D1683455

**Optimization of Polymer
Nanocomposite Properties**

Edited by
Vikas Mittal

Related Titles

Pascault, J.-P., Williams, R. J. J. (eds.)

Epoxy Polymers

New Materials and Innovations

2010
ISBN: 978-3-527-32480-4

Cosnier, S., Karyakin, A. (eds.)

Electropolymerization

Concepts, Materials and Applications

2010
ISBN: 978-3-527-32414-9

Matyjaszewski, K., Müller, A. H. E. (eds.)

Controlled and Living Polymerizations

From Mechanisms to Applications

2009
ISBN: 978-3-527-32492-7

Ghosh, S. K. (ed.)

Self-healing Materials

Fundamentals, Design Strategies, and Applications

2009
ISBN: 978-3-527-31829-2

Ghosh, S. K. (ed.)

Functional Coatings

by Polymer Microencapsulation

2006
ISBN: 978-3-527-31296-2

Astruc, D. (ed.)

Nanoparticles and Catalysis

2008
ISBN: 978-3-527-31572-7

Matyjaszewski, K., Gnanou, Y., Leibler, L. (eds.)

Macromolecular Engineering

Precise Synthesis, Materials Properties, Applications

2007
ISBN: 978-3-527-31446-1

Optimization of Polymer Nanocomposite Properties

Edited by Vikas Mittal

WILEY-VCH

WILEY-VCH Verlag GmbH & Co. KGaA

The Editor

Dr. Vikas Mittal
BASF SE
Polymer Research, G201
67056 Ludwigshafen
Germany

All books published by Wiley-VCH are carefully produced. Nevertheless, authors, editors, and publisher do not warrant the information contained in these books, including this book, to be free of errors. Readers are advised to keep in mind that statements, data, illustrations, procedural details or other items may inadvertently be inaccurate.

Library of Congress Card No.: applied for

British Library Cataloguing-in-Publication Data
A catalogue record for this book is available from the British Library.

Bibliographic information published by the Deutsche Nationalbibliothek
The Deutsche Nationalbibliothek lists this publication in the Deutsche Nationalbibliografie; detailed bibliographic data are available on the Internet at <http://dnb.d-nb.de>.

© 2010 WILEY-VCH Verlag GmbH & Co. KGaA, Weinheim

All rights reserved (including those of translation into other languages). No part of this book may be reproduced in any form – by photoprinting, microfilm, or any other means – nor transmitted or translated into a machine language without written permission from the publishers. Registered names, trademarks, etc. used in this book, even when not specifically marked as such, are not to be considered unprotected by law.

Typesetting   Toppan Best-set Premedia Limited
Printing   betz-druck GmbH, Darmstadt
Binding   Litges & Dopf GmbH, Heppenheim
Cover Design   Grafik-Design Schulz, Fußgönheim

Printed in the Federal Republic of Germany
Printed on acid-free paper

ISBN: 978-3-527-32521-4

Contents

Preface XV
List of Contributors XIX

1 **Polymer Nanocomposites: Synthesis, Microstructure, and Properties** 1
Vikas Mittal
1.1 Introduction 1
1.2 Means of Synthesis and Microstructure 3
1.3 Importance of Thermogravimetric Analysis and X-Ray Diffraction for Filler and Nanocomposite Microstructure Characterization 6
1.4 Polar and Nonpolar Polymer Systems 9
1.5 Advances in Filler Surface Modifications 14
1.6 Prediction of Composite Properties 15
References 17

2 **Morphology Development in Thermoset Nanocomposites** 21
Peter J. Halley
2.1 Introduction 21
2.2 Epoxy Nanocomposite Systems 22
2.3 Effects of Processing and Aging 27
2.4 Other Thermoset Nanocomposite Systems 30
2.5 Recent Advances in Thermoset Nanocomposites 33
2.5.1 Epoxy-HBP Nanostructured Systems 33
2.5.2 Ternary Nanostructured Systems and Multiscale Composites 34
2.5.3 Novel Characterization Methods 36
2.5.4 Modeling Thermoset Nanocomposite Systems 36
2.6 Summary 38
References 38

Optimization of Polymer Nanocomposite Properties. Edited by Vikas Mittal
Copyright © 2010 WILEY-VCH Verlag GmbH & Co. KGaA, Weinheim
ISBN: 978-3-527-32521-4

3	**Morphology and Interface Development in Rubber–Clay Nanocomposites** 41
	Yong-Lai Lu and Li-Qun Zhang
3.1	Introduction 41
3.2	Melt Compounding 42
3.2.1	Mechanism and Influencing Factors 42
3.2.1.1	The Organic Modification 43
3.2.1.2	The Features of Rubber and Compatibilizers or Coupling Agents 44
3.2.1.3	Melt-Compounding Conditions 44
3.2.2	Evolution of Morphology and Interface during Vulcanization of RCNs 44
3.2.2.1	Changes in the Local Microstructure of Clay Particles 44
3.2.2.2	Change in the Spatial Distribution of Clay Particles 45
3.3	Latex Compounding 57
3.3.1	Mechanism and Influencing Factors 57
3.3.2	Interface Enhancement 60
	References 65

4	**Morphology Development in Polyolefin Nanocomposites** 67
	Mitsuyoshi Fujiyama
4.1	Introduction 67
4.2	Intercalation, Exfoliation, and Dispersion of MMT 68
4.2.1	Manufacturing Processes 68
4.2.2	Dispersion (Exfoliation) State of Nanoclays 69
4.2.3	Exfoliation Process of Nanoclays 72
4.2.4	Control of Exfoliation/Dispersion of Nanoclays 75
4.2.4.1	Raw Materials 75
4.2.4.2	Mixing Methods 78
4.2.4.3	Mixing Conditions 82
4.2.5	Morphology of Base Polymers 82
4.3	Crystallization and Crystalline Structure of Matrix Polymers 83
4.3.1	Crystallization 83
4.3.1.1	Quiescent Crystallization 83
4.3.1.2	Flow-Induced Crystallization 84
4.3.2	Crystalline Structure 84
4.3.2.1	Quiescent Crystallization 84
4.3.2.2	Flow-Induced Crystallization 86
4.4	Morphology Development in Processing 86
4.4.1	Injection Molding 87
4.4.1.1	Conventional Injection Molding 87
4.4.1.2	Dynamic Packing Injection Molding 88
4.4.2	Sheet Extrusion 89

4.4.3	Film Extrusion Casting	90
4.5	Conclusions	90
	References	91

5	**Rheological Behavior of Polymer Nanocomposites**	**93**
	Mo Song and Jie Jin	
5.1	Introduction	93
5.2	Rheological Behavior of Polymer Nanocomposites in Solution State	95
5.3	Rheological Behavior of Polymer Nanocomposites in Melt State	107
5.4	Conclusions	118
	References	119

6	**Mechanical Property Enhancement of Polymer Nanocomposites**	**123**
	Nourredine Aït Hocine	
6.1	Introduction	123
6.2	Material Stiffness	124
6.2.1	Experimental Investigations	124
6.2.2	Analytical Modeling	125
6.3	Ultimate Mechanical Properties	129
6.3.1	Experimental Investigations	129
6.3.2	Analytical Modeling	131
6.3.2.1	Yield Stress	131
6.3.2.2	Properties at Break	132
6.4	Conclusions	135
	References	136

7	**Stress Transfer and Fracture Mechanisms in Carbon Nanotube-Reinforced Polymer Nanocomposites**	**139**
	Bhabani K. Satapathy, Martin Ganß, Petra Pötschke, and Roland Weidisch	
7.1	Introduction	139
7.2	Experimental Studies	142
7.2.1	Fabrication of Composites	142
7.2.2	Morphology Characterization	142
7.2.2.1	Transmission Electron Microscopy (TEM)	142
7.2.2.2	Atomic Force Microscopy (AFM)	143
7.2.2.3	2-D Wide-Angle X-Ray Diffraction	143
7.2.3	Thermal Characterization	143
7.2.3.1	Differential Scanning Calorimetry	143
7.2.3.2	Dynamical Mechanical Analysis	143
7.2.3.3	Melt Rheological Investigations	144
7.2.4	Mechanical and Fracture Mechanical Investigations	144
7.2.4.1	Tensile Testing	144
7.2.4.2	Essential-Work-of-Fracture Approach	144

7.2.4.3	Kinetics of Crack Propagation Measurement Using a Single-Specimen Technique *145*	
7.3	Mechanical Behavior of Polymer Nanocomposites and Stress Transfer *145*	
7.3.1	Amorphous Thermoplastic *145*	
7.3.2	Semi-Crystalline Thermoplastic *149*	
7.4	Fracture Mechanics of CNT-Polymer Nanocomposites *156*	
7.4.1	Amorphous Thermoplastic *156*	
7.4.2	Semi-Crystalline Thermoplastic *162*	
7.5	Concluding Remarks *168*	
	Acknowledgments *169*	
	References *169*	

8 Barrier Resistance Generation in Polymer Nanocomposites *173*
Vikas Mittal

8.1	Introduction *173*
8.2	Theory of Permeation *174*
8.3	Barrier Generation in Polar Nanocomposites *176*
8.4	Barrier Generation in Nonpolar Nanocomposites *183*
8.5	Modeling of Barrier Properties of Composites *189*
	References *192*

9 Mechanisms of Thermal Stability Enhancement in Polymer Nanocomposites *195*
Krzysztof Pielichowski, Agnieszka Leszczyńska, and James Njuguna

9.1	Introduction *195*
9.2	The Mechanisms of Thermal Stability Improvement by Different Nanofillers *196*
9.2.1	Clay Minerals *196*
9.2.1.1	Barrier Effect *196*
9.2.1.2	Restricted Thermal Motions *198*
9.2.1.3	Char Forming and Catalytic Effects *198*
9.2.1.4	Radical Trapping and Sorption Mechanisms *201*
9.2.2	Carbonaceous Nanofillers *202*
9.2.2.1	Carbon Nanotubes and Carbon Nanofibers *202*
9.2.2.2	Carbon Black *205*
9.2.2.3	Fullerenes *205*
9.2.2.4	Graphite *205*
9.2.3	Silica-Based Nanofillers *205*
9.2.3.1	Silica Oxide *205*
9.2.3.2	Polyhedral Oligomeric Silsesquioxane *206*
9.2.4	Metals and Metal Oxides *206*
9.2.5	Other Fillers *207*
9.3	Concluding Remarks *207*
	References *208*

10	**Mechanisms of Tribological Performance Improvement in Polymer Nanocomposites** *211*
	Ga Zhang and Alois K. Schlarb
10.1	Introduction *211*
10.2	Nanoparticle Reinforcements *213*
10.2.1	Improvement of Wear Performance by Using Nanoparticles *214*
10.2.2	Roles of Nanoparticles on Transfer Film Formation *214*
10.2.3	Structure–Tribological Property Relationships *215*
10.2.3.1	Effect of Grafting Treatment of Nanoparticles on Tribological Improvement of Epoxy Nanocomposites *217*
10.2.3.2	Role of Nano-SiO_2 Particles on the Mechanical and Tribological Behaviors of PEEK *218*
10.3	Carbon Nanotubes *223*
10.4	Synthetic Roles of Nanoparticles with Traditional Fillers *226*
10.4.1	Tribological Behavior of Traditional and Nanofillers (or Sub-Micro)-Filled Epoxy *226*
10.4.2	Roles of Nanoparticles on the Tribological Behavior of SCF/PTFE/Graphite-Filled PEEK *227*
	References *233*

11	**Mechanisms of Biodegradability Generation in Polymer Nanocomposites** *235*
	Mitsuhiro Shibata
11.1	Introduction *235*
11.2	PBAT Nanocomposites *237*
11.2.1	Preparation and Morphology of PBAT Nanocomposites *237*
11.2.2	Mechanical Properties of PBAT Nanocomposites *241*
11.2.3	Thermal Properties of PBAT Nanocomposites *243*
11.2.4	Biodegradability of PBAT Nanocomposites *244*
11.3	PBS Nanocomposites *245*
11.3.1	Preparation and Morphology of PBS Nanocomposites *245*
11.3.2	Mechanical Properties of PBS Nanocomposites *248*
11.3.3	Thermal Properties of PBS Nanocomposites *252*
11.3.4	Biodegradability of PBS Nanocomposites *253*
11.4	Conclusions *256*
	References *258*

12	**Self-Healing in Nanoparticle-Reinforced Polymers and other Polymer Systems** *261*
	Stephen J. Picken, Steven D. Mookhoek, Hartmut R. Fischer, and Sybrand van der Zwaag
12.1	Introduction *261*
12.2	Microstructured Self-Healing Polymer Structures *264*
12.2.1	Liquid-Based Self-Healing Thermosetting Polymers *264*
12.2.2	Liquid-Based Self-Healing Thermoplastic Polymers *265*

12.2.3	Geometric Aspects in Encapsulation 265
12.3	Nanoparticle-Reinforced Self-Healing Polymer Systems 270
12.3.1	Modeling the Modulus of Nanoparticle-Filled Polymers 270
12.3.2	Experimental Validation for Non-Self-Healing Systems 273
12.3.3	Design of a Self-Healing Nanoparticle Composite 274
12.4	Concluding Remarks 277
	Acknowledgments 277
	References 277
13	**Crystallization in Polymer Nanocomposites** 279
	Jyoti Jog
13.1	Introduction 279
13.2	Nanofillers 280
13.2.1	Silicates 280
13.2.2	Carbon Nanotubes 281
13.2.3	Exfoliated Graphite 281
13.2.4	Other Nanoparticles 282
13.3	Isothermal and Nonisothermal Crystallization in Polymers 282
13.3.1	Polypropylene (PP) 283
13.3.1.1	Crystallization 283
13.3.1.2	Polymorphism in PP 285
13.3.2	Poly-1-Butene (PB) 285
13.3.3	Polybutylene Terephthalate (PBT) 286
13.3.4	Polyethylene Terephthalate (PET) 287
13.3.5	Poly Trimethylene Terephthalate (PTT) 288
13.3.6	Polyethylene Naphthalate (PEN) 288
13.3.6.1	Crystallization 288
13.3.6.2	Polymorphism in PEN 289
13.3.7	Polylactic Acid (PLLA) 289
13.3.8	Polyhydroxy Alkonate (PHA) 290
13.3.9	Polyether Ether Ketone (PEEK) 290
13.3.10	Nylon 6 291
13.3.10.1	Crystallization 291
13.3.10.2	Polymorphism in Nylon 6 292
13.3.11	Nylon 66 293
13.3.11.1	Crystallization 293
13.3.11.2	Polymorphism in Nylon 66 293
13.3.12	Nylon 11 293
13.3.13	Nylon 10,10 294
13.3.14	Polyvinylidene Fluoride (PVDF) 294
13.3.14.1	Crystallization 294
13.3.14.2	Polymorphism in PVDF 294
13.3.15	Syndiotactic Polystyrene (sPS) 295

13.3.15.1	Crystallization 295
13.3.15.2	Polymorphism in sPS 296
13.4	Conclusions 296
	References 297

14 Prediction of the Mechanical Properties of Nanocomposites *301*
Qinghua Zeng and Aibing Yu

14.1	Introduction 301
14.1.1	Nanocomposites 301
14.1.2	Some Issues in Nanocomposites 301
14.1.2.1	Dispersion of Nanoparticles 301
14.1.2.2	Interface 303
14.1.2.3	Crystallization 303
14.1.3	Property Predictions 304
14.2	Analytical and Numerical Techniques 305
14.2.1	Analytical Models 305
14.2.1.1	Rule of Mixtures 305
14.2.1.2	Halpin–Tsai Model 306
14.2.1.3	Mori–Tanaka Model 306
14.2.1.4	Equivalent-Continuum Approach 307
14.2.1.5	Self-Similar Approach 307
14.2.2	Numerical Methods 307
14.2.2.1	Molecular Dynamics 308
14.2.2.2	Monte Carlo 309
14.2.2.3	Brownian Dynamics 309
14.2.2.4	Dissipative Particle Dynamics 310
14.2.2.5	Lattice Boltzmann 310
14.2.2.6	Time-Dependent Ginzburg–Landau Method 311
14.2.2.7	Dynamic Density Functional Theory 311
14.2.2.8	Finite Element Method 312
14.2.2.9	Boundary Element Method 312
14.2.3	Multiscale Modeling 313
14.2.3.1	Challenges 313
14.2.3.2	Sequential and Concurrent Approaches 313
14.2.3.3	Applications in Polymer Nanocomposites 314
14.3	Prediction of Nanocomposite Properties 314
14.3.1	Mechanical Properties 316
14.3.1.1	Stiffness and Strength 316
14.3.1.2	Stress Transfer 320
14.3.1.3	Mechanical Reinforcement 321
14.3.1.4	Interfacial Bonding 323
14.3.1.5	Viscoelasticity 323
14.3.2	Mechanical Failure 324
14.3.2.1	Buckling 324

14.3.2.2	Fatigue	325
14.3.2.3	Fracture	325
14.3.2.4	Wear	326
14.3.2.5	Creep	327
14.4	Conclusions	327
	Acknowledgments	328
	References	329

15 Morphology Generation in Polymer Nanocomposites Using Various Layered Silicates 333
Kenji Tamura and Hirohisa Yamada

- 15.1 Introduction 333
- 15.2 Aspects of Layered Silicates 334
- 15.2.1 General Structure 334
- 15.2.2 Various Types of Layered Silicates 335
- 15.3 Conventional Layered Silicate Polymer Nanocomposites using Smectite and Expandable Synthetic Fluoro-Mica 338
- 15.3.1 Relationship Between Morphology and Properties 338
- 15.3.2 Properties of Conventional Layered Silicate/Polymer Nanocomposites 339
- 15.4 Aspect Ratio Variation Using Various Layered Silicates 344
- 15.4.1 Exfoliation of High Crystallinity Nonexpandable Mica 344
- 15.4.2 Controlling the Number of Nanolayers (in the Dispersed Platelets): Interstratified Layered Silicate/Polymer Nanocomposites 347
- 15.5 Summary 348
- References 349

16 Thermomechanical Properties of Nanocomposites 351
Lucia Helena Innocentini-Mei

- 16.1 Introduction 351
- 16.2 Thermomechanical Analysis 352
- 16.3 Dynamic Mechanical Analysis and the Principle of Time-Temperature Superposition 354
- 16.4 Nanoclays and Their Influence on the Thermomechanical Properties of Polymer Composites: Some Case Studies 355
- 16.5 Conclusions 366
- References 367

17 Effect of Processing Conditions on the Morphology and Properties of Polymer Nanocomposites 369
Michele Modesti, Stefano Besco, and Alessandra Lorenzetti

- 17.1 Introduction 369
- 17.2 Melt-Intercalation of Polymer Nanocomposite Systems 370
- 17.2.1 Melt Intercalation of Polymer/Clay Systems 370
- 17.2.1.1 Effects of Temperature, Shear, and Residence Time 370

17.2.1.2	Effects of Extruder Configuration and Screw Profiles	*374*
17.2.1.3	Effect of Processing Route *377*	
17.2.2	Melt-Intercalation of Polymer/CNT Systems *378*	
17.3	Solution-Intercalation of Polymer Nanocomposites *380*	
17.4	Progress in Polymer Nanocomposites Processing *385*	
17.4.1	Water Injection-Assisted Melt-Compounding *385*	
17.4.2	Supercritical CO_2-Assisted Melt-Compounding *388*	
17.4.3	Ultrasound-Assisted Melt-Compounding *391*	
17.5	Processing of Thermoset Nanocomposites *394*	
17.6	Conclusions *399*	
	References *400*	

Index *407*

Preface

During the past two decades, research into nanocomposites has led to the development of materials with properties that are far superior to those of not only the parent materials but also of conventional microcomposites. Nanocomposites are organic inorganic hybrid materials where the inorganic filler has at least one dimension in the nanometer scale. The nanoscale dispersion of the filler within the polymer matrices leads to tremendous interfacial contacts between the organic and inorganic phases, which in turn generates an interfacial material that has an altogether different morphology, and also has properties that are superior to those of the bulk polymer phase. As a result of this, significant improvements in the properties of nanocomposites may be realized at much lower filler concentrations. Most importantly, these properties cover the entire spectrum of the polymer nanocomposites' potential applications, from automobile body parts to high-barrier packaging materials, and from highly scratch-resistant composites to biodegradable nanocomposites.

In recent years, nanocomposites with practically all polymer systems have been used to improve one property or another, with varying degrees of success. A range of factors that influence not only the morphology but also the final properties of composites have been identified, including interfacial interactions between the filler and the polymer phase (optimization of filler surface modification, kinetic and thermodynamic factors influencing intercalation and exfoliation, etc.), the nature of the polymer (polar or nonpolar, molecular weight, etc.), the nature of the filler (aspect ratio, size, geometry, cation-exchange capacity, etc.), the processing methodologies, and the amount of inorganic filler. Yet, these improved properties are the result of many different mechanisms at play, owing to the presence of inorganic fillers within the polymers; consequently, an enhancement of one property does not directly translate into an enhancement of the other properties. Thus, it is important to gain insights into these different factors and considerations that are responsible for enhancing the various properties, the optimization of which may – in time – lead to nanocomposites being designed according to need.

This book describes the mechanisms of enhancement for the different properties in polymer nanocomposites, with each chapter focusing on a specific characteristic property. Chapter 1 reviews the synthesis methodologies of the

Optimization of Polymer Nanocomposite Properties. Edited by Vikas Mittal
Copyright © 2010 WILEY-VCH Verlag GmbH & Co. KGaA, Weinheim
ISBN: 978-3-527-32521-4

nanocomposites with both polar and nonpolar polymer matrices, and also provides details on the characterization of the microstructure and recent advances in filler surface modifications. Chapters 2 to 4 focus on the morphology developments in various polymer systems. As the generation of intercalated/exfoliated filler morphologies in the polymer matrices requires specific considerations, depending on the nature of the polymer used, it is necessary to consider polar and nonpolar polymer systems separately, as reported in Chapters 2 and 4 respectively. In the meanwhile, Chapter 3 focuses on the mechanisms of morphological development in rubbers, based on their completely different nature and their interactions with fillers other than polar (e.g., epoxy) or nonpolar (e.g., polyolefin) polymers. Chapters 5 details the rheological properties of the polymer nanocomposites, both in solution and in the melt state, and describes the relationships between the generated morphology and the rheological behavior of the composite materials that have been established to optimize these properties. Chapter 6 then focuses on the factors that must be considered in order to enhance the mechanical properties of nanocomposites, while Chapter 7 explains the various mechanisms of stress transfer and fracture in polymer nanocomposites. Chapter 8 considers the optimization of the nanocomposites' barrier properties; these are highly sensitive to minute changes at the interface between organic and inorganic components, and nanocomposites in which polar or nonpolar polymer systems are used for barrier applications differ totally in this respect. Following a similar theme, the mechanisms whereby the thermal stability of nanocomposites may be enhanced are explained in Chapter 9, while Chapter 10 details the optimization of polymer nanocomposites for use in tribology. Owing to the need to generate environment-friendly nanocomposites, it is important to optimize biodegradability in these materials, and this topic is dealt with in Chapter 11. Another subject of much recent interest is that of self-healing, the generation of which in polymers, in the presence of nanofillers, is described in Chapter 12. In Chapter 13, the optimization of polymer crystallization in nanocomposites is outlined, by examining the molecular-level interactions between the filler and the polymer phases. The ability to predict the properties of nanocomposites is equally important when optimizing their design and properties, and this is described in Chapter 14, using both analytical and numerical methods. A nanocomposite's morphology and interfacial interactions, as well its ultimate properties, are strongly affected by the aspect ratio of the filler; hence, the role of fillers with different aspect ratios, and how they affect both microstructure and properties, is considered in Chapter 15. Finally, Chapter 16 reviews the criteria applied to the optimization of the thermomechanical properties of nanocomposites, while Chapter 17 outlines the role of the various processing conditions on the resultant composite morphology and properties. In the latter case, considerations of both thermoplastic and thermoset polymer systems have been included.

It gives me immense pleasure to thank those people without whose help this project may not have become reality. I am indebted to Wiley-VCH for providing the opportunity to publish this book. I am equally thankful to my dear wife, Preeti, who coedited the book and whose continuous support and positive criticism throughout the project was immensely helpful. I dedicate this book to my family, and especially to my mother, who has been an ever-increasing source of inspiration for me.

Ludwigshafen, November 2009   *Vikas Mittal*

List of Contributors

Nourredine Aït Hocine
Université Européenne de Bretagne
LIMATB/Equipe Rhéologie
UFR Sciences et Techniques
6 avenue Victor Le Gorgeu
CS 93837-29238, Brest, Cedex 3
France

Stefano Besco
University of Padova
Department of Chemical Process Engineering
35131 Padova
Italy

Hartmut R. Fischer
TNO Science and Industry
P.O. Box 6235
5600 HE Eindhoven
The Netherlands

Mitsuyoshi Fujiyama
Fujiyama Polymer Research
Natajima 2670
Yamaguchi-shi
Yamaguchi-ken 754-0892
Japan

Martin Ganß
Friedrich-Schiller-University of Jena
Mechanics of Functional Materials Group
Institute of Materials Science and Technology
Jena, 07743
Germany

Peter J. Halley
The University of Queensland
Centre of High-Performance Polymers
School of Engineering/Australian Institute for Bioengineering and Nanotechnology
Brisbane 5
QLD 4072
Australia

Lucia Helena Innocentini-Mei
Universidade Estadual de Campinas (UNICAMP)
Faculdade de Engenharia Quimica
Caixa Postal 6066
Av. Albert Einstein, 500
CEP 13083-852
Campinas, SP
Brazil

Optimization of Polymer Nanocomposite Properties. Edited by Vikas Mittal
Copyright © 2010 WILEY-VCH Verlag GmbH & Co. KGaA, Weinheim
ISBN: 978-3-527-32521-4

List of Contributors

Jie Jin
Loughborough University
Department of Materials
Loughborough, LE11 3TU
UK

Jyoti Jog
National Chemical Laboratory
Polymer Science and Engineering
Division
Pune 411 008
India

Agnieszka Leszczyńska
Cracow University of Technology
Department of Chemistry and
Technology of Polymers
ul. Warszawska 24
31-155 Kraków
Poland

Alessandra Lorenzetti
University of Padova
Department of Chemical Process
Engineering
35131 Padova
Italy

Yong-Lai Lu
Beijing University of Chemical
Technology
Key Laboratory for Nanomaterials
Ministry of Education
Beijing 100029
China

Vikas Mittal
BASF SE
Polymer Research
67056 Ludwigshafen
Germany

Michele Modesti
University of Padova
Department of Chemical Process
Engineering
35131 Padova
Italy

Steven D. Mookhoek
University of Technology Delft
Faculty of Aerospace Engineering
Kluyverweg 1a
2629 HS Delft
The Netherlands
and
Dutch Polymer Institute (DPI)
P.O. Box 902
5600 AX Eindhoven
The Netherlands

James Njuguna
Cranfield University
Department of Sustainable Systems
Bedfordshire, MK43 0SU
UK

Stephen J. Picken
University of Technology Delft
Department of Chemical Technology
Julianalaan 136
2628 BL Delft
The Netherlands

Krzysztof Pielichowski
Cracow University of Technology
Department of Chemistry and
Technology of Polymers
ul. Warszawska 24
31-155 Kraków
Poland

Petra Pötschke
Leibniz Institute of Polymer Research
Dresden e.V.
Hohe Str. 6
Dresden 01069
Germany

Bhabani K. Satapathy
Centre for Polymer Science and
Engineering
Indian Institute of Technology
Delhi, Hauz Khas
New Delhi, 110016
India

Alois K. Schlarb
University of Kaiserslautern
Chair of Composite Engineering
Kaiserslautern 67663
Germany

Mitsuhiro Shibata
Faculty of Engineering
Chiba Institute of Technology
Department of Life and
Environmental Sciences
2-17-1, Tsudanuma
Narashino
Chiba 275-0016
Japan

Mo Song
Loughborough University
Department of Materials
Loughborough, LE11 3TU
UK

Kenji Tamura
National Institute for Materials
Science
Photocatalytic Material Center
1-1 Namiki
Tsukuba
Ibaraki 305-0044
Japan

Roland Weidisch
Friedrich-Schiller-University of Jena
Mechanics of Functional Materials
Group
Institute of Materials Science and
Technology
Jena, 07743
Germany

Hirohisa Yamada
National Institute for Materials Science
Photocatalytic Materia Center
1-1 Namiki
Tsukuba
Ibaraki 305-0044
Japan

Aibing Yu
The University of New South Wales
School of Materials Science and
Engineering
Laboratory for Simulation and
Modeling of Particulate Systems
Sydney, NSW 2052
Australia

Qinghua Zeng
The University of New South Wales
School of Materials Science and
Engineering
Laboratory for Simulation and Modeling of Particulate Systems
Sydney, NSW 2052
Australia

Ga Zhang
University of Kaiserslautern
Institut für Verbundwerkstoffe GmbH
Kaiserslautern 67663
Germany

Li-Qun Zhang
Beijing University of Chemical
Technology
Key Laboratory for Nanomaterials
Ministry of Education
Beijing 100029
China
and
Beijing University of Chemical
Technology
Key Laboratory of Beijing City on
Preparation and Processing of Novel
Polymer Materials
Beijing 100029
China

Sybrand van der Zwaag
University of Technology Delft
Faculty of Aerospace Engineering
Kluyverweg 1a
2629 HS Delft
The Netherlands

1
Polymer Nanocomposites: Synthesis, Microstructure, and Properties[1]

Vikas Mittal

1.1
Introduction

Polymer–silicate nanocomposites are hybrid organic inorganic materials, in which mixing of the filler phase is achieved at the nanometer level, so that at least one dimension of the filler phase is less than 100 nm. During recent years, these nanocomposites have generated much research interest owing to remarkable enhancements in the various composite properties at very low volume fractions [1–10]. This morphology of nanoscale dispersion of the filler phase in the polymer matrix leads to tremendous interfacial contacts of the nanoparticles with the polymer matrix, and subsequently to confined polymer chains in-between the nanometer-thick delaminated elementary clay layers. This leads to synergistic improvements in the composite properties, where the achieved properties are superior to those of the individual components. The fillers generally used for such composites are layered aluminosilicates, and most commonly montmorillonites (MMTs) from the family of aluminosilicates. The properties of the composites are directly affected by the filler volume fraction, the aspect ratio, alignment in the composite, and other geometric considerations.

The layered aluminosilicates such as MMT are plate-like particles and belong to the family of 2:1 phyllosilicates. A 2:1 layer consists of two tetrahedral silica sheets sandwiching an alumina octahedral sheet [11, 12]. The physical dimensions of one such layer may be 100 nm in diameter and 1 nm in thickness. Due to isomorphic substitutions in the octahedral and tetrahedral sheets, the layers have a net negative charge. The most common substitutions are Al^{3+} for Si^{4+} in the tetrahedral sheet and Mg^{2+} for Al^{3+} in the octahedral sheet. The negative charges are counterbalanced by the interlayer alkali or alkaline earth metal cations, and as a result of this the 2:1 layers are held together in stacks by electrostatic and van der Waals forces. Most of these inorganic minerals have

---

1) These studies were carried out at the Institute of Chemical and Bioengineering, Department of Chemistry and Applied Biosciences, ETH Zurich, 8093 Zurich, Switzerland.

high energetic hydrophilic surfaces, which make them incompatible with the hydrophobic polymer matrices. MMTs are known to swell easily in water, and consequently can be delaminated in water to give nanosized platelets, the inorganic surface cations of which can then be exchanged with organic cations. An exchange of inorganic cations with organic cations renders the clay organophilic and hydrophobic, and lowers the surface energy of the clay layers. It then becomes possible for the organic polymer to diffuse between the clay layers and to delaminate the clay platelets to individual layers. This technology has been widely developed, as reported by Theng [13], Lagaly [14], Pinnavaia [15], and Giannelis [16]. Long-chain alkyl ammonium salts have been widely used for exchanging the inorganic cations because they increase the basal spacing of the clay to a large extent, apart from lowering the surface energy, which can further be helpful in achieving exfoliation of the clay layers in the polymer matrix. Figure 1.1 shows a representation of the surface modification process. Although, originally, the alkyl ammonium ions were the modifications conventionally used, in recent years more advanced surface modifications for the fillers have been developed; these include surface modifications with reactive groups, modifications with initiator molecules or with monomer molecules, and so on. Based on the basal plane spacing, it can also be predicted that either the alkyl chains lie flat on the silicate surface in monolayer, bilayer or pseudo-trimolecular arrangement, or radiate away from the surface giving paraffin-type geometry [17–20]. This may also help in predicting their possible interactions with the polymer matrix. It should be noted that the filler platelets are still partially polar, even after the surface modification with the long-chain alkyl ammonium ions, and therefore after drying would again form thick stacks of platelets joined

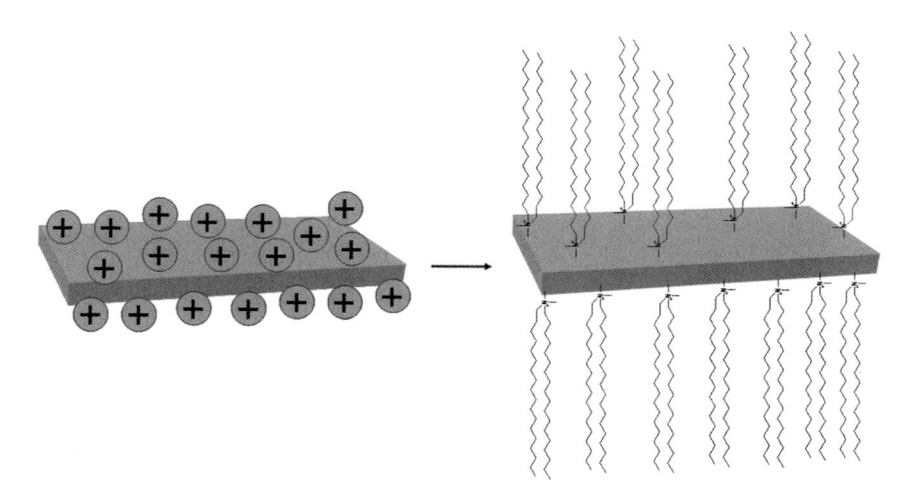

Figure 1.1 Surface modification of the aluminosilicate surface, which includes the exchange of cations present on the surface with long-chain alkyl ammonium ions.

together by electrostatic forces. Thus, it is of immense importance to achieve the uniform and nanoscale dispersion of these platelets in the polymers matrix by the action of thermodynamic and kinetic forces. It is also possible to use the other varieties of aluminosilicates such as vermiculite or mica. These minerals differ from MMT in terms of their layer charge; this is high in mica and vermiculite, which makes them difficult to swell in water and hence to achieve complete surface modification. However, owing to the higher charges present per unit area as compared to the MMT surface, a greater degree of surface modification can be exchanged on the surface. This, in turn, leads to a more straight positioning of the modification molecules on the surface in the case of vermiculite and mica, thus creating a much higher basal plane spacing as compared to MMT for the same cation exchanged on the surface. The MMTs themselves also have wide range of layer charges or charge densities, depending on their source.

1.2
Means of Synthesis and Microstructure

Polymer nanocomposites have been generated in several different ways. In cases where the monomers are present in liquid form, a bulk polymerization of the monomer is often carried out in the presence of clay. This leads to much better interfacial contacts between the organic and inorganic phases, and the system does not suffer from the challenge of intercalation of high-molecular-weight polymer chains inside the clay interlayers. Otherwise, a solvent can also be used as the reaction medium in order to reduce the viscosity of the bulk medium and to distribute the heat more uniformly. In such cases, a solvent is chosen in which the polymer and monomer are soluble, and the solvent also is able to swell the clay. After polymerization and precipitation the precipitate can be collected and dried. Water may also be used as the reaction medium, while emulsion and suspension polymerization methods can also be used to generate polymer nanocomposites using monomers such as styrene and methylmethacrylate. In this case, the modified clay can be dispersed in the water phase by shearing; the emulsifier and monomer can then be added, followed by subsequent polymerization.

Melt compounding is the most commonly used approach for generating polymer nanocomposites. Here, high-molecular-weight polymers such as polypropylene (PP), polyethylene (PE) and polystyrene (PS) can be melted at high temperature, after which the modified clay powder is added to the melt. The filler is then kneaded and compounded thoroughly with the polymer melt in order to achieve a uniform dispersion and distribution of the filler. Although the melt compounding approach requires the use of a high temperature, this can sometimes cause concern with regards to surface modifications. Ammonium-based surface modifications have an onset of degradation close to 200 °C, which is a common temperature used for melt compounding of the polymers. Any degradation of the surface modification may have a serious impact on the composite properties, and so should be avoided by using either a slightly lower temperature or more advanced thermally stable modifications. The time during which the organic inorganic phases are mixed at high temperature should also be optimized in order to minimize any thermal degradation of the ammonium modification.

The microstructure of the polymer nanocomposites is ideally classified as *intercalated, exfoliated,* and *unintercalated* or *microcomposites*. The composite microstructure is classified as exfoliated when the filler platelets are completely delaminated into their primary nanometer-scale size; moreover, the platelets should be far apart from each other so that the periodicity of the platelet arrangement is totally lost. This occurs when the electrostatic forces of interaction between the platelets have completely been overcome by the polymer chains. Figure 1.2 shows a series of transmission electron microscopy (TEM) images depicting the various morphologies of the polymer nanocomposite structures. Figure 1.2a represents the exfoliated morphology, where the black lines are the cross-section of the aluminosilicate platelets. The platelets can be seen as single and uniformly dispersed, although they are completely misaligned. On many occasions, bending and folding of the platelets has also been observed. For example, when a single (or occasionally more than one) extended polymer chain is intercalated into the clay interlayers, the periodicity of the clay platelets remains intact; such a microstructure is termed *intercalated*. Whilst this structure indicates that the organic–inorganic hybrid has been formed, the electrostatic forces of interaction between the clay platelets could not be totally dissolved. Figure 1.2b shows the micrograph with intercalated platelets; here, the microstructure represents a well-ordered multilayer morphology consisting of alternate polymer and inorganic layers. Such a periodicity produces a signal in the X-ray diffractograms, such that the degree of intercalation can be quantified by comparison with the basal plane spacing of the modified filler. The exfoliated morphology does not generate any diffraction signal, owing to a loss of periodicity; an absence of any diffraction peak is taken as proof of the generation of completely exfoliated nanocomposites. Based on the interfacial interactions and mode of mixing of the organic and inorganic phases, it is possible that the two phases do not intermix at all, but rather a microcomposite or unintercalated composite is formed. Such a morphology is shown in Figure 1.2c. This type of structure is not a nanocomposite but, like conventional composites, will require a large amount of filler to achieve any significant improvement in the composite's properties, which otherwise could be achieved at much lower levels of filler in the case of nanocomposites. It should be noted that a classification of composite microstructure as either exfoliated or intercalated is unrealistic since, in reality, a variety of morphologies is generally present. Different extents of both intercalation and exfoliation are generally observed, whilst only a qualitative classification of morphology as more or less intercalated or exfoliated can generally be observed.

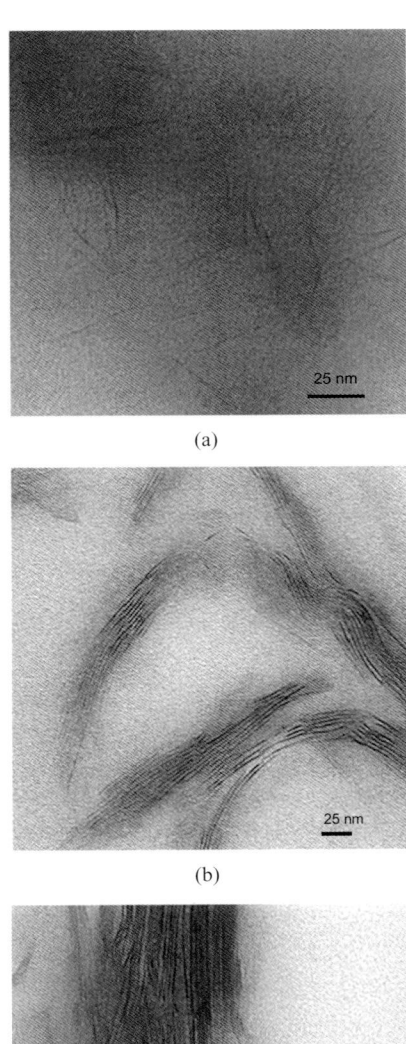

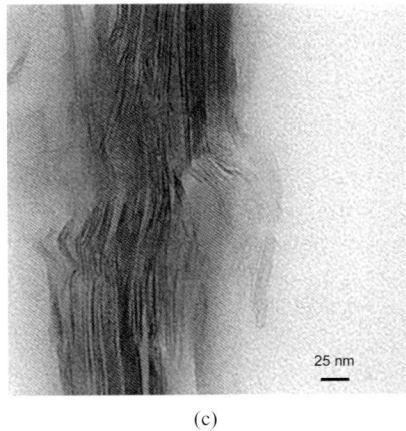

Figure 1.2 Transmission electron microscopy images indicating various possible morphologies in the composites as a function of the filler distribution. (a) Exfoliated; (b) Intercalated; (c) Unintercalated.

1.3
Importance of Thermogravimetric Analysis and X-Ray Diffraction for Filler and Nanocomposite Microstructure Characterization

Both, XRD and TGA, are important techniques used to characterize the microstructure of nanocomposites. In general, TGA is used to assess the amount of organic matter exchanged on the clay surface during the surface modification process. High-resolution TGA can also be used to ascertain if there is presence of any excess of surface modification molecules present as a pseudo bilayer, but unbound to the surface. XRD is also used to quantify increases in basal plane spacing in the filler following surface modification, and also after composite generation. The increased basal plane spacing that occurs after surface exchange leads to information regarding the possible alignment of the modification molecules inside the clay interlayers, whereas the presence or absence of diffraction peaks in the composites is used to provide information concerning the microstructure of the composites. Although XRD also provides information relating to the amount of organic matter present in the clay interlayers, it cannot provide information on the excess of surface modification molecules present on the clay surface, as does TGA. Information relating to excess modification is very important, especially when the filler must be compounded with the polymer at high temperature, as any free modification present on the surface would have a much lower thermal degradation temperature and may impact negatively on the composite's properties [21]. Figure 1.3 shows a comparison of the XRD and TGA of the fillers, which were modified with octadecyltrimethylammonium, dioctadecyldimethylammonium and trioctadecylmethylammonium. The modifications differed in terms of the increasing number of octadecyl chains in the molecule. As shown in Figure 1.3a, the basal plane spacing of the filler increased as the chain density in the molecule increased. The cation-exchange capacity (CEC) of the clay also has a major impact on the basal plane spacing of the filler, with a low-CEC filler generally having a lower basal plane spacing as compared to a high-CEC counterpart. The fillers in Figure 1.3a had CEC-values of 680 and $880 \mu Eq\, g^{-1}$. Figure 1.3b shows the TGA thermograms of the same fillers modified with the above-mentioned surface modifications, and using MMT with a CEC of $880 \mu Eq\, g^{-1}$. The increased amount of organic matter was clearly visible in the TGA thermograms when the chain density was increased in accordance with the increase in basal plane spacing in the XRD. Yet, the TGA thermograms can provide additional information on the state of the surface of the MMTs, with those of the trioctadecylmethylammonium-modified MMT showing a sharp degradation peak at low temperature. This effect is due to a small amount of the ammonium modification being present as a pseudo-bilayer, and not bound ionically to the clay surface. As noted above, these molecules degrade at much lower temperatures than those which bind ionically to the clay surface; however, such information cannot be obtained from the X-ray diffractograms. Thus, it is very important to quantify the state of the filler surface by using a combination of high-resolution TGA and XRD.

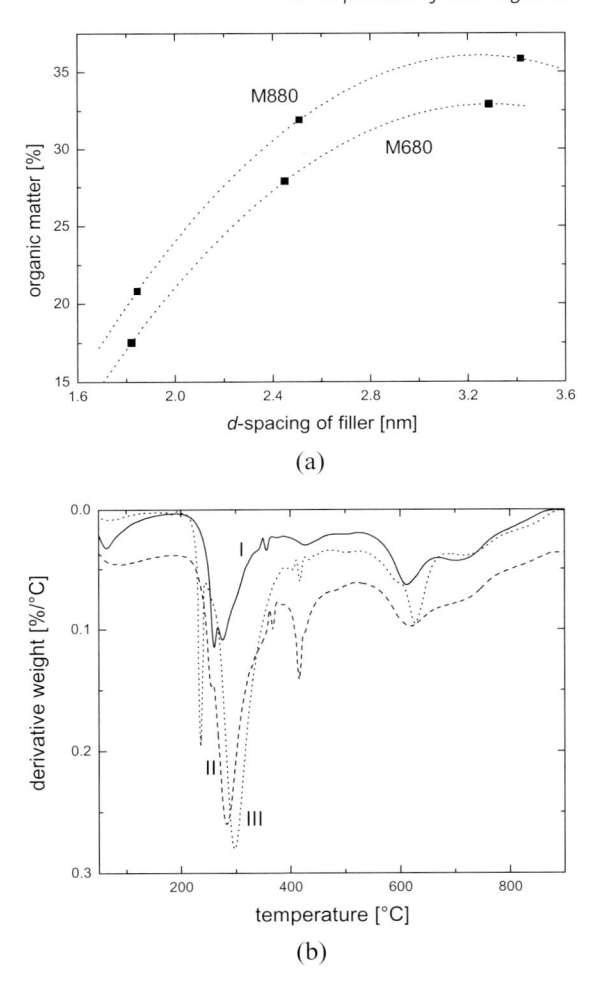

Figure 1.3 (a) Basal plane spacing in the montmorillonites with increasing chain density in the modification and as a function of filler cation-exchange capacity (CEC); (b) Thermogravimetric analysis (TGA) thermograms of the filler with a CEC of $880\,\mu Eq\,g^{-1}$ modified with (I) octadecyltrimethylammonium, (II) dioctadecyldimethylammonium, and (III) trioctadecylmethylammonium.

TGA can also be used to gain insights into the overall thermal performance of composites. As an example, Figure 1.4 shows the TGA thermograms of a pure polypropylene matrix (curve 1) and the MMT modified with dioctadecyldimethylammonium (curve 2). The TGA thermogram shown in curve 3 was obtained when these systems were formed into a composite with a 3 vol.% filler content. With regards to the thermal behavior of the nanocomposite, a synergy between the composite components is clearly visible, as thermal degradation of the composite

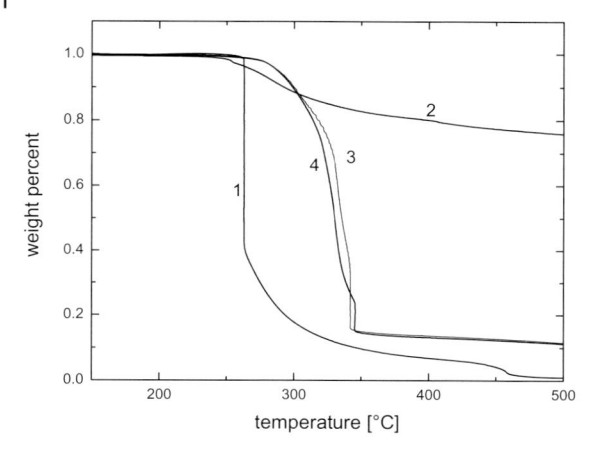

Figure 1.4 Thermogravimetric analysis thermograms of the (1) polypropylene (PP) matrix, (2) dioctadecyldimethylammonium-modified clay, (3) polypropylene composite with 3 vol% of clay of (2), and (4) polypropylene composite with 3 vol% of clay of (2) and 4 wt% of PP-g-MA compatibilizer.

begins at a higher temperature than for any of the components. The TGA thermogram of the composite with 4 wt% polypropylene-grafted maleic anhydride compatibilizer is shown as curve 4 in Figure 1.4. As these compatibilizers have very low molecular weights, it is important to observe the thermal behavior of the composite in the presence of a compatibilizer. An examination of the TGA thermograms showed the thermal behavior of the composites to be similar in the presence or absence of compatibilizer, which indicated that the compatibilizer had no negative effect on thermal performance [22].

The intensity of X-ray diffractograms is generally taken as a measure of classifying the microstructure as either intercalated or exfoliated. For example, as shown in Figure 1.5, the intensity is seen to increase as the amount of filler in the composite is increased. It is thus notable that the composites become increasingly intercalated as the filler content rises, as a greater area under the curve would indicate a greater level of intercalated material. However, it should be noted that the X-ray signals are highly qualitative in nature and strongly influenced by the sample preparation and orientation of the platelets, as well as by defects present in the crystal structure of the MMTs. Thus, a classification of nanocomposite microstructure based only on the intensity may prove to be faulty. Neither does the presence of a diffraction signal in the diffractograms of the composite mean that 100% of the microstructure is intercalated; indeed, it is quite possible for there to be a significant amount of exfoliation present in the composite. Similarly, an absence of any diffraction signal does not guarantee complete exfoliation, as small or randomly oriented intercalated platelets may still be present in the composite. The X-ray diffractograms shown as examples in Figure 1.6a are of the MMT modified with benzyldibutyl(2-hydroxyethyl)ammonium, and its composite with epoxy.

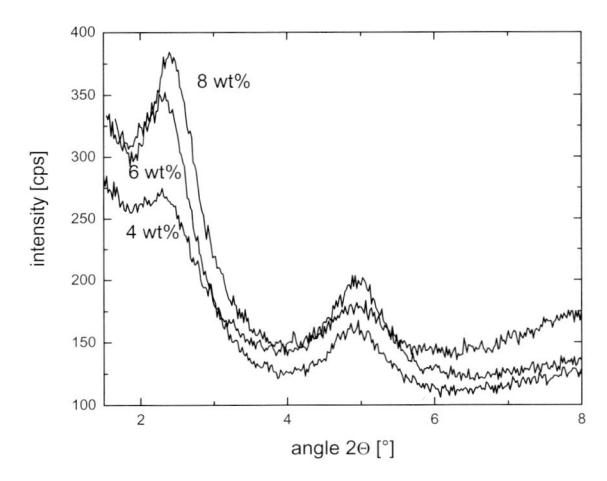

Figure 1.5 X-ray diffractograms of polyurethane nanocomposites containing different weight fractions of the filler.

Here, any increase in basal plane spacing of the filler after composite synthesis was minimal, while the diffraction signal of the filler in the composite material was quite strong in the diffractograms, and indicated an intercalated structure. However, when investigated with TEM, a large proportion of the filler was seen to exfoliated, and single platelets uniformly distributed in the polymer matrix (Figure 1.6b) [10]. It should be noted, therefore, that the XRD signals are qualitative in nature, and any idealized classification of the composite morphologies as interacted or exfoliated is entirely arbitrary.

1.4
Polar and Nonpolar Polymer Systems

Since the initial development of MMT/Nylon 6 [2, 23, 24] nanocomposites by the research group at Toyota during the early 1990s, investigations into reinforcing polymers by incorporating surface-treated layered aluminosilicates have been extensive. This technology of breaking the organically treated inorganic minerals into their nanoscale building blocks was successfully applied to polymer systems such as epoxies [10, 25–27], polyimides [28, 29], and polydimethylsiloxanes [30]. The intercalated and exfoliated nanocomposites achieved by this concept were found be very effective in improving the physical, mechanical and thermal properties of polymers at very low filler loadings as compared to the conventional composites. Conventional composites have the limitation that the clay tactoids form an unintercalated segregated phase in which the full potential of filler in improving the properties is never realized. For this reason, these composites require very

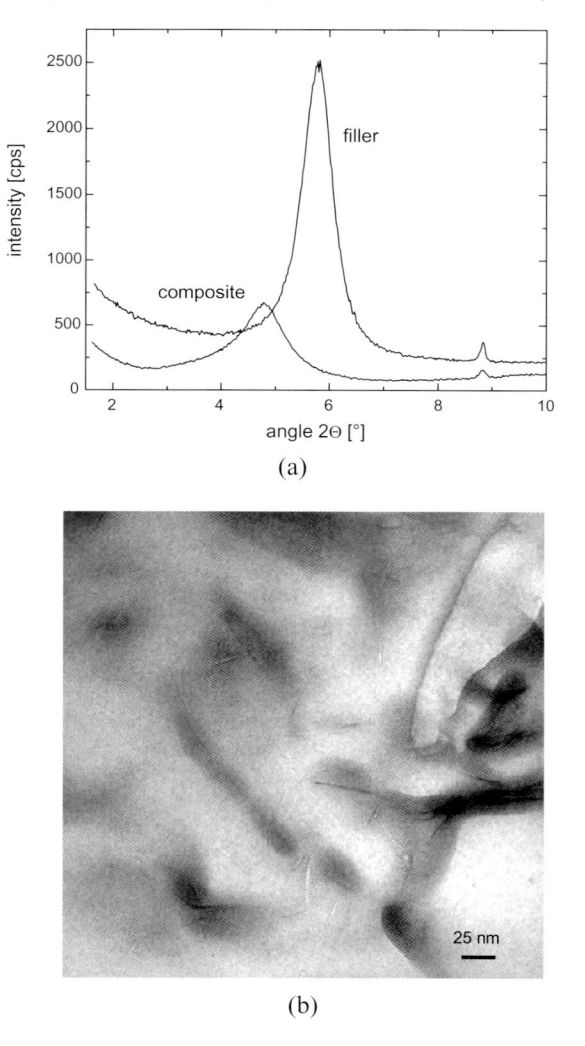

Figure 1.6 (a) X-ray diffractograms of benzyldibutyl(2-hydroxyethyl)ammonium-modified montmorillonite and its nanocomposites with epoxy polymer containing 3 vol% filler; (b) Transmission electron microscopy image of the epoxy nanocomposite.

high filler loadings in order to achieve impressive property improvements, but this makes them very bulky and opaque.

The mode of filler delamination in polar and nonpolar polymers has been observed to be very different. Polar polymers are generally observed to have more filler exfoliation as compared to nonpolar polymers, owing to the better match of polarity of polar polymers with the partially polar surface of the MMT. Thus, in the case of polar polymers, it is more likely the interfacial interactions between the organic and inorganic phases which lead to delamination of the filler. The

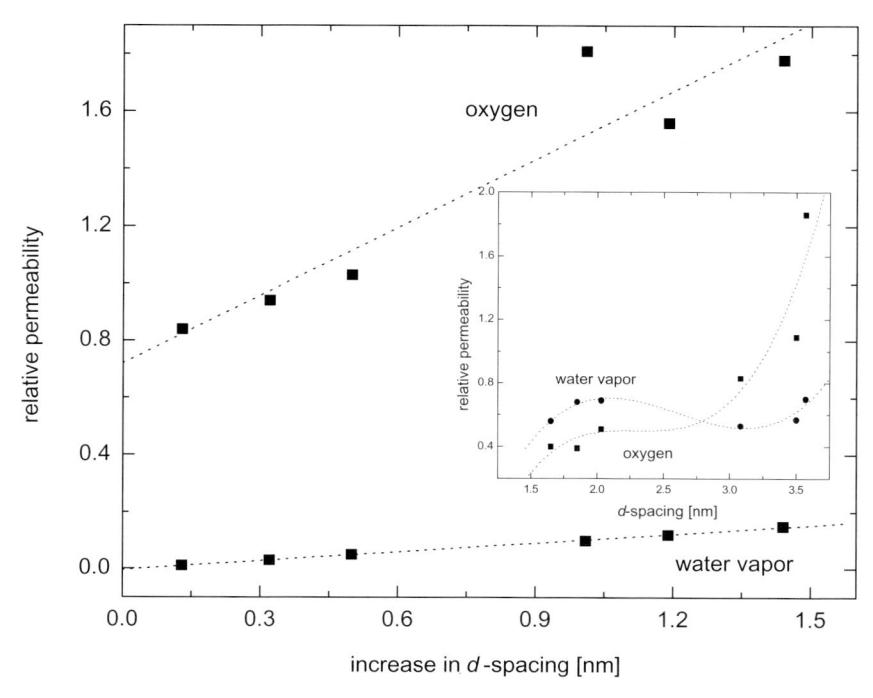

Figure 1.7 Permeation through the epoxy composites containing filler with different surface modifications as a function of filler basal plane spacing (see inset) or an increase in filler basal plane spacing in the composite.

basal plane spacing of the filler is not the prime factor to achieve exfoliation, as was also observed in Figure 1.6 for epoxy composites. The same notion is further reinforced in Figure 1.7, where oxygen and water vapor permeation though the epoxy nanocomposites is plotted as a function of the filler basal plane spacing in the composites, or as the increase in basal plane spacing of the filler after composite synthesis. In both cases, there is an indication of an increased permeation through the composites as the basal plane spacing is increased, which in truth should not be the case if the permeation were really influenced by the basal plane spacing of the filler. This confirms that it is not the basal plane spacing, but rather the exfoliated platelets that are generated, owing to the positive interfacial interactions which lead to improved composite properties in polar composites. There are generally also reactive modifications exchanged on the surface in such polymers, so that the polymer can be chemically tethered to the filler surface. As an example, benzyldibutyl(2-hydroxyethyl)ammonium was modified on the clay surface to generate epoxy nanocomposites, in the hope that the hydroxy groups of the modification would react chemically with the epoxy polymer chains. However, such bonds would be difficult to prove using infrared (IR) spectroscopy studies, as shown in Figure 1.8. The signal from the OH groups on the filler surface was observed at

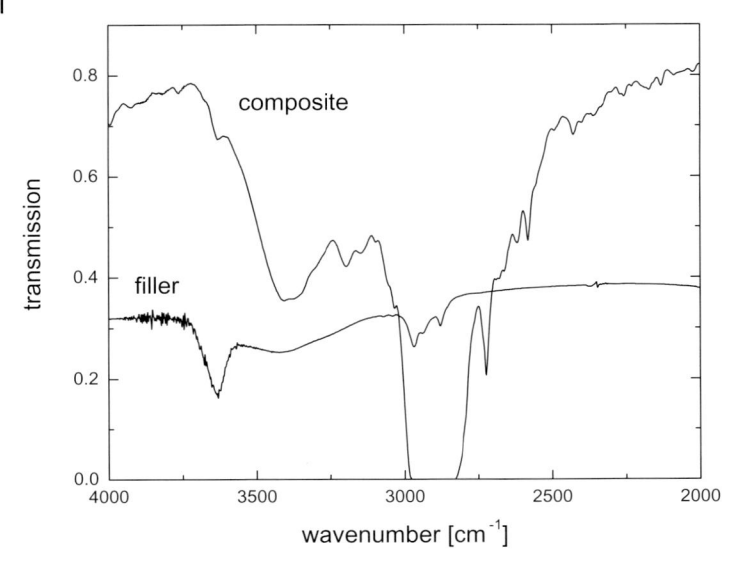

Figure 1.8 Infrared spectroscopy plots of the benzyldibutyl(2-hydroxyethyl)ammonium-modified clay, and its 3 vol% epoxy nanocomposite.

approximately 3300 cm^{-1}, an absence of which in the composite would confirm chemical tethering of the clay with the polymer matrix. However, there was a strong signal in this wavenumber range in the IR spectrum of the composite, owing to the generation of a large number of hydroxy groups on the polymer chains due to opening of the epoxy rings. Hence, whilst it is difficult to confirm the chemical reaction at the interface, the interfacial interactions in the studied systems were sufficiently positive to lead to a extensive delamination of the filler in the polymer matrix.

The nonpolar polymers, however, demonstrate different factors that affect delamination of the clay contained in them. In the absence of any positive interactions at the interface between the organic and inorganic phases, it is generally the basal plane spacing of the filler that determines the potential of the filler to be delaminated within the polymer matrix. A higher basal plane spacing in the filler leads to reduced electrostatic forces of interaction between the platelets, indicating that the platelets are more loosely held. This, in turn, increases the platelets' chances of exfoliation in the polymer matrices when mixed at high temperature in the compounder. This phenomenon is shown graphically in Figure 1.9, where PP composites with octadecyltrimethylammonium-, dioctadecyldimethylammonium- and trioctadecylmethylammonium-modified MMTs, using MMTs of two different CEC-values (680 and 880 μEq g^{-1}) were analyzed for their tensile strength and oxygen permeation [31, 32]. On increasing the chain density, the basal plane spacing was seen to increase, the effect of which was also reflected in an increasing

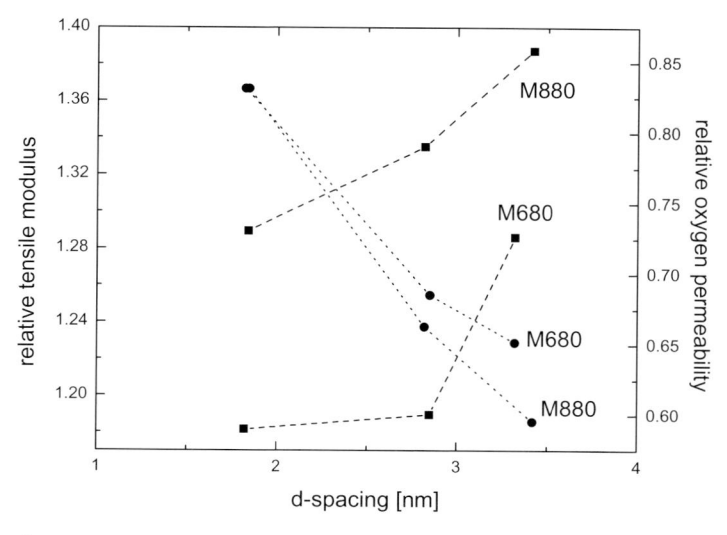

Figure 1.9 Improvements in the tensile modulus and oxygen permeation through polypropylene nanocomposites as a function of the cation-exchange capacity of the clay, and of the basal plane spacing of the modified clay.

value of the tensile modulus or a decreasing value of oxygen permeation. The data indicated clearly that an increased basal plane spacing in the filler rendered the platelets more susceptible to delamination, owing to reduced attractive forces among them.

It has also been shown important to ascertain the effect of a modified filler on the crystallinity of the polymer matrix. Occasionally, the filler may lead to changes in the crystallinity of the polymer, such that the composite properties become a result not only of the filler but also to crystallinity effects caused by the presence of the filler. The differential scanning calorimetric thermograms of the PP nanocomposites with octadecyltrimethylammonium-, dioctadecyldimethylammonium- and trioctadecylmethylammonium-modified MMTs are shown in Figure 1.10. Here, the thermograms were plotted as a comparison to the differential scanning calorimetry (DSC) thermogram of the pure PP matrix. The melting temperature and extent of crystallinity were observed to be unaffected by the change in the filler surface modification; this indicated that, under the conditions employed for synthesis of the composites, the modified filler had no effect on the crystallinity of the filler. However, Maiti *et al.* reported a decrease in spherulite size with an increase in clay content [33]. The presence of tactoids, owing to a poor dispersion of the filler, was also reported to cause a decrease in spherulite size [34]. Kodgire *et al.* reported that, in the presence of clay, PP showed an advanced crystallization and a fibrous morphology rather than the usual spherulite behavior [35]. Similarly, a decrease in crystallinity and an

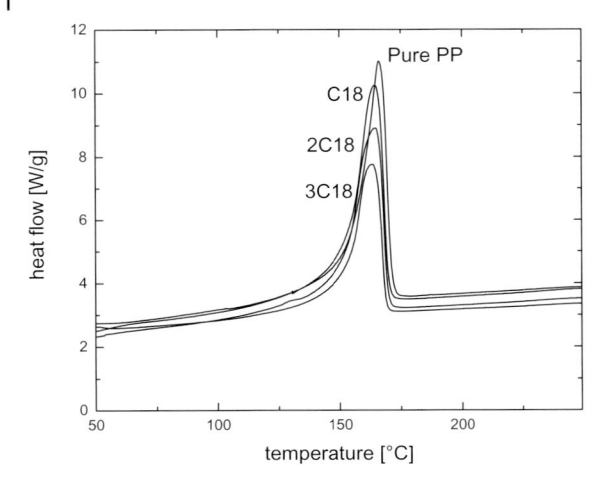

Figure 1.10 Differential scanning calorimetry thermograms of the polypropylene nanocomposites containing 3 vol% of octadecyltrimethylammonium- (C18), dioctadecyldimethylammonium- (2C18), and trioctadecylmethylammonium- (3C18) modified montmorillonites.

increase in nucleus density was observed, owing to a nucleation effect of the clay platelets [36, 37].

1.5
Advances in Filler Surface Modifications

Apart from the conventional long-chain alkyl ammonium modifications, many developments have been introduced as surface modifications for inorganic fillers. Typically, these modifications have aimed at achieving a higher organophilization of the filler surface, in order to render them more susceptible to exfoliation when compounded with polymers (especially polyolefins). Surface reactions on the clay have been studied extensively so as to generate polymer chains and achieve much higher basal plane spacing values; this would otherwise be difficult to achieve via the exchange of preformed long chains, owing to problems of solubility and steric hindrance. The two forms of polymerization reaction are represented in Figure 1.11a: (i) polymerization "to" the surface, which is generally achieved by exchanging a monomer on the filler surface, followed by its polymerization with the external monomer; and (ii) polymerization "from" the surface, in which generally an initiator is bound ionically to the filler surface, which then is used to initiate the polymerization of an externally added monomer [38]. Other than these two options, a number of other possibilities of surface reactions exist, including *surface esterification*. Here, the reactive surface modifications are first exchanged on the clay surface, which can then be used to undergo simple esteri-

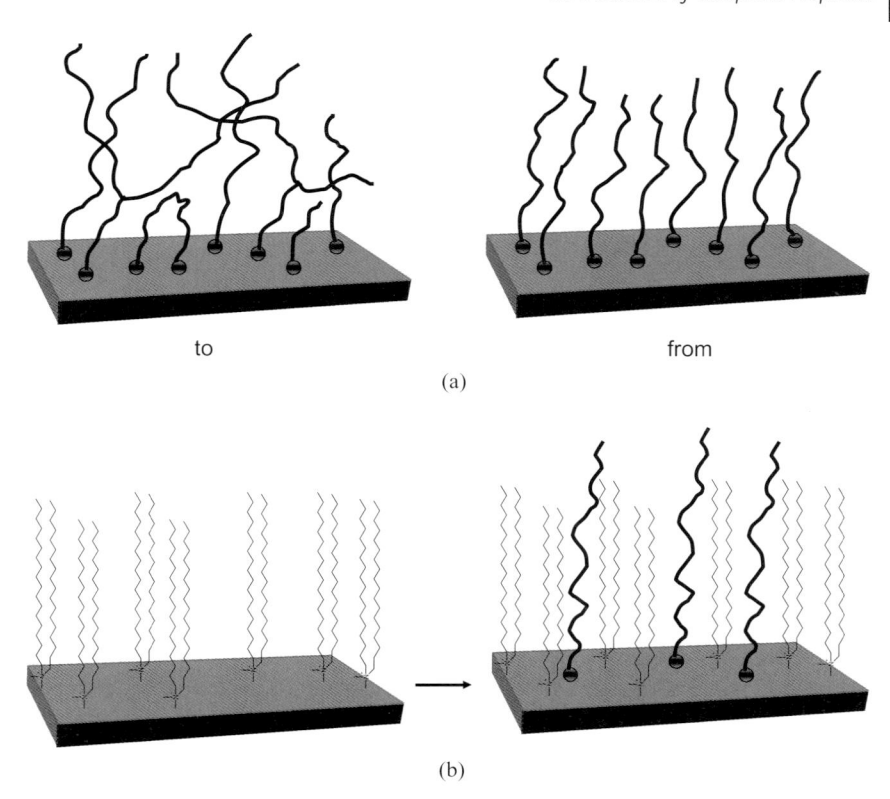

Figure 1.11 (a) Representation of the polymerization "to" the surface and polymerization "from" the surface; (b) Representation of physical adsorption onto the clay surface.

fication reactions leading to a higher basal plane spacing in the fillers [39]. The physical adsorption of organic molecules into vacant spaces on the clay surface after surface modification can also be used to achieve organophilization of the clay surface (Figure 1.11b). As the area available per cation is generally larger than the area of the cation exchanged on a clay surface, there will always be vacant spaces on the clay surface that can be targeted for occupation by organic molecules, with hydrogen bonds leading to much lesser electrostatic forces of interaction among the platelets [40].

1.6
Prediction of Composite Properties

The modeling of nanocomposite properties is important to achieve optimum improvements in composite properties, by optimizing the various factors that

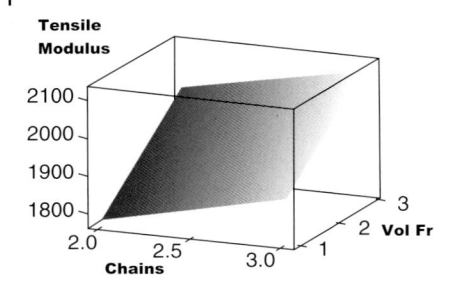

Figure 1.12 Prediction of tensile modulus of the polypropylene nanocomposites as a function of number of octadecyl chains in the ammonium modification and the inorganic volume fraction.

affect composite behavior. Although, in the past, many micromechanical models have been used traditionally to predict nanocomposite properties, these models have assumed the presence of: (i) a perfect alignment of the filler platelets in the composite; (ii) complete exfoliation of the filler; and (iii) interfacial adhesion [41–44]. These assumptions are not true, however, in the case of nanocomposites, and especially for polyolefins. Consequently, these models have been modified during recent years to incorporate the effects of incomplete exfoliation and misalignment [45–47]. In addition to these micromechanical models, several new models have been proposed that are based on factorial and mixture designs using a design of experiments methodology [48]. One such example is shown in Figure 1.12, where the tensile modulus of the composites is predicted as a function of the octadecyl chain density in the surface modification and inorganic volume fraction. Such models do not include any unrealistic assumptions, and are thus more representative of the microstructure of the nanocomposites. These factorial designs can also help to quantify the interactions of the various factors (including volume fraction, chain density, and the CEC of the MMT) on each other, and to generate an entire spectrum of values of the composite property for different values of the factors.

A number of finite element models have also been developed to predict the properties of nanocomposites [49, 50]. These models also incorporate the effects of incomplete exfoliation as well as filler misalignment, owing to the fact that these factors have significant effects on the composite properties. An example of this is shown in Figure 1.13, where the effect of platelet misalignment on the reduction of oxygen permeation through nanocomposite films is identified as a function of the aspect ratio of the platelets and the filler volume fraction. Although the reduction in oxygen permeation is much better when the platelets are completely aligned (Figure 1.13a), permeation is improved when the platelets are misaligned (Figure 1.13b). Hence, care must be taken when selecting a model used to predict a nanocomposite's properties; notably, the model should be capable of simulating the actual behavior of a nanocomposite material.

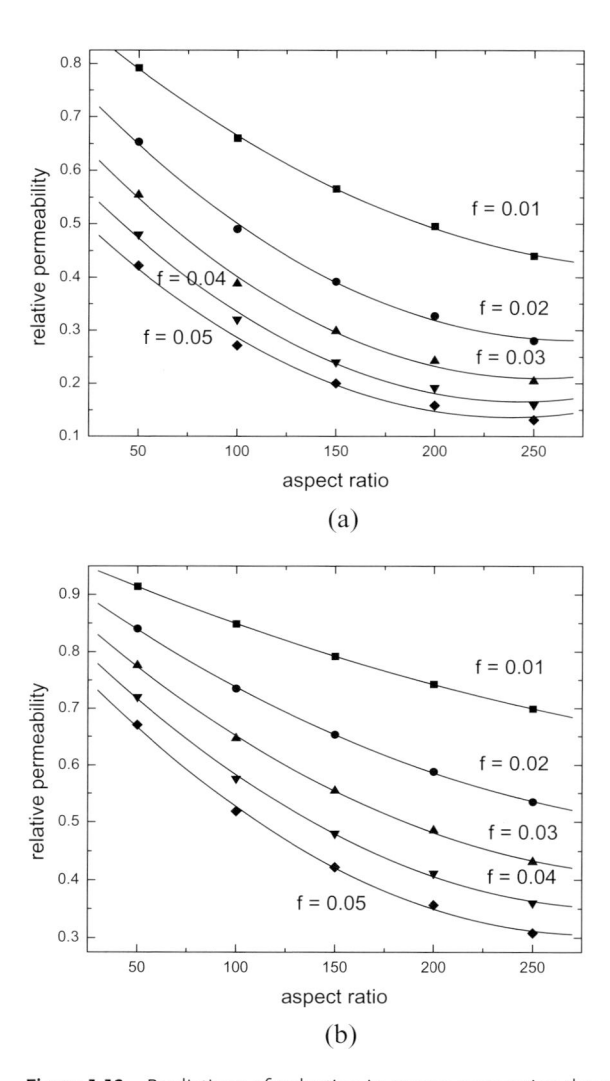

Figure 1.13 Predictions of reduction in oxygen permeation through the polymer nanocomposites as a function of aspect ratio and filler volume fraction when the filler platelets are (a) completely aligned and (b) completely misaligned [49].

References

1 Yano, K., Usuki, A., Okada, A., Kurauchi, T., and Kamigaito, O. (1993) *J. Polym. Sci., Part A: Polym. Chem.*, **31**, 2493.
2 Kojima, Y., Fukumori, K., Usuki, A., Okada, A., and Kurauchi, T. (1993) *J. Mater. Sci. Lett.*, **12**, 889.
3 Lan, T., Kaviratna, P.D., and Pinnavaia, T.J. (1994) *Chem. Mater.*, **6**, 573.
4 Lan, T. and Pinnavaia, T.J. (1994) *Chem. Mater.*, **6**, 2216.
5 LeBaron, P.C., Wang, Z., and Pinavaia, T.J. (1999) *Appl. Clay Sci.*, **15**, 11.

6 Wang, Z. and Pinnavaia, T.J. (1998) *Chem. Mater.*, **10**, 3769.
7 Messersmith, P.B. and Giannelis, E.P. (1995) *J. Polym. Sci., Part A: Polym. Chem.*, **33**, 1047.
8 Yano, K., Usuki, A., and Okada, A. (1997) *J. Polym. Sci., Part A: Polym. Chem.*, **35**, 2289.
9 Osman, M.A., Mittal, V., Mobridelli, M., and Suter, U.W. (2003) *Macromolecules*, **36**, 9851.
10 Osman, M.A., Mittal, V., Mobridelli, M., and Suter, U.W. (2004) *Macromolecules*, **37**, 7250.
11 Bailey, S.W. (1984) *Reviews in Mineralogy*, Virginia Polytechnic and State University, Blacksburg.
12 Brindley, G.W. and Brown, G. (1980) *Crystal Structures of Clay Minerals and Their X-Ray Identification* (eds G.W. Brindley and G. Brown), Mineralogical Society, London.
13 Theng, B.K.G. (1974) *The Chemistry of Clay-Organic Reactions*, Adam Hilger, London.
14 Lagaly, G. (1986) *Developments in Ionic Polymers* (eds A.D. Wilson and H.J. Prosser), Elsevier Applied Science Publishers, London/New York.
15 Pinnavaia, T.J. (1983) *Science*, **220**, 365.
16 Giannelis, E.P. (1996) *Adv. Mater.*, **8**, 29.
17 Lagaly, G. (1986) *Solid State Ionics*, **22**, 43.
18 Lagaly, G. and Benecke, K. (1991) *Colloid Polym. Sci.*, **269**, 1198.
19 Osman, M.A., Seyfang, G., and Suter, U.W. (2000) *J. Phys. Chem. B*, **104**, 4433.
20 Osman, M.A., Ernst, M., Meier, B.H., and Suter, U.W. (2002) *J. Phys. Chem. B*, **106**, 653.
21 Mittal, V. (2008) *J. Comp. Mater.*, **42**, 2829.
22 Mittal, V. (2008) *J. App. Polym. Sci.*, **107**, 1350.
23 Usuki, A., Kawasumi, M., Kojima, Y., Okada, A., Kurauchi, T., and Kamigaito, O. (1993) *J. Mater. Res.*, **8**, 1174.
24 Usuki, A., Kawasumi, M., Kojima, Y., Okada, A., Fukushima, Y., Kurauchi, T., and Kamigaito, O. (1993) *J. Mater. Res.*, **8**, 1179.
25 Lee, A. and Lichtenhan, J.D. (1999) *J. Appl. Polym. Sci.*, **73**, 1993.
26 Becker, O., Cheng, Y.-B., Varley, R.J., and Simon, G.P. (2003) *Macromolecules*, **36**, 1616.
27 Messersmith, P.B. and Giannelis, E.P. (1994) *Chem. Mater.*, **6**, 1719.
28 Wang, S., Ahmad, Z., and Mark, J.E. (1994) *Proc. ACS, Div. Polym. Mater. Sci. Eng. (PMSE)*, **70**, 305.
29 Kakimoto, M., Iyoku, Y., Morikawa, A., Yamaguchi, H., and Imai, Y. (1994) *Polym. Prep.*, **35** (1), 393.
30 Burnside, S.D. and Giannelis, E.P. (1995) *Chem. Mater.*, **7**, 1597.
31 Osman, M.A., Mittal, V., and Suter, U.W. (2007) *Macromol. Chem. Phys.*, **208**, 68.
32 Mittal, V. (2007) *J. Thermoplastic Comp. Mater.*, **20**, 575.
33 Maiti, P., Nam, P.H., Okamoto, M., Kotaka, T., Hasegawa, N., and Usuki, A. (2002) *Polym. Eng. Sci.*, **42**, 1864.
34 Svoboda, P., Zeng, C., Wang, H., Lee, L.J., and Tomasko, D.L. (2002) *J. Appl. Polym. Sci.*, **85**, 1562.
35 Kodgire, P., Kalgaonkar, R., Hambir, S., Bulakh, N., and Jog, J.P. (2001) *J. Appl. Polym. Sci.*, **81**, 1786.
36 Gopakumar, T.G., Lee, J.A., Kontopoulou, M., and Parent, J.S. (2002) *Polymer*, **43**, 5483.
37 Ma, J., Zhang, S., Qi, Z., Li, G., and Hu, Y. (2002) *J. Appl. Polym. Sci.*, **83**, 1978.
38 Mittal, V. (2007) *J. Colloid Interface Sci.*, **314**, 141.
39 Mittal, V. (2007) *J. Colloid Interface Sci.*, **315**, 135.
40 Mittal, V. and Herle, V. (2008) *J. Colloid Interface Sci.*, **327**, 295.
41 Kerner, E.H. (1956) *Proc. Phys. Soc.*, **B69**, 808.
42 Hashin, Z. and Shtrikman, S. (1963) *J. Mech. Phys. Solids*, **11**, 127.
43 Halpin, J.C. (1969) *J. Comp. Mater.*, **3**, 732.
44 Halpin, J.C. (1992) *Primer on Composite Materials Analysis*, Technomic, Lancaster.
45 van Es, M., Xiqiao, F., van Turnhout, J., and van der Giessen, E. (2001) *Specialty Polymer Additives: Principles and Application* (eds S. Al-Malaika, A.W. Golovoy, and C.A. Wilkie), Blackwell Science, Melden, CA, MA.

46 Fornes, T.D. and Paul, D.R. (2003) *Polymer*, **44**, 4993.
47 Brune, D.A. and Bicerano, J. (2002) *Polymer*, **43**, 369–387.
48 Mittal, V. (2008) *J. Thermoplastic Comp. Mater.*, **21**, 9.
49 Bharadwaj, R.K. (2001) *Macromolecules*, **34**, 9189.
50 Osman, M.A., Mittal, V., and Lusti, H.R. (2004) *Macromol. Rapid Commun.*, **25**, 1145.

2
Morphology Development in Thermoset Nanocomposites
Peter J. Halley

2.1
Introduction

Thermoset polymers form the basis of many important advanced materials, such as computer chip packaging (insulation), protective coatings, adhesives and advanced aerospace composites, based on their great strength, high-temperature stability, good processability, and good chemical resistance. However, because these desirable properties of thermoset polymers typically stem from the highly crosslinked nature of the cured polymer, these materials may have shortcomings in terms of their final properties, notably their brittleness and lack of durability. Consequently, thermoset nanocomposite materials have been developed to circumvent some of these property shortcomings, primarily by redesigning the nanostructure and underlying morphology of the system.

Some important aspects of the development of thermoset nanocomposites are highlighted in Figure 2.1, including the important steps of nanoparticle design and selection, nanoparticle modification (shape, size, and organo modifications), nanocomposite mixing and processing, nanostructure and morphology measurement and understanding, and final performance property measurements. It should be noted that this process may be cyclic in nature, as such development may require iterative steps with continual feedback, from each of the measurement stages to the design and processing stages.

In this chapter, attention will be focused on the importance of nanostructure, morphology and key property measurements in the design and performance of thermoset polymer nanocomposite material systems. Specifically, the importance of kinetic, chemo-rheology (reactive flow properties) and structure–property experiments to fully characterize effects of these thermoset nanocomposites will be highlighted.

Optimization of Polymer Nanocomposite Properties. Edited by Vikas Mittal
Copyright © 2010 WILEY-VCH Verlag GmbH & Co. KGaA, Weinheim
ISBN: 978-3-527-32521-4

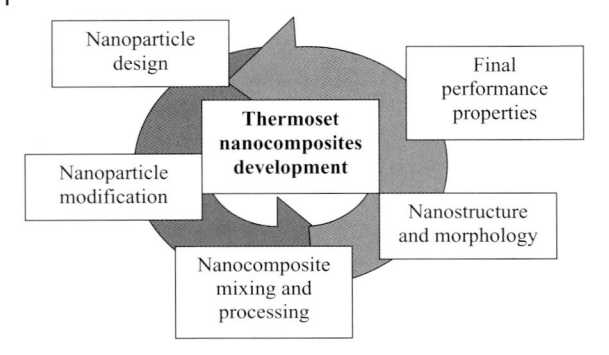

Figure 2.1 Thermoset nanocomposite development.

2.2
Epoxy Nanocomposite Systems

Epoxy nanocomposite systems, which are the most widespread thermoset nanocomposite systems studied, were first reported by Giannelis and Pinnavaia and their coworkers [1, 2].

Wang and Pinnavaia [2] examined the self-polymerization of the diglycidyl ether of bisphenol A (DGEBA) with the concurrent delamination (exfoliation) of acidic forms of montmorillonite (MMT) clays at elevated temperatures. Initially, the polymerization–clay delamination temperature (PDT) was observed to depend on the heating rate and the nature of the clay-exchange cation. It was noted acidic onium ions of the type $[H_3N(CH_2)_{n-1}COOH]^+$, $[H_3N(CH_2)_nNH_2]^+$, $[H_3N(CH_2)_nNH_3]^{2+}$, and $[H_3N\text{-}(CH_2)_{n-1}CH_3]^+$ ($n = 6$ and 12) facilitate the polymerization–delamination process. Subsequently, Messersmith and Giannelis [1] examined an epoxy-silicate nanocomposite by dispersing an organically modified mica-type silicate in an epoxy resin (DGEBA) and curing in the presence of nadic methyl anhydride (NMA), benzyldimethylamine (BDMA), or boron trifluoride monoethylamine (BTFA). Dispersion of the layered silicate within the crosslinked epoxy matrix was verified using X-ray diffraction (XRD) and transmission electron microscopy (TEM), and revealed layer spacing >10 nm and a good wetting of the silicate surface by the epoxy matrix. The curing reaction was claimed to involve the hydroxyethyl groups of the alkylammonium ions located in the galleries of the organically modified silicate, which participate in the crosslinking reaction and result in a direct attachment of the polymer network to the molecularly dispersed silicate layers. Furthermore, a final properties enhancement was observed; for example, the dynamic storage modulus of the nanocomposite containing 4 vol% silicate was approximately 58% higher in the glassy region and 450% higher in the rubbery plateau region compared to the unmodified epoxy. Based on these initial reports, it was proposed that the epoxy nanocomposite's final properties could be enhanced by appropriate nanoscale interactions, and morphological changes were also confirmed.

Brown et al. [3] examined the role of various quaternary ammonium-modified MMTs in epoxy/diamine nanocomposites formation. A hydroxyl-substituted quaternary ammonium modifier combined both catalytic functionality, which increases the intragallery reaction rate, and enhances miscibility toward both reagents. After curing, exfoliated and partially exfoliated epoxy/diamine nanocomposites were produced with an enhanced heat-distortion temperature and an increased flammability resistance.

Montserrat et al. [4] examined the cure reaction kinetics of epoxy resin, with organically modified MMT loadings of up to 20 wt% and with stoichiometric conditions by using differential scanning calorimetry (DSC). The kinetic analysis of isothermal and nonisothermal cures showed the autocatalytic model to be the more appropriate for describing the kinetics of these reactions; it was also observed that a dominant effect of the MMT was to catalyze the curing reaction. Complex effects on the reactions were also noted, especially etherification by cationic homopolymerization catalyzed by the onium ion of the organically modified MMT. As the homopolymerization reaction results in an excess of diamine in the system, the reaction in practice is off-stoichiometric; this leads to a reduction in both the heat of cure and the glass transition temperature (T_g) as the MMT content increases. Small-angle X-ray scattering (SAXS) of the cured nanocomposites showed that an exfoliated nanostructure is obtained in nonisothermal cure at slow heating rates, whereas for a nonisothermal cure at faster heating rates intercalation occurs. Resin/clay mixtures, before the addition of any crosslinking agent, were preconditioned by storage for long periods at room temperature. The development of nanostructures before the addition of any crosslinking agent was attributed to the presence of clay agglomerations in the original resin/clay mixtures, and highlights the importance of the quality of the dispersion of the clay in the resin.

Chatterjee et al. [5] examined adding nanosize TiO_2 fillers into an epoxy resin. For this, an ultrasonic mixing process (via sonic cavitation) was employed to disperse the particles into the resin system. The thermal, mechanical, morphologic and viscoelastic properties of the nanocomposite and the neat resin were monitored using thermogravimetric analysis (TGA), dynamic mechanical analysis (DMA), TEM and the Instron universal testing machine (UTM). The nanoparticles were observed to be dispersed evenly throughout the entire volume of the resin, while the nanofiller infusion improved the thermal, mechanical, and viscoelastic properties of the epoxy resin. The nanocomposite showed increases in storage modulus, T_g, tensile modulus, flexural modulus, and short beam shear strength compared to the neat epoxy resin. The mechanical performance and thermal stability of the epoxy nanocomposites were shown to depend on the dispersion state of the TiO_2 in the epoxy matrix, and correlated with its loading (0.0015–0.006% by volume). In addition, the nanocomposite showed enhanced flexural strength. Several reasons have been proposed to explain these effects in terms of reinforcing mechanisms. First, nanoparticle infusion may alter the morphology of the resin systems, which in turn increases the T_g of the bulk matrix along with the mechanical and thermal properties of the nanocomposites. Nanoparticle incorporation also enhanced the activation energy of the nanocomposites; this was seen to level off at 1% loading and then to decrease with additional loading. The integral procedure decomposition temperature (IPDT) for the epoxy TiO_2-nanocomposite also increased with nanoloading, although this trend was dissimilar from that reported elsewhere. The same trend has been observed for the IPDT and activation energy of the epoxy TiO_2-nanocomposite, with the results of studies showing that nanofiller infusion would improve the thermal, mechanical, and viscoelastic properties of the epoxy resin.

Wang et al. [6] investigated the effect of clay concentration on the structures and properties of bisphenol-A epoxy/nanoclay composites. For this, three composites with organoclay concentrations of 2.5, 5.0, and 7.5 wt% with respect to the epoxy resin were prepared by *in situ* polymerization under mechanical stirring, followed by ultrasonic treatment. The clay was found to aggregate on a microscale, indicating the absence of any fully exfoliated nanocomposites, while the layer space decreased with increases in clay concentration. Whilst the thermal decomposition temperature remained almost unchanged with increases in clay concentration, the T_g of the composites was decreased slightly and the storage modulus increased as the clay concentration was increased.

Chen et al. [7] examined an epoxy nanocomposite filled with 12 nm spherical silica particles as a function of silica loading. The nanoparticles were easily dispersed with minimal aggregation for loadings up to 25 wt% as determined using TEM and ultra-small-angle X-ray scattering (USAXS). A proportional decrease in cure temperature and T_g (for loadings of ≥10 wt%) was observed with increased silica loading. The morphology, as determined by USAXS, was consistent with a zone around the silica particles from which neighboring particles were excluded (where the exclusion zone extended to 10-fold the particle diameter). For samples with loadings <10 wt%, increases of 25% in tensile modulus and 30% in fracture toughness were obtained. More highly loaded samples continued to increase in modulus, but decreased in strength and fracture toughness.

Ingram et al. [8] examined the addition of Cloisite 30B to diglycidyl ether of bisphenol F (DGEBF) and cured with diaminodiphenylsulfone to investigate the influence of organoclay on the extent of cure. An increase in the extent of cure was found with the addition of Cloisite 30B, when lower cure temperatures were employed. Cloisite 30B, when added at 2 wt%, resulted in a 40 °C increase in T_g, while an increase in the magnitude of the bending modulus was also detected.

Kortaberria et al. [9] examined concurrent dielectric and rheological measurements in "real-time" to follow the cure of a diglycidyl ether of bisphenol-A epoxy resin with 4,4-methylene bis(2,6-diethylaniline) hardener and different amounts of organically modified nanoclay (Nanofil 919). Both, the gelation and vitrification times agreed well with rheological and dielectric experiments; moreover, the gel and vitrification times for the nanocomposite systems were lower than for the neat resins, indicating a catalytic effect of the clays on the curing reaction.

Abdalla et al. [10] examined the interrelationships between carbon nanotube (CNT) surface modification, dispersion in an epoxy resin, and the final properties. Both, carboxylated and fluorinated CNTs were dispersed separately in an epoxy resin, after which a torsional dynamic mechanical analysis was applied both parallel and perpendicular to the long axis of the multiwall nanotubes (MWNTs). The epoxy/MWNT (1 wt%) nanocomposites showed higher shear moduli in the glassy state, with the highest values being observed when the nanotubes were parallel to the direction of the applied torque. These epoxy/MWNT (1 wt%) nanocomposites also exhibited higher T_g-values than the neat resin. In addition, the rubbery plateau modulus (between 150–200 °C) was higher by a factor of three for the epoxy/MWNT (1 wt%) nanocomposites. Samples containing fluorinated nanotubes exhibited the highest T_g-values, the longest relaxation times, and the highest activation energies compared to the carboxylated nanotube composite samples and the neat resin. This was indicative of stronger interactions between (the better dispersed) fluorinated nanotubes and the polymer.

Ceccia et al. [11] reported the details of an *in situ* intercalative polymerization of nanocomposites via ultraviolet (UV)-curing to produce hybrid films. Here, organophilic clays were first dispersed in UV-curable epoxy oligomers by ultrasonication, after which a diol was added to the suspension to improve the curing kinetics and final conversion. The addition of nanoclays introduced a solid-like behavior to the low-frequency region (compared to the normally Newtonian epoxy), while addition of the diol caused an increase in the dispersion of the clay within the matrix.

Dean et al. [12] examined multiscale fiber-reinforced nanocomposites which had been manufactured using a vacuum-assisted resin infusion molding (VARIM) process. The nanocomposites were epoxy resin prepolymers dispersed with layered silicates, and the effects of silicate loading on the flow, isothermal cure behavior and solid-state properties were studied. Although the addition of silicate resulted in an increase in resin viscosity (see Figure 2.2), this was still within the range of processability by VARIM. A slight decrease in resin gel times was also observed with an increase in silicate content. Subsequent XRD studies demonstrated an exfoliated morphology for the 2% silicate, and an intercalated morphology for the 4% and 6% specimens. In addition, improvements of 31% in flexural modulus and 24% in flexural ultimate strength for the 2% silicate fiber-reinforced nanocomposites were achieved.

Dean et al. [13] examined the effects of organoclay nanoparticles on the rheology, morphology development and properties for epoxy/organoclay nanocomposites. The interlayer spacing increased with the cure temperature, which in turn resulted in intercalated morphologies with varying degrees of interlayer expansion, depending on the cure temperature used. The results of rheological studies of the curing process indicated that intergallery diffusion before curing, and catalysis during cure, are essential for exfoliation. These findings were in line with previous data obtained by Wang and Pinnavaia [2], which identified the importance of balancing extragallery and intergallery polymerization.

Torre et al. [14] investigated the kinetic and thermal characterization of an epoxy resin (DGEBA) polymerized with a methyl tetrahydrophthalic anhydride reinforced with MMT-layered silicates. The nanoreinforcement was compatibilized by exchanging the cations between the silicate layers with alkylammonium salts, containing long hydrocarbon chains. Unexpected effects on the polymerization kinetics of the epoxy–anhydride system were detected, and the nanofiller was

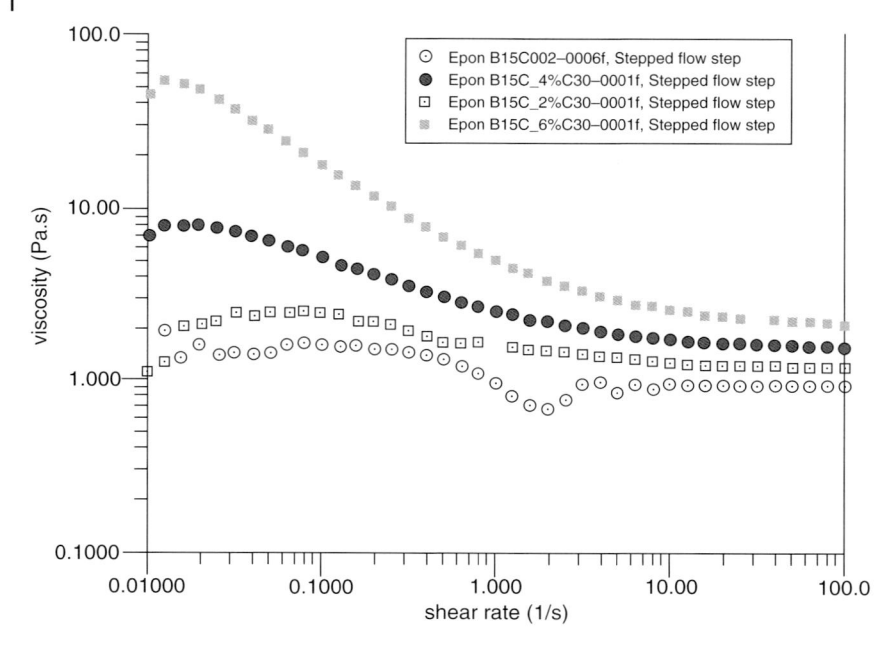

Figure 2.2 Viscosity of pure epoxy resin and samples containing 2%, 4%, and 6% silicate, respectively. Adapted from Ref. [12] with permission from Elsevier.

observed to behave as a catalyst for the crosslinking reaction, thus affecting both the activation energy and curing temperature. Subsequently obtained thermograms revealed a higher thermal stability during degradation in oxygen of the nanocomposite with respect to the pure resin.

Zilg et al. [15] studied nanocomposites based upon hexahydrophthalic anhydride-cured DGEBA and layered silicates including synthetic fluoromica (F-mica), purified sodium bentonite, and synthetic hectorite (all of which were rendered organophilic by means of ion-exchange with mono- and di-functional alkyl ammonium ions). Enhanced toughness is associated with the formation of dispersed anisotropic laminated nanoparticles consisting of intercalated layered silicates utilizing only 5% nanofiller.

Gilman et al. [16] examined the use of new, thermally stable imidazolium-treated layered silicates for the preparation of nanocomposites (in order to overcome the limited thermal stability of alkylammonium cations intercalated into smectite nanomaterials such as MMT). For this, several trialkylimidazolium salt derivatives were prepared with propyl, butyl, decyl, and hexadecyl alkyl chains attached to the imidazolium; these were then used to prepare treated layered silicates. The study results suggested that the use of 1-alkyl-2,3-dimethylimidazolium and 1-alkyl-2,3-dimethylimidazolium salts to replace the sodium in natural MMT would produce organophilic MMT with a 100 °C improvement in its thermal stability.

Mittal [17] examined epoxy nanocomposites with organically modified vermiculite via solution casting. In these studies, the modified vermiculite mineral had a much higher basal plane spacing than the MMT platelets modified with the same ammonium ions, owing to a smaller area per cation available on the surface. The basal spacing values were enhanced after nanocomposite synthesis, thus confirming intergallery polymerization. A TEM analysis indicated the presence of a mixed morphology, with tactoids of varying thicknesses present, together with a small number of single layers. In addition, the permeation properties were significantly improved, despite the pure matrix itself being resistant to oxygen and water vapor permeation.

Becker *et al.* [18] examined the influence of an organically modified clay on the curing behavior of three epoxy systems used widely in the aerospace industry, and also monitored the effect of different structures and functionalities. For this, DGEBA, triglycidyl *p*-amino phenol (TGAP) and tetraglycidyl diamino diphenylmethane (TGDDM) were each mixed with an octadecyl ammonium ion-modified organoclay and cured with diethyltoluene diamine (DETDA). As the nanofiller concentration increased, the gelation time was seen to decrease; this effect was due to the fact that the rate of reaction (but not the reaction mechanism) was increased by the addition of an organoclay.

Dean *et al.* [13] examined epoxy-layered silicate nanocomposites and showed that the interlayer spacing increased with the cure temperature. This resulted in intercalated morphologies with varying degrees of interlayer expansion, depending on the cure temperature used. The results of chemorheological studies of the curing process indicated that intergallery diffusion and catalysis before curing occurred, and was essential for exfoliation, before the morphology was frozen in by gelation and vitrification.

Sun *et al.* [19] examined the use of novel silica nanofillers as an underfill for flip-chip applications, and showed both precure rheology and postcure T_g-values to be affected by nanosilica surface treatment that may increase rates.

A recent review of epoxy nanocomposites [20] highlighted recent developments in the formation and properties of epoxy-layered silicate nanocomposites. Here, the authors examined the effects of processing conditions on cure chemistry and morphology, and discussed the relationship between these conditions and a wide range of material properties. Consequently, in the following sections attention will be focused on the importance of processing and aging on epoxy nanocomposite systems.

2.3
Effects of Processing and Aging

The effects of processing and aging on epoxy nanocomposites have become an important consideration in later studies on these materials. The results showed that optimal processing types, processing conditions and aging conditions should also be considered when designing epoxy nanocomposite systems, in addition to nanofiller type, shape, size, surface treatment, epoxy functionality, and curing conditions. One key approach to monitoring structural and property changes during processing is via chemorheology; indeed, a recent review [21] has expanded on chemorheological tests, chemorheological models and their value in a variety of processes for thermoset systems in general. In this section, however, attention will be focused specifically on processing and aging effects on thermoset nanocomposites.

Both, LePluart et al. [22] and Becker and coworkers [23], have demonstrated the effects of nanoclays on dynamic rheology and steady viscosity pre- and post-cure. The results indicated a clear increase in low-frequency elastic modulus and low-shear viscosity pre-cure, and a clear increase in mechanical properties with increasing clay content post-cure. Interestingly, these data show a direct effect of nanoclay addition on both the pre-cure and post-cure properties, thereby indicating that the nanoclay has an influence on both the pre- and post-cure nanostructures. Similar pre-cure studies were undertaken by Hadjistamov [24], who examined the effect of nanoscale silica on the rheology of silicone oil and uncured epoxy (Araldite) systems. Both, shear thickening and yield–stress-like behavior were observed, these being due to a network structure build-up of the nanocomposite phase in the pre-cure materials. Yet, such an effect was not surprising, as many previous studies have identified the effects of nanomaterials on polymer nanocomposite rheology. Such studies have highlighted a range of important effects, including increased dynamic shear elastic modulus at low frequencies in polyethylene oxide (PEO)-laponite [25], increased shear thinning and reduced high-shear melt elasticity in high-impact polystyrene (HIPS)-MMT [26], increased elongation stiffening in polyamide (PA)-MMT [27], increased relaxation times in epoxy CNTs [28], and a change of structure during annealing in PP-MAg-MMT [29].

Chen et al. [30] examined the effects of processing on nanocomposite formation. Here, details of the dispersion behavior of organic layered silicates (OLS), as a function of the processing procedure, were reported for all stages of processing: in the solvent, in the epoxy prepolymer, and in the epoxy through cure. A different processing procedure (as shown in Table 2.1) was designed for each stage and used so that the morphology of the epoxy/layered-silicate nanocomposite could be regulated. For example, a mild low-shear processing resulted in an intercalated nanocomposite with large-sized aggregates (>10 μm), while high-shear processing resulted in an intercalated nanocomposite with relatively small-sized aggregates (0.5–3 μm). In contrast, the high-shear and ultrasonication processing procedures gave rise to an exfoliated nanocomposite.

Kotsilkova [31] examined "direct" (without solvent) and "solvent" processing techniques for epoxy–smectite clay nanocomposites with three types of organic modifier. Notably, the use of a solvent during processing assists in the enhancement of clay exfoliation. Dispersions of exfoliated nanoclays at concentrations above the critical volume fraction (φ^*) were determined, and a scaling of rheological properties was observed to occur (similar to strongly flocculated suspensions theory, as shown in Figure 2.3). Whilst the curing kinetics was found to depend on both the organic modifier and solvent, the extent of curing was essentially

Table 2.1 Summary of the epoxy/layered-silicate nanocomposites with different morphologies through different processing procedures.

Processing	a	b	c	d
	Low-shear	High-shear	US (one-pot, acetone)	High-shear plus US
Interplanar spacing (Å)	~150	~150	~150	No aggregation
Aggregate size	>10 μm	0.5–3 μm	100–500 nm lateral, 10s of nanometers thick, homogeneous, and random	100–500 nm lateral, nanometers thick, homogeneous, and random
Number of nanolayers per aggregate	Hundreds	30–200	2–8	1 or 2

Adapted from Ref. [30], with permission from Wiley.

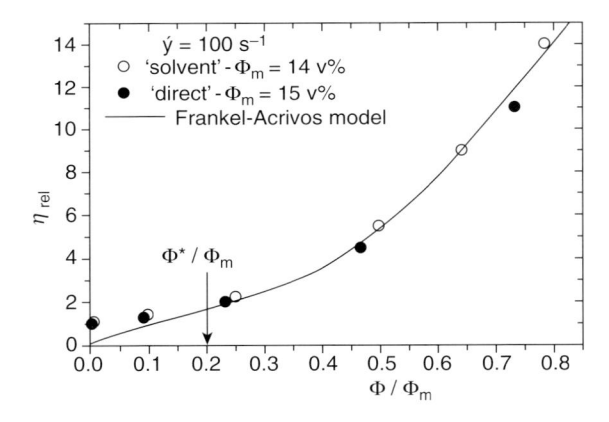

Figure 2.3 The relative viscosity (η_{rel}, at $\gamma = 100\,s^{-1}$) of epoxy suspensions plotted as a function of the reduced volume fraction of smectite (ϕ/ϕ_m) compared with the theoretical prediction by the Frankel–Acrivos model showing that, in the limit of packing ($\phi/\phi_m > 0.2$), nanofiller addition leads to flocculation. Adapted from Ref. [31], with permission from Wiley.

equivalent for both the pure epoxy resin and the nanocomposites. The structure of the nanocomposites produced by the direct processing technique could be controlled by the organic modifier. Hence, by using a solvent-processing technique, the effect of the solvent would dominate that of the organic modifier, and presumably lead to exfoliated nanocomposites. In addition, the mechanical and thermal properties would be greatly enhanced above the φ^* of smectites, and also depend significantly on the type of nanocomposite structure and the use of a solvent.

Koerner et al. [32] examined effects of mechanical processing history and chemistry at the organo-montmorillonite (OMM) surface. For this, the group specifically examined Cloisite 30A (quaternary ammonium OMM) and I.30E (primary ammonium OMM), each of which contained surfactants that had different catalytic effects on the curing chemistry of epoxy Epon 862. A poor intercalation was identified (via sonication) for either treatment, leading to "hybrid" micron-scale reinforcing particles, and no nanoscale dispersion of the individual layers. In contrast, subambient temperature (cryo)compounding had a substantial impact on the ability to reduce tactoid and agglomerate size, and to increase the homogeneity of dispersion for Cloisite 30A. The reactivity close to the Cloisite 30A surface was similar to that in the bulk; thus, any localized gelation around the layer-stacks did not retard particulate refinement. Interestingly, in all cases any alteration of the global epoxy network structure was ruled out by Fourier transform-infrared (FTIR) and nuclear magnetic resonance (NMR) measurements. The coefficients of thermal expansion and hardness were seen to depend only weakly on the morphology, whereas the T_g depended heavily on the extent of OMM dispersion and interfacial chemistry. In general, the inter-relationships between mechanical processing, the OMM surface chemistry and the desired property enhancements were found not to be linear, and so should be considered in light of a final application in order to achieve maximal benefit.

Benson Tolle and Anderson [33] examined the role of preconditioning on the morphology development of organically modified MMT–epoxy nanocomposites. For this, *in situ* synchrotron SAXS studies were performed to relate the initiation and levels of exfoliated morphologies with various silicate preconditioning processes. Significantly, exfoliation could be achieved in systems initially considered intercalated through preconditioning by further epoxy–silicate mixture aging (as shown in Table 2.2), where the resultant morphologies led to a slightly improved toughness. Thus, it was concluded that any interaction between the OLS and the epoxy prior to polymerization would play a major role in how morphology–and, specifically, exfoliation–develops. These findings were similar to those reported by Dennis et al. [34] for thermoplastic nanocomposite extrusion, and suggested that exfoliation was largely affected by the residence time (diffusion) in the extruder.

2.4
Other Thermoset Nanocomposite Systems

Recently, many studies have been conducted on other thermoset polymer nanocomposite systems, including polyester, polyimide, and polyurethane nanocomposites. For example, when Mironi-Harpaz et al. [35] investigated the addition of organo-clays to novel unsaturated polyester (UP)-alkyd chains (without styrene) crosslinked in the presence of peroxide, the induction of high shear levels for prolonged periods was found to promote intercalation and exfoliation of the silicate layers, and to result in a better dispersion of the nanoclay particles. Crosslinking of the UP-alkyd/organo-clay nanocomposites led to changes in their

Table 2.2 The d-spacings reached by I.30E/Epon 828/mPDA at an isothermal temperature of 80 °C, showing the effects of mixture aging.

Weight fraction I.30E	Fabrication process	Initial d-spacing (Å)	d-spacing at 60 min (Å)
5%	Baseline	34	42[a]
5%	Epoxy–silicate mixture aged 1 week	34	45
5%	Epoxy–silicate mixture aged 4+ weeks	36	104
5%	Epoxy–silicate mixture aged 8 weeks	35	105
5%	Epoxy–silicate mixture aged 16+ weeks	36	110
7%	Baseline	33	37
7%	Epoxy–silicate mixture aged 16+ weeks	35	106

a) Final time: 45 min; error typically within 1 Å.
Reproduced from Ref. [33], with permission from Wiley.

nanostructure that were dependent on the peroxide content used; indeed, varying the content between 1% and 6% allowed the realization of either an exfoliated or a combined intercalated/exfoliated structure.

Tugrul Seyhan *et al.* [36] studied CNT/thermoset polyester nanocomposite systems, by using a three-roll mill and sonication technique to process nanocomposites with and without amine (NH_2) functional group treatments on CNTs. The CNT/polyester suspensions were shown to exhibit a shear-thinning behavior, whereas neat polyester resins acted as a Newtonian fluid (as highlighted in Figure 2.4). The nanotubes with amine functional groups were also found to have a better tensile strength when compared to those with untreated CNTs, and to show a greater degree of dispersion of CNTs within the matrix.

McClory *et al.* [37] studied thermosetting polyurethane (PU)/multiwalled carbon nanotube (MWCNT) nanocomposites at loadings of up to 1 wt% via an addition–polymerization reaction. The morphology of the nanocomposites and degree of dispersion studies showed the nanotubes to be highly dispersed in the PU matrix, while the addition of only 0.1 wt% MWCNTs resulted in significant enhancements in stiffness, strength, and toughness. Increases in the Young's modulus, percentage elongation at break and ultimate tensile strength of 561%, 302% and 397%, respectively, were identified for the nanocomposites compared to the unfilled PU. The effect of the MWCNTs on the modulus of the PU was also evaluated using the rule of mixtures, and the Krenchel and Halpin–Tsai models. Only the Halpin–Tsai model, when applied to high-aspect ratio nanotubes, was in good agreement with the modulus values determined experimentally. A strong interfacial shear stress, of up to 439 MPa, was found between the PU chains and nanotubes; this

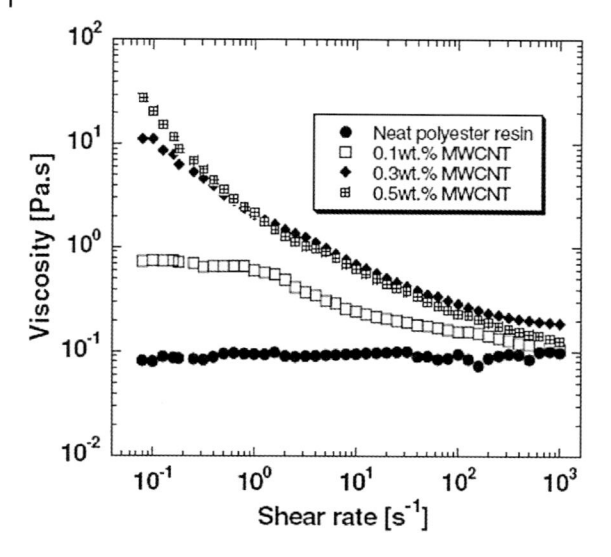

Figure 2.4 Viscosity of the polyester suspension with MWCNTs as a function of shear rate. Adapted from Ref. [36], with permission from Elsevier.

Table 2.3 Values of N (clay layers in stacks) and TEM# (average size in nm of clays) representing clay delamination and dispersion in PMR-15/clay composites.

Composite	N	TEM#	TEM# Std dev
1	3.47	8.9	2.0
1A	3.17	6.0	2.5
2	2.97	8.9	1.7
2A	2.55	7.9	2.4

Adapted from Ref. [38], with permission from Elsevier.

was calculated using a modified Kelly–Tyson model. Evidence for such strong interfacial interactions was obtained from the Raman spectra of both the precursor materials and nanocomposites.

Gintert et al. [38] investigated the addition of nanoclay to a polyimide thermoset resin (PMR). For this, a novel processing method which involved nanoclay intercalation by a lower-molecular-weight PMR monomer prior to dispersion in primary, higher-molecular-weight PMR resin and resin curing was investigated. Sonication of the clay at the time of intercalation by lower-molecular-weight PMR resin was found to help achieve a higher degree of exfoliation. Subsequently, details of clay delamination and dispersion were reported (see Table 2.3) using quantitative TEM with a system similar to that described by Dennis et al. [34] and Finnigan et al. [39]. Such an approach represents an excellent systematic and quantitative means of analyzing electron microscopy data relating to nanoparticle dispersion and delamination.

Clays obtained after ion exchange with a 50:50 mixture of N-[4(4-aminobenzyl)phenyl]-5-norborene-2,3-dicarboximide (APND) and dodecylamine (C12) showed a better exfoliation than did commercial Cloisite 30B clay. The resultant nanocomposites also showed a higher thermal stability and a higher tensile modulus.

An examination of the effects of reaction sequences on thermoset PU nanocomposite formation by Woo et al. [40] showed that, even when identical compositions were employed, any variation in the synthesis routes could affect the structure and properties of the nanocomposite product. The stoichiometry, the dispersion of the layered silicates and the chemical structure each represented principal factors inducing these changes. Among the various synthesis routes, the extent of reaction between the surfactant on the organoclay and the isocyanate precursors of the PU was varied, and this led in turn to imbalances of stoichiometry in the different samples with subsequent direct effects on their mechanical properties. The level of exfoliation of the layered silicates also showed a strong correlation with the degree of polymerization reaction occurring in the interlayer of the organoclay.

2.5
Recent Advances in Thermoset Nanocomposites

2.5.1
Epoxy-HBP Nanostructured Systems

Recently, the development of dendritic and hyperbranched polymers (HBPs) has attracted much attention [41–48]. Both, dendrimers and HBPs are globular macromolecules that have a highly branched structure with multiple reactive chain ends (the shell) that converge to a central focal point (the core). Although, the terms *dendrimers* and *HBPs* are often interchanged, these polymers are different structurally and may have quite different properties [49]. Strictly speaking, dendrimers have a precise end group multiplicity and functionality but are very tedious to synthesize, whereas HBPs mimic (but do not achieve) regular dendritic growth but are easily synthesized. Many HBPs have been synthesized based on PS, polyesters, polyethers, polyphenylene, polyamides, and various engineering polymers [43]. The key attributes of HBPs are their high potential for crosslinking and blending (due to a large number of reactive end groups) and their low viscosities (their viscosity is less than that of linear polymers of equivalent molecular weight [48], due to an elimination of intermolecular entanglements). In particular, it is these properties that have led to HBPs being used as tougheners in epoxy systems.

Recently, several reports have been made on the nanostructured design of epoxy-hyperbranched toughened polymers. In fact, it has been shown [50] that epoxy-functionalized HBPs can greatly improve polymer toughness (250% increase) without affecting the Young's modulus, T_g and/or processing viscosity of a model DGEBF/isophorondiamine epoxy matrix at very low levels of HBP (5 wt%). In this study, the shell chemistry and polarity was seen to affect the initial miscibility and kinetics of phase separation, whilst the subsequent cure conditions and phase separation could be used to control the processability and structure. Unfortunately, details of the shell chemistry and cure profiles were not disclosed by these authors, and consequently detailed relationships between phase separation and shell chemistry/cure conditions could not be determined. In contrast, it has been shown recently [51] that minimal improvements in toughness were obtained for hydroxy- and acetyl terminated-HBPs added to DGEBA/DDS epoxy systems compared to a system containing a linear analogue of the HBPs. Unfortunately, it was noted [51] that the use of epoxy-terminated end groups might promote a better interfacial adhesion, and thus an improved toughness. It was also apparent that a better control of phase separation (by varying the cure conditions) might be needed to further optimize the final structure and properties. It has been postulated that the role of the HBP in toughening would be to act in similar fashion to a core–shell particle; that is, the core of the HBP would serve to cavitate and promote shear yielding, while the shell could be tailored to control aggregation and interactivity with the epoxy matrix. Increases in the core molecular weight should promote cavitation and shell chemistry functionalization, and also increase dissolution and reactivity with the epoxy. However, unlike the core–shell particles, the inherent increase in shell sites and the low viscosity of the HBP would enable toughening to occur without any deleterious effects on other properties. Taken together, these findings represent an exciting future area for nanostructured thermosets.

2.5.2
Ternary Nanostructured Systems and Multiscale Composites

Another exciting area for future research is that of ternary nanostructured systems or multiscale composites. Theses systems involve a thermoset matrix, a filler phase, and an additional phase (either a thermoplastic phase for toughening, or a macroscale composite for strength) in order to enhance the performance properties of the composite. Although the possibility of complex interactions is clearly greater in a ternary system, this offers a greater degree of tailorability in terms of desired product performance.

Mei *et al.* [52] examined a DGEBA epoxy resin containing different shaped nanofillers (e.g., MWCNTs, clay and clay–MWCNT composites), with diaminodiphenyl sulfone (DDS) hardener being used as the curing agent. The nanocomposites were processed by shear mixing using different clay contents (2% and 5 wt%) and MWCNT concentrations (0.1–0.5 wt%). Subsequent investigations using XRD and field emission scanning electron microscopy (SEM) revealed that an exfoliated nanocomposite structure was formed in which the clay layers were randomly dispersed within the matrix. The fracture toughness test results showed the clay–MWCNT to exhibit a synergistic effect on toughening the epoxy resin.

Table 2.4 The time for the onset of phase separation, the gelation or vitrification, the end of phase separation and the relative gelation time for the filled and unfilled specimens.

T (°C)	Epoxy/PEI				Epoxy/PEI/OLS			
	t_{os} (s)	t_{gt} (s)	t_{end} (s)	$t_{get,r}$	t_{os} (s)	t_{gt} (s)	t_{end} (s)	$t_{get,r}$
120	7150	8670	13 780	0.229	4967	6340	12 130	0.192
130	4728	5605	8 902	0.210	3518	4200	7 180	0.186
140	3035	3600	5 750	0.208	2043	2428	4 050	0.192
150	2066	2420	3 749	0.210	1413	1780	3 420	0.183
160	1344	–	–	–	1060	1330	2 541	0.182
170	979	–	–	–	770	915	1 625	0.170
180	731	–	–	–	568	730	1 550	0.165
190	560	–	–	–	437	548	1 156	0.154
200	494	–	–	–	310	390	820	0.157
210	–	–	–	–	220	286	645	0.155
220	–	–	–	–	168	200	372	0.157

Adapted from Ref. [53], with permission from Elsevier.

When Peng et al. [53] incorporated modified OLSs into mixtures of epoxy and thermoplastic poly(ether imide) (PEI) to investigate reaction-induced phase separation, the onset of phase separation and the gelation or vitrification times were shown to be greatly brought forward, while the periodic distance of phase-separated structure was reduced when the OLS was incorporated (see Table 2.4). The OLS-filled hybrid nanocomposites carried out an obvious phase separation at cure temperatures ranging from 120 to 220 °C, but this was much restricted in the absence of nanocomposite addition. Such behavior was shown to be related to the preferential wettability, the chemical reaction of OLS with the epoxy oligomer, and the enhanced viscosity of the nanocomposite mixture.

Schubel et al. [54] examined the effects of two clays (Cloisite10A and Garamite 1958) and a conventional low-profile additive (polyvinyl acetate; PVAc) with calcium carbonate filler at various loadings on the volumetric shrinkage and T_g of polyester components used in the construction of cars. A series of hybrid matrices consisting of clay and PVAc were used to impregnate random E-glass preforms via resin transfer molding (RTM). The results suggested that the exfoliated clay systems worked synergistically with conventional additives to reduce shrinkage and residual volatile organic compounds (VOCs), whilst in many cases improving the final performance properties above the level of the base resin.

Bauer et al. [55] reviewed the applications of nanocomposites as fiber-reinforced polymer matrix composites for lightweight design, and as high-performance composites for traffic engineering uses. Desired improvements areas include mechanical properties, interlaminar shear strength, reinforcement in z-direction, fiber–matrix adhesion, and the obtaining of new functionalities.

2.5.3
Novel Characterization Methods

Drummy et al. [56] highlighted a new quantitative analysis of layered silicate (nanoclay)–thermoset polymer nanocomposite morphology using two characterization methods, namely electron tomography and SAXS. In these studies, it was noted that the segmentation of two-dimensional Z-contrast projection images of randomly oriented, highly anisotropic nanoparticles (such as layered silicates in polymer nanocomposites) was extremely inaccurate, on the basis that 75% of the volume of MMT layers in an epoxy matrix would not be identified in the segmentation. By using electron tomography, this number could be reduced to below 15%, and a tomographic reconstruction revealed three-dimensional (3-D) information. The corresponding 3-D fast Fourier transformation (FFT) indicated that the image volume did not contain a sufficient distribution of local environments to directly correspond to the global average, as revealed with SAXS. However, in contrast to SAXS the tomographic reconstruction provided precise details of the distribution of morphological features, as well as statistical averages over the sample volume (refer to the process in Figure 2.5). It is anticipated that SAXS and tomography should be used in tandem in order to provide more accurate information on morphology.

2.5.4
Modeling Thermoset Nanocomposite Systems

At this point, it is very important to note the growing interest in the modeling of thermoset nanocomposite interactions on a molecular level. These studies are essential to provide not only a clear understanding of the fundamental interactions but also an ability to predict the structures and properties of new nanocomposite systems.

Recently, Yu et al. [57] investigated the effect of different-sized alumina (Al_2O_3) nanoparticles on the mechanical properties of thermoset epoxy-based nanocomposites using molecular dynamics (MD) simulations combined with sequential scale bridging methods. Based on the MD simulation results, the sequential bridging method has been adopted for the efficient estimation of particle size and epoxy networking effects, while an effective interface concept has been incorporated as a characteristic phase which can describe the particle size effects. A schematic of the basic unit is shown in Figure 2.6.

Wang and Pan [58] presented a general review of numerical multiscale property models for various complex multicomposite materials. In particular, the studies conducted by Buryachenko et al. [59] were detailed, as these have focused on property modeling for thermoset nanocomposites. Also reviewed were methods of calculating nanoelement-reinforced composite mechanical properties, together with highlights of a multiscale modeling methodology to calculate elastic constants and local/interface properties for systems with a statistically homogeneous distribution of embedded nanofibers, nanoparticles,

2.5 Recent Advances in Thermoset Nanocomposites

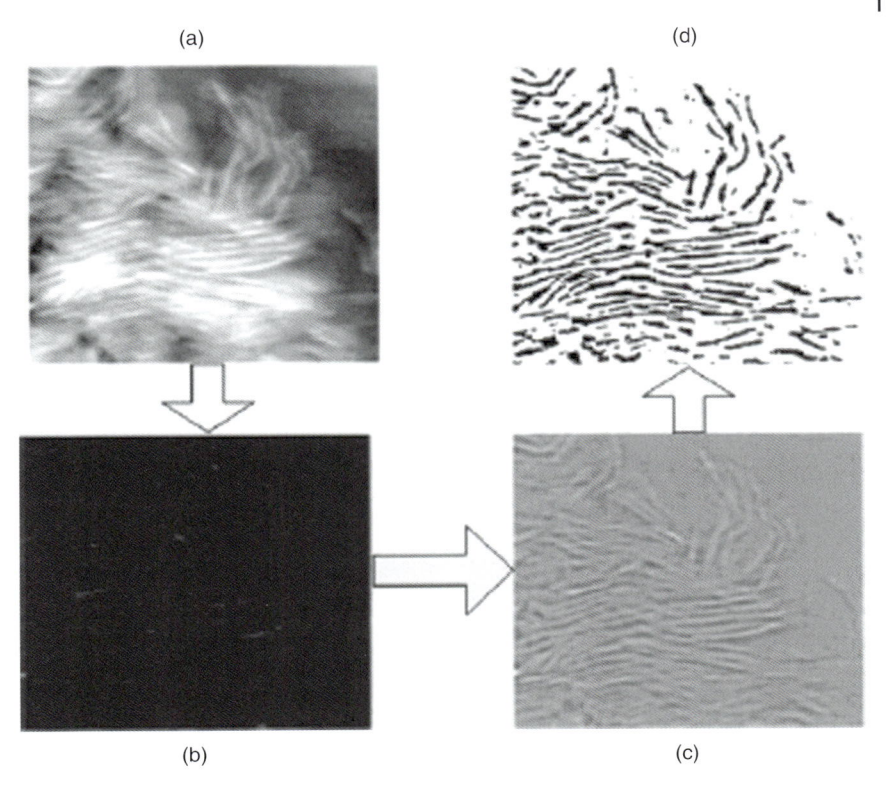

Figure 2.5 Image processing steps for segmentation. (a) A Gaussian blur and a high-pass filter are applied to the image to remove noise as well as long-range background intensity variations; (b) An opening filter is used to remove "bridges" and small particles of less than a predetermined diameter; (c) An arithmetic Laplacian filter is used to calculate the second derivative of the image, filter out low frequencies, and enhance the boundaries of objects; (d) The image is then contrast-enhanced and inverted, and a threshold applied to produce a binary image. Adapted from Ref. [56], with permission from The American Chemical Society.

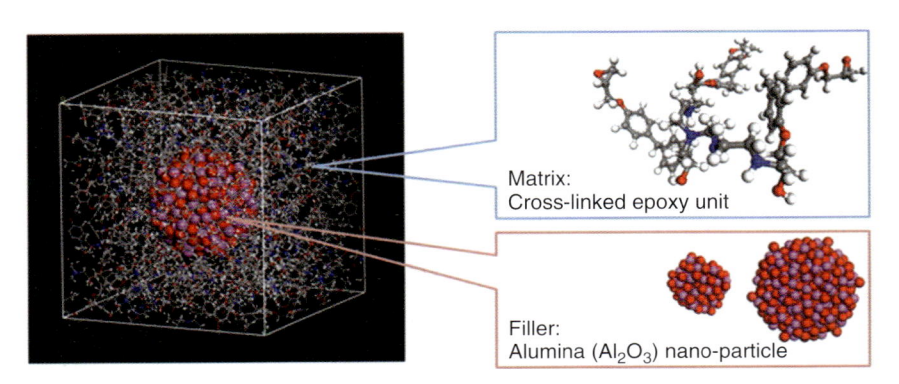

Figure 2.6 The nanocomposites unit cell consists of the crosslinked epoxy unit as matrix and alumina particles as the filler. Adapted from Ref. [57], with permission from Elsevier.

nanoplates, or other heterogeneities that are either aligned or randomly oriented.

To summarize, it is essential that future modeling studies are developed in parallel with novel experimental investigations of thermoset nanocomposites, in order to provide a greater depth of understanding with regards to current and new systems.

2.6 Summary

In this chapter, it has been shown that the organically modified inorganic filler materials will affect the cure kinetics, nanostructure, processing rheology, materials properties and ultimate performance properties of thermoset nanocomposites. An understanding of the complex inter-relationships in thermoset systems is achieved in the nanocomposites via complex chemical and physical interactions directly at the nanoscale and via secondary effects on larger microscales, especially in ternary composite systems. As the potential of thermoset nanocomposites continues to generate great excitement, based on the number of complex interactions possible, the realization of such potential will require an understanding of these complexities, obtained via state-of-the-art characterization and modeling tools that are only now beginning to come within reach.

References

1 Messersmith, P. and Giannelis, E. (1994) *Chem. Mater.*, **6**, 1719–1725.
2 Wang, M. and Pinnavaia, T.J. (1994) *Chem. Mater.*, **6**, 468–474.
3 Brown, J.M., Curliss, D., and Vaia, R.A. (2000) *Chem. Mater.*, **12** (11), 3376–3384.
4 Montserrat, S., Roman, F., Hutchinson, J.M., and Campos, L. (2008) *J. Appl. Polym. Sci.*, **108**, 923–938.
5 Chatterjee, A. and Islam, M.S. (2008) *Mater. Sci. Eng. A*, **487**, 574–585.
6 Wang, J., Kong, X., Cheeng, L., and He, Y. (2008) *J. Univ. Sci. Tech. Beijing*, **15** (3), 320.
7 Chen, C., Justice, R.S., Schaefer, D.W., and Baur, J.W. (2008) *Polymer*, **49**, 3805–3815.
8 Ingram, S.E., Pethrick, R.A., and Liggat, J.J. (2008) *Polym. Int.*, **57**, 1206–1214.
9 Kortaberria, G., Solar, L., Jimeno, A., Arruti, P., Gomez, C., and Mondragon, I. (2006) *J. Appl. Polym. Sci.*, **102**, 5927–5933.
10 Abdalla, M., Dean, D., Adibempe, D., Nyairo, E., Robinson, P., and Thompson, G. (2007) *Polymer*, **48**, 5662–5670.
11 Ceccia, S., Turcato, E.A., Maffettone, P.L., and Bongiovanni, R. (2008) *Prog. Org. Coatings*, **63**, 110–115.
12 Dean, D., Obore, A.M., Richmond, S., and Nyairo, E. (2006) *Comp. Sci. Technol.*, **66**, 2135–2142.
13 Dean, D., Walker, R., Theodore, M., Hampton, E., and Nyairo, E. (2005) *Polymer*, **46**, 3014–3021.
14 Torre, L., Frulloni, E., Kenny, J.M., Manferti, C., and Camino, G. (2003) *J. Appl. Polym. Sci.*, **90**, 2532–2539.
15 Zilg, C., Mulhaupt, R., and Finter, J. (1999) *Macromol. Chem. Phys.*, **200**, 661–670.
16 Gilman, J.W., Awad, W.H., Davis, R.D., Shields, J., Harris, R.H., Davis, C., Morgan, A.B., Sutto, T.E., Callihan, J., Trulove, P.C., and Delong, H.C. (2002) *Chem. Mater.*, **14** (9), 3776–3785.

17 Mittal, V. (2008) *J. Comp. Mater.*, **42**, 2829–2839.
18 Becker, O., Cheng, Y., Varley, R., and Simon, G. (2003) *Macromolecules*, **36**, 1616–1625.
19 Sun, Y., Zhang, Z., and Wong, C. (2006) *IEEE Transact Comp. Packag. Technol.*, **29**, 190–197.
20 Becker, O. and Simon, G. (2005) *Adv. Poly. Sci.*, **179**, 29–82.
21 Halley, P.J. and George, G.A. (2009) *Chemorheology – From Fundamentals to Reactive Processing*, Cambridge University Press, Cambridge, UK.
22 Lepluart, L., Duchet, J., Sautereau, H., Halley, P., and Gerard, J. (2004) *Appl. Clay Sci.*, **25**, 207–219.
23 Becker, O., Simon, G., Varley, R., and Halley, P. (2003) *Polym. Eng. Sci.*, **43**, 850–862.
24 Hadjistamov, D. (1999) *Appl. Rheol.*, 212–218.
25 Lingibson, S., Kim, H., Schmidt, G., Chan, C.C., and Hobbie, K. (2004) *J. Colloid Interface Sci.*, **274**, 515–525.
26 Chen, D., Yang, H., He, P., and Zhang, W. (2005) *Comp. Sci. Technol.*, **65**, 1593–1600.
27 Seong, D.G. and Yeong, J.R. (2004) Proceedings, XIVth International Congress of Rheology, Seoul, Korea.
28 Kotsilkova, R. and Pissis, P. (2005) *J. Poly. Sci., Part B: Polym. Phys.*, **43**, 522.
29 Li, J., Zhou, C., Wang, G., and Zhao, D. (2003) *J. Appl. Polym. Sci.*, **89**, 318–323.
30 Chen, C., Benson-Tolle, T., Baur, J.W., and Sputthanarat (2008) *J. Appl. Polym. Sci.*, **108**, 3324–3333.
31 Kotsilkova, R. (2005) *J. Appl. Polym. Sci.*, **97**, 2499–2510.
32 Koerner, H., Misra, D., Tan, A., Drummy, L., Mirau, P., and Vaia, R. (2006) *Polymer*, **47**, 3426–3435.
33 Benson Tolle, T. and Anderson, D.P. (2004) *J. Appl. Polym. Sci.*, **91**, 89–100.
34 Dennis, H., Hunter, D., Chang, D., Kim, S., White, J., and Cho, J. (2001) *Polymer*, **42**, 9513–9523.
35 Mironi-Harpaz, I., Narkis, M., and Siegmann, A. (2006) *Macromol. Symp.*, **242**, 201–207.
36 Tugrul Seyhan, A., Gojny, F.H., Taniglu, M., and Schulte, K. (2007) *Eur. Polym. J.*, **43**, 374–379.
37 McClory, C., McNally, T., Brennan, G.P., and Erskine, J. (2007) *J. Appl. Polym. Sci.*, **105**, 1003–1011.
38 Gintert, M.J., Jana, S.C., and Miller, S.G. (2007) *Polymer*, **48**, 4166–4173.
39 Finnigan, B., Jack, K., Campbell, K., Halley, P., Truss, R., Casey, P., Cookson, D., King, S., and Martin, D. (2005) *Macromolecules*, **38** (17), 7386–7396.
40 Woo, T., Halley, P., Martin, D., and Kim, D.S. (2006) *J. Appl. Polym. Sci.*, **102**, 2894–2903.
41 Chu, F. and Hawker, C. (1993) *Polym. Bull.*, **30**, 265.
42 Feast, W. and Stanton, N. (1995) *J. Mater. Chem.*, **5**, 405.
43 Kim, Y. (1998) *J. Poly. Sci., Part A: Polym. Chem.*, **36**, 1685.
44 Malmstrom, E., Johansson, M., and Hult, A. (1995) *Macromolecules*, **28**, 1698.
45 Newkome, G., Yao, Z., Baker, G., and Gupta, V. (1985) *J. Org. Chem.*, **50**, 2003.
46 Tomalia, D. (1985) *Polym. J.*, **17**, 117.
47 Webster, O. (1991) *Science*, **251**, 887.
48 Wooley, K., Frechet, J., and Hawker, C. (1994) *Polymer*, **35**, 4489.
49 Frechet, J., Hawker, C., Gitsov, I., and Leon, J. (1996) *J. Macromol. Sci.: Pure Appl. Chem.*, **A33**, 1399.
50 Boogh, L., Pettersson, B., and Manson, J. (1999) *Polymer*, **40**, 2249.
51 Wu, H., Xu, J., Liu, Y., and Heiden, P. (1999) *J. Appl. Poly. Sci.*, **72**, 151.
52 Mei, Q.L., Wang, J.H., Huang, Z.X., and Yuan, L. (2008) *2nd IEEE International Nanoelectronics Conference*, vols 1–3, pp. 280–284.
53 Peng, M., Li, H., Wu, L., Chen, Y., Zheng, Q., and Gu, W. (2005) *Polymer*, **46**, 7612–7623.
54 Schubel, P.J., Johnson, M.S., Warrior, N.A., and Rudd, C.D. (2006) *Composites: Part A*, **37**, 1757–1772.
55 Bauer, M., Kahle, O., Landeck, S., Uhig, C., and Wurzel, R. (2008) Frontiers in Materials Science and Technology, in *Advanced Materials Research*, vol. 32 (eds J. Bell, C. Yan, L. Ye, and L. Zhang), Trans Tech Publications, Brisbane, Australia, pp. 149–152.
56 Drummy, L.F., Wang, Y.C., Schoenmakers, R., May, K., Jackson, M.,

Koerner, H., Farmer, B.L., Mauryama, B., and Vaia, R.A. (2008) *Macromolecules*, **41**, 2135–2143.

57 Yu, S., Yang, S., and Cho, M. (2009) *Polymer*, **50** (3), 945–952.

58 Wang, M. and Pan, N. (2008) *Mater. Sci. Eng. R*, **63** (1), 1–30.

59 Buryachenko, V.A., Roy, A., Lafdi, K., Anderson, K.L., and Chellapilla, S. (2005) *Comp. Sci. Technol.*, **65**, 15–16.

3
Morphology and Interface Development in Rubber–Clay Nanocomposites

Yong-Lai Lu and Li-Qun Zhang

3.1
Introduction

The concept of polymer/layered silicate nanocomposites (PLSNs) or polymer/clay nanocomposites (PCNs) – alternatively referred to as polymer–clay hybrids (PCHs) – was first presented in 1987 by a research team from the Toyota Central Research & Development Laboratories (TCRDL) in Japan, following their pioneer investigations into nylon-6–clay hybrids [1]. Since that time, these types of nanocomposite have attracted much interest, both in industry and in academia, the reason being that they often exhibit remarkable improvements in materials properties when compared to virgin polymer or conventional micro- and macrocomposites. Consequently, numerous investigations have been carried out on such nanocomposites, the results of which have been extensively reviewed [2, 3].

Studies on rubber–clay nanocomposites (RCNs) were started later than those of plastic-based nanocomposites [4], with relevant reports appearing in international journals only after the year 2000. In such studies, the properties of clay-reinforced rubbers were shown to depend very heavily on the dispersion state of the silicate, whilst a good interfacial interaction between the rubber and the nanosilicate layers would enhance any effects on property improvement, most notably those of mechanical strength and gas-barrier capabilities. The aim of this chapter was, therefore, to survey not only the current situation relating to research into the mechanisms and influential factors linked to the morphological development of RCNs prepared by melt compounding and latex compounding, but also some of the methods used to enhance the interface between rubber and nanosilicates.

Although *in situ* polymerization and solution compounding represent two other available methods for the preparation of RCNs, both techniques proved to be either too complex or non-environment-friendly, and consequently progress in these areas will not be included in this chapter. However, these investigations have been reviewed extensively elsewhere [4].

Optimization of Polymer Nanocomposite Properties. Edited by Vikas Mittal
Copyright © 2010 WILEY-VCH Verlag GmbH & Co. KGaA, Weinheim
ISBN: 978-3-527-32521-4

3.2
Melt Compounding

3.2.1
Mechanism and Influencing Factors

Among the many currently available rubber-compounding techniques, the melt-compounding method (MCM) is probably the most widely applied for preparing RCNs, with related R&D studies having profited greatly from the general rules deduced for thermoplastic-based nanocomposites. The successful preparation of RCNs via melt compounding requires that the clay is first organically modified in order to: (i) reduce its hydrophilicity; and (ii) facilitate intercalation of the rubber chains into the gallery of clay layers.

Vaia and Giannelis were the first to establish the thermodynamic principles for polymer melt intercalation in organically modified clays (OMCs) [5, 6]. Here, the free energy change (ΔF) upon intercalation was separated into an energy change (ΔE) due to new intermolecular interactions and a combinational entropy change (ΔS) associated with the conformational change of the polymer ($\Delta S^{polymer}$) and the modifier's aliphatic chains (ΔS^{chain}); this could be expressed by following two equations:

$$\Delta F = \Delta E - T\Delta S \tag{3.1}$$

$$\Delta S \approx \Delta S^{chain} + \Delta S^{polymer} \tag{3.2}$$

If ΔF is below zero, then the intercalation would be thermodynamically favorable; otherwise, it should be thermodynamically forbidden. For a nonpolar polymer, ΔE caused by the change in intercalated structures would be small, so that the intercalation should be dominantly determined by ΔS. According to both theoretical calculations [5] and experimental validation [6], there is a critical gallery height h_c (~2.4 nm; the corresponding basal spacing is ~3.4 nm) for the compounding system containing clay modified by octadecylammonium salt (the initial basal spacing is ~2.3 nm) and an arbitrary aliphatic polymer (nonpolar), below which the penalty for polymer confinement is compensated by entropy gains of the tethered surfactant chains associated with interlayer, and the overall entropy change is near zero. For a nonpolar polymer having weak interactions with surfactant chains and silicate surfaces, ΔF for intercalation of the polymer into the silicate gallery is always below zero if $h < h_c$ (where h is the gallery height of the layered silicates); in contrast, ΔF is larger than zero if $h > h_c$. In some nonpolar rubbers, however, intercalation structures with a gallery height far larger than h_c (i.e., 2.4 nm) were obtained by melt blending [see Figure 3.1, which shows the wide-angle X-ray diffraction (WAXD) pattern and transmission electron microscopy (TEM) image of isoprene-isobutadiene rubber (IIR)/clay]. In order to interpret this phenomenon, the theory for melt intercalation was slightly modified by considering the special features of the rubber [7]. In comparison with common thermoplastic polymers, rubber has a much greater molecular weight, so that the viscosity and shear stress for rubber compounds during melt-blending remain quite high. Rubber chains are oriented by shear stress to a large extent, especially at the gap

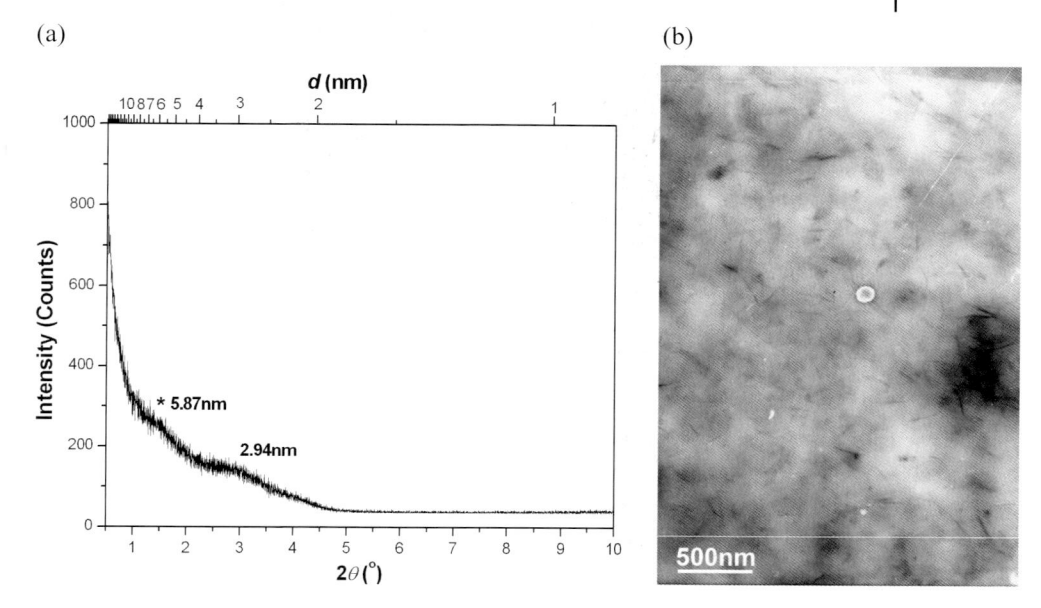

Figure 3.1 Microstructure and morphology of IIRCN (IIR/OMC = 100/10) prepared by melt compounding. (a) WAXD patterns (the basal spacing of OMC is ~2.4 nm); (b) Transmission electron microscopy images. Reproduced from Ref. [7] with permission from John Wiley & Sons Ltd.

region between two rolls, and this may decrease the entropy of the chains before their intercalation. Therefore, $\Delta S^{polymer}$ related to polymer confinement would be reduced during melt intercalation, so that h_c for rubber in this dynamic state may be far larger than h_c(~2.4 nm) estimated based on thermoplastics. These analyses may also explain, from the viewpoint of melt intercalation, the influence of mixing shear magnitude on the dispersion of OMCs.

At present, a number of primary factors determining the morphology and interface of RCNs prepared with melt compounding have been identified, as summarized in following sections.

3.2.1.1 The Organic Modification

The results of studies on nitrile-butadiene rubber (NBR)/OMC nanocomposites [8, 9] have indicated that increasing the alkyl length of the surfactant would not only improve the spatial dispersion state (i.e., the dispersion dimension reduced but distribution homogeneity increased; see Ref. [8]), but also change the local structure in terms of intercalation and exfoliation (see Ref. [9]). Apart from the alkyl length of the surfactant, Zhang and coworkers also recognized that the intercalation extent increased with increasing surfactant dosage in the ethylene propene diene monomer rubber (EPDM)/OMC nanocomposites [10]. A study involving the styrene butadiene rubber (SBR)/OMC system also confirmed the above-described point that longer alkyl chains and a larger dosage of the modifier would facilitate the formation of a better dispersion state [11].

3.2.1.2 The Features of Rubber and Compatibilizers or Coupling Agents

As with thermoplastic-based clay nanocomposites, it is relatively straightforward to obtain a good dispersion of clay particles in the rubber matrix with high polarity, examples being either NBR [8, 9, 12] and hydrogenated nitrile butadiene rubber (HNBR) [13]. In this case, polar rubber could be used as the compatibilizer for improving the dispersion state of OMC, as well as for the interaction with nonpolar rubber. For instance, epoxidized natural rubber (ENR) [14] and maleic anhydride-grafted EPDM (EPDM-g-MAH) [10, 15, 16] were used for NR/OMC and EPDM/OMC, respectively. Kim *et al.* also disclosed that the addition of a silane coupling agent could enhance the dispersion of OMC (C18-MMT) in NBR [17].

3.2.1.3 Melt-Compounding Conditions

Based on the results of their research into SBR/OMC and EPDM/OMC, both Schön *et al.* [11] and Gatos *et al.* [16] drew the same conclusions, respectively, that a high-shear mixing force could improve the dispersion of OMCs in a rubber matrix. However, whilst Schön and colleagues suggested that compounding in an open roll mill was more effective than using an internal mixer, Gatos and coworkers arrived at an opposing conclusion. In fact, Gatos *et al.* proposed that a high temperature of compounding might even improve the dispersion of the clay.

3.2.2
Evolution of Morphology and Interface during Vulcanization of RCNs

For most rubber-based composites, the curing process must be conducted under high temperature and high pressure after compounding, so as to obtain the crosslinked nanocomposites. It was also shown that that the morphology might even change greatly during the vulcanization process, which could be classified as two types, namely changes in the local microstructure and in the spatial distribution of the clay particles.

3.2.2.1 Changes in the Local Microstructure of Clay Particles

This microstructural change could be deduced from WAXD patterns of RCNs before and after curing, and can be further divided into three subclasses:

- **Further intercalation:** The position of the (001) reflection peak of the OMC is shifted to a lower 2θ angle, or the peak may even disappear. This implies the formation of exfoliated structures or intercalated structures with a low degree of order.

- **Confinement or collapse:** The position of the (001) peak is obviously shifted to a higher angle after curing, but the basal spacing is still larger than that for a pure OMC used in this compound. This means that the originally interacted rubber chains moved outside the silicate gallery during the vulcanization process.

- **Deintercalation:** The position of the (001) peak is larger than that of the pure OMC after curing. This suggests that not only the rubber chains moved outside the silicate gallery, but also that there was deadsorption of the organic surfactant. In some cases, totally inorganic clays might even be formed.

3.2.2.2 Change in the Spatial Distribution of Clay Particles

This type of change includes the dimension of dispersed clay particles and their distribution homogeneity on a large scale, with such information normally being acquired by using TEM. With regards to reinforcement with OMC, their spatial distribution homogeneity is more important than whether exfoliated clay layers exist. Hence, TEM observations at low magnification may be more suitable for detecting this type of morphological change.

Recently, Lu *et al.* observed the local microstructures and spatial distribution of OMC particles during the different stages of vulcanization of a variety of RCNs [18]. Beyond all of Lu *et al.*'s and other research groups' previous expectations, the results revealed that obvious microstructural changes at both local (Figure 3.2) and spatial scales (Figure 3.3) could occur over the entire vulcanization course, this being attributed to the high mobility of rubber segments in crosslinked network structures. The magnitude of change in morphology is most likely dependent on the matrix rubber polarity, with the morphological changes in nonpolar rubber-based nanocomposites (such as EPDM) being considerably greater than those in their polar rubber-based counterparts (such as BIIR). It should be noted here that the spatial dispersion state of clay particles in EPDMCNs was improved to some extent during the early stage (i.e., T_{10}), but became far worse on completion of the entire vulcanization course (i.e., T_{100}).

A number of studies have been conducted to identify the reasons and mechanisms for these morphological changes, the ultimate aim being to take advantage of these types of changes in order to tailor the microstructures of the RCNs. At present, the study results have suggested that two classes of factor—namely *chemical* and *physical*—influence the morphological evolution of the RCNs during vulcanization.

Chemical Factors The chemical factors further include two aspects, notably: (i) reactions between rubber and curatives at the initial stage of the curing process; and (ii) reactions between curatives and amine-type intercalants within the silicate galleries. When Usuki *et al.* investigated the influence of the curing accelerator type on microstructures of crosslinked EPDM/clay nanocomposites (EPDMCNs) [20], the silicate layers of the clay were shown to be exfoliated and almost to disperse as monolayers in the cured EPDMCNs when a thiuram (TS) or a dithiocarbamate (PZ) -type vulcanization accelerator was used. It was also assumed that radicals produced by the thermal dissociation of TS or PZ could combine with carbon atoms in EPDM chains to polarize the EPDM molecules; this in turn would result in a further intercalation of EPDM molecules into the clay galleries through hydrogen bonds between the polarized EPDM and the clay surface. Other studies of natural rubber (NR) [21], ENR [22], EPDM [23, 24] and HNBR [25] -based clay nanocomposites, as conducted by Karger-Kocsis *et al.*, revealed that the confinement, deintercalation and further intercalation could occur simultaneously during vulcanization of the RCNs. Hence, these authors speculated that a Zn complex

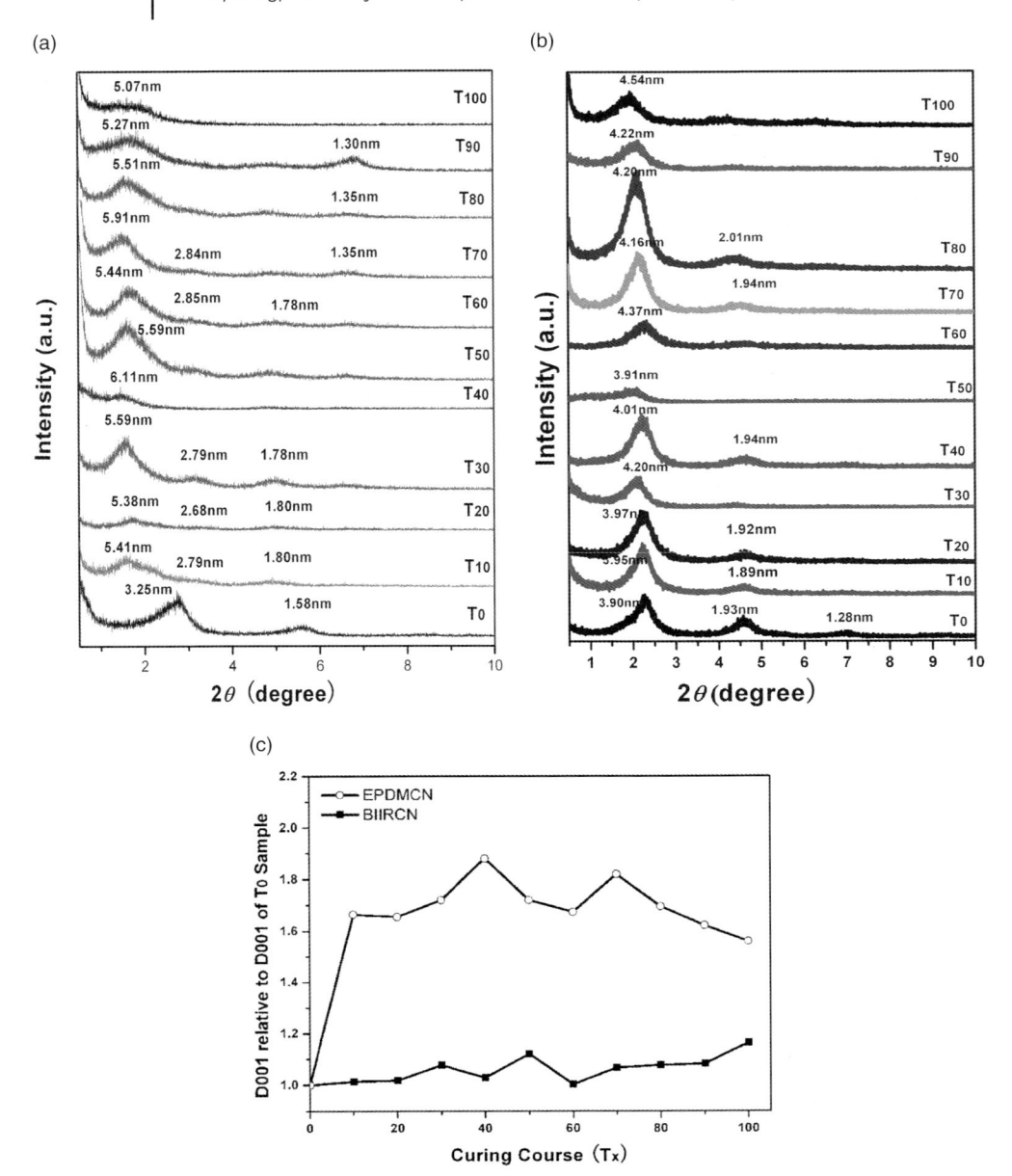

Figure 3.2 Evolution of local microstructures of EPDMCNs and BIIRCNs undergoing different stages of vulcanization. (a) WAXD patterns for EPDMCNs; (b) WAXD patterns for BIIRCNs; (c) D001 changing trends with curing course. The T0 sample represents the uncured RCNs. T10, T20...T90, and T100 represent the time when the curing course reached 10%, 20%...90%, and 100% of the full vulcanization course, respectively. Reprinted from Ref. [18] with permission.

(a)

| T0 | T10 | T50 | T100 |

(b)

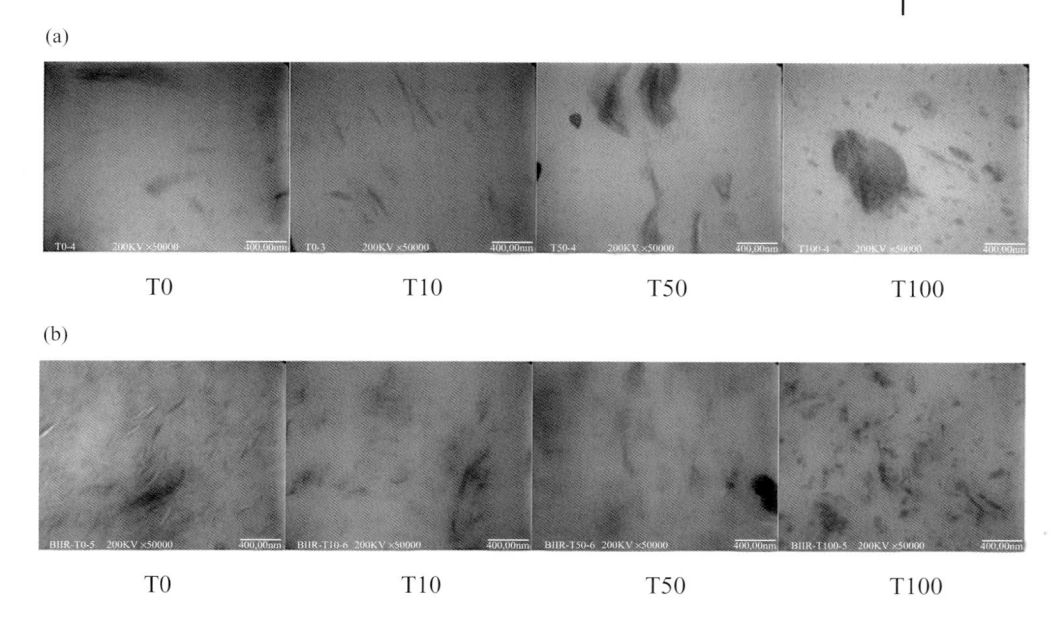

| T0 | T10 | T50 | T100 |

Figure 3.3 Transmission electron microscopy images of (a) EPDMCNs and (b) BIIRCNs at different curing stages. Reprinted from Ref. [18] with permission from John Wiley & Sons Ltd.

containing sulfur and amine groups from the organic amine-type intercalants might be formed during vulcanization. This Zn complex would serve as an intermediate for sulfur vulcanization, and react with rubber. However, if it were to migrate into the rubber matrix to take part in the vulcanization process, then the intercalants would be extracted from the clay galleries, with a resultant confinement and even deintercalation. But, if the complex were to cause rubber crosslinking inside the galleries, then the rubber molecules would be further inserted into the clay galleries, resulting in further separation of layers or delamination/exfoliation. It was further disclosed that the different types of amine intercalant played important roles in the structural changes of clay layers. For example, when a primary amine (e.g., octadecylamine, ODA) was used as the intercalant, then confined and even deintercalated structures were always found in the crosslinked RCNs. In contrast, when a quaternary amine (which has less reactivity with curatives) was used, then confinement and deintercalation was almost nonexistent [24, 25]. When Karger-Kocsis and colleagues used peroxide as a substitute for sulfur curatives to cure HNBR/OMC composites, the RCNs obtained demonstrated a well-ordered intercalated structure, despite a primary amine having been used as the intercalant [26].

Lu et al. reported a dramatic impact of curing temperature on microstructural changes in both the local and spatial scales of sulfur-cured IIRCNs [19]. Here, it was found that significant deintercalation occurred only when the curing temperature was above a certain level (i.e., 140 °C in this system; Figure 3.4). Moreover, a

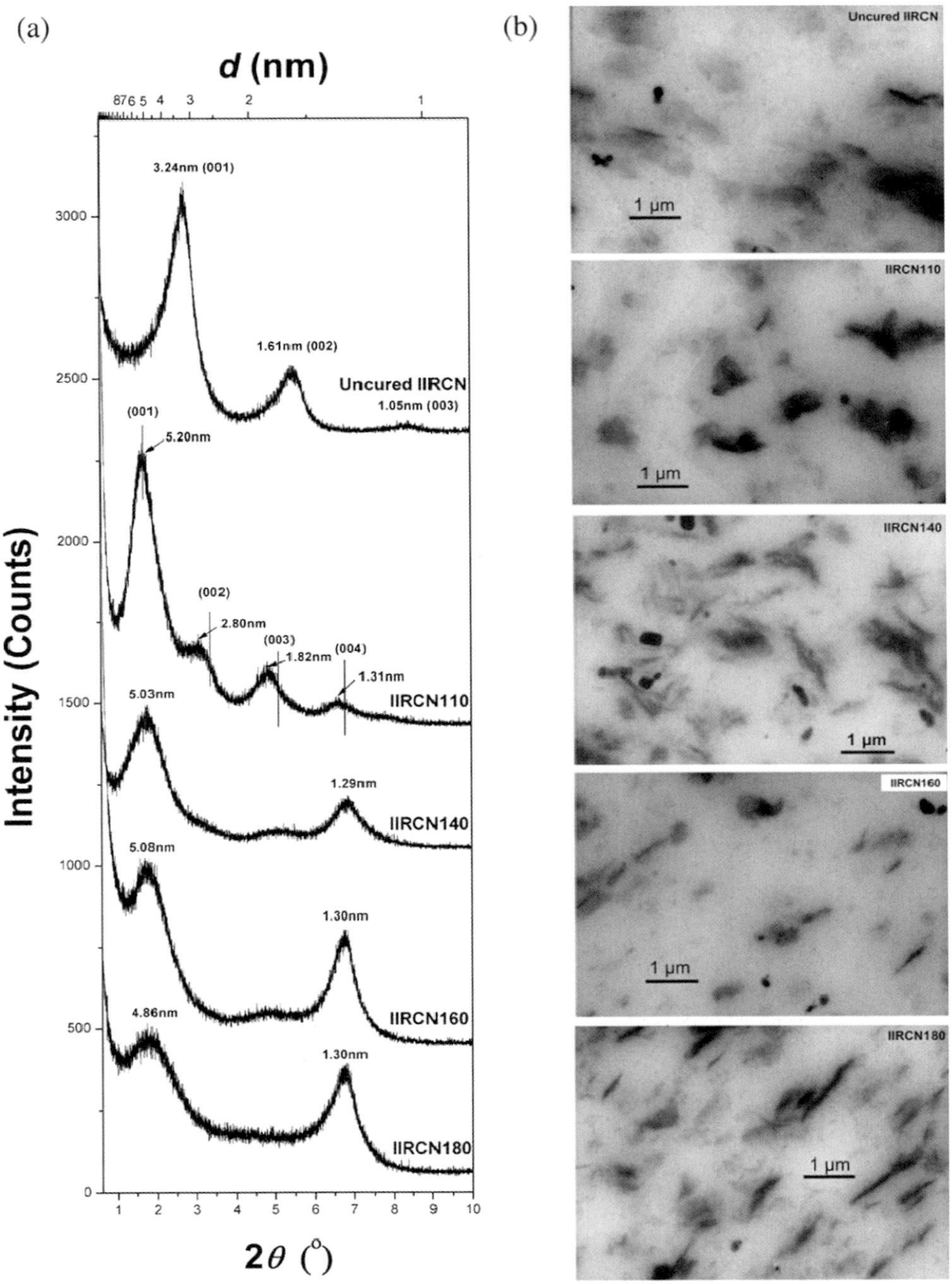

Figure 3.4 (a) WAXD patterns and (b) TEM images of IIRCNs (IIR:OMC = 100:10) cured at different temperatures. The WAXD curves were stacked vertically for clarity. Reprinted from Ref. [19] with permission from John Wiley & Sons Ltd.

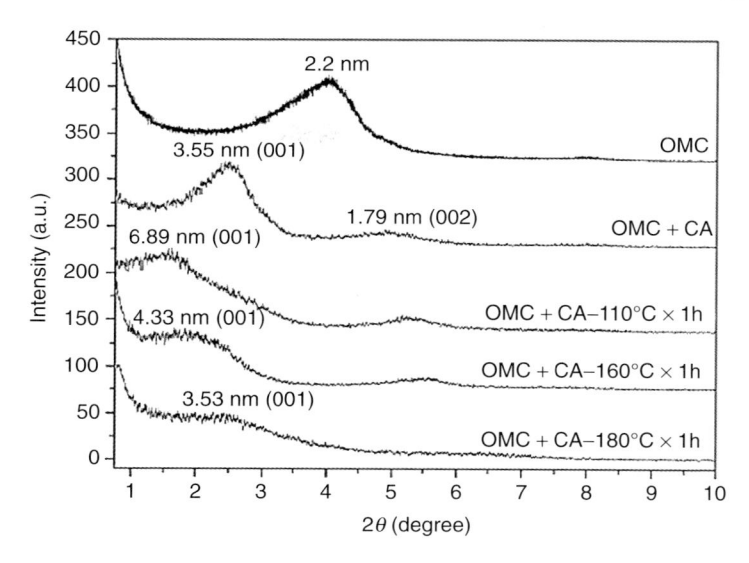

Figure 3.5 Influence of treating conditions on WAXD patterns of the mixture of OMC and curing agent (CA). The WAXD patterns of pure OMC and untreated mixture of OMC and CA were also plotted for comparison. The mixture containing: OMC 10 parts, zinc oxide (ZnO) 5 parts, stearic acid (SA) 2 parts, tetramethyl thiuram disulfide (accelerator TMTD) 1.0 part, 2-mercapto benzothiazole (accelerator M) 0.5 part, sulfur (S) 1.8 parts, N-phenyl-α-naphthylamine (antioxidant A) 1.0 part. Reprinted from Ref. [19] with permission from John Wiley & Sons Ltd.

WAXD investigation (Figure 3.5) of the mixture containing OMC and sulfur curatives after thermal treatment at different temperatures suggested that the reactions between curatives and amine intercalants that occurred at high temperature resulted in only layer separation and delamination rather than deintercalation in the absence of rubber.

Lu and coworkers also showed that organic amine intercalants could accelerate the vulcanization rate of rubber (Table 3.1), which in turn implied that the curing rate of rubber intercalated into the silicate gallery might be higher than that of the rubber outside; consequently, more rubber could be driven to intercalate into the silicate gallery during vulcanization (Y.L. Lu et al., unpublished results). These recently obtained data suggested that the hypothesis defining the role of the curing reaction in microstructural changes, as proposed by Karger-Kocsis and colleagues, might indeed be valid.

Physical Factors The two physical factors influencing microstructural changes among RCNs were identified as the *high temperature* and *high pressure* applied during the vulcanization process. As noted in Section 3.2.1, thermodynamically unstable intercalated structures can be produced by melt compounding and, at room temperature, these structures would persist due to the high viscosity of the

3 Morphology and Interface Development in Rubber–Clay Nanocomposites

Table 3.1 Influence of clay and amine intercalant on curing parameters of EPDM compounds.

Compound	T_{10} (s)	T_{90} (s)	CRI (1000∗s^{-1})
Pure EPDM compound	344	1070	1.38
EPDM/Na-MMT (100/7)	134	1071	1.07
EPDM/amine intercalant (100/3)	129	779	1.54
EPDM/OMC (100/10)	137	893	1.32

Note: CRI (curing rate index) was defined as $1/(T_{90} - T_{10}) \times 1000$, and represents the vulcanization speed. The EPDM compound has a sulfur vulcanization system. The OMC was I.30P (Nanomer®) and contained about 70% clay and 30% amine intercalant, which was determined by TGA.

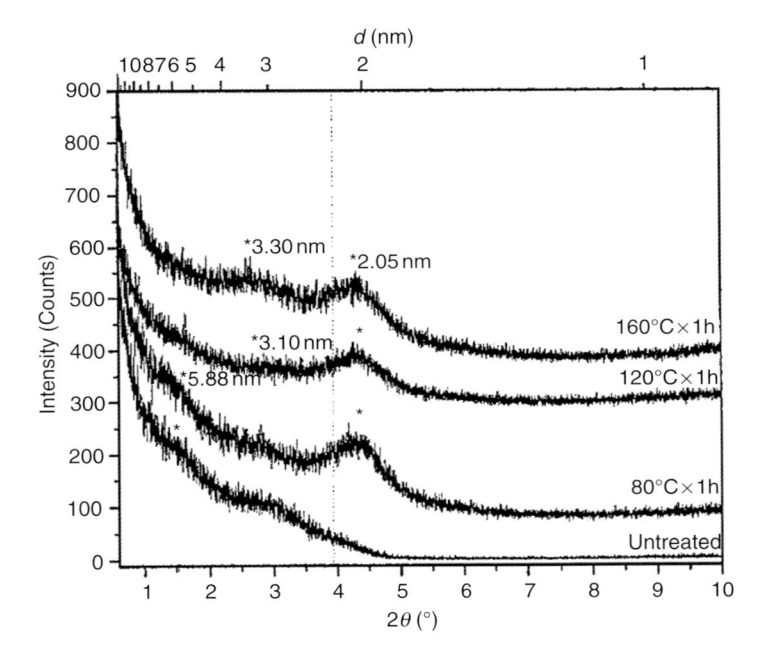

Figure 3.6 WAXD patterns of untreated IIRCN and IIRCN (IIR:OMC = 100:10) thermally treated at different temperatures and atmospheric pressures (AP) for 1 h. The asterisks indicate the (001) peak for OMC dispersed in the IIR matrix. The vertical dotted line indicates the location of the silicate (001) reflection of pure OMC. The curves were stacked vertically for clarity. Reproduced from Ref. [7] with permission from John Wiley & Sons Ltd.

compound. At high temperature, however, they would transform spontaneously to thermodynamically stable structures (as shown in Figure 3.6 [7]). High pressure was also found to enhance and accelerate this transformation (see Figure 3.7).

Perhaps more importantly, high pressure was shown to be the critical factor causing the aggregation of originally well-dispersed OMC particles [27]. As shown

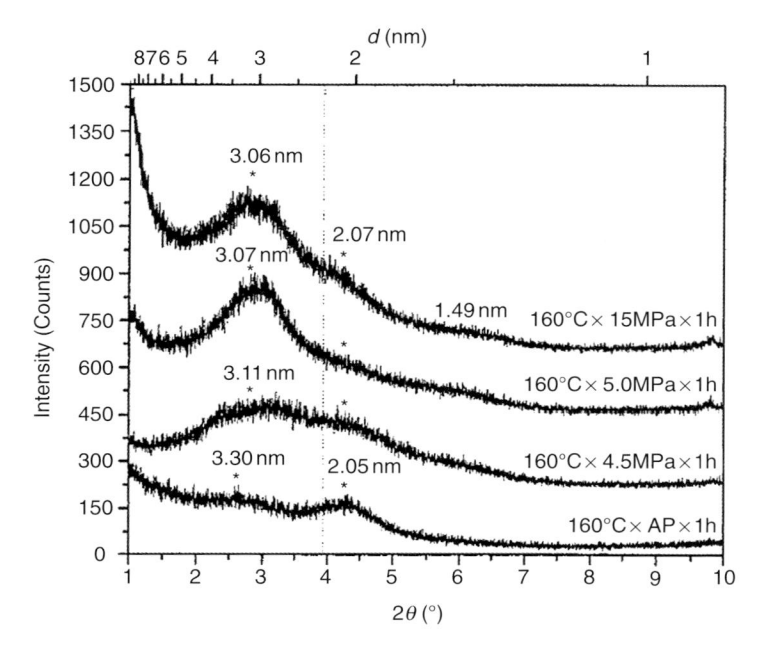

Figure 3.7 Influence of pressure on WAXD patterns of IIRCN treated at 160 °C for 1 h. The asterisks indicate the (001) peak for OMC dispersed in the IIR matrix. The vertical dotted line indicates the location of the silicate (001) reflection of pure OMC. Reproduced from Ref. [7] with permission from John Wiley & Sons Ltd.

in Figure 3.8, the dimension of the dispersed OMC particles was clearly increased, and their spatial distribution homogeneity reduced dramatically after high-pressure vulcanization, when compared to uncured compounds. In contrast, the spatial dispersion state would be greatly improved after vulcanization under atmospheric pressure (AP) (see Figure 3.9).

Various methods to reduce the negative effect of high pressure have been considered in order to optimize the properties of RCNs, the most direct approach being to perform the vulcanization under a lower pressure. The data listed in Table 3.2 show that not only the mechanical but also the gas-barrier properties of IIRCN, when cured at 3 MPa pressure, are clearly superior to those of IIRCN cured at 15 MPa [28]. These results indicate the essential role of the clay dispersion state on the properties of the RCNs.

Another approach would be to adjust the vulcanization temperature [19] and, indeed, results have shown that reactions during the initial curing period lead to a further intercalation of rubber chains into the clay galleries, thus improving the spatial dispersion of the clay particles. High pressure, in turn, leads to the aggregation of the clay layers to form larger clay agglomerates, although increasing the curing rate by raising the temperature can lead to a reduction in such aggregation.

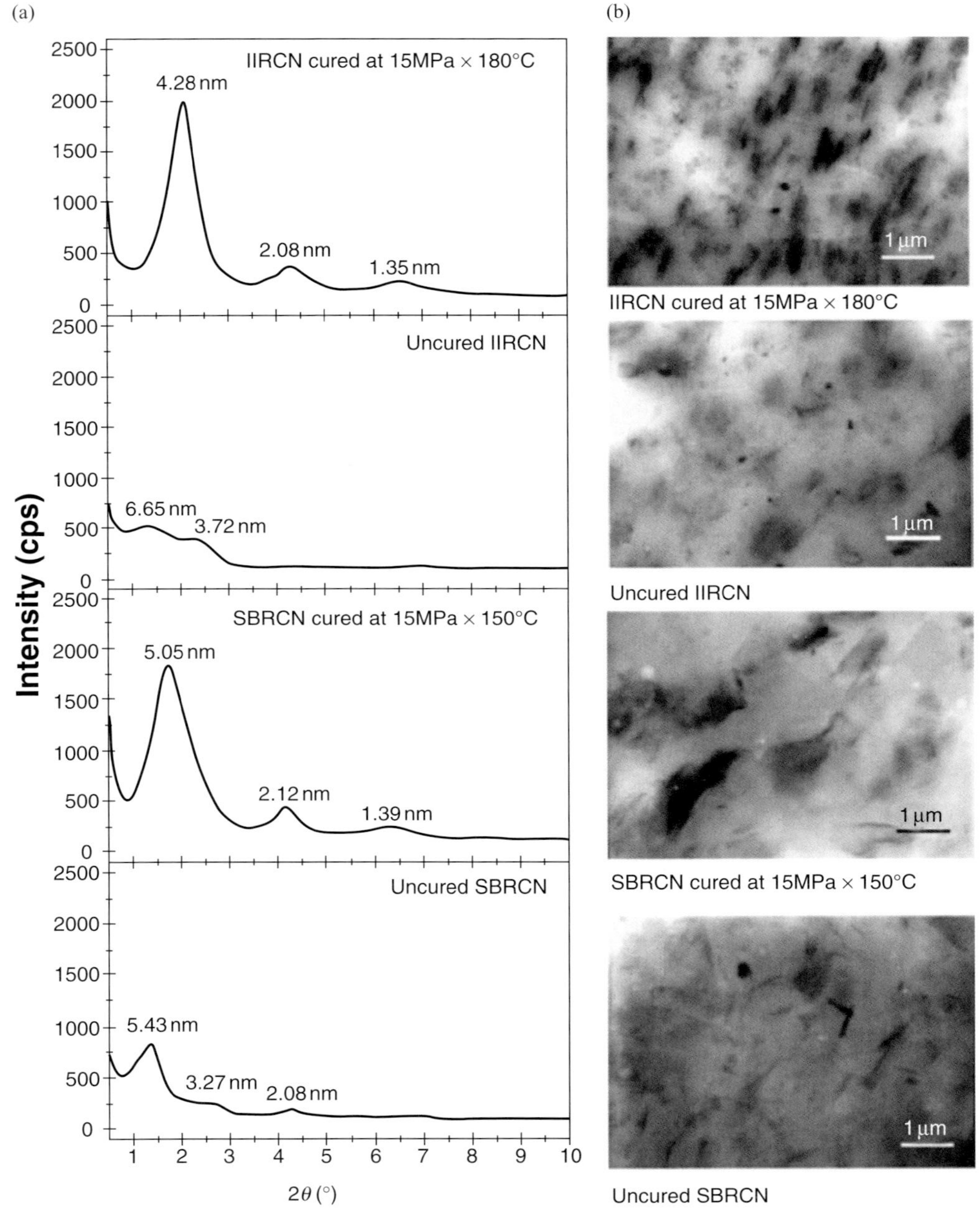

Figure 3.8 (a) WAXD patterns and (b) TEM images of cured and uncured IIRCN and SBRCN prepared by melt intercalation. Reprinted from Ref. [27] with permission from John Wiley & Sons Ltd.

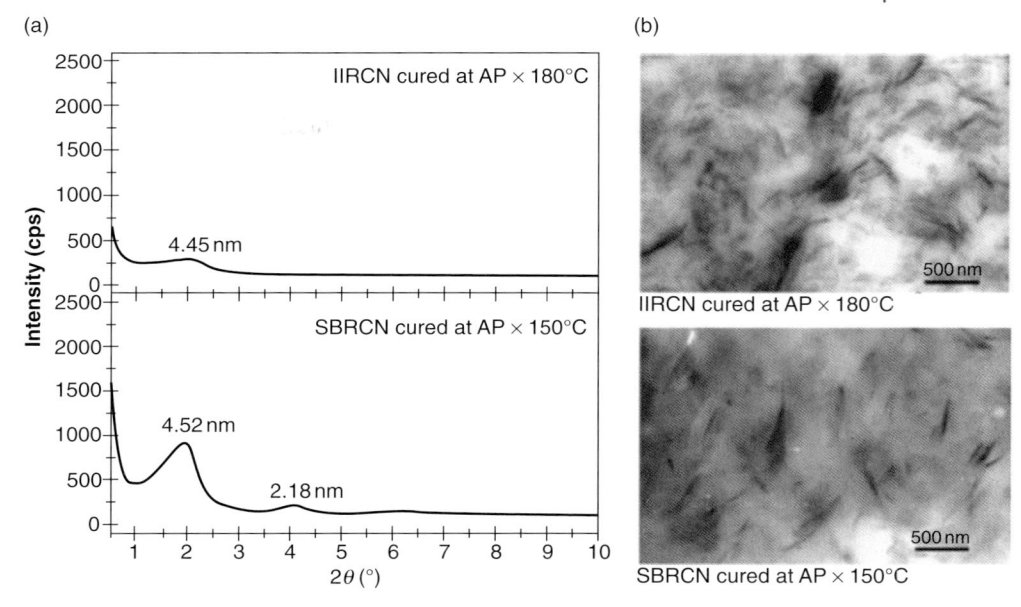

Figure 3.9 (a) WAXD patterns and (b) TEM images of IIRCN and SBRCN cured at atmospheric pressure. The vertical scales of the WAXD patterns are the same as that for Figure 3.8a, in order to allow comparison. Reproduced from Ref. [27] with permission from John Wiley & Sons Ltd.

Table 3.2 A comparison of properties between IIRCNs cured at 15 MPa pressure and 3 MPa pressure.

Property	Curing pressure (MPa)	
	15	3
Shore A hardness (degree)	43	47
Elastic modulus, G' (MPa)	4.9	8.0
Tensile strength (MPa)	14.8	20.0
Elongation at break (%)	657	631
Tear strength (kN m^{-1})	19	23
Relative permeability $(P_c/P_p)^a$	0.70	0.54

a) P_c and P_p are the permeabilities of the composite and pure polymer, respectively.

However, extreme curing temperatures (e.g., 160–180 °C for this system) also resulted in a deadsorption of intercalants and the formation of larger amounts of inorganic clays, thus weakening the interface interaction between clay and rubber. Hence, the optimum curing temperature range for obtaining a better dispersion was identified as approximately 140 °C for the IIRCN system investigated (as shown in Figure 3.4). Because the crosslinking density, filler dispersion state and filler–rubber interactions are greatly influenced by the curing temperature, the gas-barrier properties and mechanical performance of IIRCN cured at different temperature would be quite different (see Figures 3.10 and 3.11).

Subsequently, it was found accidentally that the vaporization of residual solvents occurring under high-temperature treatment at AP could expand and delaminate the clay galleries, and that this would result in a major improvement in the spatial distribution of clay particles in the RCNs [29] (see Figure 3.12). Enlightened by these results, a new strategy for the preparation of RCNs by using a melt-blending process was developed. For this, a certain amount of an organic solvent and OMC were first mixed to obtain a pre-swelled OMC (PSOMC) that could be incorporated into the rubber by melt blending. During the curing process, the solvents within the silicate galleries were vaporized, which in turn caused exfoliation and prevented any aggregation of the silicate layers. The RCNs prepared in this way showed a better dispersion morphology, and consequently their mechanical and gas-barrier properties were superior to those of RCNs

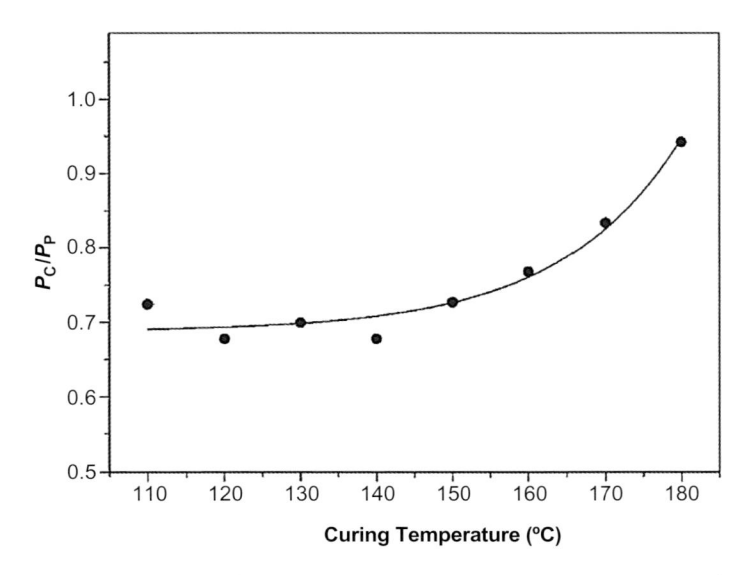

Figure 3.10 Impact of the curing temperature on the gas-barrier property of cured IIRCN (IIR:OMC = 100:10). P_c and P_p represent the gas permeability of the composite and pure matrix rubber, respectively. Reproduced from Ref. [19] with permission John Wiley & Sons Ltd.

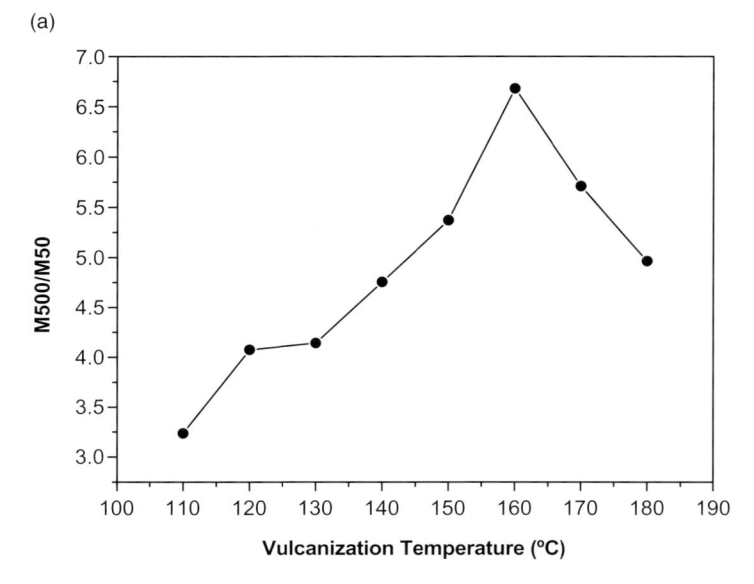

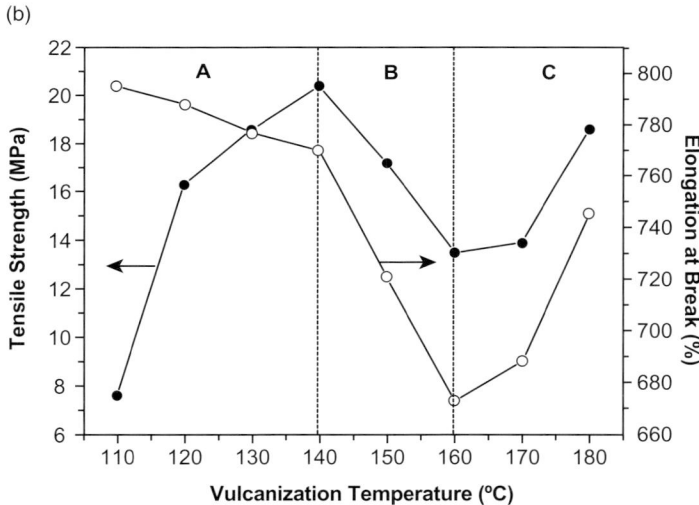

Figure 3.11 Influence of vulcanization temperature on (a) reinforcement index (M500/M50) of IIRCN (IIR:OMC = 100:10) and (b) tensile strength and elongation at break. The reinforcement index can represent the interface strength between filler and rubber. Reproduced from Ref. [19] with permission John Wiley & Sons Ltd.

prepared by traditional melt blending. As an example, the tensile strength and gas-barrier properties of IIRCN (100/5) created using this new method were ~75% and 15% higher, respectively, than those of a counterpart prepared via traditional melt blending [30, 31].

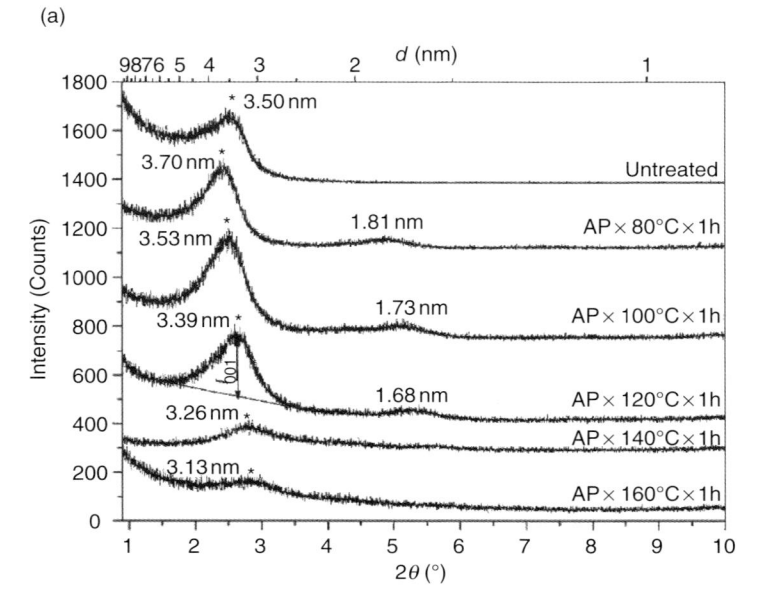

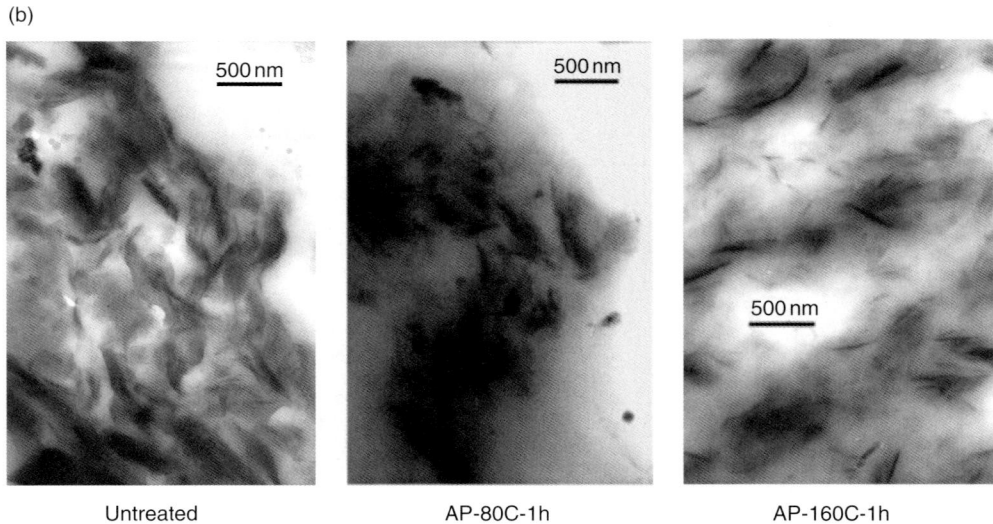

Figure 3.12 (a) WAXD patterns and (b) TEM images of IIRCN (IIR/OMC = 100/10) prepared by solution compounding with AP thermal treatments under different temperatures. Reprinted from Ref. [29] with permission John Wiley & Sons Ltd.

3.3
Latex Compounding

3.3.1
Mechanism and Influencing Factors

Layered pristine silicates can be dispersed in water, which acts as a swelling agent owing to hydration of the intercalary cations (usually Na^+ ions). It should be noted that most rubbers are available in latex form, which is a more or less stable aqueous dispersion of fine rubber particles (in the submicron to micron range). During the late 1990s, Zhang *et al.* conducted a series of pioneering studies on the preparation of RCNs by mixing the rubber latex with a pristine clay water suspension, followed by coagulation [32–34]. Until now, many types of rubber-based clay nanocomposites, including NR [35, 36], SBR [33, 34], carboxylated NBR [37], NBR [38], polyurethane (PU) [39], polybutylene rubber [41], and polychloroprene (CR) [4], have been successfully prepared via a latex compounding method (LCM). Compared with the melt, solution or *in-situ* polymerization methods, LCM (when pristine clay is employed) has demonstrated great promise for industrial application, due to the low cost of the pristine clay, the simplicity of the preparation process and the superior cost:performance ratio.

The mechanism by which the nanocomposite structure is formed via the latex route differs totally from that of melt compounding. A schematic representation of the mixing and co-coagulating process is shown in Figure 3.13a. A WAXD study revealed that the clay could be exfoliated in water when its content was very low in aqueous suspension (e.g., 10 wt% for Na^+-MMT suspension; see Ref. [40]). At the stage of mixing the rubber latex with the dilute clay suspension, the clay was also dispersed as individual silicate layers, as indicated by WAXD patterns (see Figure 3.13b). When a flocculant was added, it coagulated the rubber latex and the silicate layers simultaneously, although in most cases the rubber macromolecules did not exactly intercalate into the galleries of clay (Figures 3.13b and 3.14). This phenomenon was mainly attributed to competition between the separation of rubber latex particles and the reaggregation of silicate layers following flocculant addition [40].

Since the rubber latex particles are composed of several macromolecules, the existence of latex particles and reaggregation of single silicate layers in the water medium should result in completely separated (exfoliated) silicate layers; this is apparent in the TEM images of various RCNs prepared using the LCM (Figure 3.15). However, cations of the flocculant caused the separated silicate layers to reaggregate, so that the rubber latex particles between the silicate layers might be expelled. As a result, there are some nonexfoliated layers in the nanocomposites. In the meantime, due to the fact that the amount of latex is more than that of silicate layers and the latex particles agglomerate rapidly, the reaggregation of silicate layers is evidently obstructed to some extent by the agglomerated latex particles around the silicate layers. Consequently, the size of aggregates of silicate layers is at the nanometer level, and the thus-obtained nanocomposites contain both the

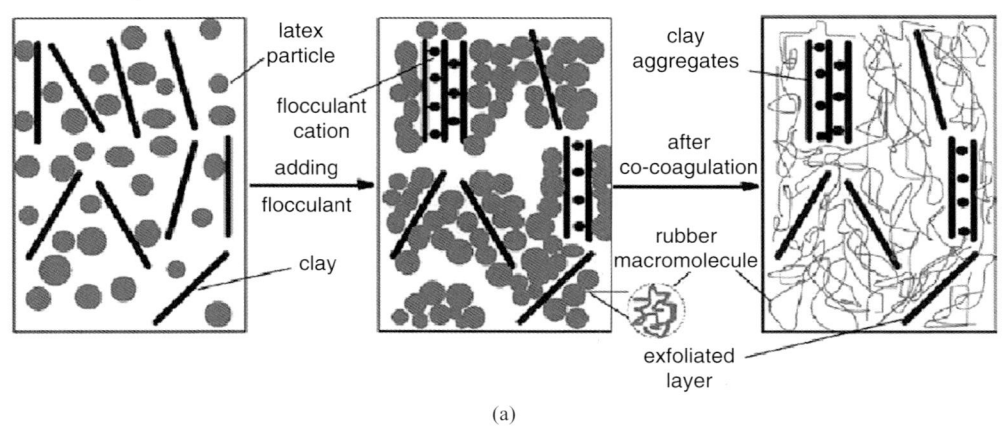

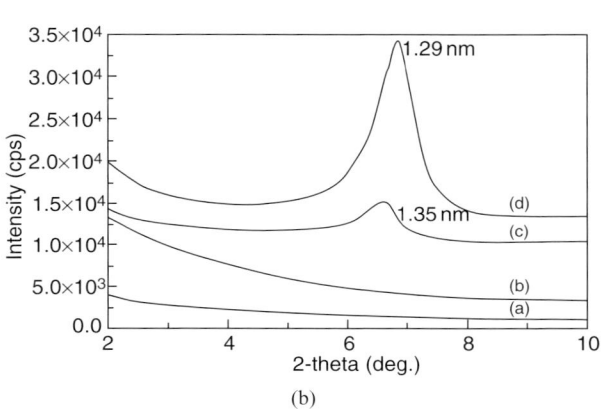

Figure 3.13 (a) Schematic illustration of the mixing and co-coagulating process, and (b) WAXD patterns for different stages of the process. Stage 1, 3% clay aqueous suspension; Stage 2, when the SBR latex mixes with the clay aqueous suspension; Stage 3, after addition of flocculant; Stage 4, the dried SBR–clay nanocomposite. Reproduced from Ref. [40] with permission from Elsevier.

exfoliated silicate layers and nonexfoliated (not intercalated) aggregates of nanometer thickness in the rubber matrix. This special morphology was referred to as a "separated" structure for presenting the difference from partly exfoliated structure prepared by melt blending [40].

The factors affecting the final dispersion level of filler in nanocomposites mainly include: (i) the size of the rubber latex, the interaction state between the rubber latex and the pristine clay; (ii) the ratio of rubber latex to clay suspension; and (iii) the speed of co-coagulation. Clearly, the smaller the size of the latex particle and the higher the latex content, the better dispersion would be achieved. Nevertheless, these two factors could not be adjusted flexibly when the RCNs were

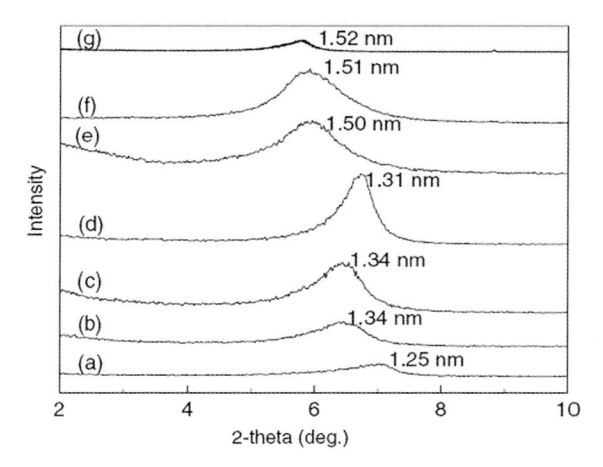

Figure 3.14 WAXD patterns of various RCNs prepared by LCM and clays. (a) Na$^+$-MMT; (b) SBR-clay; (c) NR-clay; (d) Organically modified clay prepared by a cation-exchange reaction between Na$^+$-MMT and excess ammonium cations from triethylenetrammonium chloride; (e) NBR-clay; (f) CNBR-clay; (g) Ca^{2+}-MMT. Reproduced from Ref. [40] with permission from Elsevier.

prepared with LCM. A stronger interaction between the latex particles and silicate would prevent the rubber latex particles between the silicate layers from being expelled. For instance, the fully exfoliated nanocomposites could be prepared by utilizing pyridine groups contained on rubber latex, as shown in Figure 3.16 (unpublished results). For most rubber systems, however, the interaction between the rubber latex particles and the pristine silicate layers could not be much stronger.

A decreasing mobility of rubber macromolecules and an increasing coagulation speed can also depress the reaggregation of silicate layers and improve the dispersion state of the RCNs obtained. In considering this mechanism, Lu *et al.* recently developed a modified latex-compounding method, where crosslinked rubber latex was used and the coagulation process was performed by spray-drying the mixture of rubber latex and clay water suspension, instead of adding the flocculants. Consequently, in the rubber–clay compound obtained, the pristine clay layers were fully exfoliated. By incorporating this compound with other rubbers (using melt blending), it was possible to prepare rubber/pristine nanocomposites with an exfoliation morphology, even in EPDM (nonpolar rubber, as shown in Figure 3.17) [41].

A strong shear force during mixing of the curatives and other agents by melt blending would break down the silicate layers and clearly reduce the aspect ratio of dispersion of the RCN; this would be harmful to the material's properties, notably those relating to the gas barrier. However, if the curatives are mixed in aqueous solution, and the rubber latex and clay water suspension mixture are used

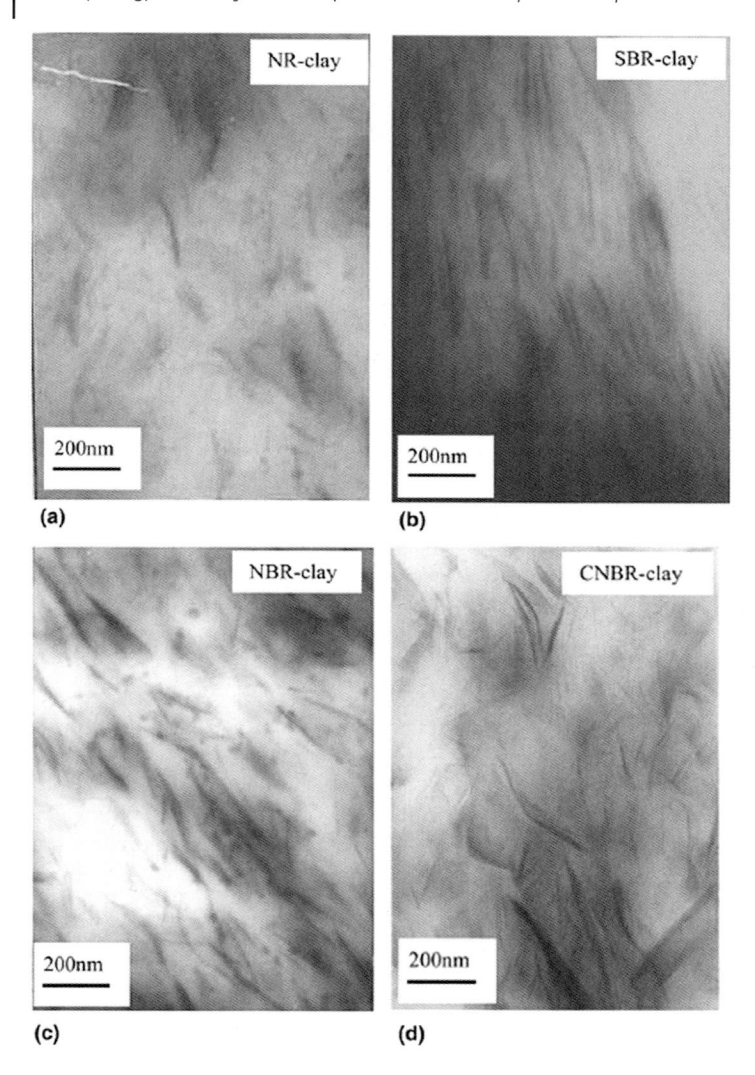

Figure 3.15 TEM micrographs of four rubber–clay nanocomposites. (a) NR–clay; (b) SBR–clay; (c) NBR–clay; (d) CNBR–clay. Reproduced from Ref. [40] with permission from Elsevier.

directly as a coating application, then this problem may be avoided. Indeed, the relative gas permeability (P/P_0) of a thin coating RCN membrane has been reported to be as low as ~0.02 when the clay content was 30 wt% [42].

3.3.2
Interface Enhancement

As most rubbers are hydrophobic, and the pristine clay is hydrophilic, any interfacial interaction between the two will always be poor – a situation confirmed by

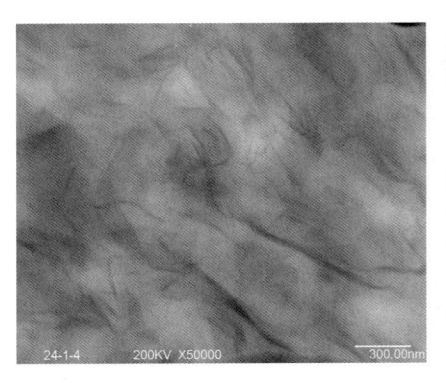

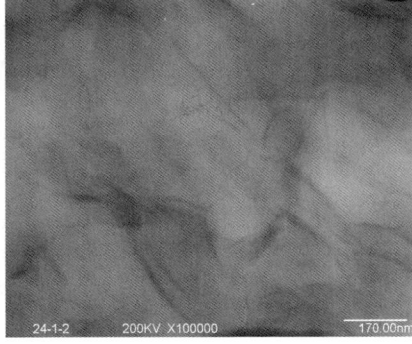

Figure 3.16 TEM images of butadiene-styrene-vinyl pyridine rubber/Na$^+$-MMT (100/10) nanocomposite prepared by LCM. Left: low-magnification image; Right: high-magnification image.

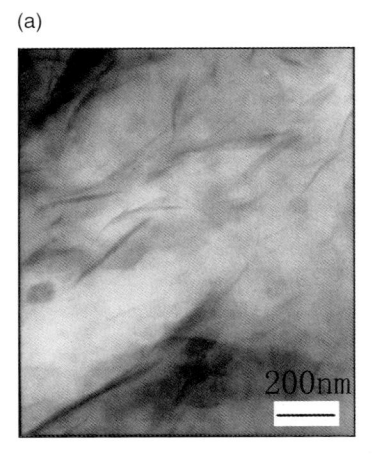

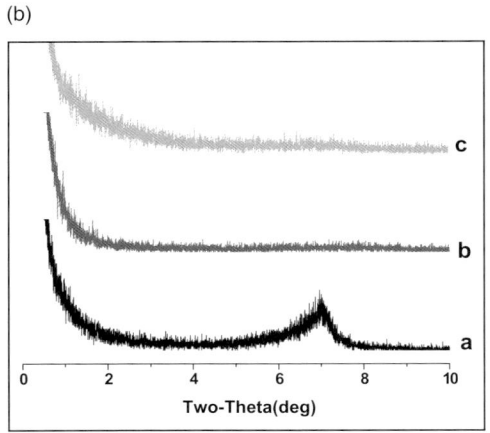

Figure 3.17 Microstructure of EPDM/clay nanocomposites prepared by modified LCM. (a) TEM image; (b) WAXD patterns. Curve a = pristine Na$^+$-MMT; curve b = cured ultrafine powdered NBR(UFPNBR)/clay compound (4:1); curve c = EPDM/UFPNBR/clay (5:4:1). Reproduced from Ref. [41] with permission.

the use of positron annihilation lifetime spectroscopy [43]. For this reason, most rubber/pristine clay nanocomposites prepared using LCM have one major shortcoming, namely that the modulus at high elongation (e.g., 300%) normally remains at a very low level, and so cannot meet the requirements of many engineering applications. In order to overcome this disadvantage, a number of studies [44–48] on interface enhancement have been conducted. For this, three types of surfactant – triisopropanolamine (TA), *m*-xylylenediamine (MXD), and allylamine (AA) – were used to first modify the pristine clay through cation-exchange reactions; the clay used to prepare RCNs via LCM was then replaced with a surfactant-modified clay [44]. As a result of this, exfoliation was achieved with the SBR/AA-modified clay, and enhancement of the mechanical properties was more prominent than when using carbon black. Jia et al. [45] chose an unsaturated organic ammonium chloride (UOAC) to modify pristine clay *in situ*. For this, they added UOAC directly to a clay water suspension (before mixing with rubber latex) in the hope that, during vulcanization, a chemical interfacial interaction would form through reaction between the double bonds of UOAC and the rubber. In the nanocomposites thus obtained, an exfoliated, flocculant ions intercalated structure and a rubber intercalated structure were seen to coexist. The use of this approach was shown to improve the interfacial compatibility between clay and rubber, and to result in a major improvement in tensile strength. However, if a strong chemical interface could not be formed, the modulus at high strain remained at low level (Figure 3.18a). Jia and coworkers subsequently developed a relatively simple method for enhancing the interface, which included the addition of an organic modifier, hexadecyl ammonium bromide (C16) and 3-aminopropyl triethoxy silane (KH550), by melt blending after the latex compounding of rubber and pristine clay [46]. The results obtained (Figure 3.18b) showed that this approach could be used to tailor the interfacial interaction state to some extent by changing the modifier type and the flocculant system, and thus adjusting the shape of the strain–stress curve of the nanocomposites. Because the modulus at high strain was increased dramatically by the addition of KH550, it was assumed that a strong chemical interface might be formed. Very recently, a two-step method for interface enhancement was reported [47, 48] in which KH550 was introduced to a clay water suspension before mixing with rubber latex; the bis(triethoxsilpropyl) tetra sulfide (Si69) was then added by melt blending after latex compounding. The results obtained showed that with a combined modification of KH550 and Si69, the SBR/clay nanocomposites exhibited a maximum improvement in their mechanical properties (Figure 3.18c). The mechanical properties of SBR/clay nanocomposites with different interface enhancements (using optimum conditions for each approach) are summarized in Table 3.3. It can be seen from these data that the value of M300/M100 (which qualitatively may represent the interface state) for SBRCN with the two-step interface enhancement was highest. Based on these findings, Jia et al. proposed the possible mechanism shown schematically in Figure 3.19. Following co-coagulation, a KH550-modified (intercalated) nanoclay should be generated which contains many active hydroxyl groups originating from the hydrolysis of Si–(OR)$_3$ of KH550 in aqueous suspension. During subsequent vulcanization, the ethoxysilylpropyl groups of Si69 would react with the Si–OH groups of nanodispersed clay in rubber, after which the sulfide groups of Si69 could react with the rubber molecules, assisted by accelerators of the curing system of rubber. As a result, a chemical bonding could be built between the clay phase and the rubber phase.

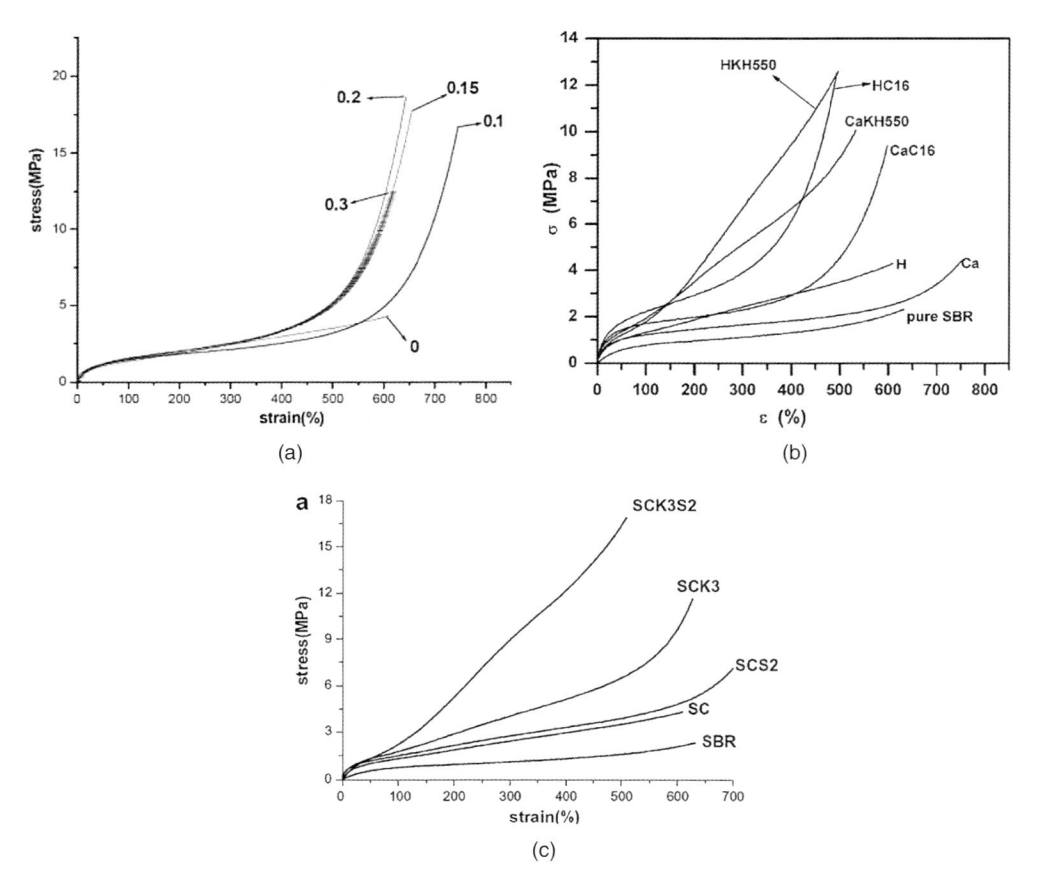

Figure 3.18 Strain–stress diagram of SBR/clay (100/10) nanocomposites prepared by LCM. (a) Combining *in situ* modification with UOAC. Here, 0, 0.1, 0.15, 0.2, and 0.3 represent the ratios of UOAC to MMT, respectively. Reproduced from Ref. [45] with permission from John Wiley & Sons Ltd. (b) Organic modifier was mixed by melt blending after latex compounding. Here, KH550 and C16 denote hexadecyltrimethyl ammonium bromide, and 3-aminopropyltriethoxy silane, respectively; H and Ca represent H^+ and Ca^+ flocculated systems, respectively. Reproduced from Ref. [46] with permission from John Wiley & Sons Ltd. (c) Enhancing interface by a two-step method. Here, SC, K and S denote the SBR/clay nanocomposite, KH550, and Si69 silane coupling agents, respectively. The number indicates the dosage (phr) for corresponding silanes. Reproduced from Ref. [48] with permission from Elsevier.

Table 3.3 Mechanical properties of interface-enhanced SBR/clay (100/10) nanocomposites by different approaches.

SBRCN	M100 (MPa)[a]	M300 (MPa)[a]	M300:M100	Tensile strength (MPa)	Elongation at break (%)	Reference(s)
UAOC-0.2	1.5	2.5	1.67	18.7	638	[45]
HKH550	1.8	6.8	3.78	12.4	504	[46]
SCK3S2	2.3	9.6	4.17	16.9	509	[47, 48]

a) M100 and M300 represent moduli at 100% and 300% strain, respectively. The ratio of these (M300:M100) was used to indicate the interface strength between the filler and the rubber.

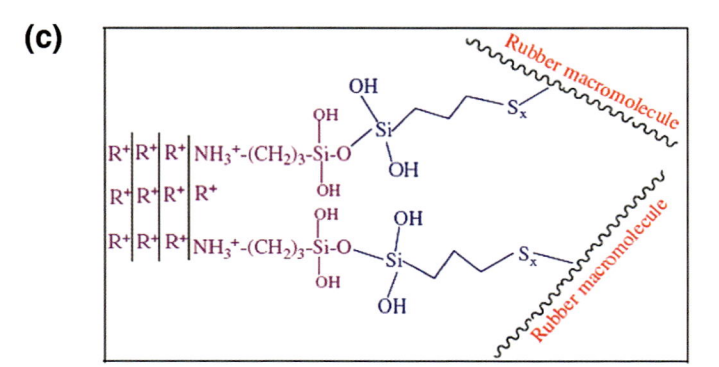

Figure 3.19 Proposed schematic of modification mechanism with silane coupling agent. (a) Cation exchange in clay suspension with KH550; (b) After co-coagulating, the structure of SCK flocculate; (c) After adding Si69 and sulfur curing ingredients on a two-roll mill, and curing the compounds at 150 °C, the interaction among KH550, Si69 and rubber macromolecules in SCKS. Reproduced from Ref. [48] with permission from Elsevier.

References

1. Kawasumi, M. (2004) *J. Polym. Sci., Part A: Polym. Chem.*, **42**, 819–824.
2. Alexandre, M. and Dubois, P. (2000) *Mater. Sci. Eng.*, **28**, 1–63.
3. Ray, S.S. and Okamoto, M. (2003) *Prog. Polym. Sci.*, **28**, 1539–1641.
4. Karger-Kocsis, J. and Wu, C.-M. (2004) *Trends Polym. Eng. Sci.*, **44**, 1083–1093.
5. Vaia, R.A. and Giannelis, E.P. (1997) *Macromolecules*, **30** (25), 7990–7999.
6. Vaia, R.A. and Giannelis, E.P. (1997) *Macromolecules*, **30** (25), 8000–8009.
7. Liang, Y.R., Ma, J., Lu, Y.L., Wu, Y.P., Zhang, L.Q., and Mai, Y.W. (2005) *J. Polym. Sci.: Part B, Polym. Phys.*, **43** (19), 2653–2664.
8. Kim, J., Oh, T., and Lee, D. (2003) *Polym. Int.*, **52** (7), 1203–1208.
9. Tim, J., Oh, T., and Lee, D. (2003) *Polym. Int.*, **52** (7), 1058–1063.
10. Zheng, H., Zhang, Y., Peng, Z., and Zhang, Y. (2004) *J. Appl. Polym. Sci.*, **92** (1), 638–646.
11. Schön, F., Thomann, R., and Cronski, W. (2002) *Macromol. Symp.*, **189**, 105–110.
12. Nah, C., Ryu, H.J., Kim, W.D., and Chang, Y.W. (2003) *Polym. Int.*, **52** (8), 1359–1364.
13. Gatos, K.G., Sawanis, N.S., Apostolov, A.A., Thomann, R., and Karger-Kocsis, J. (2004) *Macromol. Mater. Eng.*, **289** (12), 1079–1086.
14. Teh, P.L., Mohd Ishak, Z.A., Hshim, A.S., Karger-Kocsis, J., and Ishiaku, U.S. (2004) *J. Appl. Polym. Sci.*, **94** (6), 2438–2445.
15. Li, W., Huang, D., and Ahmadi, S.J. (2004) *J. Appl. Polym. Sci.*, **94** (2), 440–445.
16. Gatos, K.G., Thomann, R., and Karger-Kocsis, J. (2001) *Polym. Int.*, **53** (11), 1191–1197.
17. Kim, J.T., Lee, D.Y., Oh, T.S., and Lee, D.H. (2003) *J. Appl. Polym. Sci.*, **89** (10), 2633–2640.
18. Li, Y., Mao, L.-X., Lu, Y.-L., Wang, C.-L., and Zhang, L.-Q. (2009) *China Synth. Rubber Ind.*, **32** (in press).
19. Lu, Y.L., Li, Z., Mao, L.X., Li, Y., Wu, Y.P., Yu, R.L., and Zhang, L.Q. (2008) *J. Appl. Polym. Sci.*, **110** (2), 1034–1042.
20. Usuki, A., Tukigas, A., and Kato, M. (2002) *Polymer*, **43** (8), 2185–2189.
21. Varghese, S. and Karger-Kocsis, J. (2004) *J. Appl. Polym. Sci.*, **91** (2), 813–819.
22. Varghese, S., Karger-Kocsis, J., and Gatos, K.G. (2003) *Polymer*, **44** (14), 3977–3983.
23. Gatos, K.G., Thomann, R., and Karger-Kocsis, J. (2004) *Polym. Int.*, **53**, 1191–1197.
24. Gatos, K.G. and Karger-Kocasis, J. (2005) *Polymer*, **46**, 3069–3076.
25. Gatos, K.G., Sawanis, N.S., Apostolov, A.A., Thomann, R., and Karger-Kocsis, J. (2004) *Macromol. Mater. Eng.*, **289**, 1079–1086.
26. Gatos, K.G., Százdi, L., Pukánszky, B., and Karger-Kocsis, J. (2005) *Macromol. Rapid Commun.*, **26**, 915–919.
27. Liang, Y.-R., Lu, Y.-L., Wu, Y.-P., Ma, Y., and Zhang, L.-Q. (2005) *Macromol. Rapid Commun.*, **26**, 926–931.
28. Lu, Y.L., Liang, Y.R., Wu, Y.P., and Zhang, L.Q. (2005) *National Conference for Polymer Science of China Symposium*, Beijing, p. 260.
29. Lu, Y.L., Liang, Y.R., Wu, Y.P., and Zhang, L.Q. (2006) *Macromol. Mater. Eng.*, **291**, 27–36.
30. Zhang, L.Q., Liang, Y.R., et al. (2004) Chinese Patent ZL200410008452.0.
31. Liang, Y.R., Cao, W.L., Li, Z., Wang, Y.Q., Wu, Y.P., and Zhang, L.Q. (2008) *Polym. Testing*, **27** (3), 270–276.
32. Zhang, L.Q., Wang, Y.Z., Yu, D.S., Wang, Y.Q., and Sun, Z.H. (1998) Chinese Patent ZL98101496.8.
33. Zhang, L.Q., Wang, Y.Z., Wang, Y.Q., Sui, Y., and Yu, D.S. (2000) *J. Appl. Polym. Sci.*, **78**, 1873–1878.
34. Wang, Y.Z., Zhang, L.Q., Tnag, C.H., and Yu, D.S. (2000) *J. Appl. Polym. Sci.*, **78**, 1879.
35. Varghese, S. and Karger-Kocsis, J. (2003) *Polymer*, **44**, 4921.
36. Wang, Y.Q., Zhang, H.F., Wu, Y.Q., Yang, J., and Zhang, L.Q. (2005) *Eur. Polym. J.*, **41**, 2776–2783.
37. Wu, Y.P., Zhang, L.Q., Wang, Y.Q., Liang, Y., and Yu, D.S. (2001) *J. Appl. Polym. Sci.*, **82**, 2842.

38 Wu, Y.P., Jia, Q.X., Yu, D.S., and Zhang, L.Q. (2003) *J. Appl. Polym. Sci.*, **89**, 3855.
39 Varghese, S., Gatos, K.G., Apostolov, A.A., and Karger-Kocsis, J. (2004) *J. Appl. Polym. Sci.*, **92**, 543.
40 Wu, Y.P., Wang, Y.Q., Zhang, H.F., Wang, Y.Z., Yu, D.S., and Zhang, L.Q. (2005) *Compos. Sci. Technol.*, **65**, 1195–1202.
41 Wang, H.T., Lu, Y.L., Liu, L., and Zhang, L.Q. (2009) *China Synth. Rubber Ind.*, **32**, 250.
42 Takahashi, S., Goldberg, H.A., Feeney, C.A., Karim, D.P., Farrell, M., O'Leary, K., and Paul, D.R. (2006) *Polymer*, **47**, 3083–3093.
43 Wang, Y.Q., Wu, Y.P., Zhang, H.F., Zhang, L.Q., Wang, B., and Wang, Z.F. (2004) *Macromol. Rapid Commun.*, **25**, 1973–1978.
44 Ma, J., Xiang, P., Mai, Y.W., and Zhang, L.Q. (2004) *Macromol. Rapid Commun.*, **25**, 1692–1696.
45 Jia, Q.X., Wu, Y.P., Xu, Y.L., Mao, H.H., and Zhang, L.Q. (2006) *Macromol. Mater. Eng.*, 291–226.
46 Jia, Q.X., Wu, Y.P., Wang, Y.Q., Lu, M., Yang, J., and Zhang, L.Q. (2007) *J. Appl. Polym. Sci.*, **103**, 1826–1833.
47 Jia, Q.X., Wu, Y.P., Lu, M., He, S.J., Wang, Y.Q., and Zhang, L.Q. (2008) *Compos. Interface*, **15**, 193–205.
48 Jia, Q.X., Wu, Y.P., Wang, Y.Q., Lu, M., and Zhang, L.Q. (2008) *Compos. Sci. Technol.*, **68**, 1050–1056.

4
Morphology Development in Polyolefin Nanocomposites
Mitsuyoshi Fujiyama

4.1
Introduction

Recently, research and development (R&D) into polymer nanocomposites (PNCs) has been a highly active topic, with the latter having already been put to wide practical use in areas such as automobile components and wrapping materials. In PNC materials, the nanofillers have sizes (at least in one direction) of nanometer order – that is, 1 to 100 nm – and are dispersed homogeneously in polymers. Initially, PNCs were prepared by penetrating a monomer into a layered silicate (e.g., montmorillonite; MMT); this would be followed by polymerization of the monomer and exfoliation of the layered silicate. Recently, it has become common practice to prepare PNCs by the simple use of a compounding method, where the MMTs are directly intercalated and exfoliated with a polymer during a melt-mixing process. Although PNCs with MMTs prepared by a direct melt-mixing method were first developed for nylon, this was quickly followed by active R&D relating to polyolefin nanocomposites, with attention focused in particular on polypropylene (PP).

The property changes that occur in PNCs compared to the base polymer are generally an enhanced stiffness and strength, and a slight deterioration in toughness; as a result, the friction coefficient is reduced and the abrasion resistance increased. Furthermore the heat resistance of the material is generally much improved, with the thermal expansion coefficient, mold shrinkage and dimensional change rate each decreased, leading to improved dimensional accuracy and stability. Functional properties, including gas-barrier properties and flame retardancy, are also improved.

Whilst not restricted to PNCs, the quality (i.e., the properties) of the processed article of a polymer is determined not only by its primary structures but also its morphology; accordingly, the quality design of a processed article can – and should – be achieved by controlling its morphology. For any processed article, the morphology is determined by the raw material's characteristics, the processing methods used, and the processing conditions, which naturally determine also the article's quality.

Optimization of Polymer Nanocomposite Properties. Edited by Vikas Mittal
Copyright © 2010 WILEY-VCH Verlag GmbH & Co. KGaA, Weinheim
ISBN: 978-3-527-32521-4

The aim of this chapter is first to outline the development of morphology in polyolefin/MMT nanocomposites. Indeed, the morphology of composites, and the intercalation, exfoliation and dispersion of the MMT, are the most important factors in general polymer/filler composites. Yet, the crystalline structures and the orientation and crystallization behavior of the base polymer may also affect the quality of the nanocomposites. Thus, the effects of the raw materials' characteristics, and also of the mixing methods and mixing conditions on the intercalation, exfoliation and dispersion of MMTs, and on the morphologies (e.g., crystallization, crystalline structure and crystal orientation) of the base polymer in nanocomposites, are first outlined. Subsequently, the development of morphology that occurs during processes such as injection molding, sheet extrusion and T-die film extrusion casting, is described.

4.2
Intercalation, Exfoliation, and Dispersion of MMT

Because, in PNCs, the nanofillers are very fine, their total surface area is enormous whereas the distance between their particles is extremely small – generally on the order of a few nanometers in the case of a low filler content. Accordingly, the nanofillers are not only highly prone to agglomerate and difficult to disperse, but are also apt to reagglomerate even after dispersion. For these reasons, dispersion technology is particularly important during the manufacture of PNCs.

Dispersion difficulties represent a problem of degree, with improvements in dispersion being carried out based on the basic theory of dispersion; namely, long-term mixing under high shear and extensional stresses, but with the interfacial tension reduced. In other words, improvements can be made by unbinding the filler lumps under a strong force for long periods of time, whereupon the affinity between the filler and polymer will be improved.

The above-mentioned physical properties of PNCs depend heavily on the dispersion of the MMT filler, with such properties being generally improved when the layered silicates are better exfoliated and dispersed.

4.2.1
Manufacturing Processes

The manufacturing processes of PNCs using nanoclays are shown in Figure 4.1 [1], where the raw materials used (nanoclays, monomer, organic cation, and polymer) are shown at the left-hand side, and the obtained PNCs on the right-hand side. The latter include an *intercalated nanocomposite*, where the polymer is intercalated between the nanoclay layers, and an *exfoliated nanocomposite*, where the nanoclay layers are exfoliated and randomly dispersed. The appearance of an intercalated or exfoliated nanocomposite is a problem of degree; in fact, the intercalated nanocomposite, with enhanced intercalation, becomes an exfoliated nanocomposite.

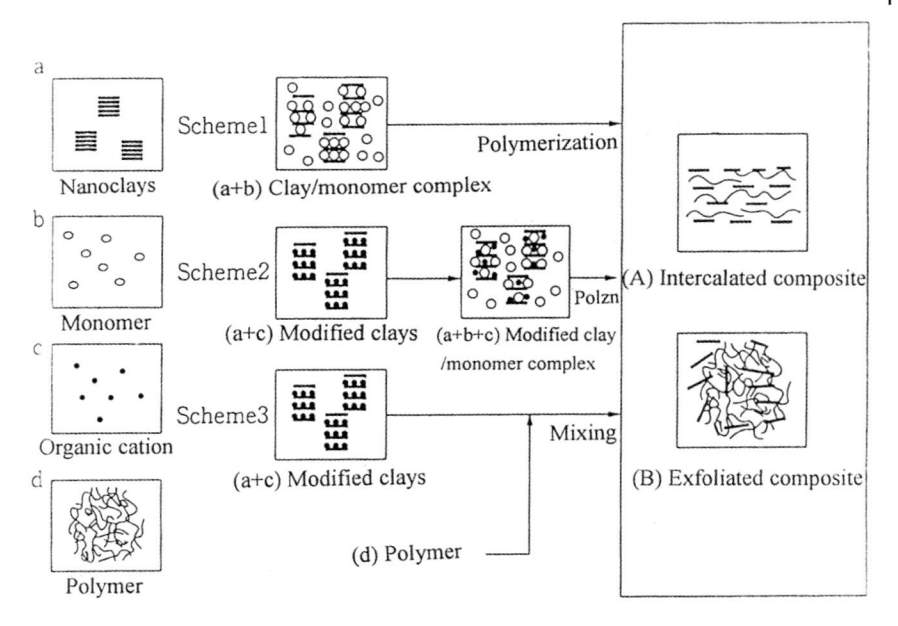

Figure 4.1 Preparing processes of PNCs with nanoclays [1]. Reprinted with permission from Plastics Age Co., Ltd.

Penetration of the nanoclays with the monomer, followed by its polymerization to produce the PNC, is shown in scheme 1 in Figure 4.1. Because the unmodified nanoclays have poor affinity with the monomer or the polymer due to the hydrophilicity, the monomer penetrates into the nanoclays and is polymerized after the clays have been treated with the organic cation and rendered hydrophobic (scheme 2 in Figure 4.1). In scheme 3 in Figure 4.1, a PNC is prepared by melt-mixing the polymer with the organophilized nanoclays (OMMTs), and shows the compounding method mainly described in this chapter.

4.2.2
Dispersion (Exfoliation) State of Nanoclays

As shown in the right-hand column of Figure 4.1, the nanoclays in PNCs with layered silicates (nanoclays) take on the dispersion states shown in Figure 4.2 [2]. The case in Figure 4.2a is where the nanoclays and polymer are immiscible and intercalation does not occur, and the nanoclays exist as lumps. Figure 4.2b shows the usual miscible case, where the nanoclays are unbound to some extent but intercalation of the polymer between the nanoclay layers has not yet occurred. In Figure 4.2c, although intercalation has occurred on the nano level, its extent is restricted, while in Figure 4.2d the intercalation has progressed and the layered silicates have been exfoliated into a near-monolayer state. As the physical properties of PNCs are generally improved when passing to the succeeding case, the

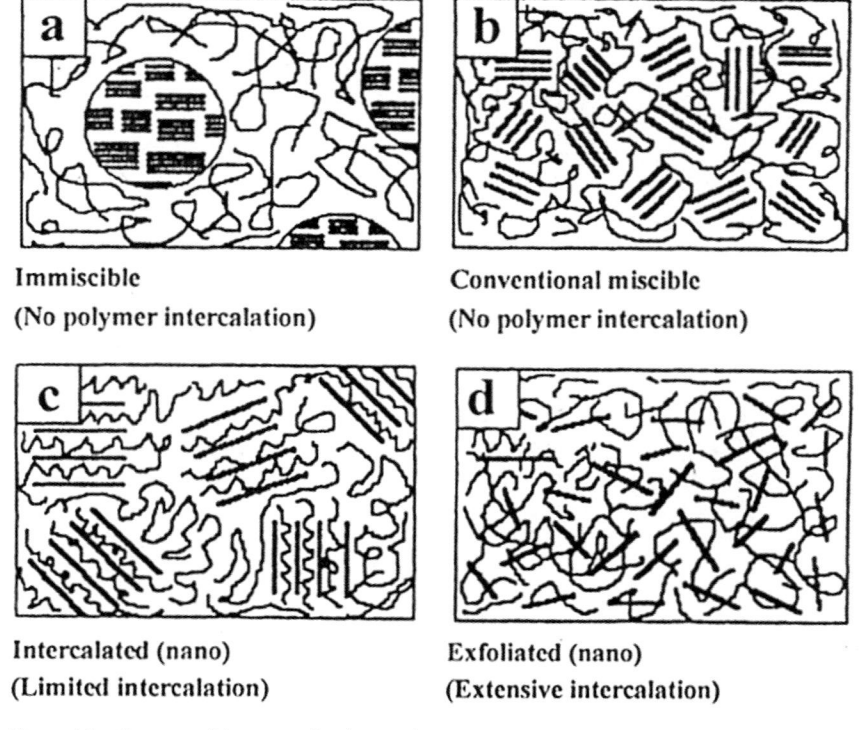

Figure 4.2 Four possible types of polymer/clay nanocomposite [2]. Reprinted with permission from Wiley-Blackwell.

Captions under figure panels:
- a: Immiscible (No polymer intercalation)
- b: Conventional miscible (No polymer intercalation)
- c: Intercalated (nano) (Limited intercalation)
- d: Exfoliated (nano) (Extensive intercalation)

manufacture of high-quality PNCs with layered silicates depends largely on how the nanoclays are brought to the dispersion state (as in Figure 4.2d).

Figure 4.3 shows the transmission electron microscopy (TEM) image and intercalation state of a polypropylene (PP)/OMMT nanocomposite prepared using the melt-mixing method [3]. The "A" regions in the TEM image correspond to the state in Figure 4.2c, with the PP molecular chains being intercalated between the OMMT layers. This situation is shown schematically using a molecular model at the right-hand side of Figure 4.3, where the "C" regions correspond to the perfect exfoliation state in Figure 4.2d. In Figure 4.3, the "B" regions are in an intermediate state between the "A" and "C" regions, while intercalation of the OMMTs has progressed further than the "A" state. Notably, the OMMT layers are no longer parallel to each other but rather exist in a disorderly state.

Kim *et al.* evaluated the length and thickness of OMMT sheets in a PP/maleated PP (MAPP)/OMMT nanocomposite prepared via a melt-mixing method from TEM observations [4]. The results are shown in Figure 4.4, where the thickness of a single OMMT sheet is seen to be approximately 1 nm. Because the OMMTs are penetrated and modified with organic matter, the mean thickness per single sheet

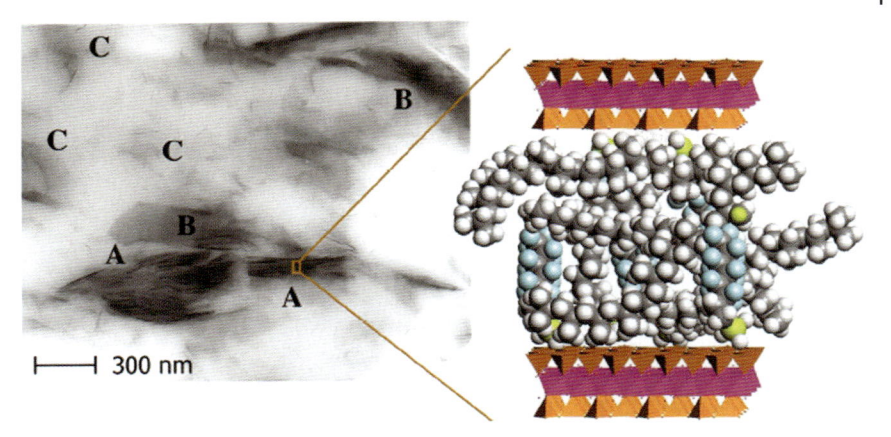

Figure 4.3 Bright-field transmission electron micrograph of PP/clay structure; the black lines correspond to the clay layers and gray regions to the PP matrix. Note the coexistence of (A) intercalated (parallel-stacked and separated by 2–3 nm of PP) and (B) disordered (with comparable PP separations, but without parallel registry) tactoids – bunches of clay layers – with (C) exfoliated formations of one to three clay layers (separated by tens of nanometers of polymer). A, B, and C refer to the way the dark lines (clay layers) are dispersed in the gray background (polymer) [3]. Reprinted with permission from The Materials Research Society.

is between 2 and 5 nm [5]. Accordingly, it can be shown that the majority of OMMT sheets are single, and that their amounts decrease rapidly as the number of sheets increases. Typically, the mean number of sheets is between 1.5 and 2.0, with the OMMT sheets being broken and dispersed at a length of approximately 200 nm. Fornes and Paul showed that the majority of OMMT sheets in nylon 6 (PA6)/OMMT nanocomposites prepared with melt-mixing were also single sheets, with their thickness decreasing rapidly with increasing the sheet number (the mean sheet number was ca. 1.4–1.7) [6]. These facts indicate that OMMTs are well exfoliated, even in PNCs prepared using a melt-mixing method.

The exfoliation state of layered silicates can be evaluated using both X-ray diffraction (XRD) and melt rheometry. The wide-angle X-ray diffraction (WAXD) diagrams of samples in various states are shown in Figure 4.5 [7]. Here, the pure clay (MMT) shows the (001) plane reflection, corresponding to an interlayer distance at $2\theta \approx 7°$. This reflection shifts to a lower angle at $2\theta \approx 4°$ for the O-clay (OMMT), which means that its interlayer distance is increased. This reflection is barely changed for the tactoid nanocomposite in which a polymer (PP) is simply mixed with the O-clays (which means that the O-clays are present as they initially were). The interlayer distance is further opened and the reflection shifted to a lower angle for the intercalated nanocomposite prepared by melt-mixing with a compatibilizer, such as MAPP. The (001) reflection disappears for the exfoliated nanocomposite in which O-clays are exfoliated and perfectly dispersed, which means that the layer structure has vanished.

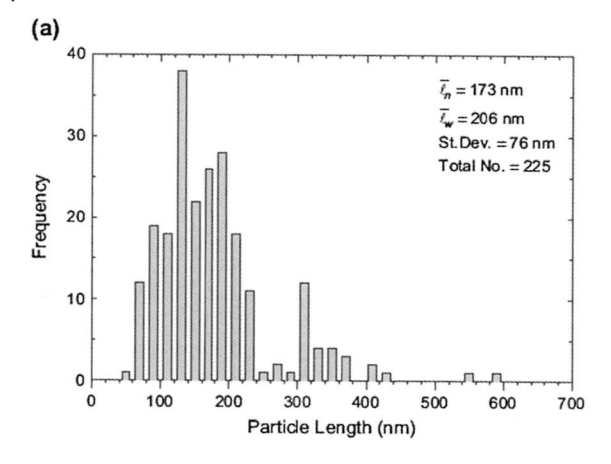

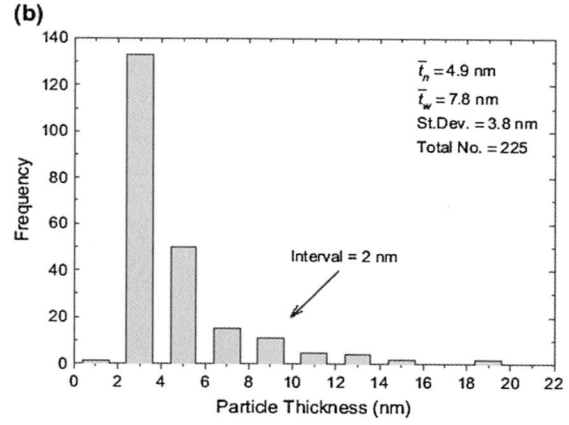

Figure 4.4 Histograms of MMT particle length (a) and thickness (b) of PP/PP-g-MA/MMT nanocomposites at a fixed PP-g-MA/organoclay ratio of 1.0, viewed parallel to the transverse direction (MMT content = 5 wt%) [4]. Reprinted with permission from Elsevier.

4.2.3
Exfoliation Process of Nanoclays

As mentioned above, in order to improve the affinity between the polyolefin and MMT, the latter is usually modified with organic matter and used as OMMT in polyolefin/clay (MMT) nanocomposites prepared using the melt-mixing method. Owing to the presence of surface modifications in the interlayer, the OMMT has a higher interlayer spacing than the MMT.

Mittal studied the relationship between the interlayer distance (d-spacing) of OMMT and the amount of organic matters ionically attached to the MMT surface by using different surface modifications [8]. Thus, the interlayer dis-

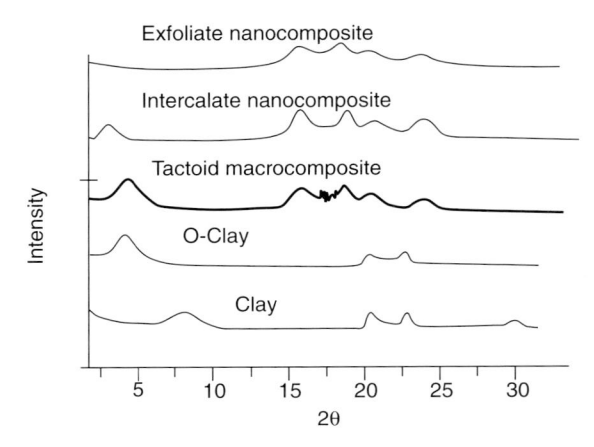

Figure 4.5 Typical X-ray patterns of different nanocomposites obtained using melt-mixing process [7]. Reprinted with permission from John Wiley & Sons Ltd.

tance was shown to increase linearly with the amount of adherent organic matter.

Toth *et al.* simulated the change of interlayer distance of MMT by the modification with organic modifiers, by employing a three-component model [9].

When OMMT is melt-mixed with a polyolefin (e.g., PP) in the presence of a compatibilizer such as MAPP, the latter intercalates between the layers of OMMT, which are then exfoliated by the action of the shear force upon further mixing and randomly dispersed in PP (see Figure 4.6 [10]).

Li *et al.* performed a model experiment on the exfoliation process of a PP/MAPP/OMMT nanocomposite during melt-mixing [11]. A PP melt-mixed with 9 wt% OMMT and a MAPP were respectively compression-molded at 190 °C into 0.4 mm-thick films. Four PP/OMMT composite films and three MAPP films were overlaid one after the other, and then pressed into a laminated sheet at 140 °C. The sheet was held in a parallel-plate rheometer maintained at a temperature above the melting point of PP, after which the dynamic viscoelasticity as a function of time was measured. The results are shown in Figure 4.7. With the lapse of time, the second plateau feature of storage modulus $G'(\omega)$ and the yield flow feature of the absolute value of complex viscosity $|\eta^*|(\omega)$ at low ω become notable. These features are due to the formation of a percolation network structure of OMMT particles caused by the advanced exfoliation/dispersion of OMMTs.

Using the storage modulus $G'(\omega)$ and the absolute value of complex viscosity $|\eta^*|(\omega)$ from Figure 4.7, the time change of exfoliation fraction $\varphi(t)$ of OMMT tactoid layers was calculated. This showed that the exfoliation/dispersion of OMMT had progressed on the order of several hundreds minutes under a 'no-mixing' quiescent state. The exfoliation rate was also seen to increase with increasing temperature.

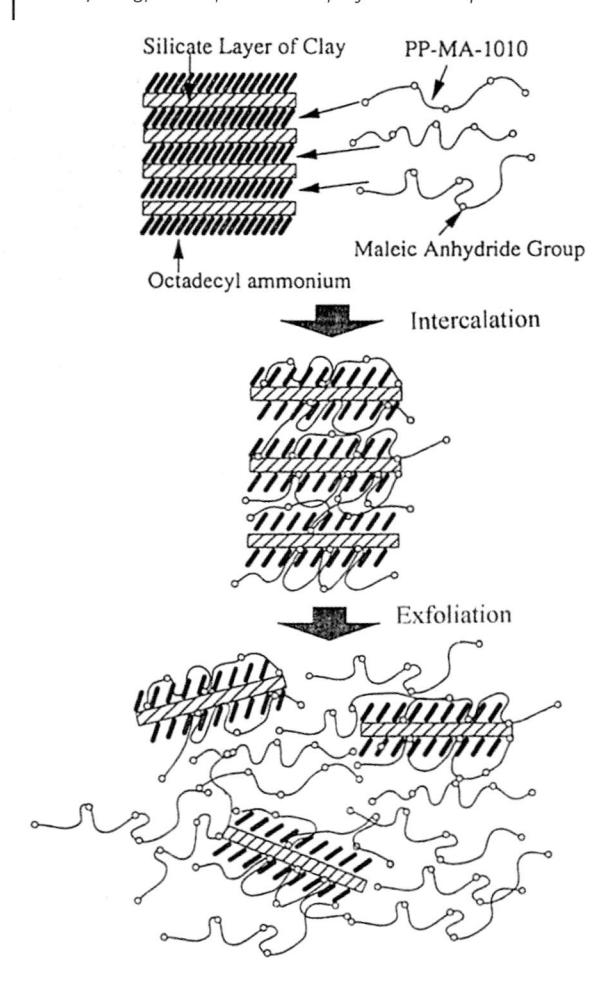

Figure 4.6 Schematic representation of the intercalation process of PP-MA1010 into the organized clay [10]. Reprinted with permission from John Wiley & Sons Ltd.

Based on the dependence of exfoliation rate on temperature, the activation energy was evaluated as $84 \, kJ \, mol^{-1}$; this was of the same order (albeit a little higher than) as $55 \, kJ \, mol^{-1}$, the activation energy of diffusion or zero-shear viscosity of MAPP. This is assumed because an interaction between MAPP and OMMT works towards the diffusion of MAPP during the exfoliation process of OMMT layers. The estimated schematic exfoliation process of OMMT is similar to that shown in Figure 4.6. Here, MAPP – which is the compatibilizer between PP and OMMT – plays an important role.

In the case of melt-mixing, the exfoliation/dispersion rate is considered to be accelerated more by the action of a mixing force caused by the mixer, than that in the case of the quiescent state.

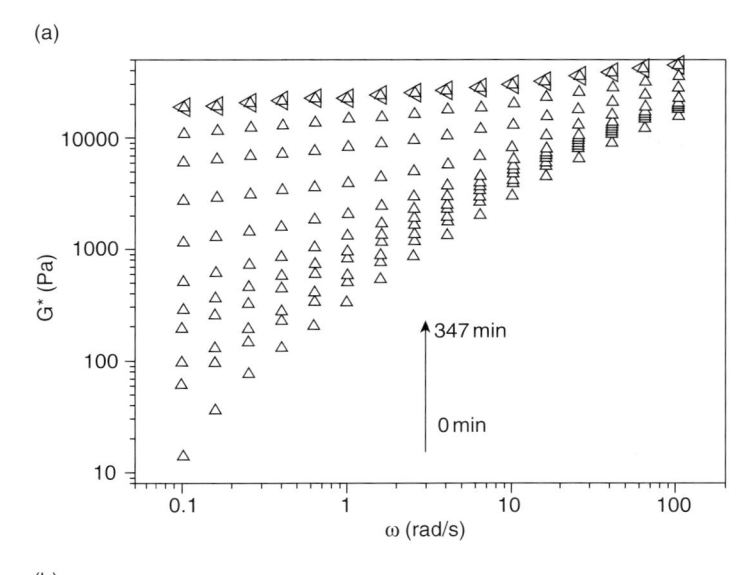

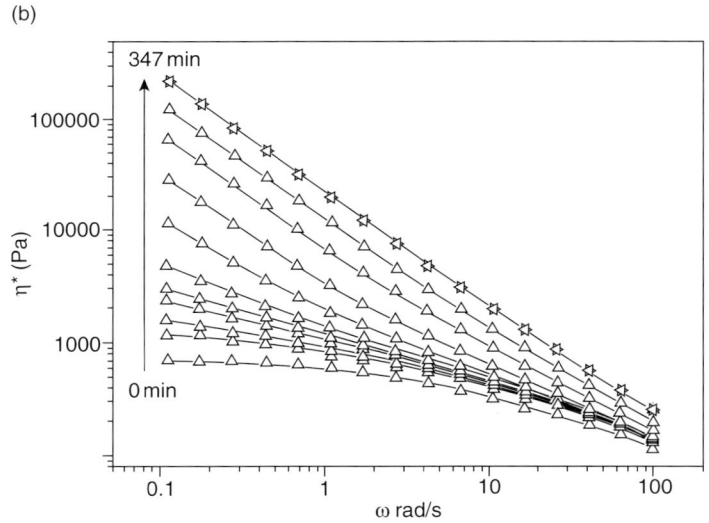

Figure 4.7 Development of dynamic modulus $G'(\omega)$ and dynamic viscosity $|\eta^*|(\omega)$ with annealing time [11]. Reprinted with permission from John Wiley & Sons Ltd.

4.2.4
Control of Exfoliation/Dispersion of Nanoclays

4.2.4.1 Raw Materials

Compatibilizers The exfoliation/dispersion of layered silicate (nanoclay) is better, as the interfacial tension between it and the polymer is lower, and their affinity is better. Generally, the surface tension of a layered silicate is high, while that of polymer would be low. Both, PA6 and PP, as polymers are frequently used for PNCs. Since PP has a lower surface tension than PA6, and hence has a higher interfacial tension with layered silicate than PA6, the exfoliation/dispersion of layered silicate in melt-mixed PP/nanoclay nanocomposite is worse than that in PA6/nanoclay nanocomposite. (An example of this is provided in Figure 4.10.)

When preparing nanocomposites using nonpolar and low-surface tension polymers (e.g., PP), these are usually added with polar materials such as MAPP, amino group-containing PP, epoxy monomer, and poly(dimethylsiloxane) in order to improve compatibility with the nanoclays.

In 2005, Lertwimolnun and Vergnes measured the WAXD diffractograms of PP/MAPP/OMMT (5 wt%) nanocomposites prepared by melt-mixing with a twin-screw extruder (TSE) [12]. The pure OMMT (Cloisite 20A™) showed the (001) plane reflection peak at $2\theta \approx 3.5°$ and its secondary reflection peak at $2\theta \approx 7°$. As with the 95/0/5 sample without MAPP, the peak positions were scarcely changed from those of the pure OMMT, although they did shift to lower angle with increasing MAPP content, which indicated an increased interlayer distance of the OMMT. The interlayer distance was seen to increase almost linearly with the increasing MAPP content [13].

The same authors also measured the frequency characteristics of the storage modulus $G'(\omega)$ in a molten state of the same PNCs [12]. On increasing the MAPP content, the second plateau feature became more remarkable, resembling the plateau with increasing annealing time in Figure 4.7a. This occurred due to the formation of a percolation network structure of OMMT sheets, and became more notable with the increasing MAPP content, due to an advanced exfoliation/dispersion of OMMT.

As noted above, the degree of intercalation/dispersion of layered silicates can be evaluated using both WAXD and melt viscoelasticity, with low and high degrees of exfoliation being detected more sensitively by the former and the latter, respectively. Whilst WAXD cannot be used to detect the degree of exfoliation/dispersion following exfoliation, the melt viscoelasticities can be used to measure this to a certain extent.

During the melt-mixing process of MMT nanocomposites, the intercalation/exfoliation of MMT lumps and the breakage of MMT sheets were shown to occur simultaneously. For example, Kim *et al.* studied the dependencies of length and thickness of MMT particles in a melt-mixed PP/MAPP/MMT nanocomposite on MAPP content [4]. Both parameters decreased rapidly with increasing MAPP content up to an MAPP/MMT ratio of about 0.5, and tended to level-off with further increases in the MAPP content. The aspect ratio was shown to increase rapidly with an increasing MAPP content, which means that the reduction in thickness of MMT particles was far larger than that in length [4].

Perrin-Sarazin *et al.* studied the effects of molecular weight and the maleic anhydride (MA) content of MAPP on the intercalation/exfoliation state of OMMT in melt-mixed PP/MAPP/OMMT nanocomposites [14]. The nanocomposite prepared using the MA9k MAPP with a low molecular weight ($M_w = 9.1 \text{ kg mol}^{-1}$) and a high MA content (3.8%) caused a uniform and good intercalation but did not lead to any further exfoliation. In contrast, the nanocomposite prepared using the MA330k MAPP with a high molecular weight (M_w = 330 kg mol^{-1}) and a low MA content (0.5%) caused only nonuniform intercalation but advanced exfoliation.

A polyolefin containing ammonium groups can be used as a compatibilizer for polyolefin nanocomposites. For this, Wang et al. proposed a schematic representation of dispersion structure of OMMTs in melt-mixed PP/OMMT nanocomposites using ammonium group-containing PPs as compatibilizers [15]. One compatabilizer containing ammonium groups in the molecular chains caused the intercalation of OMMT, but did not lead to exfoliation. In contrast, a compatabilizer containing ammonium groups at the ends of the molecular chains advanced the intercalation to the exfoliation, providing well-dispersed OMMT sheets.

Molecular Weight and Molecular Weight Distribution of Polymers Both, the molecular weight (M_w) and molecular weight distribution (MWD) of polymers affect the dispersion of fillers in polymer/filler composites, mainly through their viscosities. This situation is true also for PNCs; for example, when the M_w of a polymer is low (and hence its viscosity is low), it will easily penetrate the openings of fillers, promote the wetting fillers, and easily co-incorporate the fillers. In contrast to this, when the M_w and viscosity of a polymer are high, then the incorporation may be difficult as the fillers break easily due to the strong shear and elongational forces during mixing, leading to good dispersion of fillers. These circumstances hold true also for the MWD, with a narrow MWD showing a low melt elasticity, leading to an easy incorporation of the fillers. Conversely, a wide MWD causes high stress, leading to an enhanced dispersion of fillers.

By using PPs with a variety of M_w-values, prepared using Ziegler–Natta (ZN) and metallocene (Met) catalysts, it was possible to prepare PNCs with OMMT and MAPP as a compatibilizer, and subsequently to monitor the WAXD of the PNCs produced [7]. Pure MMT was shown to possess a diffraction peak by MMT layers at $2\theta \approx 8°$, which shifted to $2\theta \approx 4°$ for OMMT, which meant that the interlayer distance had been expanded. This dependence on M_w of the PP of the PNC suggested that the diffraction peak would shift to a lower angle and its intensity would be reduced with decreasing M_w, indicating a better exfoliation of OMMT. The Met-PP with a narrower MWD than ZN-PP showed a lower diffraction angle and hence a higher extent of exfoliation of OMMT than did the ZN-PP. In other words, the lower the M_w and the narrower the MWD, then the greater the exfoliation of the OMMT layers. This occurred because the diffusion/penetration of PP molecular chains between OMMT layers was easier in such cases. In fact, this proved to be related to the same phenomenon as that of the effect of temperature on the exfoliation rate of OMMT layers (see Section 4.2.3 and Figure 4.7), where the higher the temperature, the faster was the exfoliation rate.

MMT Content The dependence, on the crystallization temperature T_c, of the interlayer distance $d_{(001)}$ of OMMT particles in a melt-mixed and then compression-molded PP/MAPP/OMMT nanocomposite sample crystallized at various T_c-values after melting at 190 °C was studied by Maiti *et al.* [16]. The interlayer distance was seen to increase linearly with T_c, while the PPCN4 sample with a lower OMMT content of 4 wt% showed a greater interlayer distance than the PPCN7.5 with a higher OMMT content of 7.5 wt% when compared at the same T_c.

The aspect ratio of MMT particles after melt-mixing was shown to decrease as the MMT content was increased, which implied a more difficult exfoliation of the MMT particles [4].

Schematic illustrations for the morphology of nanocomposites with different OMMT contents have shown that the PPCN7.5 sample with a higher OMMT content has a shorter length L_{clay}, a higher thickness d_{clay}, and a smaller aspect ratio of OMMT than does PPCN2 with a lower MMT content [17]. It also showed a shorter correlation length χ_{clay} between the dispersed OMMT particles, and a shorter interlayer distance $d_{(001)}$, indicating a less advanced intercalation. The crystal long period $L_{lamellae}$ and lamellar thickness $d_{lamellae}$ were near-equivalent for both samples.

4.2.4.2 Mixing Methods

Masterbatch Method PNCs are usually manufactured by first preparing a masterbatch with a high concentration of layered silicate, and then by diluting this with diluent resins. Both, the preparation of the masterbatch and the dilution, are occasionally carried out using TSEs. When Treece *et al.* studied the effect of TSE type on the exfoliation/dispersion of OMMT layers from measurements of WAXD and melt viscoelasticity [18], the exfoliation states of the OMMT layers were shown to be delicately different depending on the combinations of extruder used for compounding. Notably, the exfoliation/dispersion state was inferior in the absence of MAPP. The masterbatch method was also shown to produce a slightly weaker exfoliation of the OMMT layers than did the one-step method.

Water-Assisted Mixing As MMT is hydrophilic in nature, it is accordingly expected that water would unbind the MMT particles, such that their intercalation/exfoliation/dispersion would become straightforward if they were to be melt-mixed with a polymer in the presence of water. Based on such a concept, water-assisted mixings were subsequently studied.

Initially, a mixing method was proposed in which a water slurry suspended with MMT particles was injected into the PA6 melt during the extrusion process [19]. The MMT particles were exfoliated and uniformly dispersed by the action of the evaporated water in the extruder. Although with pure MMT such exfoliation/dispersion is achieved, its stability is not good and it is apt to reagglomerate in the molten state after compounding. This can be prevented and improved by the simultaneous addition of organic modifiers at the extrusion stage.

The same method can be applied to PP nanocomposites; a schematic representation for the dispersion process of MMT is shown in Figure 4.8 [20]. Here, both PP and MMT (clay minerals) are fed into an extruder with OTM (octadecyl

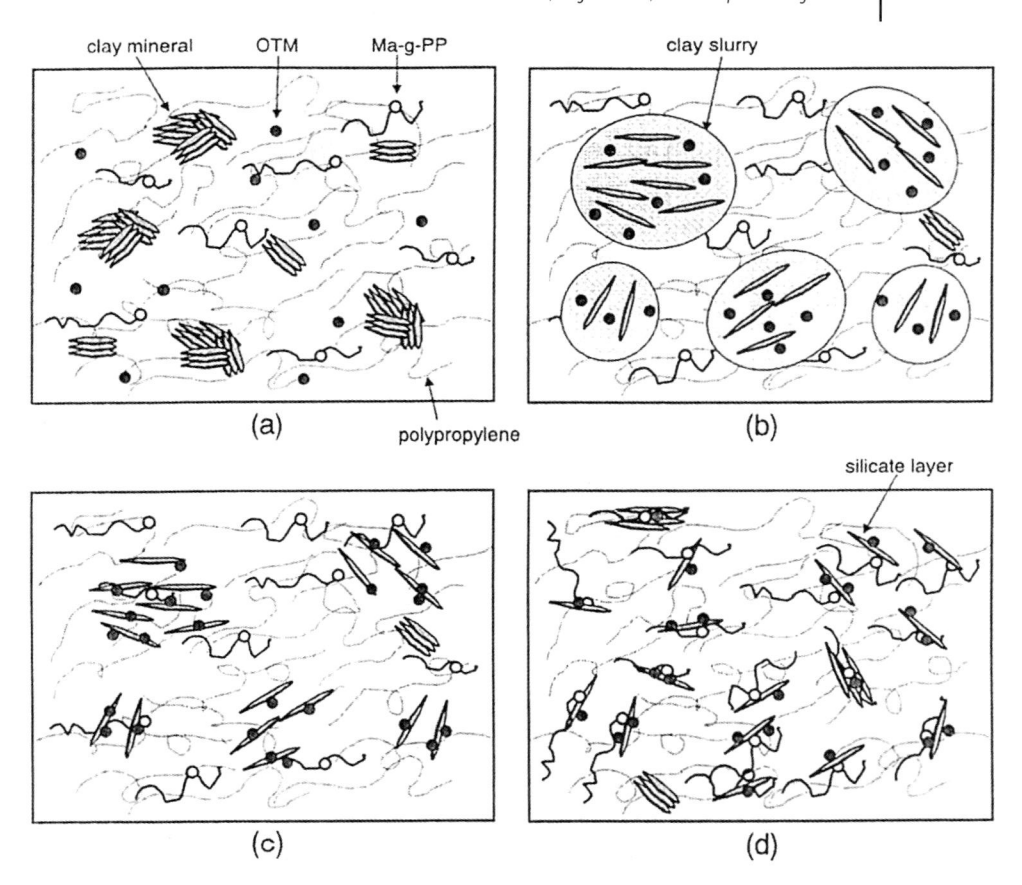

Figure 4.8 Schematic depicting the dispersion of silicate layers of clay in PP during water-assisted compounding by extruder [20]. Reprinted with permission from The Society of Plastic Engineers.

trimethyl ammonium chloride) and MAPP (Ma-g-PP) as the compatibilizers, and then melt-mixed (Figure 4.8a). Water is then injected, whereupon the clay minerals become swollen and exfoliated, so as to form a clay slurry (Figure 4.8b). The OTM then reacts with Na cations present on the silicate layers of clay in the slurry, and the hydrophilic silicate layers stabilize in the hydrophobic PP (Figure 4.8c). The clays then exfoliate and disperse by the action of shear forces (Figure 4.8d), after which any remaining water is eliminated from the extruder by venting. Since this method does not require any initial preparation of OMMT and subsequent melt-mixing with a polymer, it can be used to (at a stretch) prepare PNCs with melt-mixing. PP/clay nanocomposites prepared by this method show similar physical properties to those created using conventional methods.

Supercritical Fluid-Assisted Mixing In addition to water-assisted mixing, supercritical fluids (e.g., supercritical CO_2) can be used to improve the intercalation/exfoliation/dispersion of MMT. Figure 4.9 shows the WAXD curves of a PP/MAPP/OMMT nanocomposite prepared by melt-mixing in a TSE after injection with supercritical CO_2 [21]. Without CO_2 injection the (001) plane distance was widened for C20A (OMMT); however, the addition of a compatibilizer MAPP led to a further widening of the plane distances after CO_2 injection, with maximum benefit at a CO_2 level of 2 wt%. Any residual CO_2 was subsequently vented from the extruder. Subsequent TEM images of nanocomposites prepared with and without CO_2 confirmed the benefit of such injection in improving the dispersion of OMMT.

Mixing in Elongational Flow Utracki et al. recently developed an extensional flow mixer (EFM) and investigated its application to PNCs. The contract and divergent flows were repeated at the contraction–divergence region (c–d plate), where dispersion of the layered silicates was advanced by the action of the extensional flow deformation.

The effect of the EFM gap, h, on the degree of exfoliation of OMMT in PA6/OMMT (Cloisite 15A; 2 wt%) and PP/MAPP/OMMT (2 wt%) nanocomposites processed with the EFM, are shown in Figure 4.10a and b [22]. Here, as with the PA6 system, the value of a commercial product (Ube PA1015C2) is also indicated on the ordinate (for reference). These nanocomposites were prepared using a single-screw extruder (SSE), or a combination of a TSE and a gear pump (GP), attached with EFMs (EFM-3 or EFM-N) and with two types of c–d plate shapes, respectively, at the exit of the extruder or the extruder system. With decreasing h, the extensional stress increased, leading to an advanced degree of OMMT exfoliation. As PA6 has a better affinity with OMMT than with PP, the

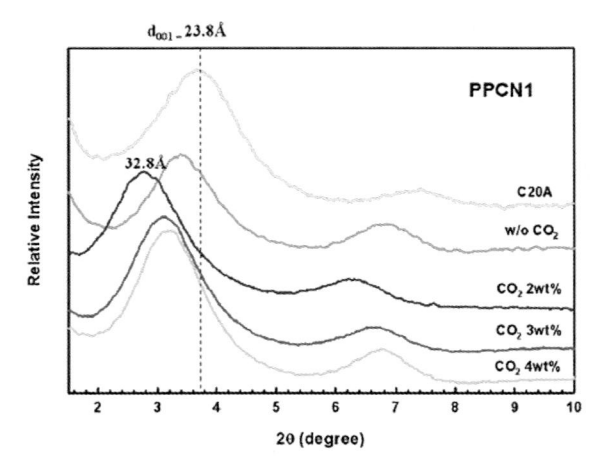

Figure 4.9 WAXD patterns of PPCN1 as a function of supercritical CO_2 concentration at a screw speed of 55 rpm [21]. Reprinted with permission from The Society of Plastic Engineers.

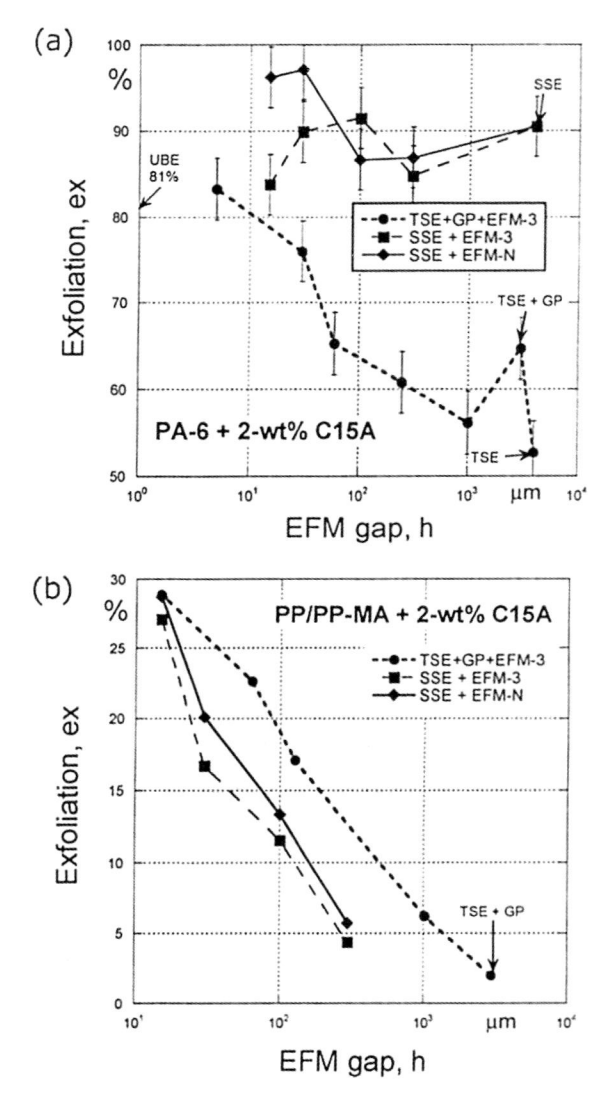

Figure 4.10 Dependence of exfoliation degree of clay (C15A) in PNCs on the EFM gap, h [22]. Reprinted with permission from Hanser Verlag.

PA6 nanocomposites generally showed a higher degree of OMMT exfoliation than did the PP nanocomposites.

Mixing Under an Electric Field Kim et al. recently reported on the compounding of PP/OMMT nanocomposites under an electric field [23]. In the apparatus used, an electric melt pipe equipped at the exit of a TSE caused an electric field E to be applied to the melt-mixed composites. It has been reported that polymer nanocomposites prepared using this method were superior with regards to the exfoliation of OMMT layers.

4.2.4.3 Mixing Conditions

Mixing Time The effect of mixing time on the frequency characteristics of the storage modulus $G'(\omega)$ for PP/MAPP/OMMT (5 wt%) nanocomposites prepared using a melt-mixing method with an internal mixer was investigated by Lertwimolnun and Vergnes [12]. By increasing the mixing time, the second plateau feature in the storage modulus $G'(\omega)$ and the yielding flow feature in the absolute value of complex viscosity $|\eta^*|(\omega)$ at low angular frequency region became more notable, with such changes saturating at mixing times in excess of 20 min. These changes, which progress with mixing time, were considered due to the advanced exfoliation/dispersion of OMMT layers. The WAXD measurements also demonstrated shifts of the (001) plane diffraction peak to lower angles, indicating that the OMMT layer distances had been widened [12].

Rotation Speed of Mixing In an internal mixer, the above-mentioned tendencies become more remarkable as the rotor rotation speed is increased [12]. Increasing the rotation speed over a similar mixing time increases the total shear strain and specific energy applied to the material; this not only produces a better mixing but also advances the exfoliation/dispersion of the OMMT layers, leading to changes in rheological characteristics and WAXD (as noted above).

Mixing Temperature The tendencies noted above become more apparent at a lower mixing temperature [12], when the diffusion rate of the PP molecular chains is decreased and the viscosity increased, as also is the shear stress applied to the OMMT particles. In this case, it is assumed that any effect of shear stress will dominate the diffusion rate, whereas the exfoliation/dispersion of OMMT layers will be more advanced at a lower mixing temperature.

Feed Rate The above-mentioned tendencies for PP/MAPP/OMMT (5 wt%) nanocomposite prepared by melt-mixing with a TSE are known to increase as the feed rate is reduced, indicating an advanced exfoliation/dispersion of the OMMT layers [12]. A lower feed rate would prolong the time for the material to be mixed, and hence produce the same effect on the exfoliation/dispersion of OMMT layers as would a long mixing time.

4.2.5
Morphology of Base Polymers

In the case of polymer nanocomposites in which the base polymers are blends and alloys, the phase morphology of the base polymers was shown to be of equal importance as the dispersion state of the MMT particles. Thermoplastic polyolefins (TPOs) are normally PPs that contain elastomer particles, such as ethylene-α-olefin copolymers. The size and deformation of the elastomer particles in a melt-mixed TPO/MAPP/MMT nanocomposite were altered by the presence of MAPP and MMT. For example, when Kim *et al.* investigated the dependence of elastomer particle size on the MMT and MAPP (PP-g-MA) contents [24], the size was seen to be reduced by the addition of MMT or MAPP. Notably, the aspect ratio or deformation of elastomer particles increased with increasing MMT or MAPP content. For example, as the MAPP content increased, the dispersion of MMT advanced and elastomer particles became smaller and more slender.

4.3
Crystallization and Crystalline Structure of Matrix Polymers

Although polyolefins used for nanocomposites are mainly PPs, high-density polyethylenes (HDPEs) have been studied to a lesser extent, while studies of other polyolefins are few in number. As these polyolefins are crystalline in nature, their crystallization behaviors and crystalline structures may greatly affect the processing and properties of the ultimate products. Typically, the presence of MMT in a polyolefin enhances its crystallization rate and reduces its crystallite (spherulite) size, since it will function as a nucleating agent. As MMT particles are also oriented in the processed article of a polyolefin nanocomposite, the polyolefin crystallites formed around nucleating agents of MMT particles may also become oriented.

4.3.1
Crystallization

4.3.1.1 Quiescent Crystallization
Maiti *et al.* studied changes of spherulite diameter D with time t for an MAPP (PP-MA) and its nanocomposites with 2 and 7.5 wt% OMMT crystallized at $T_c = 135\,°C$ [16]. Here, D was shown to increase linearly with t following an induction time, t_0. Compared to the pure MAPP, those nanocomposites with OMMT showed shorter crystallization times and faster crystallization rates.

The dependence of the crystallization temperature T_c on the MMT content of PP/MMT nanocomposites during nonisothermal crystallization showed that T_c rises with the MMT content, and then tends to level-off [25]. This T_c-rising effect was weaker for the nanocomposite with MMT (the Ca^{2+} of which were exchanged with Na^+) than for a nanocomposite with an unmodified MMT. A composite with OMMT modified with *N*-cetylpyridinium chloride showed scarcely any T_c-rising effect.

4.3.1.2 Flow-Induced Crystallization

Here, the crystallization of a polymer is affected by the shear flow, and the crystallization during flow or after the cessation of flow is more pronounced than in the usual, quiescent state.

Rozanski *et al.* studied isothermal crystallization behaviors under quiescent conditions and after the cessation of shear flow at 136 °C for PP/MMT and PP/MAPP/MMT nanocomposites [26]. Whilst the crystallization behavior of nanocomposite without MAPP was scarcely affected by the shear flow, the behavior of the nanocomposite with MAPP was largely promoted by the shear flows.

4.3.2 Crystalline Structure

4.3.2.1 Quiescent Crystallization

The polarized micrographs (POMs) of crystallized PP-MA, PPCN2, and PPCN7.5 (see Section 4.3.1.1) are shown in Figure 4.11 [16]. The addition of OMMT causes the spherulites to become smaller, with such tendency increasing as the OMMT content is increased. The spherulite growth rate (see Section 4.3.1.1 and Figure

Figure 4.11 Optical micrographs of PP-MA and PPCNs crystallized after full solidification at 130 °C. (a) PP-MA; (b) PPCN2; (c) PPCN7.5 [16]. Reprinted with permission from The American Chemical Society.

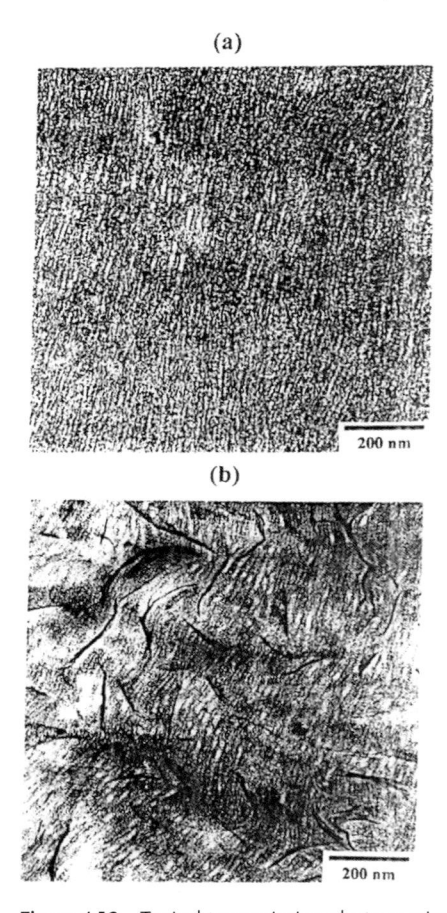

Figure 4.12 Typical transmission electron microscopy images of (a) PP-MA and (b) PPCN4 crystallized at 130 °C (after full solidification) [27]. The samples were stained with RuO_4. Reprinted with permission from The Society of Plastic Engineers.

4.11) show that OMMT functions as a nucleating agent for PP, whilst reducing its crystallite size and promoting the crystallization rate.

The optical micrographs in Figure 4.11 display spherulitic structures which are structures of considerably macroscopic order, while Figure 4.12 shows TEM images in which the spherulitic structures have been observed in greater detail [27]. Here, crystalline lamellae 10–20 nm thick can be observed in the PP-MA sample. Notably, a "house-of-cards structure", where OMMT layers with similar thicknesses to those of the PP-MA crystalline lamellae are randomly dispersed in the PP-MA crystalline lamellae, is formed in the nanocomposite with 4 wt% OMMT.

The dependency of the crystallization enthalpies ($\propto$ crystallinities) on MMT content [25] showed a similar tendency to that of crystallization temperatures, whilst the crystallization enthalpies increased with increasing MMT content and tended to level-off. MMT was shown to increase the crystallinity of PP as well as to enhance its crystallizationnate.

Among the crystalline polymorphs of PP, α-, β-, γ-, and smectic forms have been identified, while the proportion of polymorphs formed in nanocomposites differs from that in pure PP. When Maiti *et al.* investigated the dependence of γ-phase content f_γ on the crystallization temperature (T_c) for pure MAPP (PP-MA), MAPP/OMMT (4 wt%) (PPCN4) and MAPP/OMMT (7.5 wt%) (PPCN7.5) nanocomposites [27], γ-form crystals were shown to be created at high temperatures and f_γ to increase with increasing T_c. Likewise, f_γ was seen to increase as the OMMT content was increased, when compared at the same T_c.

4.3.2.2 Flow-Induced Crystallization

Not only the crystallization behavior but also the formed crystals are affected by the shear flow. The micrographs of crystal structures of PP/MAPP/MMT nanocomposite observed by Rozanski *et al.* showed that, with increasing the flow shear rate before the crystallization, the spherulites would become finer [26].

Both, OMMT and MAPP (added as a compatibilizer) also affect the crystalline orientation state of the polymer (PP). Rozanski *et al.* measured the WAXD patterns of samples crystallized after the cessation of shearing flow at a shear rate of $2\,s^{-1}$ [26]. While PP/MAPP without OMMT and PP/MMT without MAPP showed almost no crystalline orientation, PP/MAPP/MMT with both OMMT and MAPP showed a crystalline orientation where the (010) planes were oriented parallel to the shear plane.

These authors proposed a schematic diagram for the crystalline orientation of flow-induced crystallized PP nanocomposite, presumed from the experimental results, such as WAXD patterns, pole figures, and small-angle X-ray scattering (SAXS) patterns [26]. Here, the (010) planes of the mother lamellae orient parallel to the shear direction, while the (010) planes of cross-hatched lamellae epitaxially overlap almost orthogonally on the (010) planes of the mother lamellae, so that the c-axes of the mother lamellae cross orthogonally to the a*-axes of the cross-hatched lamellae. Such a crystalline orientation state closely resembles that in the injection moldings of talc-filled PPs [28, 29].

4.4
Morphology Development in Processing

To date, studies on morphology development and morphology in processed polyolefin articles are few in number. Nonetheless, some results on injection molding, sheet extrusion, and film extrusion casting will be introduced in the following sections.

4.4.1
Injection Molding

4.4.1.1 Conventional Injection Molding

Some POM images of thin sections sliced normal to the flow direction from the injection moldings of PP and PP/MMT nanocomposite are shown in Figure 4.13 [30]. The PP injection moldings usually form skin/core structures, which are observed also for both samples. As the skin layer is rapidly cooled and crystallized under a high shear stress, it is composed of fibrillar structures such as "shish-kebab," and appears featureless and homogeneous within the order of optical micrographs. In contrast, as the core layer is slowly crystallized under a low shear stress, it is usually composed of spherulites and demonstrates a heterogeneous structure of spherulites under optical microscopy. An injection-molded PP shows such a heterogeneous structure of spherulites in the core layer, whereas an injection-molded PP/MMT nanocomposite displays an almost featureless homogeneous structure at this magnification. This effect is due to MMT functioning as a nucleating agent for PP and reducing its spherulites size (see Figure 4.11 and Section 4.3.2.1).

The addition of MMT affects not only the crystalline morphology of injection-molded PPs but also their crystalline orientation. The comparison between the distributions of the degree of crystalline molecular chain orientation of PP (as monitored using infrared dichroism by Fourdrin *et al.* [30]), along the thickness direction of the injection moldings in Figure 4.13, shows that the degrees of orientation show positive values, which means that the crystalline molecular chains orient along the flow direction. The figures show two peaks of degree of orientation around the surface and inner regions, and the lowest value around the center in the thickness direction. The nanocomposite shows higher values of degree of crystalline molecular chain orientation than the pure PP throughout the thickness of the molded articles.

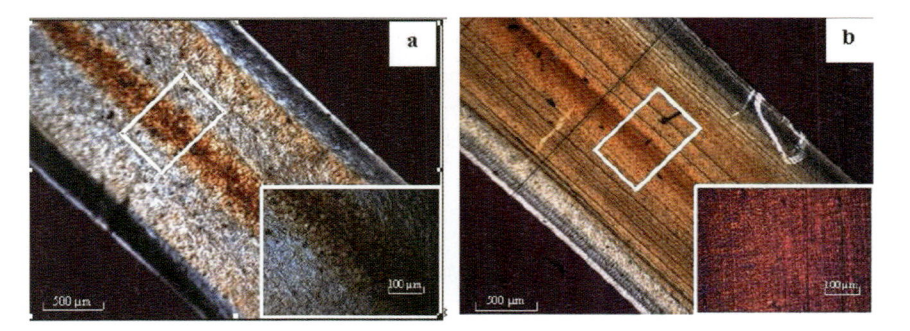

Figure 4.13 Through-the-thickness microstructure of an injection-molded part.
(a) Heterogeneous structure for PP; (b) Homogeneous structure for PP/nanoclay [30]. Reprinted with permission from Prof. Krawczak.

4.4.1.2 Dynamic Packing Injection Molding

Fu et al. recently studied, on an energetic basis, the structure and properties of injection-molded PPs by using a dynamic packing injection molding (DPIM) method. The polymer nanocomposites prepared via this technique were also studied [31–34]. DPIM operates on a similar principle to shear-controlled orientation injection molding (SCORIM). For this, a molten resin is injected into a mold cavity and then cooled and solidified under a reciprocating vibrational flow. A bar molding prepared with the DPIM has a three-layered structure composed of a surface skin layer, an intermediate oriented layer, and an inner core layer.

Figure 4.14 shows the change (in the thickness direction) of the WAXD curve of a PP/MAPP/OMMT (5 wt%) nanocomposite molding processed with DPIM [32]. The figure shows that the exfoliation of OMMT layers advances on passing inwards from the skin to the core layers. In DPIM, a reciprocating vibrational flow is applied continuously to the molten resin following its injection into a mold cavity until its solidification. Clearly, the surface layer will cool and solidify more quickly than will the inner layer; accordingly, the more innermost the layer the longer would be the time for the molten resin to be applied the vibrational flow, and the greater extent of exfoliation/dispersion of OMMT layers within it. As proof, the TEM images of the surface, intermediate and core layers confirmed that the more inner the layer, the greater the extent of exfoliation.

DPIM has been seen to affect not only the dispersion of OMMT particles but also the orientation of the PP crystalline molecular chains. Wang et al. measured WAXD patterns of skin, oriented, and core layers in PP and PP/MAPP/OMMT (10 wt%) nanocomposite molded with DPIM [33]. As with pure PP, whilst the crystalline molecular chains were oriented along the flow direction in the skin and oriented layers, they were scarcely oriented in the core layer. On the other hand,

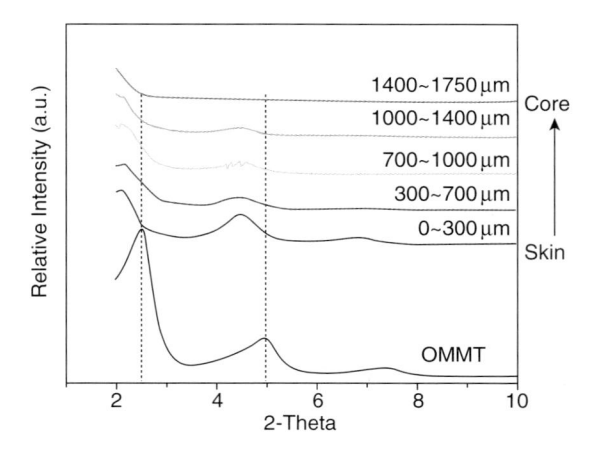

Figure 4.14 WAXD profiles representing morphology development of dynamic packing injection-molded PP/MAPP/OMMT (5 wt%) nanocomposite from the skin to the core [32]. Reprinted with permission from John Wiley & Sons Ltd.

the nanocomposite showed a high degree of orientation, and a considerable orientation even in the core layer. The orientation order parameters of PP were: for the skin layer −0.20; for the oriented layer −0.15; and for the core layer 0.0. Those of the nanocomposite were: for the skin layer −0.26; for the oriented layer −0.25; and for the core layer −0.19.

Wang et al. proposed schematic illustrations for the orientation states of OMMT particles and PP lamellae in a PP/MAPP/OMMT nanocomposite molding prepared with DPIM, presumed from electron micrographs and WAXD patterns taken from various directions [34]. The PP crystals showed an orientation where the c- and a*-axes were mixedly oriented to the flow direction, while the b-axes were oriented along the thickness direction of molding. The OMMT particles were oriented parallel to the molding surface in the skin layer, whilst tilted particles also coexist in the oriented and core layers. Such orientation states of OMMT particles and PP crystals were very similar to those of talc particles and PP crystals in injection moldings of talc-filled PPs [28, 29].

The reason why the degree of molecular chain orientation of PP crystals in an injection-molded PP nanocomposite is greater than in an injection-molded pure PP was demonstrated by Wang et al. [33]. It was proposed that, in the case of PP, the shear-induced melt molecular chain orientation would freely relax until the crystallization/solidification approach an unoriented state. In contrast, in the case of a nanocomposite containing sheet-like OMMT particles, as the shear-induced and oriented melt PP molecular chains were restrained by the OMMT particles and difficult to relax until crystallization/solidification, the crystallization of PP would occur with the molecular chains being oriented, and result in a high degree of orientation of PP crystalline molecular chains in the molded article.

4.4.2
Sheet Extrusion

As PP crystallizes/solidifies under or after the shear or elongational flow also in sheet extrusion, the OMMT particles and PP molecular chains in a sheet will be oriented.

WAXD patterns taken from the various directions of PP/MAPP and various PP/MAPP/OMMT nanocomposites sheets are very alike those of the injection moldings described above [35]. A schematic illustration of the orientation state of OMMT particles and PP crystals, presumed from the WAXD patterns, is shown in Figure 4.15 [35]. The OMMT plates are oriented parallel to the sheet surface, while the b-axes of PP crystals are oriented in the thickness direction, and the c-axes and a*-axes are mixedly oriented in the flow direction (MD); this is also very similar to that of injection moldings of talc-filled PPs. The dependence of the degree of b-axis orientation, $I(040)/I(110)$, on the OMMT content showed that by increasing the OMMT content, $I(040)/I(110)$ would be decreased and then level-off, indicating an increasing degree of b-axis orientation [35].

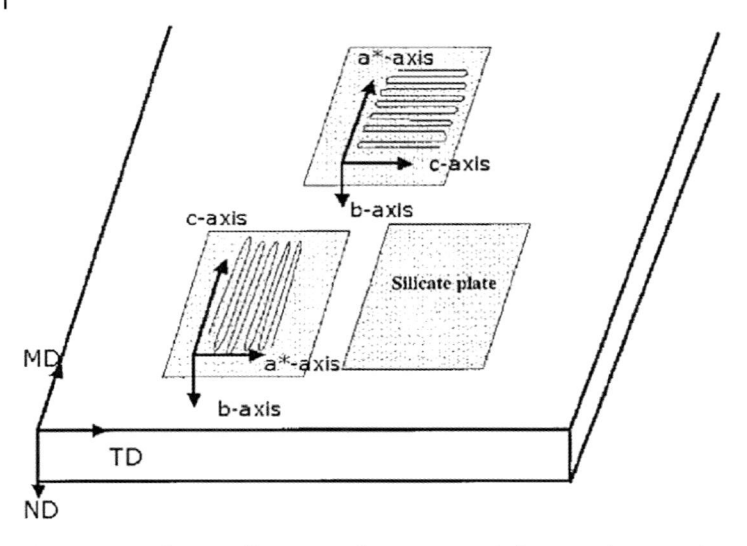

Figure 4.15 Schematic illustration of orientations of silicates and PP crystals [35]. TD = transverse direction; MD = machine direction; ND = normal direction. Reprinted with permission from John Wiley & Sons Ltd.

4.4.3
Film Extrusion Casting

Film extrusion casting is regarded as a sheet extrusion where the thickness of sheet is reduced. As a polymer also crystallizes/solidifies under shear or elongational stress, orientations of the MMT particles and polymer molecular chains also occur. Due to rapid cooling/solidification, the crystallinity of film is usually less than that of sheets.

Bafna *et al.* monitored SAXS and WAXD patterns from various directions of an HDPE/maleated polyethylene (MAPE)/MMT (6 wt%) nanocomposite and evaluated the orientation states of MMT particles and HDPE crystals by using the Wilchinsky triangle representation [36]. The HDPE crystalline lamellae were seen to be oriented normal to the flow direction (MD). As with MMT, the modified MMT (OMMT) sheets were oriented parallel to the film surface, while unmodified MMT sheets were rotating around an axis of MD.

4.5
Conclusions

In the case of PNCs containing nanoclays (MMTs), the performance of a polymer is largely improved while its functionalities – such as gas-barrier properties and flame retardancy – can be endowed and improved by the addition of very small amounts of MMT. Accordingly, in recent years the R&D of PNCs has accelerated, especially during the past decade. Recently, polyolefin nanocomposites have attracted much attention due to advances in the methods of their manufacture. Among these materials, great expectation has been directed towards PP, and indeed it is not an exaggeration to suggest that investigations into PP nanocomposites have occupied the major proportion of those studies relating to PNCs. Moreover, as PP is a resin that is well balanced in its processability, product properties and cost, the preparation of its nanocomposites is currently of major interest.

By adding small amounts of MMT into a polymer (polyolefin) so as to prepare a nanocomposite, both its melt characteristics (processabilities) and solid characteristics (product properties) are greatly altered. The main melt characteristic is that non-Newtonian flow behavior at low shear rates becomes more notable and shows a yield flow feature. A second plateau feature appears at low angular frequencies, which is more of a solid-like characteristic. Taken together, these melt characteristics will affect the flow processabilities of the polymer.

In this chapter we have described the morphology developments governing the solid characteristics of PNCs, and details have been provided of the morphologies – as intercalation/exfoliation/dispersion of MMT – that most affect the products' properties. Although investigations into composites have been commenced only relatively recently, the objectives have been rather basic in nature, and studies aimed at processing PNCs for practical purposes and identifying the morphologies of the processed materials remain few in number.

Clearly, today investigations with PNCs – and in particular of polyolefin nanocomposites – will continue to be actively pursued, and as the importance of R&D in this area becomes increasingly recognized major advances are to be expected. Today, whilst clay nanocomposites have been recognized as hugely important, other PNCs that contain nanofillers, including CNTs, carbon nanofibers, silica, carbon black, nanocalcium carbonate, polyhedral oligomeric silsesquioxane (POSS), and others continue to attract a great deal of attention. In this respect, studies of PP/CNT nanocomposites have been particularly vigorous over the past few years, and will surely continue to do so in the future.

References

1 Tamura, K. and Nakamura, J. (1999) *Plastics Age*, **45** (11), 106.
2 Zhang, Q., Fu, Q., Jiang, L., and Lei, Y. (2000) *Polym. Int.*, **49**, 1561.
3 Manias, E. (2001) *MRS Bull.*, 862.
4 Kim, D.H., Fasulo, P.D., Rodgers, W.R., and Paul, D.R. (2007) *Polymer*, **48**, 5308.
5 Ciardelli, F., Coiai, S., Passaglia, E., Pucci, A., and Ruggeri, G. (2008) *Polym. Int.*, **57**, 805.
6 Fornes, T.D. and Paul, D.R. (2003) *Polymer*, **44**, 4993.
7 Moncada, E., Quijada, R., and Retuert, J. (2007) *J. Appl. Polym. Sci.*, **103**, 698.
8 Mittal, V. (2007) *J. Thermoplast. Compos. Mater.*, **20**, 575.
9 Toth, R., Coslanich, A., Ferrone, M., Fermeglia, M., Pricl, S., Miertus, S., and Chiellini, E. (2004) *Polymer*, **45**, 8075.
10 Kato, M., Usuki, A., and Okada, A. (1997) *J. Appl. Polym. Sci.*, **66**, 1781.

11 Li, J., Zhou, C., Wang, G., and Zhao, D. (2003) *J. Appl. Polym. Sci.*, **89**, 318.
12 Lertwimolnun, W. and Vergnes, B. (2005) *Polymer*, **46**, 3462.
13 Százdi, L., Árányi, Á., Pukánszky, B., Jr, Vancso, J.G., and Pukánszky, B. (2006) *Macromol. Mater. Eng.*, **291**, 858.
14 Perrin-Sarazin, F., Ton-That, M.-T., Bureau, M.N., and Denault, J. (2005) *Polymer*, **46**, 11624.
15 Wang, Z.M., Nakajima, H., Manias, E., and Chung, T.C. (2003) *Macromolecules*, **36**, 8919.
16 Maiti, P., Nam, P.M., Okamoto, M., Hasegawa, N., and Usuki, A. (2002) *Macromolecules*, **35**, 2042.
17 Nam, P.H., Maiti, P., Okamoto, M., Kotaka, T., Hasegawa, N., and Usuki, A. (2001) *Polymer*, **42**, 9633.
18 Treece, M.A., Zhang, W., Moffitt, R.D., and Oberhauser, J.P. (2007) *Polym. Eng. Sci.*, **47**, 898.
19 Hasegawa, N., Okamoto, H., Kato, M., Usuki, A., and Sato, N. (2003) *Polymer*, **44**, 2933.
20 Kato, M., Matsushita, M., and Fukumori, K. (2004) *Polym. Eng. Sci.*, **44**, 1205.
21 Han, J.H., Lee, S.M., Ahn, Y.J., Kim, H., Kim, J.G., and Lee, J.W. (2005) *SPE Technical Paper, 63rd ANTEC*, **51**, 1928.
22 Utracki, L.A., Sepehr, M., and Li, J. (2006) *Int. Polym. Process.*, **21**, 3.
23 Kim, D.H., Park, J.U., Cho, K.S., Ahn, K.H., and Lee, S.J. (2006) *Macromol. Mater. Eng.*, **291**, 1127.
24 Kim, D.H., Fasulo, P.D., Rodgers, W.R., and Paul, D.R. (2007) *Polymer*, **48**, 5960.
25 Pozsgay, A., Fráter, T., Papp, L., Sajó, I., and Pukánszky, B. (2002) *J. Macromol. Sci. Part B: Phys.*, **B41**, 1249.
26 Rozanski, A., Monasse, B., Szkudlarek, E., Pawlak, A., Piorkowska, E., Galeski, A., and Haudin, J.M. (2009) *Eur. Polym. J.*, **45**, 88.
27 Maiti, P., Nam, P.H., Okamoto, M., Kotaka, T., Hasegawa, N., and Usuki, A. (2002) *Polym. Eng. Sci.*, **42**, 1864.
28 Rybnikář, F. (1989) *J. Appl. Polym. Sci.*, **38**, 1479.
29 Fujiyama, M. and Wakino, T. (1991) *J. Appl. Polym. Sci.*, **42**, 9.
30 Fourdrin, S., Soulestin, J., Lafranche, E., Lacrampe, M.F., and Krawczak, P. (2007) *Proceedings of the 23rd Polymer Processing Society*, PPS-23, P03-040.
31 Wang, K., Liang, S., Du, R., Zhang, Q., and Fu, Q. (2004) *Polymer*, **45**, 7953.
32 Wang, K., Liang, S., Zhang, Q., Du, R., and Fu, Q. (2005) *J. Polym. Sci., Polym. Phys. Ed.*, **43**, 2005.
33 Wang, K., Xiao, Y., Na, B., Tan, H., Zhang, Q., and Fu, Q. (2005) *Polymer*, **46**, 9022.
34 Wang, K., Zhao, P., Yang, H., Liang, S., Zhang, Q., Du, R., Fu, Q., Yu, Z., and Chen, E. (2006) *Polymer*, **47**, 7103.
35 Koo, C.M., Kim, J.H., Wang, K.H., and Chung, I.J. (2005) *J. Polym. Sci., Part B: Polym. Phys.*, **43**, 158.
36 Bafna, A., Beaucage, G., Mirabella, F., and Mehta, S. (2003) *Polymer*, **44**, 1103.

5
Rheological Behavior of Polymer Nanocomposites
Mo Song and Jie Jin

5.1
Introduction

Over the past two decades, polymer materials reinforced with nanofillers, such as layered silicate, graphite nanosheets, carbon nanotubes (CNTs), metal oxide nanoparticles and layered titanate, have attracted much attention from both the scientific and the industrial communities as a result of the significant property enhancement obtained from relatively low nanofiller loading [1–10]. In particular, polymer nanocomposites based on layered silicates have shown a number of impressive property enhancements, including increased tensile strength, modulus, hardness and fracture toughness, decreased gas permeability and flammability, and improved thermal stability and specific heat resistivity [11–13]. CNTs are ideal reinforcing fillers for a polymer matrix, because of their nanometer size, high aspect ratio and, more importantly, their excellent mechanical strength, and electrical and thermal conductivity [14–17]. CNT-filled polymer materials enjoy the additional functionality of exceptionally high electrical and thermal conductivity while maintaining the polymer matrix properties (including elasticity, strength, and modulus) [18–20]. The main reason for these property improvements is the strong interfacial interactions between polymer matrix and nanofillers. A variety of polymer/clay or polymer/CNTs nanocomposites, including poly(propylene) [21, 22], polystyrene [23, 24], polyurethane [25, 26], epoxy [27, 28] and nylon [29, 30], have been successfully synthesized to date. Such novel polymer nanocomposites open opportunities to new multifunctional materials with broad commercial and industry applications. Polymer/clay nanocomposites have also been used in a number of applications, ranging from high-barrier packing for food and electronics to strong, heat-resistant automotive components. Polymer/CNT nanocomposites have been mainly used in the two major areas of electronics and automotives. In electronics, CNTs are used to dissipate the build up of unwanted static charges. This dissipative effect is achieved by thoroughly dispersing CNTs in the polymeric material to form an interconnecting network. In the automotive industry, CNTs are applied as a conducting agent in electrostatic painting.

Optimization of Polymer Nanocomposite Properties. Edited by Vikas Mittal
Copyright © 2010 WILEY-VCH Verlag GmbH & Co. KGaA, Weinheim
ISBN: 978-3-527-32521-4

Three common methods have been adopted to fabricate polymer/clay or polymer/CNTs nanocomposites, namely: (i) *in situ* polymerization; (ii) solution casting; and (iii) melt blending. The layered silicates commonly used for preparing polymer nanocomposites belong to the same general family of 2:1 layered- or phyllosilicates [31]. The layer thickness is around 1 nm and the lateral dimensions of these layers may vary from 30 nm to several microns. On the basis of the particular characteristics of layered silicates, two structurally different types of polymer/clay nanocomposites can be fabricated: (i) intercalated polymer/clay; and (ii) exfoliated polymer/clay nanocomposites. Solution and melt intercalation usually result in intercalated nanocomposites (unless special additives or processes are followed), where the polymer chains are inserted between clay layers but the well-ordered multilayer morphology of the clay remains. The exfoliated clay structure, where the clay layers are completely delaminated into single sheets and randomly oriented in the polymer matrix, can be achieved through *in situ* polymerization. In terms of the mechanical properties of nanocomposites, the elastic modulus is reported to be significantly enhanced [32]. The better the silicate platelets are exfoliated, the stronger the reinforcement effect on the resulting nanocomposite at a given clay content will be. The degree of exfoliation or of intercalation of the clay in the polymer matrix has been found to have implications for the final properties of the nanocomposites [32].

It is believed that the preparation of polymer/clay or polymer/CNTs nanocomposites with homogeneous dispersion of nanofillers in the matrices is a crucial step to developing high-performance polymer nanocomposites. This is because a few weight percent of nanoparticles that are properly dispersed throughout the matrix can create a much larger surface area for polymer filler interactions. The homogeneous dispersion of nanofillers is, however, difficult to achieve due to their nanometer size and strong intrinsic van der Waals attractions, especially for CNTs. Furthermore, the incorporation of nanofillers into the polymer matrix also leads to an increase in the viscosity of the mixture, which in turn opposes the nanoparticle flow in the matrix [33]. Numerous investigations have demonstrated that, if nanofillers are not properly dispersed in the polymer matrix, then the properties of the polymer will not be improved and may even worsen [33, 34]. The compatibility and optimum interactions of polymer molecular chains with the surface of nanofillers are very important for realizing efficient load transfer from the matrix to the nanoparticles. Interactions between the polymeric matrix and nanofillers are usually enhanced by chemical bonding. Good adhesion and favorable interactions between polymer chains and nanoparticles will determine the enhanced properties of the final products.

From the point of view of processing and application, a key objective in preparing polymer nanocomposites is to achieve an homogeneous dispersion of the nanofillers in the polymer matrices, which in turn dictates the nature of the morphology and the physical properties of the nanocomposites. *Rheology* is the study of flow matter. One of the properties often dealt with in rheology is viscosity, which may provide some useful guidance to overcome the difficulties resulting from the large changes in melt viscoelastic properties observed in the nanocomposites.

Rheometry has also been recognized as a powerful tool for investigating the inner microstructures of nanocomposites [35, 36], such as the state of dispersion of clay and the confinement effect of silicate layers on the motion of polymer chains. It can provide a more complete picture of the shear response of sample, which in turn offers a fundamental understanding of the processability of these materials and the structure–property relationship in nanocomposites. In this chapter, some of the current studies in both the solution- and melt-state rheological behavior of polymer nanocomposites based on layered silicates and CNTs, will be reviewed.

5.2
Rheological Behavior of Polymer Nanocomposites in Solution State

In situ polymerization is one method used to prepare exfoliated and intercalated polymer/clay nanocomposites. Generally, *in situ* polymerization consists of two stages: (i) mixing of the polymer precursor or monomer and the clay; and (ii) *in situ* polymerization in the presence of clay. *In situ* polymerization is understood to be responsible for the exfoliation and intercalation of clay in the polymer matrix. The viscosity of the polymer/clay dispersions is important for the *in situ* formation of polymer nanocomposites. Processing conditions such as mixing temperature and time, shearing speed, molecular weight of prepolymer and incorporation of nanofillers are all important factors which influence the flow nature of the mixture. A lower viscosity means a better flow ability and more homogeneous mixing. Furthermore, low viscosity is very helpful in removing bubbles before any chemical reactions, which is a key step in polymer preparation, especially in the preparation of polyurethane (PU) nanocomposites. Reflecting their widespread application, a variety of intercalated and exfoliated PU/clay nanocomposites have been produced to date. The rheological behavior of PU nanocomposites in solution state has been widely discussed.

Song and coworkers [34] reported that in processing PU nanocomposites, the viscosity of the polyol/clay mixture was strongly influenced by the amount of clay loading. The viscosity was increased slightly when the clay loading was increased from 0 to 3 wt%, but with a further increase of clay loading to 5 wt% the viscosity of the mixture increased by three-fold of that at 3 wt% clay (see Figure 5.1 [34]). This means that the critical viscosity of the mixture is reached when the clay content is about 3 wt%. This information provides helpful guidance for the preparation of PU/clay nanocomposites. In higher clay content systems, the higher viscosity of the mixture means that good dispersion of the clay is difficult to maintain, and improved physical properties of the final product are difficult to achieve [33, 34].

Xia and Song [37] studied the rheological behavior of the polyol/clay mixture during the *in situ* polymerization of PU/clay nanocomposites. The variation of the viscosity of polyol/clay (5 wt%) dispersions with mixing time at various temperatures was examined for different types of clay system. The results showed that, at 20 °C, the viscosity of polyol/clay C30B dispersions increased slightly from 1.94 to

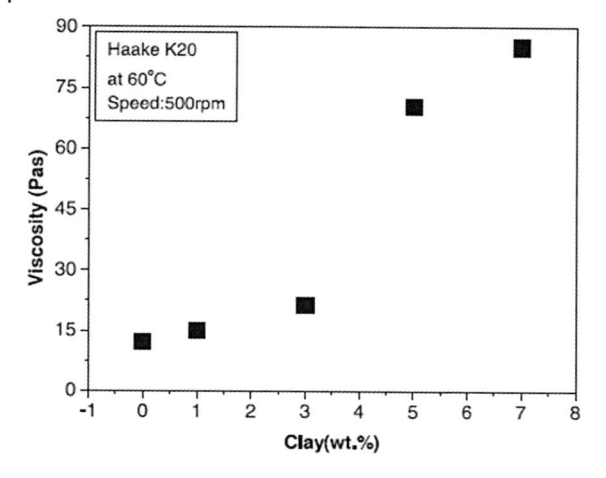

Figure 5.1 Viscosity of polyol/clay versus clay concentration. Reprinted from Ref. [34]; © John Wiley & Sons Ltd.

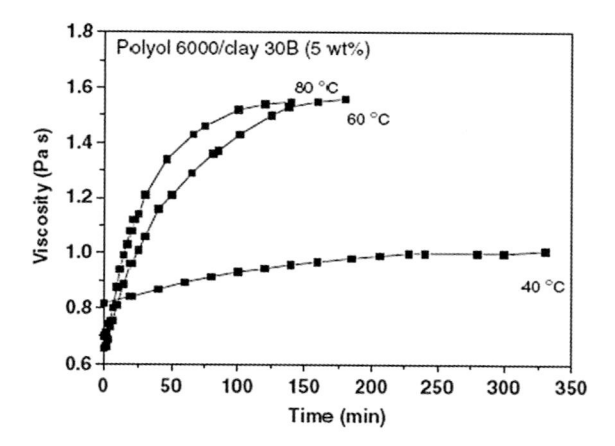

Figure 5.2 Variation of viscosity of polyol/clay C30B dispersions with mixing time at various temperatures. Reprinted from Ref. [37]; © John Wiley & Sons Ltd.

2.21 Pa·s after 4h of mixing, but that at higher temperatures the viscosity of the polyol/clay C30B dispersions increased with the mixing time. As shown in Figure 5.2, after mixing for 5.5h at 40 °C, the viscosity increased to only 1 Pa·s, but after mixing for 2h at 80 °C it reached 1.55 Pa·s. Figure 5.3 shows the variation of the viscosity of polyol/clay 20A dispersions with mixing time at 60 and 80 °C. At 20 °C, the viscosity of polyol/clay 20A dispersions practically did not change with mixing time; however, at 60 and 80 °C the viscosity of these dispersions increased slightly with mixing time. At 80 °C, and after 90 min, the viscosity reached a plateau value

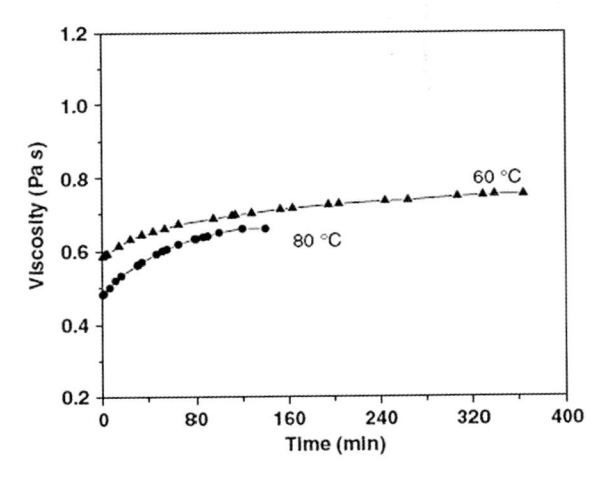

Figure 5.3 Variation of viscosity of polyol/clay 20A dispersions with mixing time at 60 and 80 °C. Reprinted from Ref. [37]; © John Wiley & Sons Ltd.

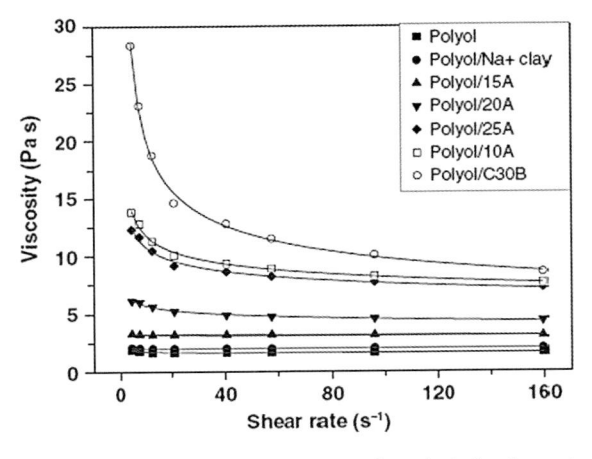

Figure 5.4 Viscosity versus shear rate for polyol/clay dispersions with various clays (viscosity determined at 20 °C). Reprinted from Ref. [37]; © John Wiley & Sons Ltd.

of 0.65 Pa·s, which was much lower than that of the polyol/C30B system. For the original unmodified Na^+ clay, the viscosity of polyol/Na^+ clay dispersions remained almost unchanged with mixing time at various temperatures.

The viscous behavior was different for the three types of clay. A rapid increase in viscosity at higher temperatures for the clay C30B system was noted, and in order to understand this phenomenon a further investigation was undertaken. Figure 5.4 shows the rheological behavior of blank polyol and polyol/clay dispersions with different types of clay determined at 20 °C, while Figure 5.5 shows the

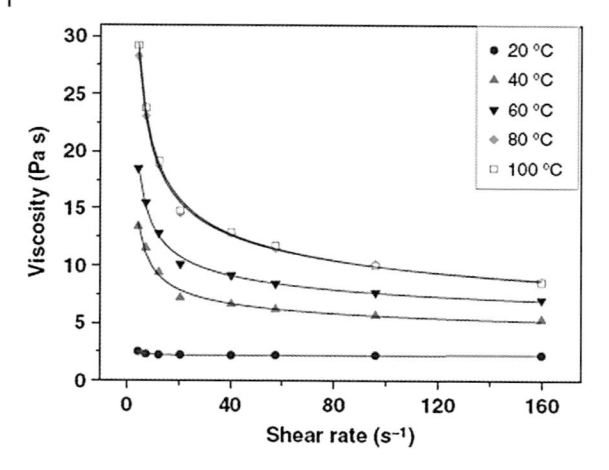

Figure 5.5 Viscosity versus shear rate for polyol/clay C30B dispersions after mixing for 4 h at 20, 40, 60, 80, and 100 °C. Reprinted from Ref. [37]; © John Wiley & Sons Ltd.

viscosity versus shear rate for polyol/clay C30B dispersions prepared at various temperatures.

The Herschel–Bulkley model [38] is commonly used to describe materials such as mud, clay suspensions, oil and drilling fluids. The Herschel–Bulkley equation is given by:

$$\sigma = \sigma_y + k\gamma^n \tag{5.1}$$

The viscosity, η, of Herschel–Bulkley suspensions can be calculated according to the following equation [39]:

$$\eta = \tau_0/\gamma + k\gamma^{n-1} \tag{5.2}$$

where σ is the shear stress, σ_y is the yield stress, γ is the shear rate, k is a constant, and n is the flow behavior index. When $n > 1$ the fluid exhibits a shear-thickening behavior; when $n = 1$ the fluid exhibits a Bingham plastic behavior; and when $n < 1$ the fluid exhibits a shear-thinning behavior. Here, n can denote the shear-thinning exponent. When τ_0 is equal to zero, Equation 5.2 is equal to a power law model. It is believed that the n, k, and τ_0 values are related to the clay dispersion, intercalation, or exfoliation states.

In Figures 5.4 and 5.5, the scatter points are the experimental data and the solid lines are the fitted curves according to Herschel–Bulkley models. The related rheological data are listed in Tables 5.1 and 5.2. For polyol/Na⁺ clay and polyol/15A dispersions, the viscosity was low. The shear-thinning parameter, n, was found to be 1.00, which is similar to that of the blank polyol dispersions. For polyol/20A, polyol/25A and polyol/10A dispersions, the viscosity was relatively high, and n decreased to ~0.9. For polyol/C30B dispersions, the viscosity was the highest, and n was found to be 0.77. When combined with the X-ray

Table 5.1 Relationship between rheological data and clay state, evaluated from XRD.

Dispersion	η_L (Pa·s)[a]	η_H (Pa·s)[a]	n	Clay state (XRD evaluation)
Blank polyol	1.87	1.63	0.99	–
Polyol/Na$^+$ clay	2.10	2.03	1.00	Low intercalation
Polyol/15A	3.30	3.09	0.99	Low intercalation
Polyol/20A	6.15	4.39	0.90	High intercalation
Polyol/25A	12.30	7.24	0.87	High intercalation
Polyol/10A	13.80	7.67	0.87	High intercalation
Polyol/30B	28.30	8.60	0.77	Exfoliation

a) η_L and η_H are the viscosity values at low shear rate (4.45 s^{-1}) and at high shear rate (159.8 s^{-1}), respectively. The shear-thinning parameter n was obtained according to Herschel–Bulkley models. Reprinted from Ref. [37]; © John Wiley & Sons Ltd.

Table 5.2 Relationship between rheological data and clay state, evaluated from XRD.

Mixing temperature (°C)	η_L (Pa·s)[a]	η_H (Pa·s)[a]	n	Clay state (XRD evaluation)
20	2.53	2.22	1.02	No intercalation or exfoliation
40	13.4	5.3	0.86	Medium exfoliation
60	18.5	7.04	0.85	Medium exfoliation
80	28.3	8.60	0.77	High exfoliation
100	29.2	8.60	0.76	High intercalation

a) η_L and η_H are the viscosity values at low shear rate (4.45 s^{-1}) and at high shear rate (159.8 s^{-1}), respectively. The shear thinning parameter n was obtained according to Herschel-Bulkley models. Reprinted from Ref. [34]; © John Wiley & Sons Ltd.

diffraction (XRD) data, the state of clays in polyol could be evaluated as shown in Tables 5.1 and 5.2.

From Tables 5.1 and 5.2, it can be seen that there was a good correlation between the rheological data and the dispersion behavior of clay. In particular, the shear-thinning exponent, n, could be used to semi-quantitatively characterize the state of the clay layers. A lower n value corresponded to a high shear-thinning behavior and a high exfoliation state of clay layers in the polymer matrix. If there was no intercalation or exfoliation, the viscosity increased slightly owing to the increase in the concentration of the dispersed phase, and the shear-thinning behavior should be similar to that of the blank polyol, as shown in Tables 5.1 and 5.2. When the clay layers were intercalated, clay tactoids of a larger size were formed, and the viscosity was increased markedly compared with the blank polyol. This was due to increases in the size and concentration of the dispersed phase, together with increases in the interactions between the polyol and clay. When the clay was exfoliated, many separated layers of clay formed; these exfoliated individual clay layers with large surface areas formed a network gel structure, leading to a significant increase in viscosity. This is also called the "house of cards" effect [40]. Once the network structure is formed, at a relatively high shear rate, the viscosity decreases owing to the alignments and disentanglements of clay layers, and thus the shear-thinning behavior will appear more obviously. Clearly, this behavior only occurred for the clay C30B system, indicating that the interactions between clay and polyol play a key role in clay exfoliation. Hence, there are different rheological behaviors for intercalated and exfoliated polyol/clay dispersions. The viscosity and the shear-thinning exponent n can be used to characterize the dispersion state of clay in polyol.

The interaction between the polyol and clay and the mixing temperature plays an important role in the occurrence of exfoliation and intercalation. The relationship between rheological data of polyol/clay dispersion and the intercalation or exfoliation state of the clay was established. This provides a convenient and efficient way to evaluate the dispersion state of the clay. Based on the experimental results, a possible layer-by-layer exfoliation mechanism has been proposed.

Recently, Song's research group prepared a series of PU/CNT nanocomposites including PU/MWNT, PU/SWNT, PU/MWNT-g-PU, PU/SWNT-g-PU and MWNT/water-based PU nanocomposites. The rheological behavior of polyol/CNT dispersions for these PU/CNT nanocomposites was reported [26]. A key issue in producing superior CNT nanocomposites for applications is the ability to control the dispersion of CNT in polymeric matrices due to its smaller diameter and strong van der Waals interactions [41, 42].

The viscosity of polyol/MWNT and polyol/SWNT dispersions with shearing rate was determined at 20 °C, 40 °C, and 60 °C, respectively, as shown in Figure 5.6. If the CNTs are in a state of good dispersion, high shearing will mean that the nanotubes align with the flow direction and cause a decrease in viscosity. If the CNTs exist in an aggregation state, the shear-thinning behavior will be very weak. For the MWNT and SWNT series, increases in the concentration of CNTs lead to decreases in the shearing thinning exponent, which means a more significant shear-thinning behavior. Furthermore, on increasing the temperature, the shear-thinning behavior decreases, which means a better dispersion state of CNTs at a higher temperature. These are common features for SWNT-polyol and MWNT-polyol. If the different dimensions of SWNTs and MWNTs are considered, it can be inferred that the SWNTs dispersion should have a more significant shearing thinning behavior due to its high aspect ratio, but this is not the case. The probable reason for this is that an individual dispersion of SWNTs is difficult to attain. In most cases, the SWNTs still existed in an aggregate state, especially in an organic medium [40]. Further, the variation of viscosity of MWNT-polyol and SWNT-polyol dispersions at a shear rate of $444.6 \, s^{-1}$ and at 60 °C with CNTs concentration was analyzed; these data are shown in Figure 5.7. The viscosity can be fitted very well with the following equation [43]:

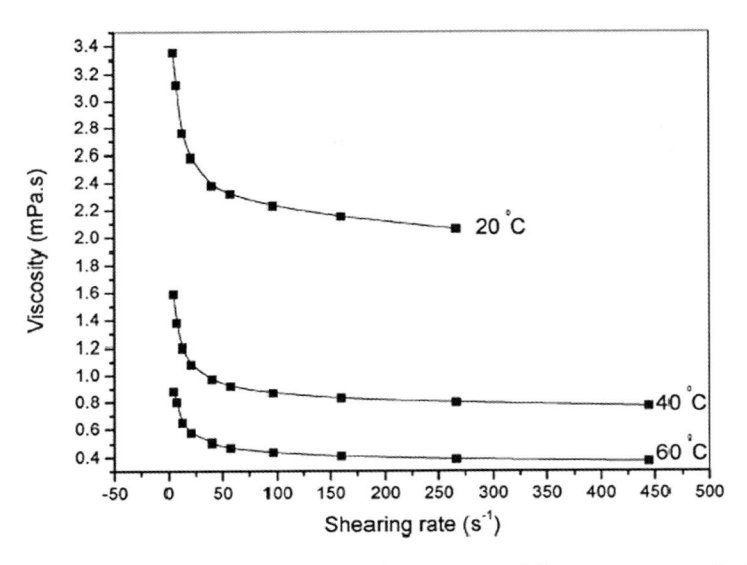

Figure 5.6 Variation of viscosity with shearing rate at different temperature for 1% SWNT-polyol dispersion. Reprinted from Ref. [26]; © Royal Society of Chemistry.

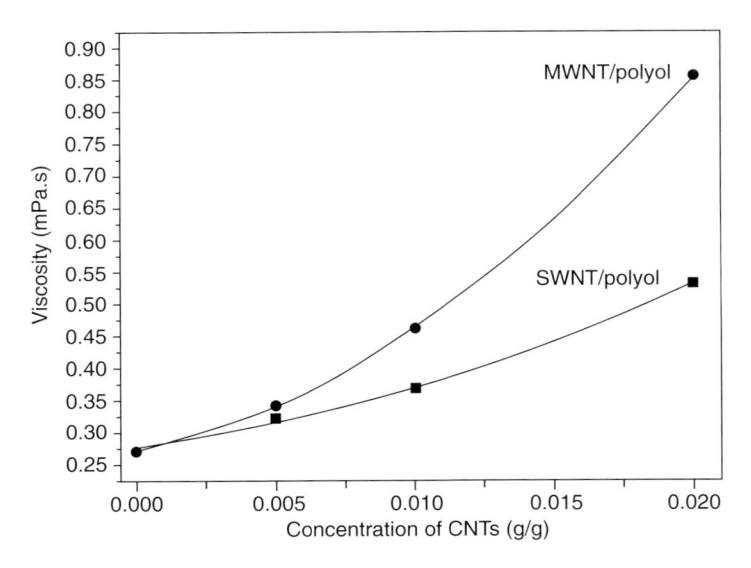

Figure 5.7 Variation of viscosity of MWNT/polyol and SWNT/polyol dispersions at a shear rate of 444.6 s^{-1} and 60 °C with CNT concentration. Reprinted from Ref. [26]; © Royal Society of Chemistry.

Table 5.3 Fitted values of η_0 and f for polyol/SWNT and polyol/MWNT dispersions.

	η_0	f
Polyol/SWNT	0.279	28.52
Polyol/MWNT	0.272	48.14

Reprinted from Ref. [26]; © Royal Society of Chemistry.

$$\eta = \eta_0\left(1 + 0.67\text{f}\,C + 1.62\text{f}^2 C^2\right) \tag{5.3}$$

where f is the shape factor, or the aspect ratio (i.e., the ratio of the longest to shortest diameter of particles), η_0 is the viscosity when the CNTs concentration is zero, and C is the concentration of CNTs. The fitted parameters η_0 and f values are listed in Table 5.3 [26]. Clearly, MWNTs have a much bigger f value than SWNTs, a result which can be related to the transmission electron microscopy (TEM) observations. The f value is dependent on the CNT morphology in polyol. A better dispersion of CNT leads to a high aspect ratio and thus a bigger f value.

It is well known that CNTs are insoluble in common solvents and have a weak interfacial interaction with the polymer matrix [44–46]. In order to prepare polymer/CNT nanocomposites, a chemical modification of CNT is necessary, and several approaches including sidewall functionalization [47], polymer encapsulation [48] and polymer grafting [49] have been applied to overcome these problems. The grafting of CNTs with polymer is one of the most important approaches due to the strong covalent bonding between the grafted polymer and the CNTs.

Poly(propylene glycol)-grafted MWNT polyurethane was synthesized based on the hydroxyl-functionalized MWNTs through a two-step reaction [50]. The grafted MWNTs can improve the rheological behavior of the polyol/MWNT dispersion. The viscosity for polyol/1% MWNT-graft-polyol dispersion was much lower than that for polyol/1% raw MWNT system, as shown in Figure 5.8. This suggests that grafting polyol on MWNT can improve the rheological behavior during the process compared to the raw MWNT. Such as improvement could result from increases in the wetting and dispersion properties of carbon MWNTs in polyol.

Polyurethane-grafted single-walled carbon nanotubes (SWNT-g-PU) were also synthesized from hydroxyl-functionalized single-walled CNTs (SWNT-OH) through a two-step reaction [51]. Based on SWNT-g-PU, PU/SWNT nanocomposites were prepared by further *in situ* polymerization. The viscosity studies suggested that SWNT-g-PU improved the dispersion of SWNT in the PU matrix and strengthened the interfacial interaction between the PU and SWNT. The variation of viscosity at 80 °C with shear rate for blank polyol, polyol/0.1% SWNT and polyol/0.1% SWNT-g-PU dispersions is shown in Figure 5.9. The viscosity for the polyol/1% SWNT-g-PU dispersion was higher than that for the 1% pristine polyol/0.1% SWNT system. The polyol/0.1% SWNT-g-PU has the lowest shearing thinning exponent, which means a more dramatically shearing thinning

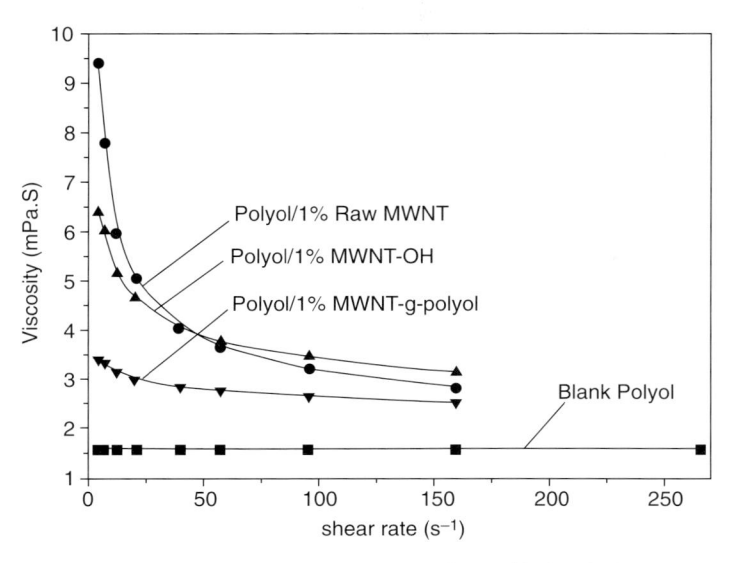

Figure 5.8 Viscosity at 20 °C versus shear rate for the blank polyol and polyol/CNT dispersions. Reprinted from Ref. [50]; © John Wiley & Sons Ltd.

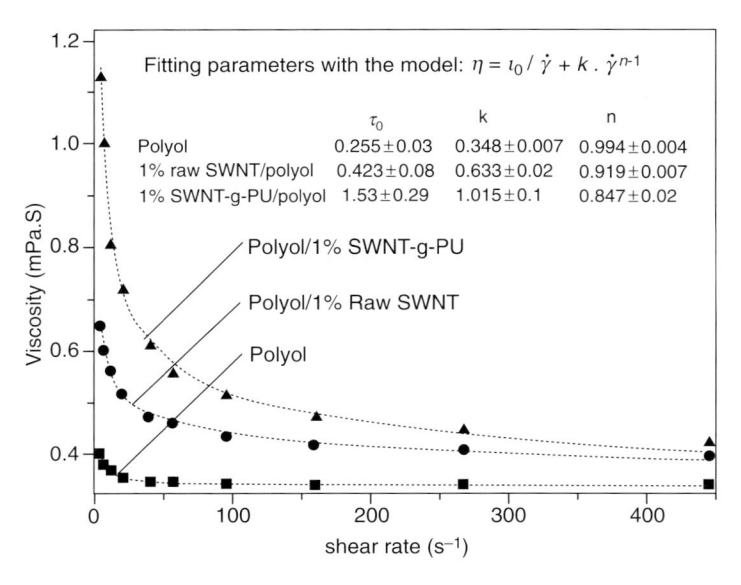

Fitting parameters with the model: $\eta = \iota_0 / \dot{\gamma} + k \cdot \dot{\gamma}^{n-1}$

	τ_0	k	n
Polyol	0.255±0.03	0.348±0.007	0.994±0.004
1% raw SWNT/polyol	0.423±0.08	0.633±0.02	0.919±0.007
1% SWNT-g-PU/polyol	1.53±0.29	1.015±0.1	0.847±0.02

Figure 5.9 The viscosity at 80 °C versus shear rate for blank polyol and polyol/01% SWNT and polyol/0.1% SWNT-g-PU dispersions. Reprinted from Ref. [51]; © Royal Society of Chemistry.

behavior, indicating a better dispersion of SWNT-g-PU in polyol. The grafted PU could act as an effective dispersion agent for SWNT in polyol.

As mentioned above, polymer/CNT nanocomposites are usually fabricated by the three traditional methods of melt blending, *in situ* polymerization, and solution casting. Chemical modification is regarded as an effective way of achieving well-dispersed CNTs in a polymer matrix [13–15], but it can result in a disruption of the tube structure that leads unavoidably to some loss in the properties of CNTs. In order to avoid this problem, colloidal physics technology has been introduced for the preparation of polymer/CNTs composites [52]. The incorporation of CNTs into a polymer matrix was achieved by obtaining a colloidal system of CNTs and polymer latex. In such as mixture the CNTs are assembled on the surface of the polymer latex particles.

Song and Cai [53] prepared MWNT/water-based PU composites through a colloidal physics technology, and investigated the effect of MWNT on the rheological behavior of the MWNT/water-based PU composites dispersions. It can be seen from the rheological curves shown in Figure 5.10 that the MWNT/water-based PU dispersions exhibit a shear-thinning behavior. At low shearing rates, the viscosity of the MWNT/water-based PU dispersions shows noticeable enhancement with increasing loading of MWNT. The rheological behavior of the PU dispersions is slightly affected by MWNT at high shearing rates. Changes in the rheological behavior revealed that the addition of CNTs increased the viscosity of the water-based PU dispersion. Changes in the flow behavior index, n, with increasing MWNT concentration are shown in Figure 5.11, which shows the n-value decreasing with increasing loading of MWCNT. This inverse relationship between flow behavior and MWNT concentration indicates that the shear-thinning behavior of MWNT/PU dispersions is stronger than that of a pure PU dispersion.

One principal disadvantage of CNTs is their high hydrophobicity, with as-prepared CNTs usually aggregating into bundles and/or becoming entangled together

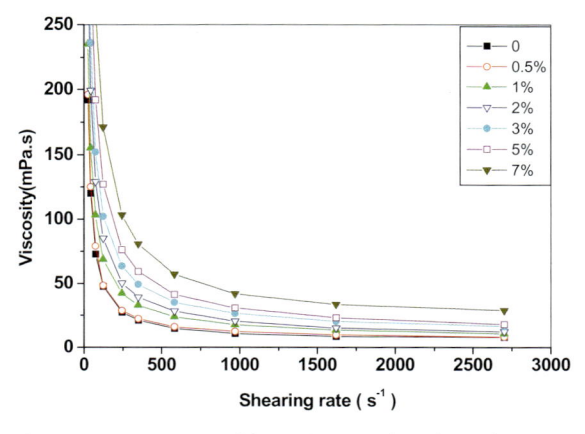

Figure 5.10 Viscosity of the MWCNT/polyurethane dispersions versus shearing rate with different loadings of MWCNTs at 20 °C. Reprinted from Ref. [53]; © John Wiley & Sons Ltd.

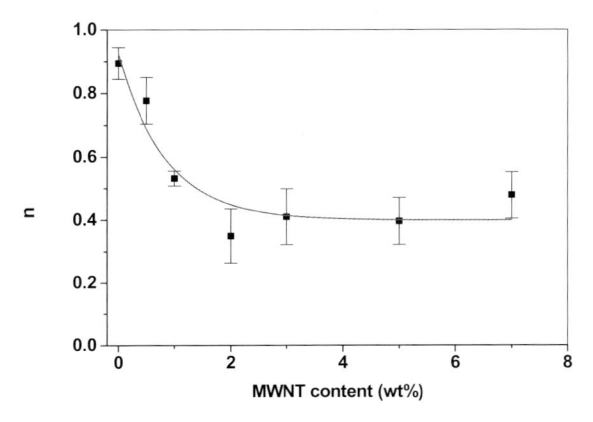

Figure 5.11 Flow behavior index versus MWNT concentration. Reprinted from Ref. [53]; © John Wiley & Sons Ltd.

because of the van der Waals attraction of the flexible nanotubes. A few polymers such as polyvinyl pyrrolidone, polystyrene sulfonate and poly(aryleneethynylene)s are known to effectively solubilize CNTs [54, 55]. However, in some cases, when CNTs are soluble in water with a linear water-soluble polymer, they are unstable and may undergo reverse solubilization, which involves their being wrapped with the polymer chains [56]. *Chitosan*, as a natural polymer, has been found to interact with CNTs and, due to the presence of the amino groups of chitosan, to induce stable chitosan solubilization and improve the interfacial strength [57, 58]. This success in dispersing CNTs in solution will greatly benefit the preparation of a variety of chitosan/CNTs composites by the solution casting method. As chitosan is an acid-soluble polymer, the solution pH will have an important effect on the dispersion state of nanofillers in the solution.

Song and coworkers (unpublished results) examined the viscosity of the chitosan/CNTs suspensions. The hydroxyl-functionalized CNTs were dispersed in chitosan solution without a dispersion agent. Figure 5.12 shows the viscosity of the chitosan/MWNT suspensions containing 0, 0.5, 1, and 2 wt% of the MWNT of suspension pH of 2.5, 4.5, and 5.5 respectively. The results indicate that the chitosan/MWNT suspension exhibits shear-thinning behavior. The increase in the viscosity of the higher-pH chitosan/MWNT suspension was significantly more than that in respect of the pure chitosan suspension prepared at the same pH value.

Chitosan is a polyelectrolyte that, due to the deprotonation of its amine groups, exhibits increasing $-NH_3^+$ charge densities with changing pH value. In low-pH solutions, the chitosan chains expand as a result of the mutual repulsions of $-NH_3^+$ ions along the chitosan backbone [59], and this expansion leads to the entanglement of the chitosan chain. With increases in the pH, the chitosan chains become aggregated because of extensive intramolecular hydrogen bonding and a decrease in the strength of the interchain electrostatic repulsion. This conforma-

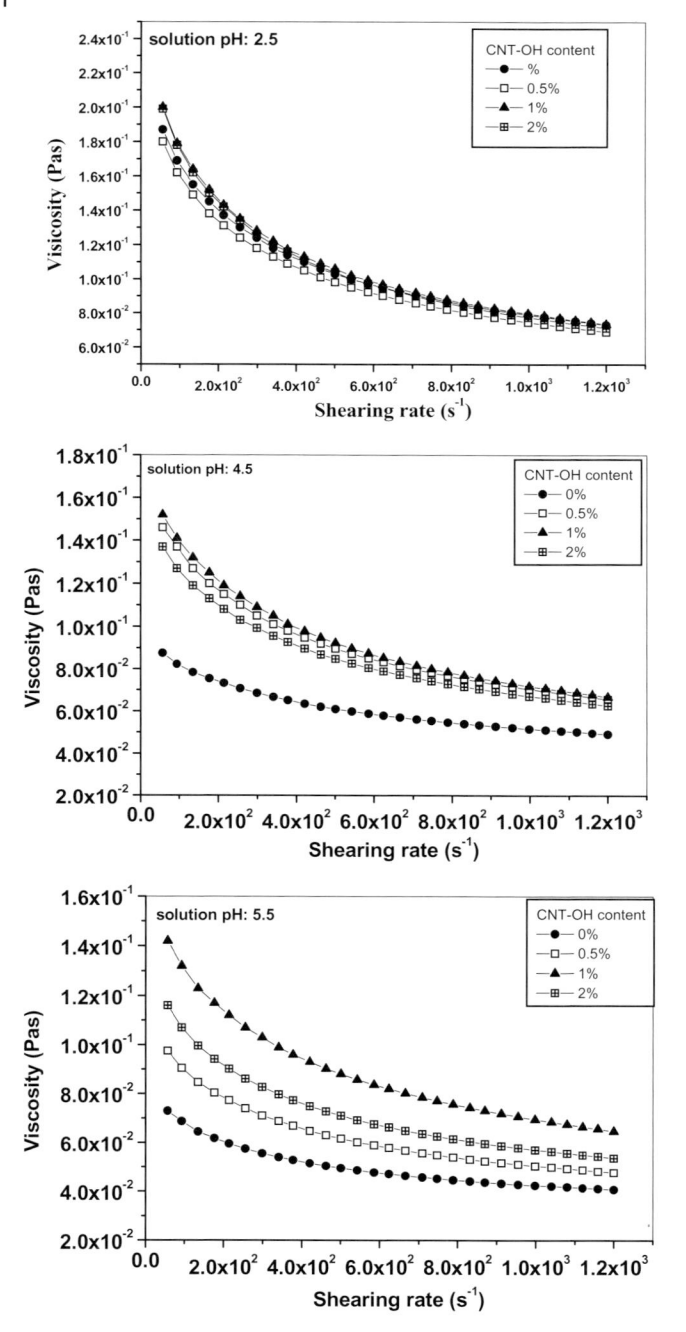

Figure 5.12 Viscosity versus shearing rate at different pH values for chitosan/MWCNT-OH solutions.

tional behavior of chitosan chains in different pH solutions also affects the interaction with the suspended nanotubes. In low-pH solutions, as a result of the electrostatic repulsion, the interaction between chitosan chain and CNTs becomes stronger, which leads to a better dispersion, showing lower viscosity. The CNTs bundles easily break up into individual tubes, which in turn could affect the state of the MWCNT dispersion in aqueous mixture and lead to different network structures of MWCNT in the matrix. These results show that the microstructure of MWCNT-filled chitosan composites can be tailored by varying the solution pH.

5.3
Rheological Behavior of Polymer Nanocomposites in Melt State

In polymer processing, such as injection molding and extrusion, the rheological properties of the nanocomposites are of vital importance. A polymer network is generally viscoelastic with a complex shear modulus having both elastic (i.e., solid-like) and viscous (i.e., liquid-like) components of similar magnitude over a large range of frequencies. The state of the dispersion of nanofiller can be studied at two levels:

- The *macroscopic* level, which involves measurements of the rheological properties of the bulk blend.
- The *microsco4pic* level, which investigates the detailed dynamics of the individual particles [60].

Over the past few years, the rheological behavior of polymer/clay nanocomposites, such as polypropylene, polystyrene, polycarbonate, polyamides, poly(ε-caprolactone), poly(ethylene oxide) and nylon 6, in melt state have been discussed [61–71].

Measurements of rheological properties under the molten state reveal that nanofiller loading leads to increases in the shear viscosity, the storage modulus (G′) and the loss modulus (G″) of nanocomposites. As with other properties, the rheological properties of nanocomposites relate directly to the degree of dispersion of nanoparticles in the polymer matrix, and to the level of interfacial interactions between nanoparticles and polymer chains. Pronounced shear thinning has been found to be a characteristic feature of truly nanodispersed composites. Under certain experimental conditions, a pseudo-solid-like rheological behavior (which possibly results from edge-to-face interaction of nanofillers) has been identified. These edge-to-face interactions help to build and mechanically stabilize the mesoscale structure of platelets within the nanocomposite.

Zhou and coworkers [61] described the melt state rheological behavior of polypropylene (PP)/montmorillonite (MMT) nanocomposites prepared by melt intercalation with maleic anhydride-modified low-isostatic PP as the compatibilizer. The linear and nonlinear rheological properties of PP nanocomposites were discussed. Rheological behavior at high frequencies is normally used to estimate the effect of the filler on processing properties. Low-frequency behavior is sensitive to the structure of the percolation state of nanofillers within the composite. The storage modulus (G′) data and loss modulus (G″) data resulting from the dynamic frequency scan measurements for PP nanocomposites with various clay loading are shown in Figure 5.13a and b, respectively. The results showed an increase in

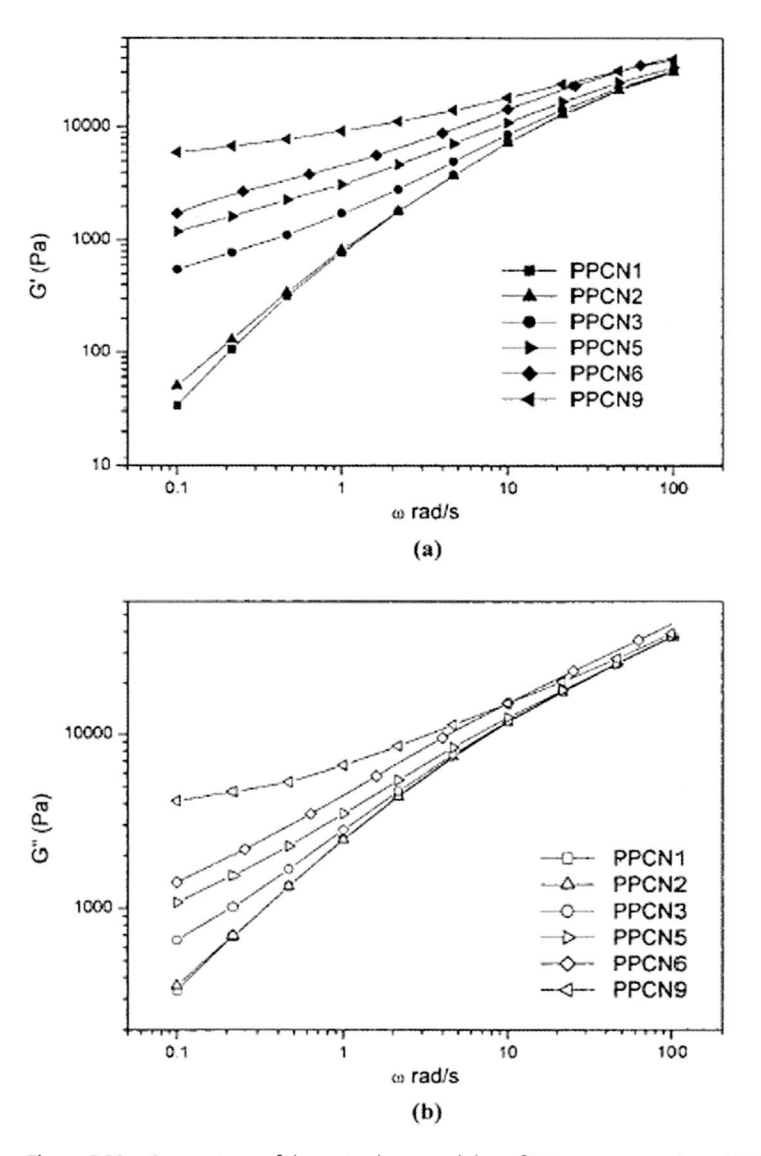

Figure 5.13 Comparison of dynamic shear modulus of PP nanocomposites. (a) Storage modulus; (b) Loss modulus. Reprinted from Ref. [61], with permission from John Wiley & Sons Ltd.

storage modulus (G′) and loss modulus (G″) of nanocomposites with clay loading. At low frequencies, the degree of dependence of G′ on the frequency was sensitive to the effect of clay on the viscoelastic properties of the nanocomposites. When the clay loading exceeded 3 wt%, the dependency of G′ of PP nanocomposites on the frequency decreased sharply and, in particular at low frequencies, the G′ curves exhibited a plateau (see Figure 5.13a). These results revealed that, when the clay loading was below 3 wt%, the viscoelastic properties of PP nanocomposites were dominated by the polymer matrix, and that once the clay loading was at or above 3 wt%, a transition from a liquid-like behavior to a solid-like behavior can be observed for PP nanocomposites. This means that, as a result, the percolation threshold for PP nanocomposites was approximately 3 wt%. The effect of clay on rheological behavior could be caused by the confinement effect of silicate layers and interparticle interactions. Such a confinement effect may lead to an alternation of the relaxing dynamic of the intercalated polymer chains. The frictional interaction between the tactoids may increase sharply when the clay loading is above the percolation threshold, which may lead to a significant enhancement of the low-frequency G′. On the other hand, if the percolation network is destroyed, the viscoelastic behavior may be changed greatly. The results from Zhou and coworkers' study [61] of the rheological behavior of PP nanocomposites showed that, having been subjected to steady pre-shear, the dynamic modulus of pre-sheared PP nanocomposite with 6 wt% clay decreased remarkably, and its low-frequency dependence was enhanced. This indicated that the tactoids were oriented preferentially in the shear direction, and that the percolation network had been ruptured.

In an effort to understand the viscoelastic behavior of PP nanocomposites at the low-frequency regions in dynamic measurements, a series of strain-scaling stress response measurements was conducted. Figure 5.14 shows the normalized transient stress response of PP/clay (6 wt%) nanocomposite in reversal flow experiments at 200 °C [61]. The magnitude of the stress overshoots observed was heavily dependent on the rest time, which was indicative of the structural evolution in the sample during the quiescent period. Initially, steady shear for 300 s resulted in rupture of the percolation network due to a preferential orientation of the tactoids, and the ruptured network could then be reorganized under quiescent conditions.

Several studies have shown that the rheological behavior of polymer/clay nanocomposites is correlated to the microstructures of the nanocomposites, and that microstructural changes cause temporal behavior. The rheological behavior of polymer/clay nanocomposite depends not only on the parent components but also on the composite phase morphology and interfacial characteristics. A different phase morphology (intercalated or exfoliated) of polymer/clay is obtained according to the interfacial characteristics between the polymer chains and clay [72]. Lim and Park [63], in their quantitative analysis, reported these effects on rheological behavior based on different characteristic polymer/clay nanocomposites. Three polymer resins with different characteristics were used to fabricate the nanocomposites with clay, namely: polystyrene (PS), polystyrene-*co*-maleic anhydride (PS-*co*-ma) and polyethylene-graft-maleic anhydride (PE-g-ma). A commercially

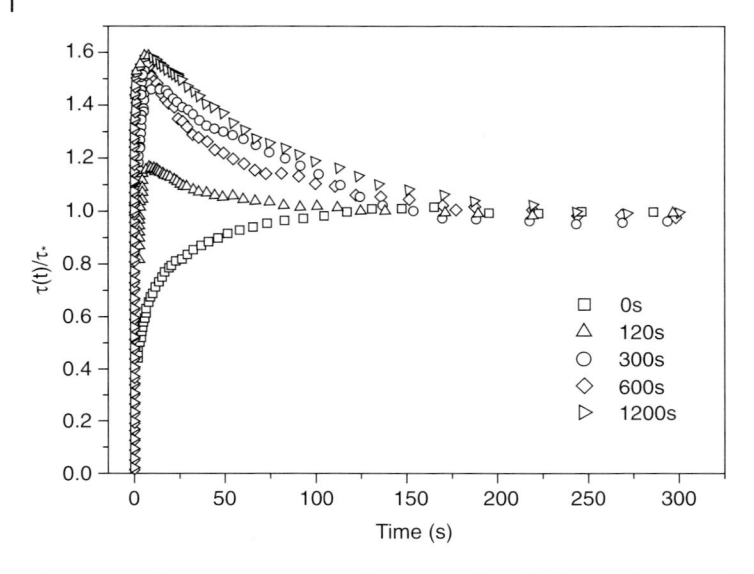

Figure 5.14 The stress response to the startup of steady shear in the reverse flow measurement. Reprinted from Ref. [61], with permission from John Wiley & Sons Ltd.

available organophilic clay M6A was used, and subsequently three different series of PS/clay, PS-co-ma/clay and PE-g-ma/clay nanocomposites were produced by melt intercalation. The final interfacial properties and phase morphology were examined using XRD and TEM. Both, PS/M6A and PS-co-ma/M6A composites had an intercalated structure, the original ordered structure of M6A was delaminated and, in the case of the PE-g-ma/M6A nanocomposite, an exfoliated structure was formed.

The dynamic mechanical properties of PS/M6A, PS-co-ma/M6A and PE-g-ma/M6A nanocomposites are shown in Figures 5.15, 5.16 and 5.17, respectively [63]. The results showed that the rheological behavior of the PS-co-ma/M6A nanocomposites at low frequency was very different from that of the PS/M6A nanocomposites. In particular, enhancement of the storage modulus of the PS-co-ma/M6A nanocomposites was very large compared to that of the PS/M6A nanocomposites and, unlike the PS/M6A nanocomposites, the PS-co-ma/M6A nanocomposites also exhibited a plateau-like behavior. The results indicated that, when the clay loading exceeded 7 wt% the liquid-like behavior of PS-co-ma/clay gradually changed to a pseudo-solid-like behavior. For the PE-g-ma/M6A nanocomposites the storage and loss moduli increased, and the frequency dependence of both moduli decreased with clay loading compared to those of the matrix polymer. The enhancement of both moduli in PE-g-ma/M6A was greater than those of PS/M6A and of PS-co-ma/M6A nanocomposites at all frequencies. These nanocomposites also exhibited a distinct plateau-like behavior at low frequency.

The interfacial interactions between the polymer chains and clay layers, in addition to the large contacting interfacial area, are responsible for the above-observed

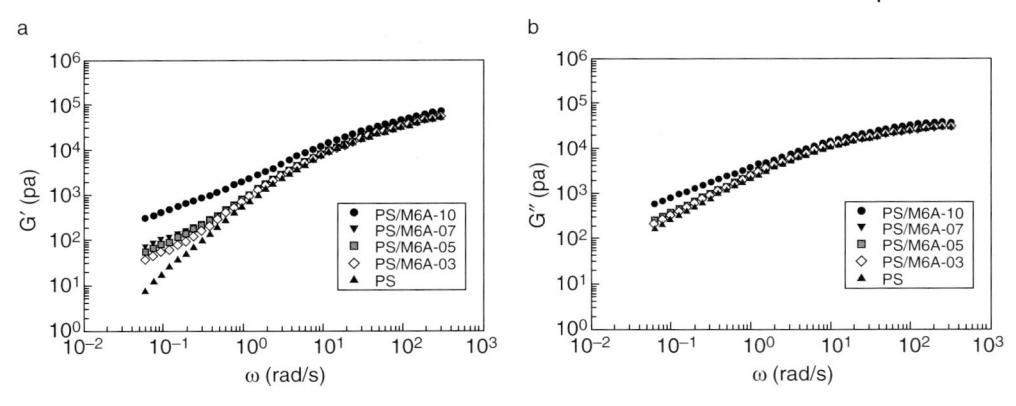

Figure 5.15 Dynamic mechanical properties of PS/M6A nanocomposites at 220 °C. (a) Storage modulus; (b) Loss modulus. Reprinted from Ref. [63], with permission from Springer.

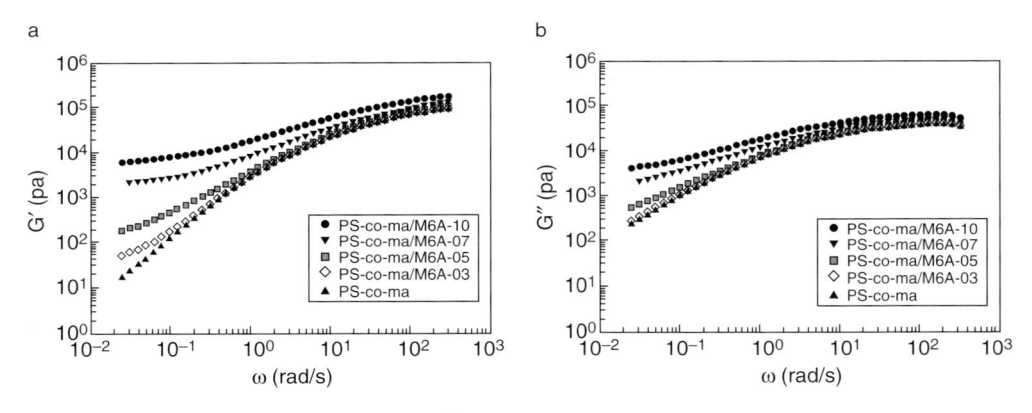

Figure 5.16 Dynamic mechanical properties of PS-*co*-ma/M6A nanocomposites at 220 °C. (a) Storage modulus; (b) Loss modulus. Reprinted from Ref. [63], with permission from Springer.

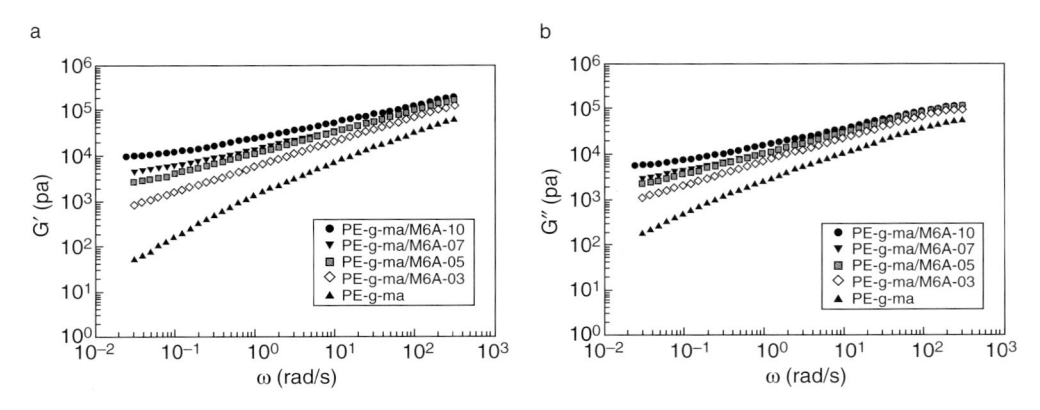

Figure 5.17 Dynamic mechanical properties of PE-*g*-ma/M6A nanocomposites at 220 °C. (a) Storage modulus; (b) Loss modulus. Reprinted from Ref. [63], with permission from Springer.

rheological behavior. In the fabrication of PS/clay nanocomposites by melt intercalation, the main driving force of PS intercalation to the gallery of clay is the weak Lewis acid–base interaction between the PS and the silicate surface [71, 73]. The PS-*co*-ma chains have a maleic anhydride group which functions as a strong active site with the silicate layer. Consequently, the PS-*co*-ma chains in the silicate gallery stick to the silicate layers more strongly than would PS chains under the same silicate loading. In addition, as detected by TEM observations, the more uniform dispersion of clay particles was achieved in PS-*co*-ma nanocomposites. As the micro-sized stacked clay particles become delaminated into smaller particles, the contacting area with polymer chains increases in the PS-*co*-ma nanocomposites. In order to fabricate PE/clay nanocomposites, a functional polar group should be introduced to the PE backbone, because the absence of any interacting moieties leads to an immiscible state between the PE and clay. When the PE-g-ma chain intercalates into the interlayer of the layered silicate due to interaction of the maleic anhydride group with the silicate surface, this forces the adjacent silicate layers to separate, and leads to a fully exfoliated structure. In exfoliated nanocomposites, when the single layers of clay are dispersed in the polymer matrix, more polymer chains come into contact with the silicate surface, whilst filler–filler interactions between clay particles are also very important, even at low clay loading. The PE-g-ma/M6A nanocomposites, which have exfoliated structures, showed a major enhancement of the storage modulus both at low and high frequency. The PS/clay nanocomposites showed a slight enhancement at low frequency because of their simple intercalated structure and weak interfacial interaction. The PS-*co*-ma/clay nanocomposites, which have a similar intercalated structure to PS/clay nanocomposites, showed a distinct plateau-like behavior at low frequency, as a result of the PS-*co*-ma having strong interactions with the clay layers. Although strong interactions between maleic anhydride and clay also exist in both PS-*co*-ma/clay and PE-g-ma/clay nanocomposites, the rheological properties of polymer/layered silicate nanocomposites are also heavily dependent on their microstructures.

The effects of phase morphology and interfacial properties on the rheological behavior of polymer/clay nanocomposites were also studied by small-amplitude oscillatory frequency tests after large-amplitude oscillatory shearing which was conducted at $\gamma = 120\%$, $\omega = 1\,\mathrm{rad\,s^{-1}}$. These measurements were carried out with 10 wt% filler sample; the results are shown in Figure 5.18. The G' of shear-aligned PS/M6A nanocomposite was lower than that of the initially unaligned sample, and close to that of the matrix polymer. A large decrease in G' was observed in the PS-*co*-ma/M6A nanocomposites, while the G' at low frequency was larger than that of PS/M6A. This indicates that strong interactions between the PS-*co*-ma and clay layer existed, and also suggests that randomly oriented silicate layers and some particle network structure were formed. For the shear-aligned PE-g-ma/M6A nanocomposite, a decrease in G' occurred only at the high-frequency region and was not large compared to that of PS/M6A or PS-*co*-ma/M6A nanocomposites. This suggests that the exfoliated structure remains in the PE-g-ma/M6A nanocomposite after large-amplitude oscillatory shearing. The results of these studies showed that a strong clay–clay interaction and a large interfacial area between the polymer

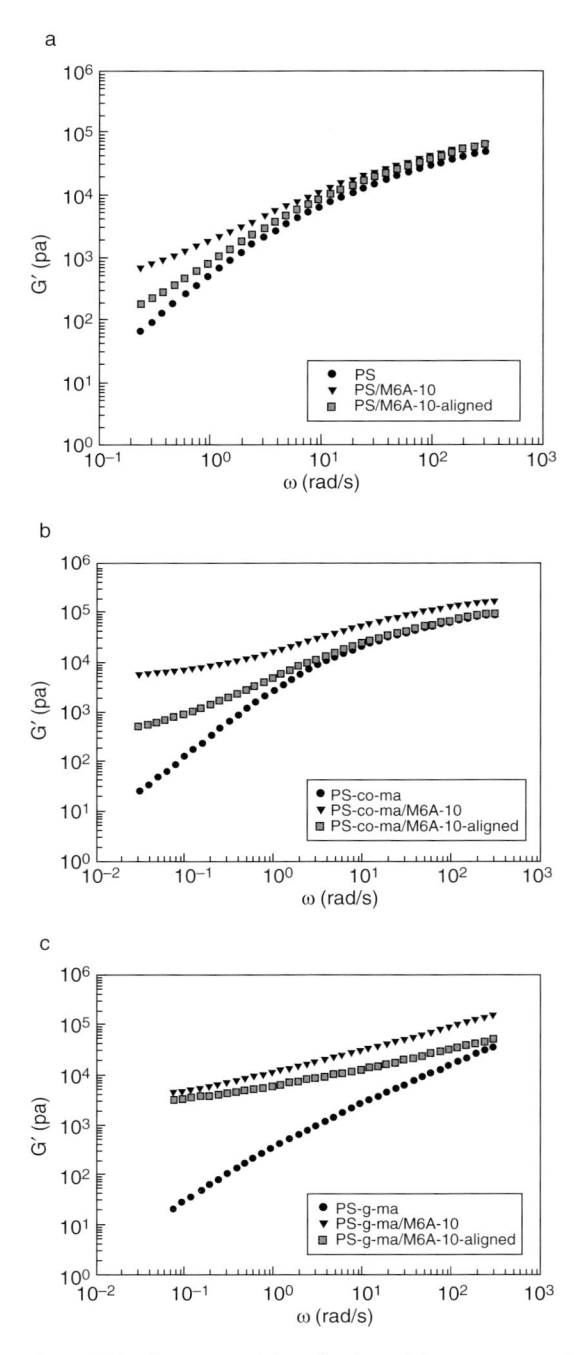

Figure 5.18 Storage modulus of polymer/clay nanocomposite after large-amplitude oscillatory shear ($\gamma = 120\%$; $\omega = 1\,\text{rad}\,\text{s}^{-1}$). (a) PS/M6A-10; (b) PS-co-ma/M6A-10; (c) PE-g-ma/M6A-10. Reprinted from Ref. [63], with permission from Springer.

chains and clay are primarily responsible for the rheological behavior of polymer nanocomposites.

Choi and coworkers [65] also described the effect of the dispersion of silicate layers and the interaction between the organoclay surface and the polymer matrix on the melt-state rheological properties of polymer/clay nanocomposites. These authors prepared a series of poly(ethylene oxide) (PEO)/organoclay nanocomposites, using three types of surfactant-modified MMT (Cloisite 15A, 20A, and 25A). First, two different organoclays were selected which had different modifier concentrations but the same alkylammonium salts. Second, the types of alkylammonium salts having the same organic modifier concentration were differentiated. All nanocomposites were fabricated using the solvent casting method. The rheological properties of these nanocomposites differed with varying modifier concentrations and surfactant size (chain lengths). The PEO/Cloisite 25A nanocomposites with Cloisite 25A loading exhibited higher zero shear rate viscosities and more rapid shear-thinning behavior than pure PEO; this was the result of the reorientation of dispersed clay particles. When the mean-field (MF) equation was applied to examine the effect of concentration of clay on viscosity, the results indicate that the rheological properties of nanocomposite systems depended on interactions between the clay platelets, and on interactions between the clay surface and polymer matrix.

Krishnamoorti *et al.* [74] examined the rheological behavior of end-tethered polymer, layered silicate nanocomposites prepared by *in situ* polymerization, which consisted of poly(ε-caprolactone) (PCL) and nylon-6 with various amounts of layered silicate. These authors used a lattice model for the interactions between polymer and organoclays, which showed that the conformation of the polymers and surfactants was decoupled [71]. Thus, the configurations of the free and tethered species were independent of each other.

The results of many studies have suggested that the melt-state linear viscoelastic characteristics (particularly the low-frequency response) of layered silicate nanocomposites are most strongly correlated to the mesoscopic structure of layered silicates [67, 75]. The mesoscopic structure is critically dependent on not only the strength of the polymer–clay interaction but also the inherent viscoelastic properties of the matrix in which the layers (or collection of layers) are dispersed [72]. Hsieh and coworkers [64] studied the rheological behavior of polycarbonate (PC)/clay nanocomposites under small-strain amplitude oscillatory shear conditions to quantify the effects of nanoclays on the linear viscoelastic response of PC nanocomposites. At high frequencies, the viscoelastic relaxation behavior was consistent with that of a well-entangled polymer, but a liquid-like relaxation with $G' \sim \omega^2$ and $G'' \sim \omega$ evident at low frequencies. For entangled homopolymers, the crossover frequency obtained from G' and G'' versus ω data (on a log-log plot) corresponds to their longest relaxation time. Figure 5.19 shows a plot of the linear dynamic viscoelastic master curves obtained for pure PC [64]. These data were shifted to a common reference temperature of 190 °C using both frequency shift factors (α_T) and modulus factor (b_T). For PC, the relaxation time, τ, was approximately 0.22 s at 190 °C.

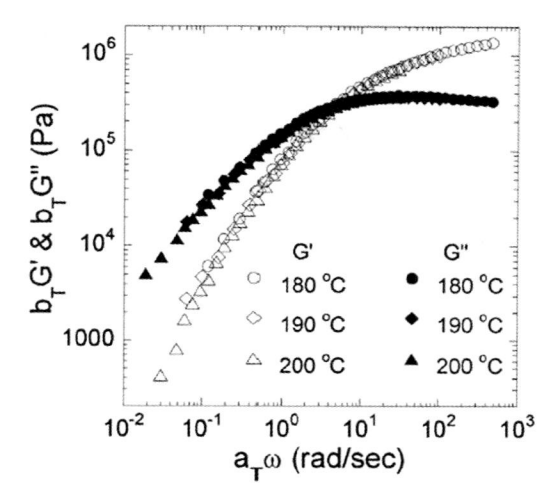

Figure 5.19 Melt-state viscoelastic master curves of storage moduli (G′) and loss moduli (G″) versus frequency (ω) data obtained for pure PC. Reprinted from Ref. [64], with permission from John Wiley & Sons Ltd.

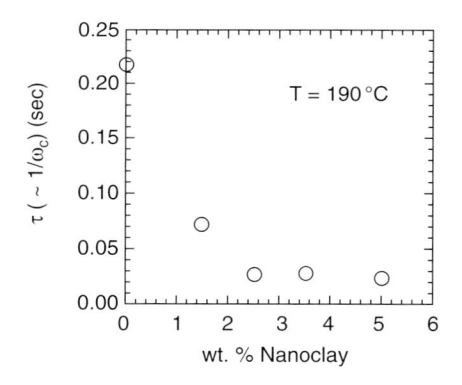

Figure 5.20 Rheological measurements of change in the longest relaxation time of PC as a function of clay content. Reprinted from Ref. [64], with permission from John Wiley & Sons Ltd.

For PC nanocomposites, the crossover frequency was noted to shift to higher frequencies, indicating a shorter relaxation time than that of the unfilled PC. Figure 5.20 shows the relaxation time, τ, as a function of the clay loading. The relaxation time decreased significantly with the incorporation of nanoclay. The divergence of viscoelastic relaxation at low frequency was presumably associated with the formation of mesoscale nanoclay structure from percolation within the PC nanocomposite.

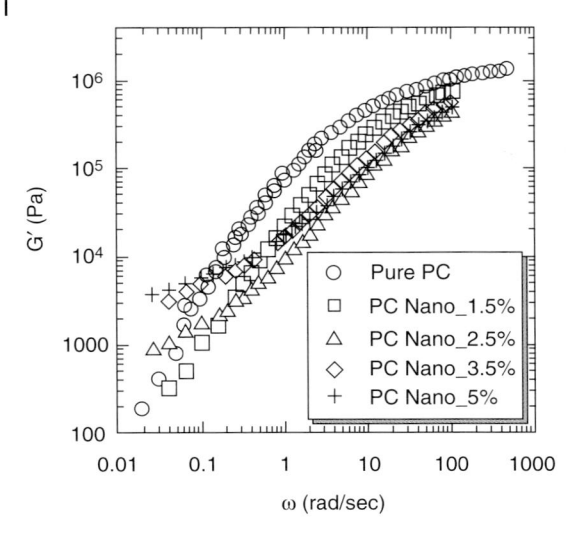

Figure 5.21 Melt-state viscoelastic data of G' versus ω for PC and PC nanocomposites consisting of 1.5, 2.5, 3.5, and 5 wt% nanoclay. Reprinted from Ref. [64], with permission from John Wiley & Sons Ltd.

The rheological measurements for PC nanocomposites containing 1.5, 2.5, 3.5, and 5 wt% nanoclays, respectively, are shown in Figure 5.21. The results indicate that G' values at low frequencies were almost independent of frequency for the 3.5 and 5 wt% clay nanocomposites, which suggested the presence of a pseudo solid-like behavior associated with a percolated network structure within these nanocomposites. The 2.5 wt% clay nanocomposite exhibited a modest dependence on G' at low frequency, but the 1.5 wt% clay nanocomposite was similar to the pure PC in exhibiting a liquid-like behavior at low frequencies. Thus, the rheological behavior of PC nanocomposites appeared to depend on the mesoscale percolated nanoclay structure, as well as an intercalated or exfoliated morphology.

Zhong and coworkers [62] studied the stress relaxation behavior of exfoliated PS/clay nanocomposites, which were produced by *in situ* polymerization. The rheological measurements were conducted using a time–temperature superposition of oscillatory shear between 150 and 300 °C and a shear amplitude of 0.5%. Figure 5.22 presents the complex viscosity of PS nanocomposites with 4, 6, and 9 wt% clay loadings, respectively, as a function of the oscillation frequency. The diverging complex viscosity with lowering frequency at all three loadings was evidence that a large-scale association of the sheet-like clay particles had occurred. The stress relaxation of PS and PS/clay nanocomposite is shown in Figure 5.23; a comparison between Figures 5.23a and b indicates that the nanocomposite has a much slower relaxation time than the pure PS. It can also be seen that, in sharp contrast to the pure PS, the stress relaxation of the nanocomposite was independ-

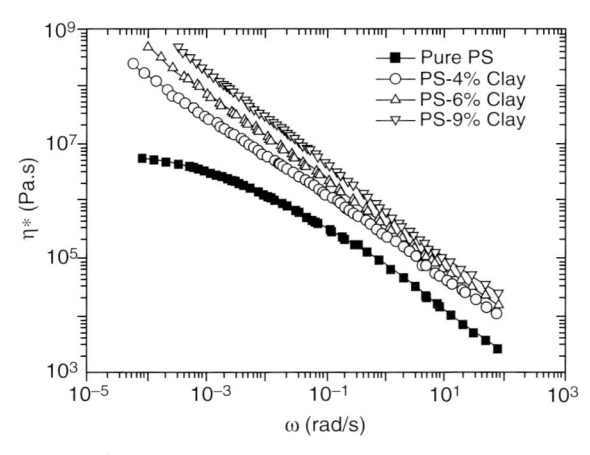

Figure 5.22 Complex viscosity as a function of frequency for the pure PS and PS/clay nanocomposites at 150 °C from oscillatory shear measurements in the temperature range of 150 to 300 °C. Reprinted from Ref. [62], with permission from Elsevier.

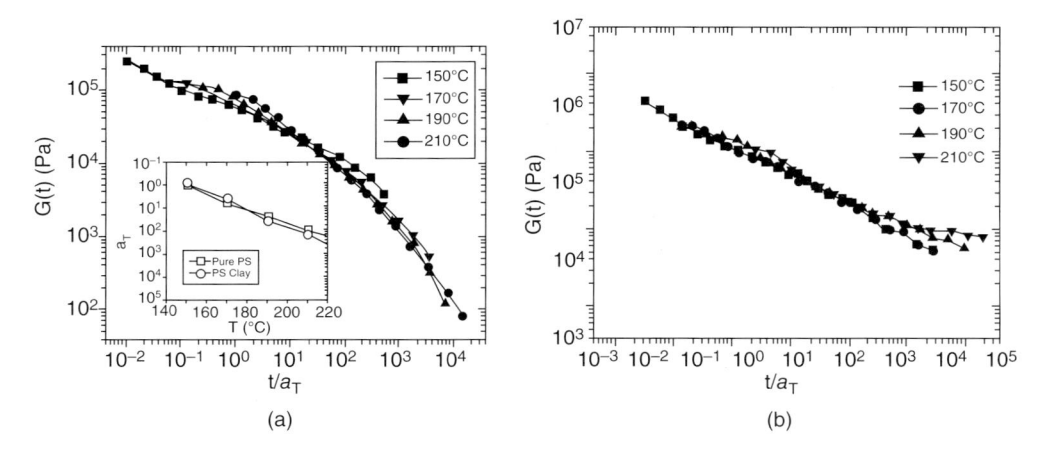

Figure 5.23 (a) Stress relaxation of the pure PS at different temperatures; (b) Stress relaxation of PS/clay 4% nanocomposite at different temperatures. Reprinted from Ref. [62], with permission from Elsevier.

ent of the temperature. Clearly, stress relaxation in the nanocomposite is not dictated by the matrix polymer relaxation but rather depends on the relaxation of the network formed by the sheet-like clay particles, which apparently is insensitive to temperature change.

Pötschke and coworkers [76] studied the dynamic flow properties of PC/MWNT composites as a function of MWNT loading, and found the weight fraction threshold to be 2 wt%. The dynamic complex viscosity η^* master curves for pure PC and

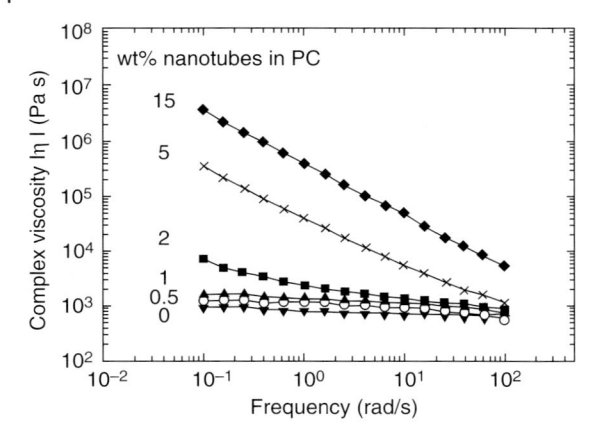

Figure 5.24 Complex viscosity of nanotube-filled polycarbonate at 260 °C. Reprinted from Ref. [76], with permission from Elsevier.

the composites are shown in Figure 5.24; from these data it was clear that the complex viscosity increased with increasing MWNT loading, and the effect of MWNT was more pronounced at low frequencies. It was also noted that the viscosity curves for 0.5 and 1 wt% nanotubes had similar frequency dependencies to the pure PC, revealing a Newtonian plateau at low frequencies. Above 2 wt% nanotubes, however, the viscosity curves showed no Newtonian plateau within the frequency range studied. At 5 wt% nanotubes the viscosity curve was almost linear over the range of frequencies shown.

The interconnected structures of anisometric fillers result in an apparent yield stress which was visible in dynamic measurements by a plateau of G' or G'' versus frequency at low frequencies [77, 78]. This effect was more pronounced in G' than in G''. As the nanotube content increased in the composite system, the nanotube–nanotube interactions began to dominate, leading eventually to percolation and the formation of an interconnected structure of the nanotubes. As shown in Figure 5.25, starting at about 2 wt% MWNT loading, G' reached such a plateau at low frequencies, which indicated that an interconnected structure had been formed.

5.4
Conclusions

The rheological properties of several polymer/clay or polymer/CNT nanocomposites in both solution and in melt state have been reviewed in this chapter. The incorporation of nanofillers into the polymer matrix dramatically alters the rheological behavior of polymer chains in the nanocomposites. Pronounced shear thinning was found to be a characteristic feature of truly nanodispersed composites. The nanocomposites were characterized by a solid-like behavior, as revealed

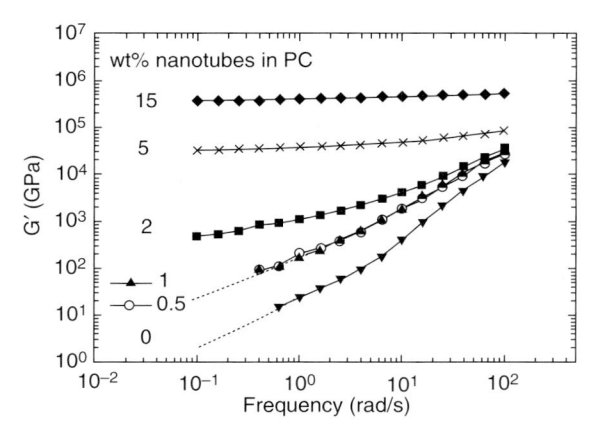

Figure 5.25 Storage modulus G′ of nanotube-filled polycarbonate at 260 °C. Reprinted from Ref. [76], with permission from Elsevier.

by the presence of a low-frequency plateau on storage modulus. As with other properties, rheological properties depend on morphology and interfacial characteristics as well as on the polymer/clay characteristics. It would be interesting to study, by means of rheometry, the process of exfoliation and intercalation of clay or nanofillers in polymer matrix, which is similar to phase separation in polymer–polymer blends.

References

1 Itagaki, T., Matsumura, A., Kato, M., Usuki, A., and Kuroda, K. (2001) *J. Mater. Sci. Lett.*, **20**, 1483–1484.
2 Chen, C. and Curliss, D. (2003) *J. Appl. Polym. Sci.*, **90**, 2276–2278.
3 Sung, J.H., Cho, M.S., Choi, H.J., and Jnon, M.S. (2004) *J. Ind. Eng. Chem.*, **10**, 1217–1219.
4 Dresselhaust, M.S. and Dresselhaust, G. (1981) *Ad. Phys.*, **30**, 139–142.
5 Cai, D. and Song, M. (2009) *Nanotechnology*, **20**, 315708(1)–315708(6).
6 Subramoney, S. (1998) *Adv. Mater.*, **10**, 1157–1171.
7 Baibarac, M. and Gómez-Romero, P. (2006) *J. Nanosci. Nanotechnol.*, **6**, 289–302.
8 Cui, X., Engellard, M.H., and Lin, Y. (2006) *J. Nanosci. Nanotechnol.*, **5**, 547–553.
9 Nussbaumer, R.J., Caseri, F.L., and Smith, P. (2006) *J. Nanosci. Nanotechnol.*, **6**, 459–463.
10 Hiroi, R., Ray, S.S., Okamoto, M., and Shiroi, T. (2004) *Macromol. Rapid Commun.*, **25**, 1359–1364.
11 Okada, A., Kawasumi, M., Usuki, A., Kojima, Y., Kurauchi, T., and Kamigaito, O. (1990) *Mater. Res. Soc. Proc.*, **171**, 45–52.
12 Lan, T. and Pinnavaia, J. (1994) *Chem. Mater.*, **6**, 2216–2219.
13 Zilg, C., Muelhaupt, R., and Finnter, J. (1999) *Macromol. Chem. Phys.*, **200**, 661–670.
14 Calvert, P. (1999) *Nature*, **399**, 210–211.
15 Shaffer, M.S.P. and Windle, A.H. (1999) *Adv. Mater.*, **11**, 937–941.
16 Qin, S.H., Oin, D.Q., Ford, W.T., Resasco, D.E., and Herrera, J.E. (2004) *J. Am. Chem. Soc.*, **126**, 171–175.

17 Andrews, R., Jacques, D., Minot, M., and Rantell, T. (2002) *Macromol. Mater. Eng.*, **7**, 395–403.
18 Saito, R., Dresselhaus, G., and Dresselhaus, M.S. (1998) *Physical Properties of Carbon Nanotubes*, Imperial College Press, London.
19 Ajayan, P.M., Schadler, L.S., Giamaris, C., and Rubio, A. (2000) *Adv. Mater.*, **12**, 750–753.
20 Dalton, A.B., Collins, S., Munoz, E., Razal, J.M., Ebron, V.H., Ferraris, J.P., Coleman, J.N., Kim, B.G., and Baughman, R.H. (2003) *Nature*, **423**, 703–704.
21 Koo, M.C., Kim, M.J., Choi, M.H., Kim, S.O., and Chung, I.J. (2003) *J. Appl. Polym. Sci.*, **88**, 1526–1535.
22 Lozano, K. and Barrera, E.V. (2001) *J. Appl. Polym. Sci.*, **79**, 125–133.
23 Hoffmann, B., Dietricha, C., Thomann, R., and Friedrich, R. (2000) *Macromol. Rapid Commun.*, **21**, 57–61.
24 Slobodian, P., Pavlínek, V., Lengálová, A., and Sáha, A. (2009) *Cur. Appl. Phys.*, **9**, 184–188.
25 Yao, K.J., Song, M., Hourston, D.J., and Luo, D.Z. (2002) *Polymer*, **43**, 1017–1020.
26 Xia, H. and Song, M. (2005) *Soft Matter*, **1**, 386–394.
27 Park, J.H. and Jana, S.C. (2003) *Macromolecules*, **36**, 2758–2761.
28 Zou, W., Du, Z.J., Liu, Y.X., Yang, X.P., Li, H.Q., and Zhang, C. (2008) *Compos. Sci. Technol.*, **68**, 3259–3264.
29 Fornes, T.D., Yoon, P.J., Keshula, H., and Paul, D.R. (2001) *Polymer*, **42**, 9929–9940.
30 Kim, H.S., Park, B.H., Yoon, J.S., and Jin, H.J. (2007) *Mater. Lett.*, **61**, 2251–2254.
31 Gilven, N. (1988) *Hydrous Phyllosilicates* (ed. S.W. Bailey), Mineralogical Society of America, Washington, DC.
32 Lan, T., Kaviratna, P.D., and Pinnavia, T.J. (1995) *Chem. Mater.*, **7**, 2144–2150.
33 Yao, K.J. (2005) High performance polyurethane Organoclay nanocomposites, Ph.D Thesis, Loughborough University, 2005.
34 Jin, J., Chen, L., Song, M., and Yao, K.J. (2006) *Macromol. Mater. Eng.*, **291**, 1411–1421.
35 Galgali, G., Ramesh, C., and Lele, A. (2001) *Macromolecules*, **34**, 852–858.
36 Lim, Y.T. and Park, O.O. (2001) *Rheol. Acta*, **40**, 220–229.
37 Xia, H. and Song, M. (2006) *Polym. Int.*, **55**, 229–235.
38 Herschel, W.H. and Bulklery, R. (1926) *Proc. Am. Test Mater.*, **26**, 621–624.
39 Matijasic, G. and Glasnovic, A. (2002) *Chem. Biochem. Eng.*, **16**, 165–169.
40 Wagener, R. and Reisinger, T.J.G. (2003) *Polymer*, **44**, 7513–7518.
41 Grossiord, N., Loos, J., and Koning, C.E. (2005) *J. Mater. Chem.*, **15**, 2349–5232.
42 Li, X.D., Gao, H.S., Scrivens, W.A., Fei, D.L., Xu, X.Y., Sutton, M.A., Reynolds, A.P., and Myrick, M.L. (2004) *Nanotechnology*, **15**, 1416–1420.
43 Cohan, L.H. (1948) *Proceedings of the Second Rubber Technology Conference*, Heffers & Sons, Ltd, Cambridge, UK.
44 Georgakilas, V., Kordatos, K., Prato, M., Guldi, D.M., Holzinger, M., and Hirsch, A. (2002) *J. Am. Chem. Soc.*, **124**, 760–765.
45 Shim, M., Kam, N.W.S., Chen, R.J., Li, Y.M., and Dai, H.J. (2002) *Nano Lett.*, **2**, 285–288.
46 Garg, A. and Sinnott, S.B. (1998) *Chem. Phys. Lett.*, **295**, 273–276.
47 Seifert, G., Kohler, T., and Frauenheim, T. (2000) *Appl. Phys. Lett.*, **77**, 1313–1316.
48 Tang, B.Z. and Xu, X. (1999) *Macromolecules*, **32**, 2569–2576.
49 Koshio, A., Yudasaka, M., Zhang, M., and Iijima, S. (2001) *Nano Lett.*, **1**, 361–363.
50 Xia, H., Song, M., Jin, J., and L. (2006) *Chen. Macromol. Chem. Phys.*, **207**, 1045–1952.
51 Xia, H. and Song, M. (2006) *J. Mater. Chem.*, **16**, 1843–1851.
52 Grossiord, N., Loos, J., and Koning, C.E. (2005) *J. Mater. Chem.*, **15**, 2349–2352.
53 Cai, D. and Song, M. (2007) *Macromol. Chem. Phys.*, **208**, 1183–1189.
54 Chen, J., Liu, H., Weimer, W.A., Halls, M.D., Waldeck, D.H., and Walker, G.C. (2002) *J. Am. Chem. Soc.*, **124**, 9034–9037.
55 O'Connell, M.J., Boul, P., Ericson, L.M., and Huffman, C. (2001) *Chem. Phys. Lett.*, **342**, 265–271.

References

56 Colvin, V.L. (2003) *Nat. Biotechnol.*, **21**, 1166–1169.
57 Moulton, S.E., Minett, A.I., Murphy, R., Ryan, K.P., McCarthy, D., Coleman, J.N., Mlau, W.J., and Wallace, G. (2005) *Carbon*, **43**, 1879–1884.
58 Spinks, G.M., Shin, S.R., Wallace, G.G., Whitten, P.G., Kim, S.I., and Kim, S.J. (2006) *Sens. Actuators*, **B115**, 678–684.
59 Roberts, G.A.F. (1992) *Chitin Chemistry*, MacMillan Press Ltd, London.
60 Solomon, M.J., Almusallam, A.S., Seefeldt, K.F., Somwangthanaroj, A., and Varadan, P. (2001) *Macromolecules*, **34**, 1864–1872.
61 Li, J., Zhou, C., Wang, G., and Zhao, D. (2003) *J. Appl. Polym. Sci.*, **89**, 3609–3617.
62 Zhong, Y., Zhu, Z., and Wang, S.Q. (2005) *Polymer*, **46**, 3006–3013.
63 Lim, Y.T. and Park, O.O. (2001) *Rheol. Acta*, **40**, 220–229.
64 Hsieh, A., Moy, P., Beyer, F.L., Madison, P., and Napadensky, E. (2004) *Polym. Eng. Sci.*, **44**, 825–837.
65 Hyun, Y.H., Lim, S.T., Choi, H.J., and Jhon, M.S. (2001) *Macromolecules*, **34**, 8084–8093.
66 Russo, G.M., Simmon, G.P., and Incarnato, L. (2006) *Macromolecules*, **39**, 3855–3864.
67 Krishnamoorti, R., Vaia, R.A., and Giannelis, E.P. (1996) *Chem. Mater.*, **8**, 1728–1734.
68 Shen, Z., Simon, G.P., and Cheng, Y.B. (2002) *Polymer*, **43**, 5011–5016.
69 Lertwimolnum, W. and Vergnes, B. (2005) *Polymer*, **46**, 3462–3471.
70 Fornes, T.D., Yoon, P.J., Keskkula, H., and Paul, D.R. (2001) *Polymer*, **42**, 9929–9940.
71 Vaia, R.A. and Giannelis, E.P. (1997) *Macromolecules*, **30**, 8000–8009.
72 Pinnavaia, T.J. and Beall, G.W. (2000) *Polymer-Clay Nanocomposites*, John Wiley & Sons, Ltd, London.
73 Vaia, R.A., Jandt, K.D., Kramer, E.J., and Giannelis, E.P. (1995) *Macromolecules*, **28**, 8080–8085.
74 Krishnamoorti, R. and Giannelis, E.P. (1997) *Macromolecules*, **30**, 4097–4102.
75 Solomon, M.J., Almusallam, A.S., Seefeldt, K.F., Somwangthanaroj, A., and Varadan, P. (2001) *Macromolecules*, **34**, 1864–1872.
76 Pötschke, P., Fornes, T.D., and Paul, D.R. (2002) *Polymer*, **43**, 3247–3255.
77 Shenoy, A.V. (1999) *Rheology of Filled Polymer Systems*, Kluwer Academic Publishers, Dordrecht.
78 Dealy, M. and Wissbrun, K.F. (1999) *Melt Rheology and Its Role in Plastic Processing Theory and Application*, Kluwer Academic Publishers, Dordrecht.

6
Mechanical Property Enhancement of Polymer Nanocomposites

Nourredine Aït Hocine

6.1
Introduction

In conventional polymer composites, fillers with dimensions in the micrometer range are added to polymers so as to enhance their mechanical properties. Such properties are related to the volume fraction, shape, and size of the filler particles [1–3]. When fillers have dimensions in the nanometer range, their mixture with a matrix leads to so-called *nanocomposites*. The first major finding in polymer nanocomposites was reported by the Toyota research group on a Nylon-6/montmorillonite (N6/MMT) blend [4]. Overviews of the different matrices used in this new class of materials have been prepared by Sur *et al.* [5] and by Tjong [6]. Because of the remarkable enhancement of their physical and mechanical properties by the addition of only a small filler volume fraction compared to traditional composites, these types of material have attracted much interest both in industry and in academia. The mechanical properties can include Young's modulus (or stiffness), yield stress, ultimate stress and strain, fracture toughness, and shock resistance [4–12]. The improvement in the mechanical performances of nanocomposites is due mainly to the nanosize dimensions of fillers (as this results in an extremely large aspect ratio), and to the strong polymer–filler interactions that may affect the effectiveness of load transfer between nanofillers and the polymer matrix. Many reports of this subject have been provided [7, 9, 12–20], in particular by the incorporation of MMT into polyamide 6 (PA6) [13], polypropylene (PP) [14–16], polyamide 66 (PA66) [17], and polystyrene [18–20]. The influence of the filler volume fraction on these properties has also been outlined [5, 12, 21–27]. In addition, the type of filler surface treatment governs the degree of particle dispersion in the matrix [28] and, thus, plays a key role in the mechanical performances of the final nanocomposite. The nanocomposite stiffness is increased by a significant factor over that of the neat matrix when a uniform dispersion is achieved in nanoscale size [29]. However, nonlinear mechanical properties such as tensile strength, elongation at break, or impact strength, generally decrease beyond a critical proportion of reinforcing particles. This may be explained by the fact that the Young's modulus is evaluated at low strains, whereas other properties are determined

Optimization of Polymer Nanocomposite Properties. Edited by Vikas Mittal
Copyright © 2010 WILEY-VCH Verlag GmbH & Co. KGaA, Weinheim
ISBN: 978-3-527-32521-4

beyond catastrophic break, where the loading transfer between matrix and fillers is important. In reality, the mechanical behavior of polymer–nanofiller composites is controlled by several microstructural parameters which are governed by, among others, the properties of the matrix and fillers, along with the processing methods employed.

Because of the wide diversity of nanocomposites and the extensive range of their properties, it is difficult to oversee all areas of interest. Consequently, this chapter will focus on the enhancement of certain macroscopic mechanical properties of thermoplastic/rigid filler nanocomposites, when compared to the virgin matrix. Particular attention is paid to models which predict these properties.

6.2
Material Stiffness

6.2.1
Experimental Investigations

The stiffness of polymer nanocomposites generally increases with the nanoparticles volume fraction, as long as a sufficient dispersion and degree of exfoliation of these particles are ensured [29, 30]. Figure 6.1 shows the typical variation in the ratio of the Young's modulus E_c of the composite to that of the neat matrix, as a function of the filler volume fraction, for polyamide nanocomposites.

Figure 6.1 highlights some interesting results, notably that although the Young's modulus increases with particle volume fraction, the increase is greater at a low filler content; this indicates that the fillers are better dispersed and/or better exfoliated. The data also show that the reinforcing effect is lower for nanocomposites with a higher filler content, owing to part of the filler being partially dispersed or

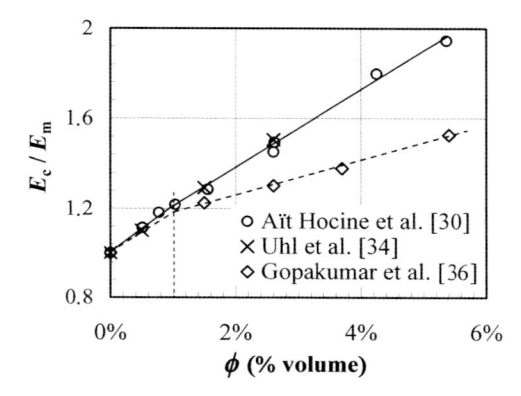

Figure 6.1 Relative Young's modulus versus filler volume fraction ϕ for polyamide nanocomposites.

exfoliated and aggregated or stacked. These results have been confirmed in other studies [5, 22, 31–37].

When Osman et al. studied the effect of the degree of nanoclay exfoliation on the tensile properties of the melt-compounded polyethylene-organomontmorillonite (PE-organoMMT) [38], they noted that the modulus progressively increased as the spacing between the individual clays was increased – that is, as the degree of exfoliation increased. The role of this powerful interaction between the matrix and fillers in providing remarkable improvements in nanocomposite rigidity was confirmed using dynamic mechanical analysis (DMA) [12]. Using a similar approach, the effect of the dimension of the dispersed clay particles on the increase in material rigidity was also determined by Sinha Ray et al. [39], who also studied the influence of compatibilizers on the morphology and mechanical properties of the nanocomposites.

6.2.2
Analytical Modeling

In order to corroborate these experimental observations, several micromechanical models have been developed to predict the macroscopic rigidity improvement of the nanocomposites. Notably, reviews of these models have been prepared by Ahmed et al. in 1990 [40] and more recently by Tjong [6]. These models are based on either a two-phase or a three-phase approach.

With regards to two-phase approaches (particle/matrix), the earliest and most popular theory was that of Einstein [41], which was developed for composites with spherical rigid inclusions in a nonrigid matrix. Since then, however, several developments have been made covering a wide variety of composites, and involving both series and parallel models [42–50]. Both, Kaplan [48] and Hansen [49] proposed models that predicted the modulus enhancement of polymer composites formed by rigid inclusions in a rigid matrix. One of the earlier phenomenological models predicting the stiffness of polymer composites, and which was more generalized in nature, was proposed by Guth in 1945 [50]. This expresses the ratio of the Young's modulus E_c of the composite to that of the neat matrix, under a second-order polynomial function of volume fraction filler ϕ:

$$\frac{E_c}{E_m} = 1 + 2.5\phi + 14.1\phi^2 \tag{6.1}$$

Unfortunately, this model suffered from lack of accuracy. In particular, it greatly underestimated the experimental stiffness of poly(vinylchloride) (PVC) reinforced with poly(methylmethacrylate) (PMMA)-encapsulated Sb_2O_3 nanoparticles [51]. In order to improve its precision, Guth's theory was modified to the following equation, which took the aspect ratio f of stiff particles into account:

$$\frac{E_c}{E_m} = 1 + 0.67 f\phi + 1.62 f^2\phi^2. \tag{6.2}$$

This led to predictions which agreed reasonably with the experimental data obtained for rubber reinforced with single-walled nanotubes (SWNTs) on the one hand, and with carbon nanofibers on the other hand [52].

Halpin and Tsai expressed the longitudinal elastic modulus E_c of continuous fiber composites as follows:

$$\frac{E_c}{E_m} = \frac{1+\xi\eta\phi}{1-\eta\phi} \quad (6.3)$$

This assumed that the fibers were discontinuous and aligned uniaxially [53–55]; the constants ξ and η are given by:

$$\xi = 2\left(\frac{l}{d}\right) \quad (6.4)$$

and

$$\eta = \frac{(E_f/E_m)-1}{(E_f/E_m)+\xi} \quad (6.5)$$

where E_f and l/d are the tensile modulus and the aspect ratio (length/diameter) of the reinforcing fibers, respectively.

Halpin and Tsai's model (Equation 6.3) was subsequently extended by Nielsen to nanocomposites with dispersed clay particles without intercalation, and was rewritten as follows [56]:

$$\frac{E_c}{E_m} = \frac{1+\xi^*\eta^*\phi^*}{1-\xi^*\psi\phi^*} \quad (6.6)$$

where:

$$\xi^* = k_E - 1, \quad (6.7)$$

$$\eta^* = \frac{(E_{clay}/E_m)-1}{(E_{clay}/E_m)+\xi^*}, \quad (6.8)$$

and

$$\psi = 1 + \left[\frac{1-\phi_{max}}{\phi_{max}^2}\right]\phi^*. \quad (6.9)$$

Here, k_E is the Einstein coefficient representing the aspect ratio of the dispersed clays, E_{clay} is the Young's modulus of the clay, ξ^* is the constant that depends on the type of nanocomposite structure and is related to the aspect ratio, and ϕ^* and ϕ_{max} are the volume fractions of clay reinforcement and the maximum packing volume fraction of clay, respectively. E_{clay} is generally assumed to be equal to 170 GPa [57].

Later, Brune and Bicerano [58] also modified the Halpin–Tsai theoretical expression to predict the tensile modulus of intercalated or incompletely exfoliated polymer/clay nanocomposites. For this, they treated the system as mélange of a matrix and pseudoparticles which were incompletely exfoliated stacks of individual platelets. In this way, the macroscopic Young's modulus of the nanocomposite could be expressed as follows:

$$\frac{E_c}{E_m} = \frac{1+\xi^{**}\eta^{**}\phi^{**}}{1-\eta^{**}\phi^{**}} \tag{6.10}$$

where η^* is given by:

$$\eta^{**} = \frac{(E_{ps}/E_m)-1}{(E_{ps}/E_m)+\xi^{**}}, \tag{6.11}$$

E_{ps} and ξ^{**} represent the modulus and the aspect ratio of the platelet stack, respectively, and ϕ^{**} is the volume fraction of platelet stacks in the matrix. Assuming that in each stack, there are N platelet layers at various distances from each others, this can be rewritten as:

$$\xi^{**} = \frac{\xi}{N}\left[\frac{1}{1+(1-1/\overline{N})(s/t)}\right] \tag{6.12}$$

$$\phi^{**} = \phi\left(1+\left(1-\frac{1}{\overline{N}}\right)\right)\frac{s}{t} \tag{6.13}$$

$$E_{ps} = E_p\left[\frac{1}{1+(1-1/\overline{N})(s/t)}\right] + \frac{(1-1/\overline{N})(s/t)}{1+(1-1/\overline{N})(s/t)} \tag{6.14}$$

$$\overline{N} = N+(1-N)\left(\frac{s}{t}\right)\left(\frac{\phi}{1-\phi}\right) \tag{6.15}$$

where s is the interplatelet spacing, and E_p and t are the modulus and thickness of an individual platelet, respectively. In the case of one platelet (i.e., $N = 1$), the model of Brune and Bicerano (Equation 6.10) reverts back to the Halpin–Tsai equation. Unfortunately, however, no comparison was made between this model predictions and this experimental data.

Takayanagi et al. [59] suggested an elaborate two-phase model which was used widely and successfully to analyze the change in the modulus of polymers, polymer blends, and composites with diverse filler morphologies. However, when this model was applied to nanocomposites it was observed that the predicted values underestimated the experimental data [60]. This divergence was related to the fact that the Takayanagi model was developed originally for traditional composites in which a dispersed particle is in the range of micrometers, such that the interfacial region contribution can be neglected. When considering nanocomposites, the interfacial zone is much larger compared to the dimensions of the particles, and so cannot be neglected owing to its dominant role. Thus, a two-phase model is irrelevant for predicting the variation in Young's modulus. Consequently, Ji et al. [60] extended the two-phase model into a fully general three-phase model which took into account the matrix, interface and volume fraction of dispersed particles of various shapes (lamellar, spherical or cylindrical) that were distributed randomly within the matrix:

$$\frac{E_c}{E_m} = \left[(1-\alpha) + \frac{\alpha-\beta}{(1-\alpha)+\alpha(k-1)/\ln(k)} + \frac{\beta}{(1-\alpha)+(\alpha-\beta)(k+1)/2+(E_f/E_m)\beta} \right]^{-1}$$
(6.16)

For nanocomposites with plate particles having thickness t_c and both length and width $= \xi_c$ (with $\xi_c \gg t_c$), $\alpha = \sqrt{[2(\tau/t_c)+1]\phi}$, $\beta = \sqrt{\phi}$, τ is the thickness of interfacial region, ϕ is the filler volume fraction, and $k = E_i(0)/E_m$ is the ratio of the interphase modulus on the surface of the particle, $E_i(0)$, to that of the matrix, E_m, assuming a linear gradient change in modulus between the matrix and the surface of particle. If the dispersed phase is in spherical form, then α is written as follows: $\alpha = \sqrt{[(r+\tau)/r]^3 \phi}$, where r is the radius of dispersed sphere. If the dispersed phase is composed of cylindrical particles, then α is given by $\alpha = \sqrt{[(r+\tau)/r]^2 \phi}$, where r represents the radius of the cylinder.

In the model of Ji et al. [60] (see Equation 6.16), the parameters controlling the material mechanical properties include: (i) the sizes of the dispersed particles t_c and ξ_c; (ii) the thickness of the interfacial region τ; (iii) the filler-to-matrix modulus ratio E_f/E_m; and (iv) the factor k. The effects of these parameters on the Young's modulus of nanocomposites were discussed by Ji et al. [60], who concluded that the role of k was not clear. Nevertheless, the point was made that the thickness of the interfacial region τ decreased when the parameter k increased, assuming that t_c remained constant. Thus, both Ji et al. [60] and Tanoue et al. [31] arbitrarily selected $k = 4$ and $k = 12$, and both groups analyzed the interaction between the matrix and fillers of different nanocomposites via the thickness τ deduced from Equation 6.16. The experimental data obtained by Tanoue et al. on polystyrene/cloisite 10A nanocomposites [31] were well described when introducing the interphase thickness $\tau = 0.4$ nm, though this was insignificant compared to $\tau = 7$ nm found by Ji et al. in the case of PA6/MMT nanocomposites [60]. Although this suggests that the quantitative evaluation of τ from Equation 6.16 is not pertinent, Ji's three-phase model appears to provide the correct tendency for the interphase contribution [31, 60], notably that the small thickness of the interphase may imply weak interactions between the clay platelets and the polymer matrix. In addition, it was shown analytically that if the particle sizes were in the range of nanometers, then the interfacial region would play an important role in the increase of the composite Young's modulus E_c.

More precisely, it was revealed that, at fixed volume fractions of the dispersed phase and the interface zone, smaller particles provided a larger increase in E_c. However, if the size of fillers was of the scale of micrometers, then the influence of the interfacial region could be neglected; in other words, $\tau = 0$ (which means $\alpha = \beta$), and the three-phase model (Equation 6.16) reverts back to Takayanagi's two-phase model [59]:

$$\frac{E_c}{E_m} = \left[(1-\beta) + \frac{\beta}{(1-\beta)+(E_f/E_m)\beta} \right]^{-1}$$
(6.17)

which was developed to describe the modulus of composites with one homogeneous rigid phase and one homogeneous matrix phase.

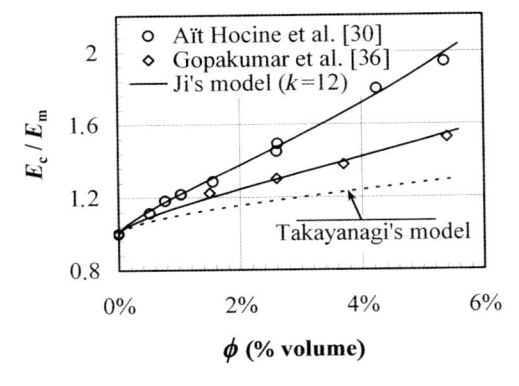

Figure 6.2 Theoretical predictions of Young's modulus versus filler volume fraction ϕ for polyamide nanocomposites.

The model of Ji et al. seemed to provide a good prediction of the tensile modulus of the various nanocomposites [30, 31, 36, 60], as shown in Figure 6.2 for polyamide nanocomposites. When modeling the Young's modulus with the three-phase model proposed by Ji et al. [60], Aït Hocine et al. recently underlined the influence of the microstructure on the linear mechanical properties of nanocomposites prepared under two processing conditions [30]. Indeed, the three-phase model highlighted that the higher the degree of exfoliation, the better the improvement in nanocomposite stiffness.

6.3
Ultimate Mechanical Properties

6.3.1
Experimental Investigations

The ultimate mechanical characteristics of nanocomposites are strongly related to their microstructures, which in turn are governed by the dispersion and/or the exfoliation of nanoparticles in the matrix. The distribution of the reinforcing fillers in the polymer depends on several parameters (type of particle and matrix, filler treatment, processing conditions, etc.) [16]. The yield stress, tensile strength and strain at break generally increase with increasing filler content up to a critical volume fraction, although beyond this limit these mechanical properties deteriorate dramatically to reach values less than those of the neat matrix [30, 61]. This is illustrated in Figures 6.3–6.5, which show the relative values of the yield stress, $\sigma_{y,c}/\sigma_{y,m}$, the stress at break, $\sigma_{b,c}/\sigma_{b,m}$, and the strain at break, $\varepsilon_{b,c}/\varepsilon_{b,m}$, as a function of loading level in the case of polyamide12/montmorillonite 'C30B' (PA12/C30B) nanocomposites. The subscripts c and m indicate that the property is that of the composite and of the matrix, respectively. As observed for the Young's modulus, the increase in the ultimate mechanical properties of nanocomposites is more

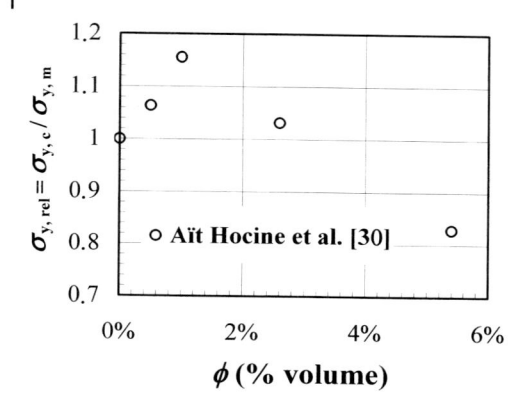

Figure 6.3 Relative yield stress versus filler volume fraction ϕ for a polyamide/montmorillonite nanocomposite.

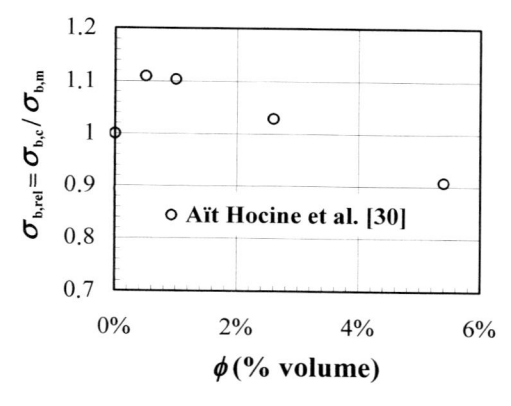

Figure 6.4 Relative stress at break versus filler volume fraction ϕ for a polyamide/montmorillonite nanocomposite.

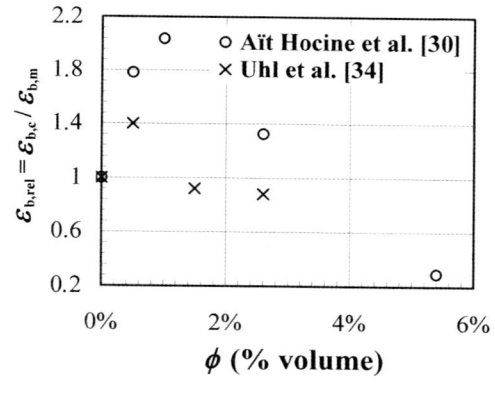

Figure 6.5 Relative strain at break versus filler volume fraction ϕ for polyamide nanocomposites.

obvious at low filler contents, which confirms that, within this range of reinforcement, the microstructure is well controlled.

Dispersion plays also key role in the improvement of such properties. In fact, a good dispersion of nanofillers in the polymer tends to improve the tensile strength and strain at break [30], although on occasion there may be significant reductions in tensile ductility and impact strength compared to the virgin polymer [62–64]. An interesting report was made by Osman *et al.* on the effect of the degree of exfoliation on tensile properties of the melt-compounded PE-organoMMT [38]. It had been observed globally that with increasing exfoliation the yield stress would increase and the yield strain decrease. A decrease in impact strength of nanocomposites was indirectly recognized by Liu *et al.* [65] who carried out dynamic mechanical testing on a polypropylene/montmorillonite (PP/MMT) system. These tests showed that the incorporation of MMT resulted in a decrease in viscosity–that is, in a decrease in shock absorption or in impact strength.

6.3.2
Analytical Modeling

Nanocomposites are developed to be used in a variety of applications where they may be submitted to loadings under low or high strain rates, and also to impact loadings. Because of their simplicity, tensile tests up to break point are used to quantify the fracture toughness of materials, while notched Charpy or Izod tests are employed to evaluate their shock resistance. Nonetheless, in order to rapidly evaluate such properties–or, at least, to better understand their variation as a function of different parameters–a range of micro- and macroscopic models was developed.

6.3.2.1 Yield Stress
Nicolais and Narkis focused their interest on predicting the yield stress of composites by proposing a phenomenological model that provided the ratio of the yield stress of the composite to that of the neat matrix, $\sigma_{y,c}/\sigma_{y,m}$, as a nonlinear polynomial function of volume fraction filler ϕ [66]:

$$\frac{\sigma_{y,c}}{\sigma_{y,m}} = 1 - 1.21\phi^{2/3}. \tag{6.18}$$

Pukanzky *et al.* developed a further robust micromechanical model which took certain microstructural aspects into account [67–69]:

$$\frac{\sigma_{y,c}}{\sigma_{y,m}} = \frac{1-\phi}{1+2.5\phi} \cdot \exp(B\phi) \tag{6.19}$$

where the fraction $[(1 - \phi)/(1 + 2.5\phi)]$ took into consideration the decrease in the effective load-bearing cross-section, while the exponential described all other effects that would result in an increase in the yield stress. From a physical point of view, the parameter B reflects the interface and the interphase properties, including the interlayer thickness, the interfacial strength, and the specific surface area of the filler particles, as underlined by the following expression:

$$B = (1 + \tau \rho_f S_f) \cdot \ln(\sigma_{y,i}/\sigma_{y,m}), \tag{6.20}$$

where τ, the thickness of the interphase, is proportional to the interfacial adhesion, defined by the parameter γ_{12}, with $\tau = \lambda \gamma_{12}$, and λ being a constant. The quantities ρ_f, S_f, and $\sigma_{y,i}$ represent the density of the filler, the specific surface area of the filler, and the yield stress of the interphase, respectively. As it is difficult to provide values for the thickness τ and the yield stress $\sigma_{y,i}$, the parameter B has been used to qualitatively quantify the effects of the interface and interphase on the yield stress. Rong et al. [70] used the model (Equation 6.19) to analyze the interfacial interactions in polypropylene nanocomposites, and showed that larger B values corresponded to a higher interfacial adhesion. Sumita et al. [71] also used this model to describe the dynamic mechanical properties of polypropylene composites filled with ultrathin particles. Recently, the experimental yield stress data of polyamide-6 nanocomposites were fitted to Equation 6.19 by Borse et al. [72], and the parameter B was evaluated. Higher values were obtained for the most compatible system, which reflected the higher interfacial adhesion. Moreover, the net effect of the low aspect ratios of reinforcing particles was reflected in lower values of B. In the same way, Aït Hocine et al. [30] analytically evaluated B from the experimental data of polyamide-12 nanocomposites, using the following expression derived from Equation 6.19:

$$B = \frac{1}{\phi} \cdot \ln\left(\frac{\sigma_{y,c}}{\sigma_{y,m}} \cdot \frac{1 + 2.5\phi}{1 - \phi}\right). \tag{6.21}$$

Here, the results were analyzed as a function of the volume fraction and the processing parameters in relationship with the microstructure. In this case, B was shown to increase up to a threshold value and then to decrease sharply for higher volume fractions of fillers. This result meant that the addition of reinforcing particles in weak proportions tended to improve the global interfacial filler/nanocomposite adhesion. At higher particle proportions, however, the concentration and number of particle aggregates was increased, and this led to a significant decrease in the average specific surface area of fillers, which was highlighted by a sharp decrease in the B parameter. Moreover, higher B-values were obtained for a better exfoliated system. In addition, an original result was underlined by the same authors which showed that the parameter B exhibited the same trend as nonlinear material properties as a function of ϕ, which in turn suggested that the B parameter governed these properties.

6.3.2.2 Properties at Break

Fracture mechanics theory makes available various parameters that allow the qualitative characterization of a material's fracture toughness. The most popular approach is based on the energy balance written for the whole specimen, which leads to the definition of a parameter, J, which represents the elastic energy required to extend a pre-existing defect of a unit area A:

$$J = -\frac{dU_{el}}{dA}\bigg|_u \qquad (6.22)$$

In Equation 6.22, U_{el} is the elastic potential strain energy (equivalent to the area under the load–displacement curve issued from the uniaxial tensile test). The subscript u indicates that the derivation is taken under a constant displacement.

When the material presents an elastoplastic behavior, J can be written as a function of the ubiquitous geometrical factor η, depending only on the geometry and the dimensions of the specimen [73–80]. More precisely, for a pre-cracked sheet of a crack length a, of a width w and of a thickness e this expression is written as a sum of an elastic part J_{el} (related to the elastic energy U_{el}) and a plastic part J_{pl} (a function of the dissipated energy U_{pl}; i.e., the plastic part of the area under the load–displacement curve):

$$J = J_{el} + J_{pl} = \eta_{el} \cdot \frac{U_{el}}{e \cdot (w-a)} + \eta_{pl} \cdot \frac{U_{pl}}{e \cdot (w-a)}. \qquad (6.23)$$

The parameter J, given by Equation 6.22, is equivalent to the well-known J – integral initially introduced by Cherepanov [81, 82] and Rice [83] in the case of small deformations:

$$J = \int_\Gamma \left(W dx_2 - \vec{t} \frac{\partial \vec{u}}{\partial x_1} ds \right) \qquad (6.24)$$

where W denotes the strain energy density, (x_1, x_2) is an orthogonal coordinate system with the x_1-axis parallel to the notch surface, Γ is a curve surrounding the notch tip (Figure 6.6), ds is a small element of Γ, $\vec{t}$ is the traction vector related to the Cauchy stress tensor components σ_{ij} by $t_i = \sigma_{ij}.n_j$ ($i = 1, 2$ and $j = 1, 2$), $\vec{u}$ is the displacement vector, and $\vec{n}$ is the outward normal on Γ. The integral assumes the same value for all paths Γ surrounding the crack tip.

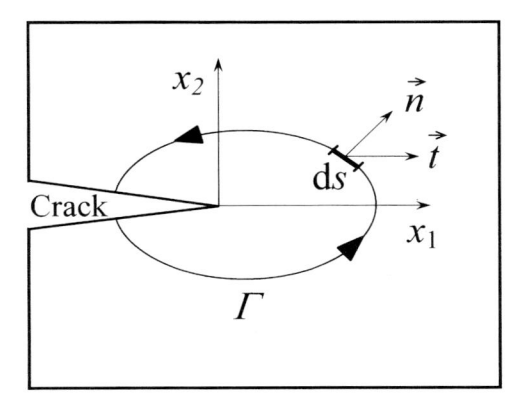

Figure 6.6 The J – integral.

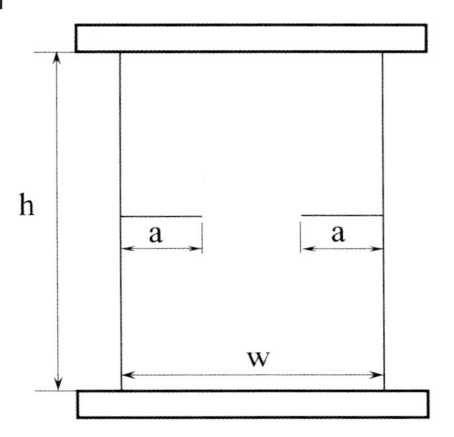

Figure 6.7 Double-edge notch in tension (DENT) specimen $h \geq 5w$.

Hence, crack growth will start at some critical value J_c of J. This value should be an intrinsic property of the material; that is, it should be independent of the crack geometry and the loading. The energy parameter J can either be determined experimentally by using the multispecimen technique developed by Begley and Landes [84], or evaluated numerically from the path integral via the finite element method (FEM).

The choc resistance of brittle materials is frequently characterized by impact energy. Moreover, an interesting procedure named the Essential Work of Fracture (EWF) has been developed and successfully applied, during the past two decades, to evaluate the fracture toughness of ductile polymers and composites or nanocomposites under quasi-static loadings [85–91]. When considering pre-cracked Double Edge Notch in Tension (DENT) specimens (Figure 6.7), the EWF concept assumes that the total fracture work, W_f, can be written as:

$$W_f = W_e + W_p \tag{6.25}$$

where:

$$W_e = \omega_e L e \tag{6.26}$$

and

$$W_p = \beta \omega_p L^2 e \tag{6.27}$$

Here, W_e is the elastic work spent to create new crack surfaces and represents the EWF. It corresponds to the qualitative total fracture energy – that is, the energy required to begin the growth of a crack initially introduced in the unloaded specimen. W_p is the part of the work spent to create a plastic zone around a pre-crack, and is called the 'nonessential work'. In addition, ω_e and ω_p are the specific essential fracture work and the specific plastic work, respectively, L is the ligament length, e the sample thickness, and β is a shape factor of the plastic zone.

Equation 6.25 can be rewritten using specific quantities:

$$\omega_f = \frac{W_f}{Le} = \omega_e + \beta \omega_p L \tag{6.28}$$

where ω_f is the specific total work.

The EWF concept is a simple method which allows evaluation of the fracture toughness of ductile materials submitted to quasi-static loadings. Several studies have been conducted to verify its relevance in the case of impact loadings by testing polymer blends [92–94] and polymer composites [95–97]. Moreover, the validity of this concept was verified by Martinatti and Ricco [98] to measure the high-rate fracture toughness of polypropylene-based materials.

It should be noted that experimental evaluations of the J – integral and the EWF require several normalized specimens to be tested.

6.4
Conclusions

This chapter has provided a succinct review on the mechanical properties enhancement of thermoplastic/rigid fillers nanocomposites, and has highlighted the fact that interest in nanocomposites is motivated by the challenge to provide lightweight, low-cost materials for industrial applications, with high mechanical performances. It has been shown that polymer/rigid filler nanocomposites can respond to such requirements, as the reinforcing particles not only have a high aspect ratio and/or large surface area but also are distributed homogenously within the matrix on the nanometer scale. Such uniform distribution will depend heavily on the type of polymer matrix, the nature and geometry of the nanofillers, and the processing conditions. During the past two decades, much effort has been expended in attempts to understand the relationship between the preparation methodologies, the microstructures and the mechanical properties of various nanocomposites, this being imperative for the development of new materials with optimized performances.

For any given polymer and reinforcing filler, nanocomposite stiffness is significantly enhanced compared to that of the neat matrix, when particles are homogeneously dispersed and/or sufficiently exfoliated. However, nonlinear mechanical properties as tensile strength and elongation at break generally decrease beyond a critical volume fraction of reinforcing particles, which are generally reached at low proportions. These problems may be explained by the fact the Young's modulus is evaluated at low strains, whereas the ultimate properties are determined at high strains where the loading transfer between matrix and reinforcing particles is important. Thus, in order to improve the nonlinear characteristics of nanocomposites, it might be possible to optimize the processing parameters or to improve the interaction between the matrix and the fillers by selecting compatible constituents of the nanocomposite and/or by chemically treating the filler surfaces. Nevertheless, it must be noted that the mechanical behavior of polymer–nanofiller composites is controlled by several factors, and in particular by those that are related directly to the microstructure.

In this chapter, particular attention was paid to micromechanical and macromechanical models that are capable of predicting the macroscopic properties of nanocomposites. Notably, several interesting theories have been proposed that can predict material stiffness as a function of the filler volume fraction. Some of these theories take not only the effect of microstructure into account, but also the aspect ratio, dispersion degree of the reinforcing fillers, and interface properties. The yield stress of nanocomposites can be predicted by the model of Pukanzky *et al.* [67–69] in relation to the thickness of the interphase, to the interfacial adhesion, the filler density, the filler specific surface area, and the yield stress of the interphase. With regards to the evaluation of nanocomposite fracture toughness, it is possible to use well-known approaches based on the *J*-integral or the EWF concepts, both of which have been applied successfully in the case of traditional materials such as metals, neat polymers, and microcomposites.

References

1 Long, Y. and Shanks, R.A. (1996) *J. Appl. Polym. Sci.*, **61** (11), 1877–1885.
2 Bartczak, Z., Argon, A.S., Cohen, R.E., and Weinberg, M. (1999) *Polymer*, **40** (9), 2347–2365.
3 Misra, R.D.K., Nerikar, P., Bertrand, K., and Murphy, D. (2004) *Mater. Sci. Eng. A*, **384** (1-2), 284–298.
4 Okada, A., Kawasumi, M., Ususki, A., Kojima, Y., Kuraushi, T., and Kamigaito, O. (1990) Synthesis and properties of nylon-6/clay hybrids in polymer-based molecular composites. *MRS Symposium Proceedings*, vol. 171 (eds D.W. Schaefer and J.E. Mark), pp. 45–50.
5 Sur, G.S., Sun, H.L., Lyu, S.G., and Mark, J.E. (2001) *Polymer*, **42** (24), 9783–9789.
6 Tjong, S.C. (2006) *Mater. Sci. Eng.*, **53**, 73–197.
7 Giannelis, E.P. (1996) *Adv. Mater.*, **8**, 29–35.
8 Giannelis, E.P., Krishnamoorti, R., and Manias, E. (1999) *Adv. Polym. Sci.*, **138**, 107–147.
9 LeBaron, P.C., Wang, Z., and Pinnavia, T.J. (1999) *Appl. Clay Sci.*, **15**, 11–29.
10 Vaia, R.A., Price, G., Ruth, P.N., Nguyen, H.T., and Lichtenhan, J. (1999) *Appl. Clay Sci.*, **15**, 67–92.
11 Biswas, M. and Sinha Ray, S. (2001) *Adv. Polym. Sci.*, **155**, 167–221.
12 Alexandre, M. and Dubois, P. (2000) *Mater. Sci. Eng.*, **28** (1–2), 1–63.
13 Okada, A. and Usuki, A. (1995) *Mater. Sci. Eng. C*, **3** (2), 109–115.
14 Hasegawa, N., Kawasumi, M., Kato, M., Usuki, A., and Okada, A. (1998) *J. Appl. Polym. Sci.*, **67**, 87–92.
15 Reichert, P., Nitz, H., Klinke, S., Brandsch, R., Thomann, R., and Mülhaupt, R. (2000) *Macromol. Rapid Commun.*, **275**, 8–17.
16 Manias, E., Touny, A., Wu, L., Strawhecker, K., Lu, B., and Chung, T.C. (2001) *Chem. Mater.*, **13** (10), 3516–3526.
17 Liu, X. and Wu, Q. (2002) *Macromol. Mater. Eng.*, **287** (3), 180–186.
18 Hasegawa, N., Okamoto, H., Kawasumi, M., and Usuki, A. (1999) *J. Appl. Polym. Sci.*, **74** (14), 3359–3364.
19 Wang, H., Zeng, C., Elekovitch, M., James Lee, L., and Koelling, K.W. (2001) *Polym. Eng. Sci.*, **41** (11), 2036–2046.
20 Park, C.I., Park, O.O., Lim, J.G., and Kim, H.J. (2001) *Polymer*, **42** (17), 7465–7475.
21 Fornes, T.D. and Paul, D.R. (2004) *Macromolecules*, **37** (20), 7698–7709.
22 Cho, J.W. and Paul, D.R. (2001) *Polymer*, **42** (3), 1083–1094.

23 Vlasveld, D.P.N., Vaidya, S.G., Bersee, H.E.N., and Picken, S.J. (2005) *Polymer*, **46** (10), 3452–3461.
24 Reichert, P., Kressler, J., Thoann, R., Mülhaupt, R., and Stöppelmann, G. (1998) *Acta Polym.*, **49** (2-3), 116–123.
25 Shelley, J.S., Mather, P.T., and DeVries, K.L. (2001) *Polymer*, **42** (13), 5849–5858.
26 Sengupta, R., Bandyopadhyay, A., Sabharwal, S., Chaki, T.K., and Bhowmick, A.K. (2005) *Polymer*, **46** (10), 3343–3354.
27 Masenelli-Varlot, K., Reynaud, E., Vigier, G., and Varlet, J. (2002) *J. Polym. Sci., Part B: Polym. Phys.*, **40** (3), 272–283.
28 Vaia, R.A. and Giannelis, E.P. (1997) *Macromolecules*, **30** (25), 8000–8009.
29 Utracki, L.A. (2004) *Clay-Containing Polymeric Nanocomposites*, vol. **2**, Rapra Technology Ltd, Shropshire, UK.
30 Aït Hocine, N., Méderic, P., and Aubry, T. (2008) *Polym. Test.*, **27** (3), 330–339.
31 Tanoue, S., Utracki, L.A., Garcia-Rejon, A., Tatibouët, J., and Kamal, M.R. (2005) *Polym. Eng. Sci.*, **45** (6), 827–837.
32 Fornes, T.D., Yoon, P.J., Keskkula, H., and Paul, D.R. (2002) *Polymer*, **43** (7), 2121–2122.
33 McNally, T., Murphy, W.R., Lew, C.Y., Turner, R.J., and Brennan, G.P. (2003) *Polymer*, **44** (9), 2761–2772.
34 Uhl, F.M., Davuluri, S.P., Wong, S.C., and Webster, D.C. (2004) *Chem. Mater.*, **16** (6), 1135–1142.
35 Thio, Y.S., Argon, A.S., Cohen, R.E., and Weinberg, M. (2002) *Polymer*, **43** (13), 3661–3674.
36 Gopakumar, T.G., Lee, J.A., Kontopoulou, M., and Parent, J.S. (2002) *Polymer*, **43** (20), 5483–5491.
37 Alexandre, M., Dubois, P., Sun, T., Garces, J.M., and Jérôme, R. (2002) *Polymer*, **43** (8), 2123–2132.
38 Osman, M.A., Rupp, J.E.P., and Suter, U.W. (2005) *Polymer*, **46** (5), 1653–1660.
39 Sinha Ray, S., Maiti, P., Okamoto, M., Yamada, K., and Ueda, K. (2002) *Macromolecules*, **35** (8), 3104–3110.
40 Ahmed, S. and Jones, F.R. (1990) *J. Mater. Sci.*, **25**, 4933–4942.
41 Einstein, A. (1956) *Investigations on the Theory of the Brownian Movement*, Dover, New York.
42 Mooney, M. (1951) *J. Colloid Sci.*, **6**, 162–168.
43 Brodnyan, J.G. (1951) *Trans. Soc. Rheol.*, **3**, 61–68.
44 Broutman, L.J. and Krock, R.H. (1967) *Modern Composite Materials*, Addison Wesley, Reading, Massachusetts.
45 Hirsch, T.J. (1962) *J. Am. Concr. Inst.*, **59**, 427–451.
46 Paul, B. (1960) *Trans. Am. Inst. Min. Metall. Eng.*, **218** (1), 36–41.
47 Counto, U.J. (1964) *Mag. Concr. Res.*, **16** (48), 129–138.
48 Kaplan, M.F. (1959) *RILEM Bull. Paris*, **1**, 58–73.
49 Hansen, T.C. (1965) *J. Am. Concr. Inst.*, **62** (2), 193–215.
50 Guth, E. (1945) *J. Appl. Phys.*, **16**, 20–25.
51 Xie, X.L., Li, R.K.Y., Liu, Q.X., and Mai, Y.W. (2004) *Polymer*, **45** (8), 2793–2802.
52 Frogley, M.D., Ravich, D., and Wagner, H.D. (2003) *Compos. Sci. Technol.*, **63**, 1647–1654.
53 Halpin, J.C. and Tsai, S.W. (1967) Air Force Materials Research Laboratory Technical Report, AFML-TR-67-423.
54 Halpin, J.C. (1969) *J. Compos. Mater.*, **3**, 732–734.
55 Halpin, J.C. and Kardos, J.L. (1976) *Polym. Eng. Sci.*, **16**, 344–352.
56 Nielsen, L.E. (1981) *Mechanical Properties of Polymer and Composites*, vol. **2**, Marcel Dekker, New York.
57 Somoza, A.M. and Tarazona, P. (1989) *J. Chem. Phys.*, **91**, 517–527.
58 Brune, D.A. and Bicerano, J. (2002) *Polymer*, **43** (2), 369–387.
59 Takayanagi, M., Uemura, S., and Minami, S. (1964) *J. Polym. Sci., Part C: Polym. Symp.*, **5**, 113–122.
60 Ji, X.L., Jing, J.K., Jiang, W., and Jiang, B.Z. (2002) *Polym. Eng. Sci.*, **42** (5), 983–993.
61 Ou, Y., Yang, F., and Yu, Z.Z. (1998) *J. Polym. Sci., Part B: Polym. Phys.*, **36** (5), 789–795.
62 Tjong, S.C. and Meng, Y.Z. (2003) *J. Polym. Sci., Part B: Polym. Phys.*, **41**, 1476–1484.
63 Lee, J.H., Jung, D., Hong, C.E., Rhee, K.Y., and Advani, S.G. (2005) *Compos. Sci. Technol.*, **65** (13), 1996–2002.

64 Tjong, S.C., Meng, Y.Z., and Xu, Y. (2002) *J. Polym. Sci., Part B: Polym. Phys.*, **40** (24), 2860–2870.
65 Liu, X. and Wu, Q. (2001) *Polymer*, **42** (25), 10013–10019.
66 Nicolais, L. and Narkis, M. (1971) *Polym. Eng. Sci.*, **11** (3), 194–199.
67 Pukanszky, B., Turcsanyi, B., and Tudos, F. (1988) *Interfaces in Polymer, Ceramics and Metal Matrix Composites* (ed. H. Ishida), Elsevier, New York.
68 Pukanszky, B. and Voros, G. (1993) *Compos. Interfaces*, **1** (5), 411–427.
69 Pukanszky, B. and Fekete, E. (1999) *Adv. Polym. Sci.*, **139**, 106–153.
70 Rong, M.Z., Zhang, M.Q., Pan, S.L., Lehmann, B., and Friedrich, K. (2004) *Polym. Int.*, **53**, 176–183.
71 Sumita, M., Tsukihi, H., Miyasaka, K., and Ishikawa, K. (1984) *J. Appl. Polym. Sci.*, **29**, 1523–1530.
72 Borse, N.K. and Kamal, M.R. (2006) *Polym. Eng. Sci.*, **49** (4), 1094–1103.
73 Sumpter, J.D.G. (1973) Elastic plastic fracture analysis and design using the finite element method, Ph.D. Thesis, University of London, 1973.
74 Rice, J.R., Paris, P.C., and Merkle, J.G. (1973) Progress in Flaw Growth and Fracture Toughness Testing, Special Technical Publication 536, American Society for Testing and Materials, Philadelphia, pp. 231–245.
75 Joyce, J.A., Ernest, H., and Paris, P.C. (1979) Special Technical Publication 700, American Society for Testing and Materials, Philadelphia, p. 222.
76 Turner, C.E. (1979) *Post Yield Fracture Mechanics* (ed. D.H.G. Latzko), Applied Science Publishers.
77 Sumpter, J.D.G. and Turner, C.E. (1976) Special Technical Publication 601, American Society for Testing and Materials, Philadelphia, pp. 3–18.
78 Turner, C.E. (1973) *Mater. Sci. Eng.*, **11** (5), 275–282.
79 Turner, C.E. (1980) Fracture Mechanics: Proceedings, Twelfth Conference, Special Technical Publication 700, American Society for Testing and Materials, Philadelphia, pp. 314–337.
80 Paris, P.C., Ernst, H., and Turner, C.E. (1980) Fracture Mechanics, Proceedings, Twelfth Conference, Special Technical Publication 700, American Society for Testing and Materials, Philadelphia, pp. 338–351.
81 Cherepanov, G.P. (1967) *J. Appl. Math. Mech.*, **31** (3), 503–512.
82 Cherepanov, G.P. (1968) *Int. J. Solids Struct.*, **4**, 811–831.
83 Rice, J.R. (1968) *J. Appl. Mech.*, **35**, 379–385.
84 Begley, J.A. and Landes, J.D. (1972) Fracture Toughness Testing, Special Technical Publication 514, American Society for Testing and Materials, Philadelphia.
85 Chan, W.Y.F. and Williams, J.G. (1994) *Polymer*, **35** (8), 1666–1672.
86 Paton, C.A. and Hashemi, S. (1992) *J. Mater. Sci.*, **27** (9), 2279–2290.
87 Hashemi, S. (1993) *J. Mater. Sci.*, **28** (22), 6178–6184.
88 Tjong, S.C., Xu, S.A., and Li, R.K.Y. (2000) *J. Appl. Polym. Sci.*, **77** (9), 2074–2081.
89 Wong, S.C. and Mai, Y.W. (1999) *Polym. Eng. Sci.*, **39** (2), 356–364.
90 Ching, E.C.Y., Li, R.K.Y., and Mai, Y.W. (2000) *Polym. Eng. Sci.*, **40** (2), 310–319.
91 Mouzakis, D.E., Stricker, F., Mulhaupt, R., and Karger-Kocsis, J. (1998) *J. Mater. Sci.*, **33** (10), 2551–2562.
92 Kudva, R.A., Keskkula, H., and Paul, D.R. (2000) *Polymer*, **41** (1), 335–349.
93 Chiou, K.C., Chang, F.C., and Mai, Y.W. (2001) *Polym. Eng. Sci.*, **41** (6), 1007–1018.
94 Wu, J.S., Mai, Y.W., and Cottrell, B. (1993) *J. Mater. Sci.*, **28** (12), 3373–3384.
95 Laura, D.M., Keskkula, H., Barlow, J.W., and Paul, D.R. (2001) *Polymer*, **42** (14), 6161–6172.
96 Tjong, S.C., Xu, S.A., and Mai, Y.W. (2003) *Mater. Sci. Eng. A*, **347** (1-2), 338–345.
97 Tjong, S.C., Xu, S.A., and Mai, Y.W. (2002) *J. Polym. Sci., Part B: Polym. Phys.*, **40** (17), 1881–1892.
98 Martinatti, F. and Ricco, T. (1995) *Proceedings of Impact and Dynamic Fracture of Polymers and Composites*, vol. **19** (eds J.G. Williams and A. Pavan), European Structural Integrity Society (ESIS).

7
Stress Transfer and Fracture Mechanisms in Carbon Nanotube-Reinforced Polymer Nanocomposites

Bhabani K. Satapathy, Martin Ganß, Petra Pötschke, and Roland Weidisch

7.1
Introduction

Over the past decade, a wide spectrum of possible functional applications of carbon nanotubes (CNTs), ranging from energy to electronics, from reinforcement to robotics, and from clinical to communications, has been proposed since the initial discovery of these materials [1–7]. This, in turn, has triggered a new dimension in novel materials research, notably in the field of nanocomposites, and the micromechanics of polymer nanocomposites has attracted considerable attention, as indicated by the huge amount of research data produced in recent years. Intensive reviews have also been prepared relating to polymer–nanotube/nanoparticle interaction, in an attempt to define the underlying mechanistic contribution to the mechanical performance of these nanocomposite materials [3–10].

Despite increasing developments in the area of CNT-filled polymer nanocomposites, the mechanism and magnitude of load transfer between polymer matrices and nanotubes remains unclear, and research in this area is still in its infancy. Evidence of stress transfer in polymer–nanotube composites is largely based on the enhancement of moduli, on fragmentation studies, and on Raman spectroscopy [2, 11–14]. The slippage of the shells in multiwalled nanotubes (MWNTs) and within single-walled nanotube (SWNT) ropes have been reported to limit stress transfer efficiency [15]. Molecular mechanics studies have revealed that the binding energies and friction forces play a minimal role in controlling the interface. Rather, helical polymer conformations facilitate polymer–nanotube interaction by the polymer wrapping around the nanotube–an effect confirmed experimentally by direct observations (using transmission electron microscopy; TEM) of polymer chains forming a sheath around a nanotube in the case of polycarbonate (PC)–nanotube systems [16–18]. The critical nanotube length for effective load transfer has also been estimated in the case of reinforcement, though this can be either enhanced or reduced via the formation of chemical crosslinks that involve less than 1% of the nanotube carbon atoms [19, 20]. It has also been reported that, when under stress, the axial stress is maximal at the mid-section of a CNT, whereas the interfacial shear stress is greatest at the tube ends. The

Optimization of Polymer Nanocomposite Properties. Edited by Vikas Mittal
Copyright © 2010 WILEY-VCH Verlag GmbH & Co. KGaA, Weinheim
ISBN: 978-3-527-32521-4

optimum aspect ratio for effective reinforcement in terms of load transfer has been found to be 1000 [21].

It has been suggested that the dynamics and stability problems of nanocomposites, and in particular fracture-related aspects, can be solved within the framework of solid mechanics [22, 23]. These proposals were adequately substantiated by the studies of Wagner *et al.* [24], in which micromechanics concepts were applied to CNT-based polymer nanocomposites to evaluate the load transfer efficiency on a semi-empirical level. A micromechanical analysis of the effective elastic properties, based on the Mori–Tanaka approach, was also reported and later compared with results from finite element modeling. Based on these findings, it was concluded that all axial and transverse elastic properties were dominated by the nanotubes and polymer matrix, respectively, whereas the interface effects greatly impacted on the transverse properties at low volume fractions [6]. Constitutive modeling has revealed that the prediction of bulk mechanical properties as a function of nanotube size, orientation and volume fraction, could be utilized, since the local molecular structure as defined by molecular dynamics approach is retained at the polymer–nanotube interface. The model also predicted that, for a nanotube length of 200 nm, the load transfer would be maximized and that the elastic properties would follow the rule of mixture [25]. Due to the nanoscale dimensions of CNTs, the interfacial regions surrounding the tubes also have nanometer dimensions. Consequently, studies of nanotube buckling have raised some interesting mechanics issues which are not found in continuum mechanics problems. *Buckling* is influenced by the forces exerted in the interfacial region, as well as by the matrix [26–28]. Multiscale modeling of CNT–polymer composites have also been developed where the stress distribution at the interface under iso-strain and iso-stress loading conditions has been examined, while simultaneously characterizing the buckling forces as a function of various aspect ratios of the nanotubes. Here, it was observed that continuous nanotubes most effectively enhance the buckling resistance of the composites [29].

For CNT-based polymer composites, several conflicting reports have been made concerning the strength of the interfaces, with some authors attributing the load transfer to be mainly through the interface while others proposed a specific structural organization which facilitated the load transfer. On the other hand, some have suggested that there is low interfacial load transfer. In addition, the stiffness of CNT-based composites has been reported to be much lower than the possible theoretical values. For example Andrews and coworkers reported normalized Young's modulus (stiffness of composite/stiffness of matrix) values in the range of 1.2–2.5 for a polystyrene (PS) matrix reinforced with 2.5 to 25 vol% of CNT. The comparable theoretical maxima and minima based on parallel (iso-strain) and series (iso-stress) models were in the range of 14 to 140 and 1.03 to 1.33 respectively, an observation which could possibly have its roots in the inadequate interfacial strength [30].

In theory, the three synergizing aspects of interaction in CNT-based nanocomposites – that is, polymer–nanotube, nanotube–nanotube, and intra-polymer interactions – have been already quantified on the basis of van der Waals force based on a truncated Lennard–Jones (6,12) type interaction (with $\varepsilon = 0.461$ KJ mol^{-1} and $\sigma = 3.65$ A°) [31], Terstoff–Brenner many body potential [32], and intra-polymer potential based on the reports of Brown *et al.*, respectively [33].

Pioneering studies in the area of the mechanics of CNT–polymer nanocomposites have been carried out by Wagner and colleagues [24, 34–37]. The detachment of nanotubes from the polymer matrix and the stress-induced fragmentation of nanotube aspects have been critically commented upon through experimental and theoretical demonstrations. It was also postulated, as a consequence of such studies, that fracture would be a complex process involving rupture of the polymer–nanotube interface, coupled with bending of the nanotube rope. Thus, such additional modes of energy dissipation could lead to high fracture strengths of polymer matrix-based nanocomposites. Later, Wagner and coworkers also measured (quantitatively) the interfacial strength and interfacial fracture energies in these composites, in an attempt to further understand the stress transfer mechanism across the polymer–nanotube interface. Both, Cox and Kelly–Tyson models were adopted to provide a qualitative appreciation of the polymer–nanotube adhesion while discussing the mechanics aspects of such nanocomposites [17, 21, 25, 29, 38]. The formation of fracture clusters, as evidence of stress transfer in consequence to fracture nucleation under the biaxial stress-induced deformation of polymer nanocomposites, has also been reported [39].

In this connection, the modeling of fracture in nanomaterials via a virtual internal bond method was already reported by Gao and Ji [40]. In materials with a characteristic size on the nanometer scale, the fracture process is often strongly dominated by the surface energy and nonlinear material properties. It was demonstrated, through the application of a virtual internal bond method, that at a critical length scale (typically on the order of nanometer scale) the fracture mechanism changes from the classical Griffith fracture to one of homogeneous failure near the theoretical strength of solids. When this transition occurs, the classical singular deformation field near a crack tip disappears, and is replaced by a uniform stress distribution with no stress concentration near the crack tip. This leads to a fracture localization which in turn could be related to intrinsic material length determined by the surface energy and the cohesive state strain energy density at the nanoscale.

The stress transfer and fracture mechanical aspects concerning nanomaterials, and especially CNT-filled polymer matrix systems, may thus fundamentally prove crucial in order to understand the overall material performance in practical situations. The mechanistic attributes that have been postulated theoretically are far from ideal, especially when nanocomposites based on polymer matrices are involved, as the molecular organization and topological dynamics of the polymer chains may be dramatically altered in the presence of CNTs. Therefore, the matrix-sensitive issues concerning molecular level interactions and their obvious consequences on macro-scale properties, such as mechanical and fracture mechanical properties, are key to the conceptualization of new-age, high-performance materials. Nanotube-induced alterations in the stress transfer mechanisms, and crack toughness aspects due to their incorporation into an amorphous thermoplastic matrix such as PC and a semi-crystalline thermoplastic matrix such as isotactic-polypropylene (i-PP), are discussed in the following sections.

7.2
Experimental Studies

7.2.1
Fabrication of Composites

The two investigated nanocomposites that are based on an amorphous polymer (PC) and semi-crystalline polymer (i-PP) were prepared via melt-mixing. The PC-MWNT nanocomposites were prepared by melt extrusion, starting from a commercially available masterbatch of PC with 15 wt% MWNT (Hyperion Catalysis Intern. Inc., Cambridge, USA). The MWNT are produced using chemical vapor deposition (CVD), and have diameter in the range of 10 to 15 nm, consisting of 8 to 15 graphite layers. The masterbatch dilution was performed using a Buss kneader at temperatures in the range of 260–280 °C, and conditions adapted to PC. The composite granules, as well as the base PC, were dried at 120 °C for 4 h in a vacuum oven and subsequently compression-molded at 260 °C into plates with a diameter of approximately 70 mm and a thickness of about 0.5 mm.

The PP-MWNT nanocomposites were melt-mixed starting from premixtures of PP powder (Moplen HF 500N, Basell Polyolefins) with powdery MWNT NC 7000 (Nanocyl S.A. Sambreville, Belgium) in 750 g batches. According to the supplier, the nanotubes have diameters in the range of 10 nm, lengths between 0.1 and 10 µm, and a carbon purity of 90%. Before processing, the materials were properly dried (PP 80 °C, 2 h, MWNT 100 °C, 2 h). Mixing was carried out using a Berstorff Twin Screw extruder ZE 25 at a screw speed of 300 rpm, and a throughput of 10 kg h^{-1}. The temperature profile ranged from 200 °C to 220 °C close to the nozzle. Pure PP was processed in the same way. The extruded granules were injection-molded to plates with the dimensions $80 \times 80 \times 1$ mm^3 using an Ergotech 100/420–310 injection-molding machine (Demag). The mold temperature was 40 °C, the injection speed was set to 70 mm min^{-1}, and the melt temperature ranged between 210 °C and 240 °C.

7.2.2
Morphology Characterization

7.2.2.1 Transmission Electron Microscopy (TEM)
For the TEM investigations, cryocuts with a thickness of approximately 90 nm were cut from a middle position of the injection-molded plates in flow direction, using a Leica Ultracut EM UC6/EM FC6 ultramicrotome (Leica Mikrosysteme GmbH, Vienna, Austria). The ultramicrotome was equipped with a diamond knife with a cut angle of 35° (Diatome, Biel, Switzerland). The TEM used was an EM 912 (LEO,

Oberkochen, Germany) operated at 120 kV, and the micrographs were taken using an energy filter in zero loss mode for an optimal contrast of the nanotubes.

7.2.2.2 Atomic Force Microscopy (AFM)

To study the bulk morphology (state of dispersion and distribution of the nanotubes), AFM specimens were prepared by cutting a strand piece along and perpendicular to the injection-molding direction in the case of PP-MWNT nanocomposites. For PC-MWNT nanocomposites, the cuts were made perpendicular to the plate axis. The freshly cut surfaces were made with a Leica RM 2155 microtome (Leica, Nussloch, Germany) equipped with a diamond knife at cryotemperature below the glass transition temperature (T_g) of i-PP and PC. The AFM measurements were performed in tapping mode using a Dimension 3100 NanoScope IIIa (Veeco, USA). For this, Pointprobe silicon-SPM-sensors (Nanosensors, Germany) were used with a spring constant of approximately $3 \, \text{N} \, \text{m}^{-1}$ and a resonance frequency of about 75 kHz, with a tip radius less than 10 nm. The scan conditions were maintained at a free amplitude of >100 nm and a set-point amplitude ratio of 0.5 in order to obtain stiffness contrast in the phase image. This means that the brighter areas in the phase image were stiffer than the darker areas, with the exception of the nanotubes which were oriented perpendicular to the surface [41].

7.2.2.3 2-D Wide-Angle X-Ray Diffraction

Wide-angle X-ray diffraction (WAXD) was carried out on the extruded granules (quasi-isotropic sample) and on pieces of the injection-molded plates to characterize crystallinity and orientation in the samples, apart from investigating the changes in the peaks corresponding to different crystal planes as a function of MWNT content. The measurements were carried out using an X-ray diffractometer P4 (Siemens AG, now: Bruker AXS Karlsruhe; 40 kV, 30 mA, Cu-Kα-Strahlung; monochromatic by primary graphite) in 360 s, with a sample-to-detector distance of 120 mm and a radial scattering range of 2–40°. The crystallinity was evaluated by applying the peak-area method while integrating in the range of $2\theta = 5\text{–}35°$ (as typical for i-PP), and applying an amorphous scattering curve which was realized by experimental and theoretical experiences.

7.2.3
Thermal Characterization

7.2.3.1 Differential Scanning Calorimetry

Differential scanning calorimetry (DSC) measurements, aimed at obtaining further information regarding the influence of MWNTs on the crystallization behavior of the PP-matrix, were conducted using a DSC Q 1000 (TA Instruments, USA) at a scan rate of $\pm 10 \, \text{K} \, \text{min}^{-1}$ over a temperature range of −60 to 210 °C.

7.2.3.2 Dynamical Mechanical Analysis

Dynamic mechanical analysis (DMA) measurements have been carried out on the test specimens with dimensions of $30 \times 8 \times$ thickness mm^3 in tensile mode on an

Q800 (TA Instruments, USA) to characterize the T_g of the nanocomposites over the temperature range of −40 to 100 °C at a frequency of 1.0 rad s^{-1} and heating rate of 3 K min^{-1}.

7.2.3.3 Melt Rheological Investigations

Melt rheological properties were obtained using an ARES oscillatory rheometer (Rheometric Scientific) at 190 °C, 210 °C, and 230 °C for PP-MWNT, and at 260 °C for PC-MWNT in a nitrogen atmosphere, using a parallel-plate geometry (plates 25 mm diameter, gap of 1–2 mm). The amplitude was selected to be within the linear viscoelastic range, which was checked by strain sweep tests.

7.2.4 Mechanical and Fracture Mechanical Investigations

7.2.4.1 Tensile Testing

Tensile tests were performed according to ISO 527-2/5A7/5 with a Tira-Test 2710 (Tira, Germany) testing machine at room temperature, with a test speed of 5 mm min^{-1}. Five tensile bars were cut in injection-molded and perpendicular to injection-molded directions, with a length of 80 mm, a thickness of 1 mm, and a gage length of 4 mm for each composition.

7.2.4.2 Essential-Work-of-Fracture Approach

The essential-work-of-fracture (EWF) method has been used to allow a distinction to be made between two terms representing the resistance to *crack initiation*, and the *crack propagation* corresponding to the inner fracture process zone and the outer plastic deformation zone. In this study, the fracture mechanical tests (test speed: 1 mm min^{-1}, room temperature) were performed on double-edge-notched-tension (DENT) specimens by a universal testing machine with mechanical grips (Z010 Zwick), and interfaced with a software-assisted video monitoring system. DENT specimens with total length of 80 mm, a thickness of 1 mm (for PP) or 0.5 mm (for PC), and a width of 20 mm, respectively, were cut from the plates. The clamp distance was 40 mm. For notching, a special device with fresh razor blades with notch tip radius of 0.2 μm, such that both notches were similarly sized. For each material, at least 10 specimens were tested with different ligament lengths ranging from about 2 mm to about 10 mm.

For plane–stress conditions, the total work of fracture W dissipated in a notched specimen can be divided into a component W_e, which characterizing the inner or fracture process zone, and W_p, which corresponds to an outer zone:

$$W = W_e + W_p = EWF \cdot B \cdot l + \beta w_p \cdot B \cdot l^2 \tag{7.1}$$

After dividing W by the ligament (notched) area $B{*}l$ (where B is the specimen thickness, l the ligament length, and β the shape factor of the plastic zone), the specific work of fracture w is obtained:

$$w = EWF + \beta w_p \cdot l \tag{7.2}$$

Based on the fact that the intrinsic fracture process takes place in the inner zone, the term EWF (or w_e), the essential work of fracture, is determined experimentally by an extrapolation of w as a function of l to zero ligament length. For this, several similar-sized specimens with different ligament lengths are monotonically loaded. Here, βw_p (nonessential work of fracture) is the slope of the linear fit extrapolation.

7.2.4.3 Kinetics of Crack Propagation Measurement Using a Single-Specimen Technique

The crack propagation behavior of the nanocomposites was studied using a single-specimen technique to determine crack resistance curves based on the elastic-plastic fracture mechanics, which is mainly applied to highly ductile materials. A video monitoring system has been used to carry out a time-dependent *in situ* image acquisition of the deformed specimens, so as to measure the crack-tip opening displacement (CTOD; δ) and the crack extension (Δa).

Parallel to the testing, images for the strain field evaluation have been acquired using an online video monitoring system at regular intervals, using a CCD camera with a resolution of 1280 × 1024 pixels (1280 pixels in tensile direction). In order to identify the strain field during the test, specimens with a ligament length of ~6 mm were spray-coated with a water-dispersed white/black dye by air brush to generate an optically active random structure. Evaluation of the acquired images could be carried out using the computation software Aramis (Gom, Germany), which essentially calculates the strain statistics and displays the strain in the specimen in terms of strain distribution images. Further details of this procedure are described elsewhere [42].

7.3 Mechanical Behavior of Polymer Nanocomposites and Stress Transfer

7.3.1 Amorphous Thermoplastic

The mechanical properties of thermosets, such as epoxy resins [34, 35, 39, 43, 44], and thermoplastic polymers, such as polystyrol (PS) [45–47], poly(methyl methacrylate) (PMMA) [48, 49] and other amorphous polymers [50–53], have been enhanced significantly with the addition of CNTs. In fact, CNT composites have also shown dramatic improvements in electrical conductivity, with a percolation threshold below 1% of the CNT concentration [53–58]. The addition of 1 wt% multi-wall nanotubes (MWNTs) to PS has been demonstrated to increase the elastic modulus by 42%, and the stress at break by 25%, in comparison to pure PS [59]. Safadi *et al.* also studied PS/MWNT composites, which showed a 40% increase of modulus with 1 wt% concentration, and a percolation threshold for conductivity at concentrations below 1 wt% [46]. In the absence of chemical bonding, the load transfer between the nanotubes and the matrix mainly arises from electrostatic and van der

Waals forces [60]. Thus, nanotubes are easily and preferentially pulled out from the matrix rather than being fractured, and so play a limited reinforcement role [61, 62]. In order to improve the solubility of the nanotubes and the nanotube–polymer load transfer, covalently functionalized SWNTs and MWNTs have been successfully used to reinforce a variety of polymers. For example, Wagner and coworkers very recently showed that a significant enhancement of the mechanical properties of poly(vinyl alcohol) (PVA) composites could also be achieved via hydrogen bonding between covalently functionalized SWNTs and PVA [53].

It was also reported that water-soluble PVA composite films, prepared using an aqueous solution of SWNTs and coated with poly(vinyl pyrrolidone) (PVP) and sodium dodecyl sulfate (SDS), exhibited significant improvements in both tensile strength and modulus [63]. Compared to the pure PVA, the tensile yield strength of a PVA/PVP/SDS/SWNT film containing 5 wt% SWNT was increased by 78%, while the Young's modulus was increased by 110% [63]. Wang and coworkers reported that a 1 wt% loading of poly(*n*-butyl acrylate)-encapsulated MWNTs led to a 31% increase in yield strength, a 35% increase in Young's modulus, but only a 6% decrease in tensile strength for nylon 6–MWNT composites [64]. Sabba and Thomas showed that a 1–3% loading of well-dispersed [(NH_2OH)(HCl)]-SWNTs led to a 6–12% decrease in tensile strength, but caused almost no change in the yield strength and Young's modulus [65]. Chen and coworkers achieved a 72% increase in tensile strength and a 28% increase in Young's modulus for the [PPE-SWNTs (2 wt%)]-Parmax composite, when using a molecular engineering method [66]. Along with the above findings, other research groups have prepared PVA nanocomposites with noncovalently functionalized MWNTs by using a simple method that achieves a dramatic enhancement in the mechanical properties. This strategy was based on the use of sodium deoxycholate (DOC; a type of bile salt) -modified MWNTs (termed DOC-MWNTs) in aqueous solution [67]. The DOC-MWNTs and PVA solution (in water), in being homogeneous, favored nanotube dispersion. However, by adding 5 wt% DOC-MWNTs, a significant increase in the tensile modulus could be achieved – a result which inherently reflected the effective nanotube dispersion and strong interfacial adhesion.

The physical interactions at the CNT–polymer interface have been elaborately studied by Wong and coworkers, where the role of the interface in controlling the mechanical properties of CNT-PS and CNT-epoxy-based systems was critically investigated. In fact, by molecular mechanics calculations, the polymer was found to adhere well to the CNT. In the absence of chemical bonding between the polymer and CNT, however, the nonbonded interactions have been reported to consist of electrostatic and van der Waals forces, which together contribute to the interfacial shear stress with magnitudes of 136 and 186 MPa for the CNT-epoxy and CNT-PS systems, respectively, this being about an order higher than the corresponding microcomposite counterparts. The local nonuniformity of CNT, and the mismatch of the coefficients of thermal expansions between CNT and polymer matrix, are believed to promote the stress-transfer ability across the interface [60]. Kashiwagi *et al.* studied the relationship between the dispersion metric which ascertains the volume of the interface, and the properties of PMMA/SWNT nanocomposites. Here, it was shown that the relative dispersion index quantified the dispersion levels, and that the physical properties followed four orders of magnitude variation in storage modulus, and almost eight orders of magnitude variation in electrical conductivity, with a 70% reduction in peak mass loss rate at the highest dispersion level. The results of these studies provided a comprehensive explanation of the correlation between the nanocomposite's properties as a function of dispersion, which in turn were dependent upon the concentration, shape, and size of the nanotubes [68]. Deformation–morphology correlations in CNT-thermoplastic polyurethane (PU) nanocomposites were studied by Koerner et al. [69]. It was found that the addition of small amount of MWNT (0.5–10 vol%) to PU causes an enhancement in mechanical properties with a very low electrical percolation threshold (0.005) that was attributed not only to the nanotubes but also to its impact on soft segment crystallization. Koerner and colleagues also showed that the large-scale deformation behavior of such nanocomposites was mainly governed by the matrix. In addition, Pötschke and coworkers showed that the rheological percolation corresponds with the electrical percolation, which fundamentally implied a possible interrelation of electrical percolation with the critical level of CNT loading required to enhance the mechanical properties in the case of a PC-MWNT system [70].

Studies with PS-SWNT-based nanocomposites conducted by Chang and coworkers have revealed, via Raman spectroscopy, that the stress transfer efficiency across the polymer–nanotube interface showed promise despite the enhancement in the moduli being nominal; however, the reason for this discrepancy was unclear [71]. Deng et al. [72] have described the stress–strain curves as a function of the MWNTs' loading at room temperature, at 100 °C, and at 200 °C. Here, the addition of MWNTs was reported to lead to an increase in the elastic modulus and yield strength at all temperatures, and to a decrease in failure strain at room temperature compared to pure polyetheretherketone (PEEK). The tensile modulus of the pure PEEK, at room temperature (i.e., 4 GPa), was increased by 89% with a MWNT content of 15 wt%, whereas at 100 °C and 200 °C such enhancement was 70% and 163%, respectively; the corresponding tensile strengths were increased by 19%, 13%, and 42%, respectively. Thus, these results imply that the enhanced stress transfer efficiency, which caused improvements in the mechanical properties due to MWNTs, was more effective above the T_g. However, these results typically remain much lower than the theoretical predictions from the 'rule of mixtures' (which takes into consideration the length and orientation factors) [50]. A simple model, which assumes perfect bonding at interfaces, would predict the modulus to be approximately 24 GPa for the composite loading 15 wt% MWNTs (assuming that the modulus of MWNTs is 450 GPa) at room temperature, which is almost threefold higher than the values found experimentally. In the case of the PC-MWNT system, a dynamic mechanical analysis was carried out in order to detect any changes in polymer mobility, and to evaluate the stiffness of the nanocomposites. Conceptually, relative differences in the molecular mobility of the polymer segments in the vicinity of the nanotubes, compared to the bulk polymer, can cause a significant difference between the viscoelastic responses of the pure

PC and the MWNT-reinforced polymer composite. Variation in the storage modulus (G') and the loss factor ($\tan\delta$) as a function of temperature are shown in Figure 7.1a. With the increase in MWNT content, an increase in the T_g of the nanocomposites (by 5 K in the case of nanocomposites with 2 and 4 wt% MWNT), in addition to a minor broadening of the $\tan\delta$ peaks, has been observed. The storage modulus was increased with the increase in MWNT content such that, at 6 wt%, the increase in modulus was accompanied by a decrease in T_g (ca. 3 K lower than PC).

The increase in T_g and broadening of the $\tan\delta$ peaks with the incorporation of MWNT may be attributed qualitatively to a confinement of the polymer chain mobility as a result of strong nanotube–polymer interactions. However, the decrease in T_g at 6 wt% MWNT content may be attributed to a molecular weight degradation of PC within the masterbatch, caused by high shear forces during the masterbatch formation process, as reported elsewhere [30]. Such an effect can lead to changes in the T_g and modulus that will counteract any effects originating from the nanotube–polymer interaction. In order to describe the mechanical behavior of the investigated PC-MWNT system, a two-phase model based on the Halpin–

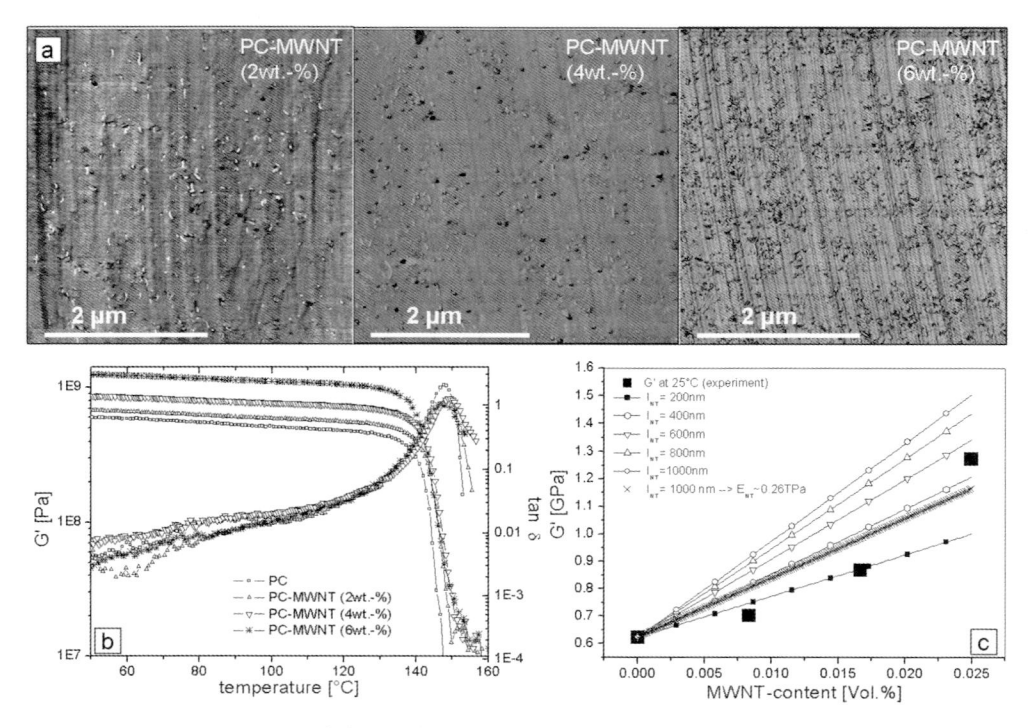

Figure 7.1 Morphology and mechanical behavior of PC-MWNT nanocomposites. (a) Atomic force microscopy phase images of the nanocomposites; (b) Dynamic mechanical analysis; (c) Modeling of the storage modulus (G') as function of the MWNT-content (for details, see the text).

Tsai approach was used, although this was later modified by Thostenson [3] for CNT-based nanocomposites. The equation can be stated as:

$$E_{11} = E_m \left(1 + 2\left(\frac{l}{d}\right)\left(\frac{(E_{NT}/E_m - (d_{NT}/4t))}{E_{NT}/E_m + (l_{NT}/2t)}\right)V_{NT}\right) \times \left(1 - \left(\frac{(E_{NT}/E_m - (d_{NT}/4t))}{E_{NT}/E_m + (l_{NT}/2t)}\right)V_{NT}\right)^{-1}$$
(7.3)

where, E_{11} is composite elastic modulus in the principal material direction, V_{NT} is the nanotube volume fraction, E_{NT} and E_m are the nanotube modulus and matrix modulus, respectively, l_{NT} is the length of the nanotubes, d_{NT} is thickness of the nanotubes, and t is the layer thickness of the nanotube (0.34 nm) [3]. The influence of the variation in nanotube length on the modulus is shown in Figure 7.1c, assuming a constant nanotube diameter of 12.5 nm (according to the supplier, 10–15 nm) and a CNT-Modulus of 0.5 TPa (the reported moduli range from 0.2 to 1 TPa for MWNT [73]). Based on these approximations, the nanotube length of the investigated system lies between 200 and 600 nm. In contrast, if a nanotube length of 1 μm was assumed, a CNT-Modulus of 0.26 TPa was calculated, which was in the predicted range. However, the strong increase in the modulus at 6 wt% of 104%, compared to increases of 13% and 39% at 2 and 4 wt% CNT, respectively, led to the assumption that the mechanical network and the interaction between the MWNTs was very efficient at high CNT-loadings. These results indicate that the mechanical percolation – that is, the mechanical contact of individual nanotubes (nanotube bundles) – together with good dispersion and distribution (as shown in Figure 7.1a) is necessary to achieve a large increase in stiffness. The electrical percolation (rheological percolation) threshold was achieved at <2 wt% MWNT, as indicated by a large increase in volume conductivity. The electrical contact of the nanotubes – that is, the direct- contact or electric-hopping effects which require a tube–tube distance of less than 5 nm [74] at 2 wt% MWNT – did not lead to any major increase in stiffness in the investigated PC-MWNT nanocomposite system. *In situ* tensile testing, via direct scanning electron microscopy (SEM) and TEM investigations, has been reported to be valuable in understanding the details of the reinforcement mechanism of MWNTs.

7.3.2
Semi-Crystalline Thermoplastic

Recently, PP-based systems have been investigated as potential flame-retardant and conducting nanocomposites; this was in addition to fundamental studies of the influence of nanotubes on crystallization behavior, and hence their crystal structure and polymorphic nature [75–79]. Nonisothermal crystallization studies conducted with this system have shown that, at low concentrations, the MWNTs act as α-nucleating agents. This finding was in agreement with previous reports, which suggested a lowering of the dimensionality of the crystal structure in PP-MWNT systems when compared to i-PP [76, 78]. The same group reported the complete absence of β-crystals and an improvement in impact strength at ~2 wt% MWNT. The increased impact strength was attributed to a lower fracture strain of the MWNTs (~10–30%), which allowed extensive bending and buckling and hence imparted flexibility to the nanotubes [77]. Dondero et al. [80] have reported that nanotubes are not efficient in resisting plastic deformation, whilst Moore et al. [81] have shown that the incorporation of MWNTs causes a reduction in toughness. A similar study with PP-SWNT nanocomposites showed no significant improvement in the mechanical properties [82].

Melt rheological investigations in the case of PP-MWNT systems have demonstrated a reinforcement effect in the molten state, as confirmed by frequency sweeps above the melting temperature of the crystals (230 °C), especially at low frequencies. An increase in the MWNT content led to an increase in both the storage modulus (see the insert in Figure 7.2a) and the complex viscosity (η^*), whereas the qualitative behavior of η^* was changed from a Newtonian plateau at low frequencies up to 1.0 wt% MWNT towards the development of a yield stress (non-Newtonian) phenomenon which started at 1.5 wt%. This change became clearer when the dependence of tan δ on frequency was taken into account, whereby a monotonic decrease was observed up to 1.0 wt% MWNT, followed by an increase in tan δ (maximum) at >1.5 wt% MWNT (see Figure 7.2a). Taken together, these results indicated the formation of an effective network structure of CNT and polymer chains between 1.0 and 1.5 wt% MWNT which were found in amorphous polymer systems within this range [70].

Dynamic mechanical studies in the solid state revealed an increase in the storage modulus (E') with the incorporation of nanotubes (see Figure 7.2b), an observation which was in accord with other results. Whilst the extent of storage modulus increase was relatively high with increasing MWNT content at lower temperatures (<0 °C), it was less pronounced at higher temperatures. For example, at room temperature the increase in E' of the nanocomposite with 5 wt% MWNT was ~23% compared to pure i-PP. Variations in the loss modulus (E'') with temperature

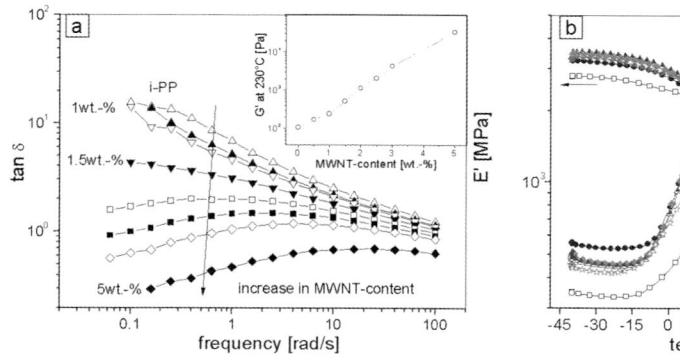

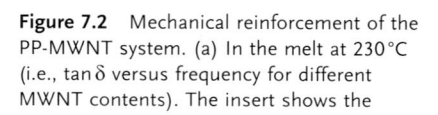

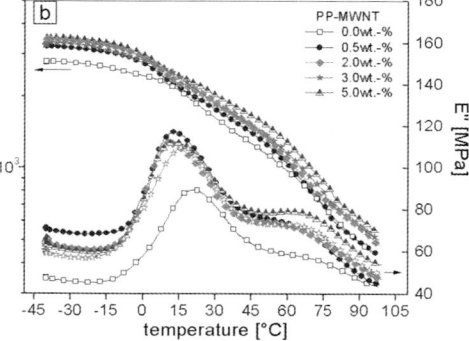

Figure 7.2 Mechanical reinforcement of the PP-MWNT system. (a) In the melt at 230 °C (i.e., tan δ versus frequency for different MWNT contents). The insert shows the storage modulus G' as a function of MWNT content at 230 °C; (b) In the solid state; that is, storage and loss modulus as a function of the temperature.

showed that the magnitude of E'' increased dramatically with the incorporation of 0.5 wt% MWNT; however, this led to a maximum which subsequently decreased with further increases in MWNT content. Over the entire composition range, E'' remained between that of the composite with 0.5 wt% MWNT and pure PP, indicating (on a qualitative basis) the enhancement in the energy-dissipating ability of the 0.5 wt% nanocomposite. This, at least in theory, implies a possible transition in terms of the response of this material to an isotropic stress situation.

The stress–strain diagrams in Figure 7.3a highlight the deformation behavior of i-PP and the PP-MWNT nanocomposites at room temperature, with respect to their processing-induced anisotropy. These stress–strain curves exhibit distinct regions of elastic, yielding, and plastic deformation accompanied by cold drawing, whereas continued deformation beyond the yield point results in necking along the gage length. It can be observed that composites in the injection-molding direction demonstrate a ductile behavior (Figure 7.3a). The deformation behavior

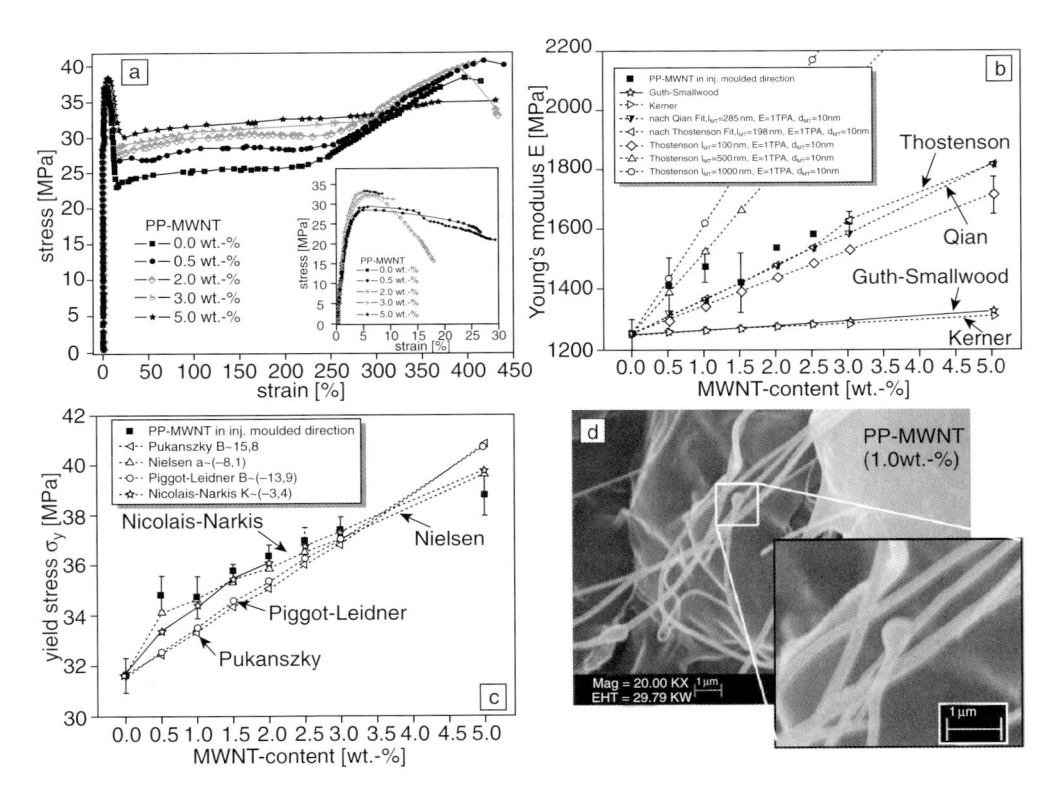

Figure 7.3 Mechanical behavior. (a) Stress–strain diagrams for nanocomposites. The insert shows the stress–strain diagram for nanocomposites with injection-molded direction perpendicular to the testing direction; (b, c) Modeling of the Young's modulus and yield stress; (d) Phenomena of "polymer sheathing" in an scanning electron microscopy image of i-PP with 1 wt% MWNT.

of samples with stress applied perpendicular to the injection-molding direction is affected by a transition from ductile/semi-ductile (for nanocomposites with a MWNT content <2.5 wt%) to a brittle (lower strain at break) behavior for composites with >3 wt% MWNT (see the insert in Figure 7.3a). The yield stress (σ_y) and Young's modulus (E), as a function of MWNT content, in the injection-molding direction parallel to the deformation direction are depicted in Figure 7.3b and c. Here, it can be seen that in case of the 0.5 wt% MWNT composite, both E and σ_y are increased (by ~13% and ~10%, respectively) under tensile stress along the injection-molding direction. The dependency of the modulus on MWNT content is observed to be nonlinear; in fact, it decreases exponentially with the increasing MWNT content. This was attributed to the increased tendency to form nanotube aggregates (reduced surface area per unit volume), as observed from AFM images (Figure 7.4a) with loosely packed collections of MWNT at higher CNT contents, which reduce the effective stress transfer between the MWNT and the polymer matrix. The dependence of mechanical properties on total surface area per unit volume (apart from dispersion and distribution) has been well reported [83, 84].

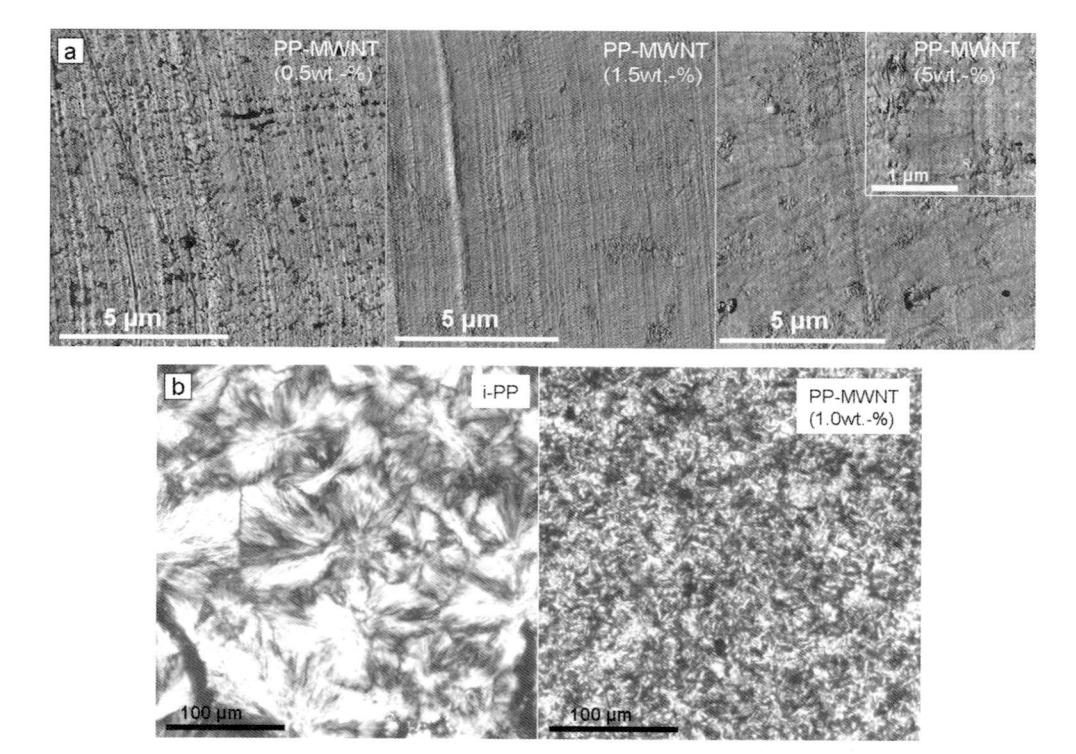

Figure 7.4 Morphology of the PP-MWNT system. (a) Atomic force microscopy phase images; (b) Polarized light microscopy images of i-PP and nanocomposite.

Several models for describing the mechanical properties of two-phase composites as a function of the filler content have been reported as successful in theoretically predicting the influence of MWNT content on Young's modulus and the yield stress. For example, the Guth–Smallwood equation (Equation 7.4), which is essentially a modified Einstein's equation, has been used as a simple model for describing E while taking into account the matrix-filler adhesion. This equation can be stated as:

$$E = E_m(1 + 2.5\phi + 14.1\phi^2) \tag{7.4}$$

where ϕ is volume fraction and E_m is modulus of the polymer matrix. Similarly, for expansible polymers with rigid spherical particles, the Kerner equation (Equation 7.5) can be used to theoretically estimate E. The equation can be given as:

$$E = E_m\left(1 + \frac{15(1-\upsilon)}{8-10\upsilon} * \frac{\phi}{1-\phi}\right) \tag{7.5}$$

where υ is the Poisson ratio, which is assumed to be approximately 0.35 [84, 85].

In contrast, Thostenson et al. [3] modified the well-known Halpin–Tsai approach in order to take into account the structural and geometrical contributions to reinforcement – that is, to predict the elastic modulus of nanocomposites as a function of the properties of both phases, the geometry of the nanotubes, and the nanotube structure. The equation is explained and discussed in Section 7.3.1.

Another suitable model to describe E, again based on the Halpin–Tsai model, was successfully applied in the case of PS-MWNT-based systems. These were assumed to be randomly oriented discontinuous fiber lamina by both Qian et al. [47] and by Guzman de Villoria and Miravete [73] for CNT-Epoxy systems, in order to evaluate the effect of dispersion. The model can be represented mathematically as:

$$E_c = E_m\left(\frac{3}{8} * \frac{1 + 2(l_{NT}/d_{NT})\eta_L V_{NT}}{1 - \eta_L V_{NT}} + \frac{5}{8} * \frac{1 + 2\eta_{NT} V_{NT}}{1 - \eta_T V_{NT}}\right) \tag{7.6}$$

$$\eta_L = \frac{(E_{NT}/E_m) - 1}{(E_{NT}/E_m) + 2(l_{NT}/d_{NT})}, \tag{7.7}$$

$$\eta_T = \frac{(E_{NT}/E_m) - 1}{(E_{NT}/E_m) + 2}, \tag{7.8}$$

The various models detailed above have been analyzed in terms of their value to PP-MWNT systems, and fits of experimentally determined E values as a function of MWNT content are shown in Figure 7.3b. Here, the theoretical values of E, when based on Kerner and Guth–Smallwood models, were found to be much lower than the experimental values, and this led inevitably to the assumption that these models were unsuitable for PP-MWNT nanocomposites. This problem may be attributed to the fact that the Young's modulus (E_{NT}) and the aspect ratio (l/d) of the filler-phase (MWNT) are ignored in these equations, despite the fact that CNTs have exceptionally high values of these parameters. The equations proposed by Thostenson et al. [3] and Qian et al. [47] included geometric parameters and the Young's modulus of the CNTs, and the proposed results based on these models have confirmed (within rational expectation) that the modulus in the nanocomposites would be significantly increased (see Figure 7.3b), with an increase in the length and a decrease in the diameter of the nanotubes. For MWCNTs, the reported moduli range from 0.2 to 1 TPa [77], while the diameter of the nanotube (as confirmed by the supplier, Nanocyl) was ~10 nm. Thus, by fitting these values into the above-described models (i.e., with a constant E_{NT} of 1 TPa and a constant nanotube thickness d_{NT} of 10 nm), the length of the nanotube (MWNT) obtained was about 200 nm, based on the approach of Thostenson et al. [3], and 285 nm based on the approach of Qian et al. [47] On the other hand, when a constant E_{NT} of 0.2 TPa was fitted to the equation adopted by Qian et al. [47], this led to an increase in nanotube length to about 400 nm. When comparing these theoretically estimated lengths of the CNTs to the MWNT lengths (as observed in the two-dimensional TEM images), most of the CNTs appeared to be shorter than 500 nm on an average. This may indicate a certain degree of nanotube breakage and, consequently, a shortening of CNTs during the mixing process. This effect was reported previously by Andrews et al. [30], following an increase in the mixing energy (mixing time) in a small-scale internal mixer. In addition, a possible reduction in the effective length of the MWNT may be due to part of the MWNT length remaining embedded in the PP-matrix, though this was limited by the thickness of the microtomed samples (i.e., 90 nm).

The yield stress (σ_y) was analyzed using the two-phase composite models, namely Nicolais–Narkis (Equation 7.9), Piggot–Leidner (Equation 7.10), Nielsen (Equation 7.11), and the Pukanszky model (Equation 7.12), to estimate the theoretical yield stress. One characteristic of all theoretical approaches is a relationship between the volume fraction (ϕ) and the projected area fraction of the particulate inclusions [85]. The Nicolas–Narkis equation (Equation 7.9) is a two-thirds power law function with K as a parameter for filler–matrix adhesion and σ_m as the yield stress of the polymer matrix:

$$\sigma = \sigma_m(1 - K\phi^{2/3}) \tag{7.9}$$

As an example, the theoretical value of K for poor adhesion is 1.21. Piggot and Leidner (Equation 7.10) used a first-power relationship with a similar parameter B, as in Equation 7.9, which is attributed to weakness in the structure due to stress concentration [85].

$$\sigma = \sigma_m(1 - B\phi) \tag{7.10}$$

The Nielsen equation (Equation 7.11) is based on a two-phase system with poor adhesion as a matrix with voids, and leads to the expression:

$$\sigma = \sigma_m \exp(-a\phi) \tag{7.11}$$

where the factor a is the stress concentration parameter, and where a high value means a high stress concentration or rather poor adhesion [85].

Similarly, the equation of Szazdi and Pukanszky *et al.* [86] can be stated as:

$$\sigma_{yt} = \sigma_{ym} \frac{1-\phi_f}{1+2.5\phi_f} \exp(B\phi_f) \tag{7.12}$$

where B is an empirical parameter that characterizes the degree of filler–matrix interaction. The value of B depends on all factors that influence the load-bearing capacity, notably the strength and size of the interface. The model was developed for composites containing spherical particles, although anisotropic fillers can lead to an increase in the load-transfer efficiency and hence to a higher value of B.

The fit of the yield stress and experimentally determined values is shown in Figure 7.3c, and provides negative K, B and a-values for the Nicolais–Narkis (Equation 7.9), Piggot–Leidner (Equation 7.10) and Nielsen (Equation 7.11) models. This leads inevitably to the assumption that, fundamentally, these models are not valid in case of the investigated nanocomposites. The inappreciable significance of the parameters based on these models is attributed to the fact that the dynamics of polymer–filler interaction is fundamentally different from that of microcomposites. In the case of nanocomposites, the immobilized layer of polymer chains that forms the interface will alter the physical significance of the polymer–filler interaction at the molecular level, since the structural organization of the matrix at the interface is largely different from that in the bulk. This is clearly evidenced by the difference in T_g of the polymer matrix as a function of MWNT content, as obtained from the loss modulus peaks based on DMA measurements.

Additionally, the Pukanszky model was used to describe the theoretical yield stress of PP-layered silicate nanocomposites [86]. For this, different types of treated layered silicates were used to compare the experimental mechanical results with the model. The B-parameter for the different nanocomposites remained between 2 and 15 for composites with an intercalated or exfoliated morphology. The model was also found to be valid for the PP-MWNT nanocomposite, with such validity being demonstrated by the linear correlation between the natural logarithm of reduced yield stress plotted against filler content. From this, a B-parameter was calculated of 15.8, which indicated an outstanding load–transfer mechanism. However, at low MWNT contents, the model fit led to a much lower σ_y than did the experimental values. Although the anisotropy and orientation of the nanofillers were not taken into account, they had a definitive influence on the extent of reinforcement. In addition, nucleation effects or other morphological changes can alter the matrix properties significantly, and may also alter the prediction of the mechanical properties with this model [86]. The observed MWNT clusters above 1.5 wt% (as shown in Figure 7.4a) may lead to a deterioration in mechanical properties compared to theoretical values, as a result of local inhomogeneities and slippage in the CNT bundles.

7.4
Fracture Mechanics of CNT-Polymer Nanocomposites

7.4.1
Amorphous Thermoplastic

The crack resistance behavior following the EWF concept postulates that EWF (w_e) and the nonessential work of fracture (βw_p) must correspond with the resistance to crack initiation, and resistance against crack propagation, respectively. In general, these nanocomposites have shown thermoplastic behavior, and the validity of the EWF methodology through the self-similarity of the load–displacement curves. The failure of DENT specimens after ligament yielding has been demonstrated via strain field analysis images, and has been discussed previously [87]. A full yielding of the ligament in the DENT specimens occurred at maximum load (F_{max}), and prior to the onset of crack propagation. It was also observed that the incorporation of MWNT into PC caused a gradual transition in the nature of the load–displacement curves, as characterized by a noticeable reduction in the maximum displacement. This provides a qualitative indication of the existence of a ductile-to-brittle transition with increasing MWNT content. As a result of the decreasing displacement, the ratio of maximum displacement (s_{max}) and maximum load (F_{max}) (the *ductility ratio*) was also decreased [87]. According to the essential studies of fracture concept, the nonessential work of fracture (βw_p) is a measure of the resistance against crack propagation. The slope of the linear-fit plot of specific work of fracture against the ligament length is represented by βw_p, and is equivalent to the J-integral concept. The similarity between the EWF concept and the crack resistance (R)-curve concept using the J-integral has been demonstrated by Mai and Cotterell [88, 89]. These authors found, that EWF = J_{Ic}, the critical J-value as a measure of the stable-crack initiation toughness, and $\beta w_p = 1/4 \times dJ/da$ (for DENT specimens), where dJ/da is the slope of the J–R curve (a is the actual crack length). Some authors confirmed that these findings corresponded experimentally to EWF and to βw_p [89], or numerically (to EWF [90]). The slope of the linear regression fit of the datum points in the plot, showing the variation in specific work of fracture as a function of ligament length (Figure 7.5a), represents the nonessential work of fracture (βw_p), which increased by 25% following the incorporation of 2 wt% MWNT as compared to pure PC. Interestingly, on further increasing the MWNT content to 4 wt% and 6 wt%, βw_p was decreased by about 25–35% compared to pure PC. The existence of such a nonlinear dependence of βw_p on the content of MWNT indicates a tough-to-brittle transition in these polymer nanocomposites. Further, it is clearly visible from Figure 7.5b, that a maximum in the nonessential work of fracture (βw_p) was observed at 2 wt% MWNT; this demonstrated an enhanced resistance to crack propagation compared to pure PC. At 2 wt% MWNT, a tough-to-brittle transition has been observed. Such a dependence of βw_p on the MWNT content might be attributed to a reduced mobility of the PC chains at higher MWNT concentrations, due to an overall increase in the interfacial area.

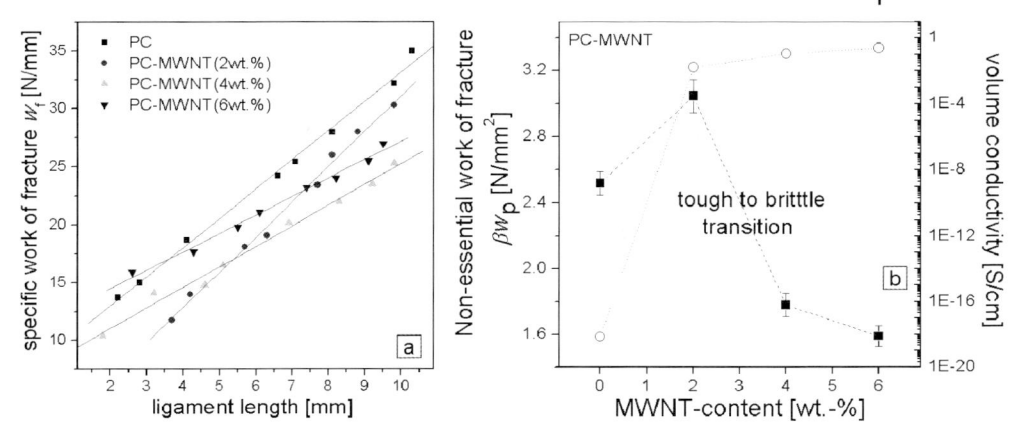

Figure 7.5 Fracture behavior of the PC-MWNT-system. (a) Linear regression plots of specific work of fracture versus ligament length; (b) Nonessential work of fracture (filled symbols) and volume conductivity (open symbols) versus MWNT content.

It has been concluded, that the reduction in molecular mobility in certain regions of the polymer chains results from an increase in the T_g and a broadening of the loss modulus peak, due to the incorporation of MWNT into a PC matrix [91], and this is also the case in the present investigations. The diameter of the CNTs are on the same size scale as the radius of gyration of the polymer chains in PC, which supports the formation of large-diameter helices around the individual nanotubes. This is also known as "polymer wrapping" [16–18, 92], and leads to a strengthening of the polymer–nanotube interface at the molecular level by entanglements [18, 36]. Such strong interface formation has also been discussed based on molecular mechanics calculations [17], and demonstrated experimentally by direct TEM observations of "polymer sheathing" in PC-MWNT composites [16]. Hence, the decrease in chain mobility at higher MWNT contents may restrict the plastic deformation of the polymer chains of PC, causing a brittle failure. At higher MWNT contents, apart from a substantial reduction in chain mobility due to the mechanical network structure, the strain energy density becomes a function of two competitive events, namely polymer wrapping around the nanotubes, and the distribution and dispersion of the nanotubes in the PC matrix. Thus, despite an efficient nanoscale interaction between the nanotubes and the polymer, an inhomogeneous distribution may cause high strain localization. Similar effects have also been discussed for styrene-butadiene rubber (SBR)/carbon black [93]. These strain localizations are also caused by an extreme modulus mismatch between the matrix and the nanotubes [94] and, as a consequence, cracks may propagate between the nanotube aggregates through the intensively strain localized zones in the PC-MWNT composite systems. This in turn causes brittle failure, such that a tough-to-brittle transition takes place in the composites at 4 wt% MWNT. Such failure modes have also been reported by Qian *et al.* [47].

7 Stress Transfer and Fracture Mechanisms

The time-dependent nature of crack growth phenomena has also been characterized based on the evaluation of individual deformation stages. For this, image acquisition has been carried out at different stages of deformation as a function of time on a single specimen with a fixed ligament length on DENT specimens from all selected nanocomposites, while assuming that the deformation fields are identical for a given ligament length. The variation in the crack propagation (Δa) and crack opening displacement (δ) with time (t) is illustrated in Figure 7.6a.

It could be clearly observed that the crack propagation occurred much higher than the crack-tip opening displacement (CTOD) after the crack-tip blunting

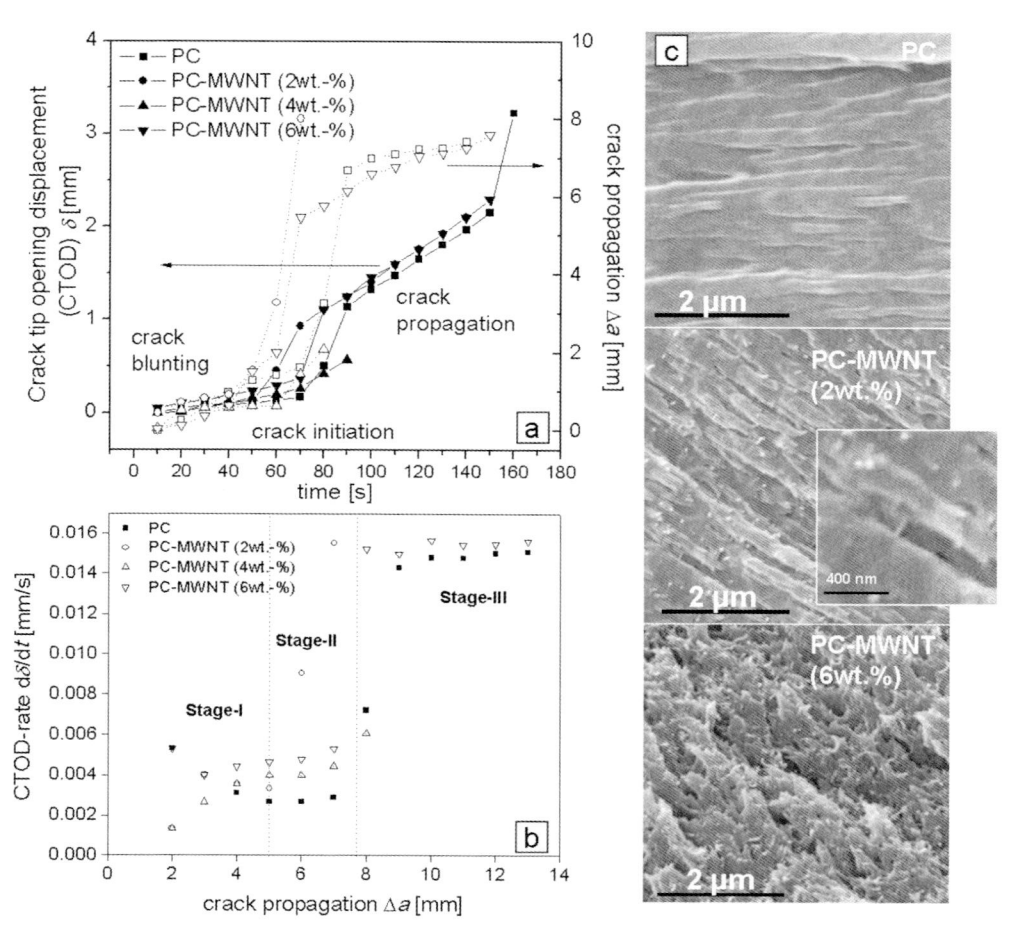

Figure 7.6 Crack kinetics and fracture surface morphology of the PC-MWNT-system. (a) Crack propagation and crack-tip opening displacement (CTOD) versus time; (b) Rate of crack opening displacement ($d\delta/dt$) versus crack propagation (Δa); (c) Scanning electron microscopy images of the fractured DENT-specimens. The magnified insert shows bridging of the nanotubes across the crack.

process, which indicated the formation of a well-localized plastic zone. Further, the rate of crack extension and CTOD were seen to undergo two distinct transitions with respect to their deformation stages. For deformation up to the first 50–70 s, depending on the material (stage-I), the values of crack speed ($\Delta a/dt$) and the corresponding CTOD rate ($d\delta/dt$) remained small (ca. 0.025 and 0.00625 mm s^{-1}, respectively) (see Figure 7.6b). However, beyond this time, the crack extension speed and speed of crack opening were much higher (i.e., 0.166 and 0.025 mm s^{-1} respectively; Stage-II). Such an acceleration with respect to the overall crack growth phenomena in these nanocomposites may be attributed to a transition in the time-dependent nature of deformation, from pure crack-tip blunting/crack initiation mode to a mixed mode of crack blunting and crack initiation/propagation. However, in the nanocomposite with 2 wt% MWNT, the crack extension speed ($\Delta a/dt$) and the CTOD speed ($d\delta/dt$) were observed as 0.35 and 0.0375 mm s^{-1} respectively, which was substantially higher (~10-fold) when compared to pure PC and the 4 wt% and 6 wt% MWNT-filled nanocomposites – that is, above the deformation stages corresponding to 50–90 s. This could be attributed to the fact that, within the range of the percolation threshold (electrical percolation), the nanocomposite undergoes only a small plastic deformation due to restrictions imposed by the immobilized polymer layers, and owing to the primary entanglement network of polymer–nanotube. Such a primary network will break down under uniaxial tensile loading, as the stress field is more concentrated in the range of the percolation threshold (i.e., ~1–2 wt% MWNT). The conventional response to the breakdown of such a primary entanglement network is the formation of a secondary filler–filler network, which could lead to a dissipation of the stress field and enhance the crack stability, as in the case of SBR-carbon black systems [94]. However, in case of a nanocomposite with 2 wt% MWNT, the formation of a stable secondary filler–filler network to sustain the stress field (unlike the highly filled nanocomposite systems) leads to large-magnitude crack propagation instability. This inevitably reaffirms the existence of a tough-to-brittle transition at 2 wt% MWNT, as discussed previously in the context of nonessential work of fracture (βw_p) and ductility. From Figure 7.6b, it is clear that the fracture with respect to $d\delta/dt$ as a function of Δa occurs in three stages. Whilst Stage-I clearly represents the crack blunting and stable crack initiation modes, Stage-II characterizes the mixed mode of crack blunting and crack initiation or propagation. However, it should be noted that a strictly defined stage corresponding to crack initiation could not be ascertained in these types of nanocomposite, as these measurements have been carried out in the post yield fracture stages. The crack resistance (R) curves as function of loading parameters (J or δ) versus stable crack growth Δa yields information about the different phases of crack growth (crack-tip blunting, stable crack initiation and propagation) as a function of time. Ideally, the physical crack initiation values can be determined based on the consideration of the kinetics of crack propagation processes. During the crack-tip blunting process in the initial stages of deformation, the stress state can be described by a yielding of the material close to the crack tip and the formation of a plastic zone. The whole deformation energy is used to calculate the J values to obtain crack propagation characteristics in terms of their stability and instability. The crack resistance curves (J versus Δa and δ versus Δa), as shown in Figure 7.7a, demonstrate the switch-over from Stage-I (crack-tip blunting/crack initiation mode) to Stage-II (mixed mode of crack blunting and crack initiation/propagation), apparently above a crack extension (Δa) of 0.5–1.5 mm depending on the amount of MWNT. In an effort to understand the crack initiation process, $J_{0.5}$ has been evaluated as the technical crack initiation resistance value at $\Delta a = 0.5$ mm. At the percolation threshold, the decrease in ductility restrains any positive effects of reinforcement, followed by a pronounced drop in the crack initiation toughness ($J_{0.5}$) (as shown in Figure 7.7). However, for volume fractions of filler much higher than the critical fraction (i.e., the percolation threshold range), the breakdown of the filler network to initiate a

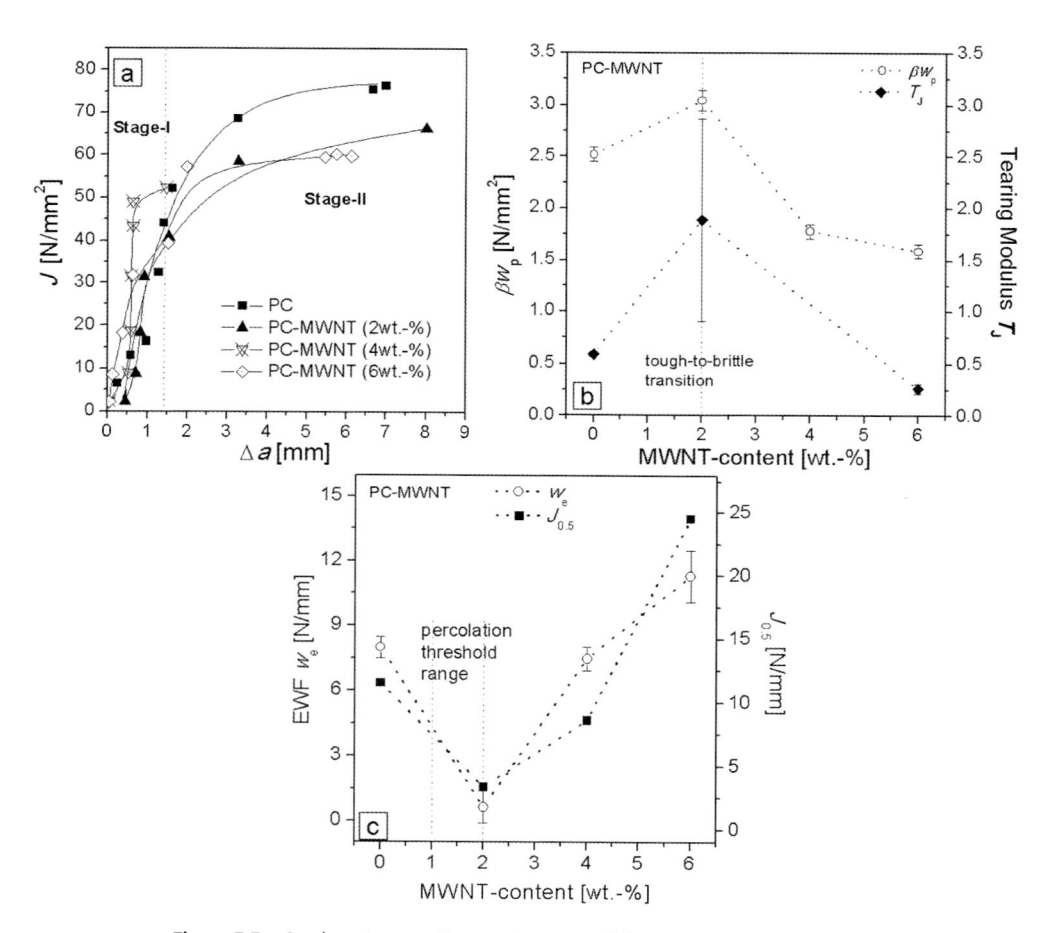

Figure 7.7 Crack resistance (R-curves) curves of the PC-MWNT system. (a) J-integral (J) versus crack propagation (Δa); (b) Variation of βw_p and T_J with MWNT content; (c) Variation of EWF and $J_{0.5}$ with MWNT content.

crack consumes increasing amounts of energy, while the resistance to crack initiation ($J_{0.5}$) is again increased as a function of the filler volume fraction. Qualitatively, such a trend is also in agreement with the w_e (EWF) values. In the case of nanocomposites with a higher nanotube fraction than the percolation threshold, the breakdown of the polymer network and the formation of a filler–filler network causes a sharp decrease in the ductility of the nanocomposites (as discussed above). As found for other polymer-based composites reinforced with active-fill fillers, which form a surrounding layer of immobilized polymer (e.g., SBR/carbon black [94] and PMMA/silica [95]), the crack initiation toughness (as resistance against stable or unstable crack initiation) is increased as a function of the filler volume fraction primary due to increasing energy dissipation capacity of the composite; this is evident from the increasing loss modulus (G'') in DMA. Unfortunately, the range between zero filler (pure PC) and the percolation threshold could not be investigated, and consequently the behavior found for EWF and $J_{0.5}$ in the said region could not be clearly understood. The maximum slope of the R-curves, dJ/da (and $d\delta/da$), as the resistance against crack propagation has been determined from Stage II in Figure 7.7a (for 4 wt%, Stage II does not exist). It has been shown in Figure 7.7b that EWF and βw_p (values from EWF concept) correlate very well with the crack initiation resistance ($J_{0.5}$ measured at $\Delta a = 0.5$ mm) and propagation resistance (dJ/da) (not shown here). Further, the tearing modulus values ($T_J = \frac{1}{4} dJ/da \times E/\sigma_y^2$ and $T_\delta = d\delta/da \times E/\sigma_y$) are determined to characterize the resistance against stable crack propagation. The dependence of tearing modulus values on the MWNT content, as calculated from the slope of the two types of R-curve, have shown opposing trends [87]. This type of inverse correlation between the nature of the slopes has been reported previously, when it was attributed to material-dependent notch sensitivity [96]. Interestingly, it was shown that T_J and T_δ have a close correspondence to βw_p and EWF, respectively [87]. Thus, a correlation was established (at least conceptually) between the nonessential work of fracture (βw_p) as a morphology-sensitive parameter related to the J-controlled crack growth process, as is the case for hetero-phase polymers and metals [97, 98] unlike EWF. Further, the correspondence of $J_{0.5}$ with EWF and T_J with βw_p (Figure 7.7b, c) structurally explains the correlation of the resistance against crack initiation $J_{0.5}$ with matrix behavior (state of percolation), while the resistance against crack propagation T_J, on the other hand, is correlated with the morphology. Thus, T_J (the tearing modulus) is controlled by morphological characteristics such as the dispersion and distribution of the nanotubes in the polymer matrix. In addition, it emphasizes the correspondence between single-specimen fracture evaluation (SSFE) and multi-specimen fracture evaluation (MSFE) methods. Several investigations of the fractured surfaces were conducted in order to analyze the micro deformation and crack propagation mechanism, using SEM. The images corresponding to pure PC and nanocomposites based on PC with 2 wt% and 6 wt% of MWNT are shown in Figure 7.6c. In the case of pure PC, a homogeneous deformation was observed during the crack propagation, with the damage initiation on the surface due mostly to the simultaneous influence of tearing and shear fracture. The advancing cracks on the surface were seen to be followed by cracks with narrower crack-tips, which would presumably undergo coalescence to form macro cracks.

In contrast, in the case of the PC-MWNT (2 wt%) nanocomposite (Figure 7.6c), the surface was seen to undergo less-intensive fracture (with smaller crack lengths) because the cracks were spanned by the nanotubes, which caused an enhanced resistance to the crack propagation process (see the insert in Figure 7.6c). Apparently, the crack dimensions in pure PC remained within the range of 0.1–0.6 µm, whilst the incorporation of 2 wt% MWNT caused the dimension to be reduced to 0.09–0.27 µm. An investigation of the fracture surface has revealed that narrower crack-tips beneath the advancing cracks are more efficiently bridged by the nanotubes at 2 wt% MWNT, which results in an increased resistance against crack propagation. Bridging of the radial cracks on the surface has been observed to be spanned across a width of ~0.07 µm, while the width of the cracks underneath (transverse to the direction of the tensile force) remained in the range of 0.05–0.12 µm. Bridging of the nanotubes as a mechanism of inhibiting crack initiation in polymer- and ceramic-based nanocomposites has been widely reported [92, 99, 100]. In contrast, at 4 wt% and 6 wt% MWNT (see 6 wt% in Figure 7.6c), the fracture surface was seen to be very rough and accompanied by extensive tearing zones that had been formed at the final deformation stages. The presence of apparently softer and isolated fibril-like structures demonstrated that a viscoplastic flow behavior existed in the shear region during loading process. This was attributed to the less homogeneous dispersion of the nanotubes at higher MWNT contents, which facilitates PC domains that are comparatively less occupied by nanotubes to undergo faster plastic deformation. The transition in the nature of the fracture surface morphology, from shear-like at 2 wt% MWNT to dimple-like structures at 4 wt% MWNT, indicated that a change from shear deformation to craze-like deformation was in agreement with the observed tough-to-brittle transition and decreased T_j and βw_p [87].

7.4.2
Semi-Crystalline Thermoplastic

The preconditions for validity of the EWF concept—that is, the self-similarity of force–displacement diagrams, plastification of the ligament before crack initiation, and plane–stress conditions—were clearly met in the case of PP-MWNT-based nanocomposites, the details of which are described in Refs [101, 102]. The specific work of fracture in this system was measured from the area under the force–displacement curve limited to F_{max}, since the PP-based nanocomposites have shown extensive yielding, causing noticeable data scatter. The resistance to crack propagation βw_p was increased by ~15 % on the incorporation of 0.5 wt% MWNT as compared to pure i-PP; however, this was decreased by ~20–30% when further increasing the MWNT content to 1.5 wt% and 5 wt%, compared to the 0.5 wt% MWNT nanocomposite, and by ~7–10% compared to pure PP (see Figure 7.8). In contrast, a sharp drop in w_e in the 0.5 wt% MWNT composite, followed by an increase in w_e up to 1.5 wt% MWNT, was also in agreement with a nonlinear

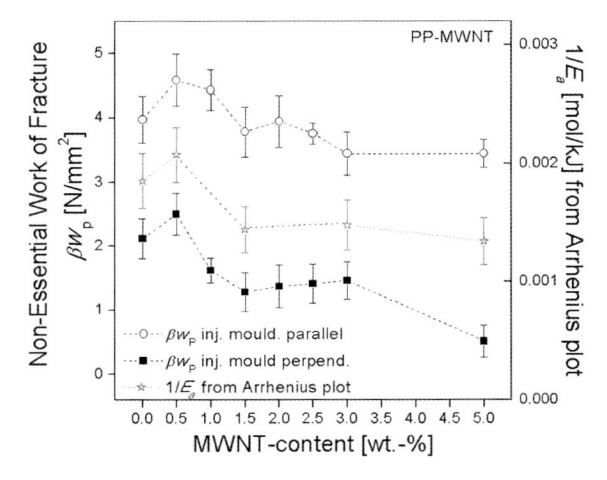

Figure 7.8 Fracture behavior and activation energy of the PP-MWNT system. Nonessential work of fracture and inverse activation energy as function of MWNT content.

dependence of crack growth prior to failure under uniaxial tensile loading. Interestingly, such a nonlinear dependence of βw_p and w_e has also been observed to be insensitive to processing-induced anisotropy. The nanocomposites have been shown to undergo ductile yielding, although the nature and speed of the ductile response varied with respect to the MWNT content. The existence of a transition in the nature of deformation mode prior to failure is in close agreement with the DMA data, where a similar transition in the magnitude of $\tan \delta$ at 23 °C (room temperature) was observed over a content composition range of 0.5 to 1.5 wt% MWNT [102].

It has been reported that the most effective way to reduce plastic resistance in the case of unoriented polymers is to quasi-uniformly cavitate the solid, thus incorporating equiaxial particles that can be forced to undergo cavitation or debonding, so as to convert the polymer into a cellular solid [103]. In PP-MWNT-based systems, the incorporation of 0.5 wt% nanotubes led to a reduction in plastic resistance [with regards to the inner fracture process zone which, according to the EWF approach based on the post-yield fracture mechanics (PYFM) concept, indicates a resistance to deformation of the plastic zone of the DENT specimens used for fracture investigations] as compared to the matrix PP. This was evident from the larger crack extension at a given dimension of crack length when compared to PP, despite this being followed subsequently by an increase in plastic resistance as the amount of MWNT was increased to 1.5 wt%, leading to a semi-ductile mode of failure.

The dependency of T_g on frequency can be used to estimate the apparent activation energy of the glass transition relaxation process, which is the energy required to activate the molecular mobility of the polymer chains (rotational carbon–carbon bonds, segmental chain bond motion, and intermolecular separation of the chains)

[11]. The relationship between glass temperature T_g ($T_g = T$ at maximum G'') and the frequency f is given by:

$$\log f = \log f_0 + \frac{\Delta E_a}{2.303RT} \qquad (7.13)$$

where R is the gas constant and ΔE_a the activation energy.

The activation energy can be calculated from the slope of the Arrhenius plot and the inverse E_a ($1/E_a$) for i-PP and nanocomposites with 0.5, 1.5, 3, and 5 wt% MWNT are shown in Figure 7.8. Interestingly the nanocomposite with 0.5 wt% MWNT showed the lowest activation energy (larger free volume) with respect to the pure PP, while nanocomposites with 1.5, 3, and 5 wt% MWNT had the highest activation energies, with $\Delta E_a > 675$ kJ mol^{-1}. This means that the molecular mobility of the polymer chains is hindered in the case of nanocomposites with >1.5 wt% MWNT. However, at low MWNT contents (0.5 wt%), chain mobility is increased due to an alteration in the amorphous phase that may be generated due to a reduction in spherulite sizes with the incorporation of nanotubes, as observed with polarized light optical microscopy (see Figure 7.4).

The reason for such a transition from low to high E_a above 0.5 wt% MWNT may be attributed to the network formation of the CNTs, which hinders the polymer chain mobility. Such a percolation (rheological percolation) was observed in the present system above 1.0 wt% MWNT, as confirmed via melt rheological investigations. These observations led to the conclusion that the crack resistance behavior of such CNT-reinforced semi-crystalline polymer matrix-based nanocomposites was primarily controlled by the structural reorganization of the matrix and the formation of a nanotube-network at higher contents. The maximum in the nonessential work of fracture (βw_p) was observed at 0.5 wt% MWNT, demonstrating an enhanced resistance to crack propagation compared to pure i-PP. This was not only in accordance with DMA results but may also be attributed to the formation of large number of smaller spherulites rather than larger spherulitic domains in the nanocomposite with 0.5 wt% MWNT, compared to a virgin PP matrix which, in turn, leads to smaller inter-spherulitic boundaries. The smaller inter-spherulitic boundary so formed through the α-nucleating effect of MWNT, whilst accompanied by a "finely dispersed" amorphous phase, facilitates the process of crack arrest when it is initiated from a highly stress-localized region; this is achieved through hindrance of the slippage of the crystal planes, and hence restricts extensive ductile yielding. Recently, SEM studies have also revealed that the network-density of fibrils is decreased in nanocomposites with 0.5 wt% MWNT (Figure 7.9d), when compared to pure PP [101, 102, 104]. This indicates a drastic hindrance of the yielding process by MWNT, despite plastic deformation, while still partly retaining the ductile nature of PP.

With a further increase in MWNT content, the nanotube-rich areas become larger and tend to show the characteristics of forming an interconnecting network. When this network-density becomes higher at higher MWNT concentrations, it

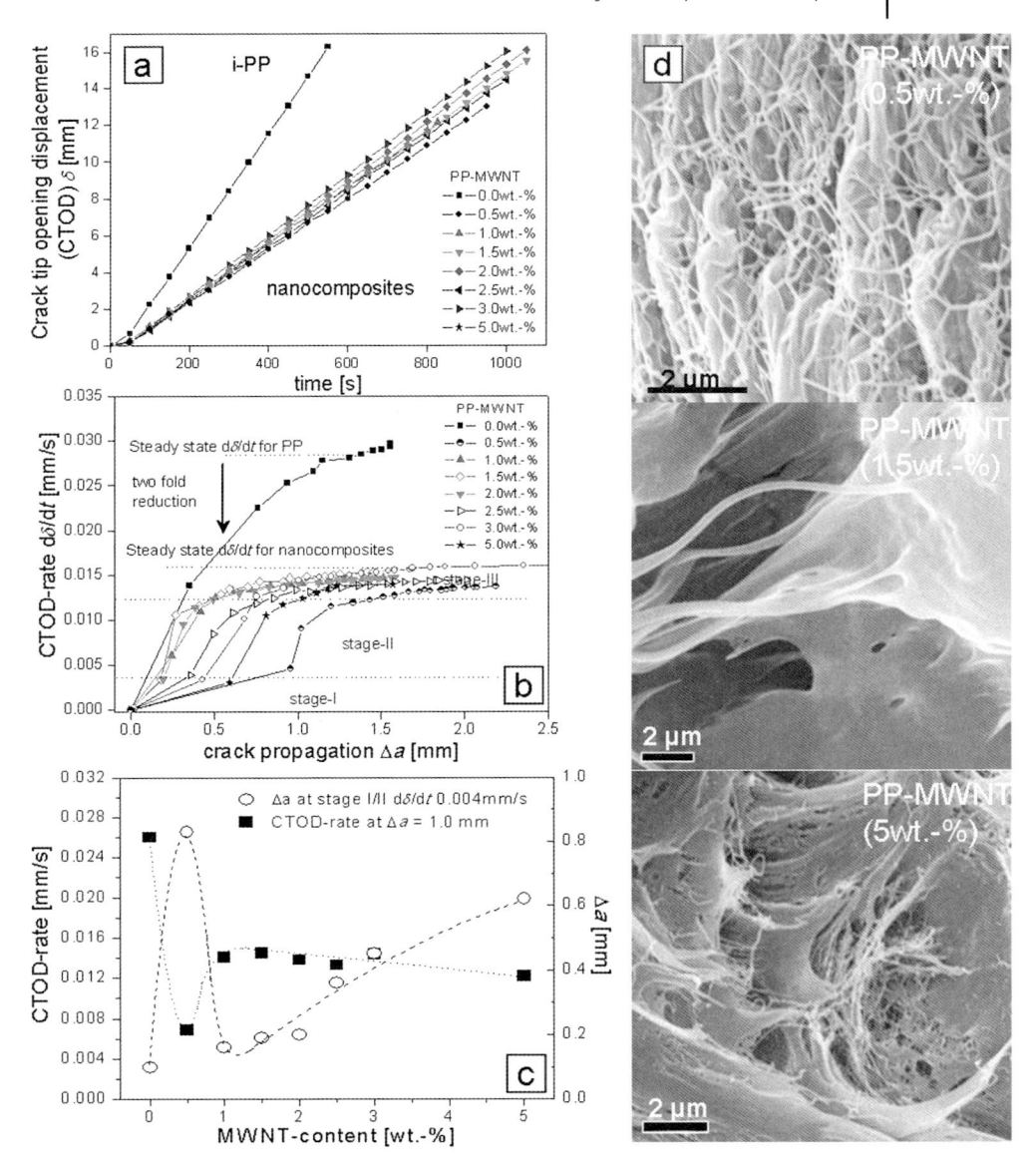

Figure 7.9 Crack kinetics and fracture surface morphology of the PP-MWNT-system. (a) Crack-tip opening displacement (CTOD) versus time; (b) Rate of crack opening displacement (dδ/dt) versus crack propagation (Δa), indicating the stages of crack growth; (c) CTOD rate and Δa as function of MWNT content (see text for details); (d) Scanning electron microscopy images of the fractured DENT-specimens.

can potentially lead to an intense strain-localization as a result of hindered plastic deformation/strain of the matrix. The strain-localization around the dense network of nanotubes ultimately causes matrix cracking, due to a severe modulus mismatch between the polymer and nanotubes; as a consequence, the ductile yielding behavior is substantially reduced, such that it becomes semi-ductile behavior. Strikingly, when the MWNT content was increased further to 1.5 wt%, the network-density of fibrils was almost negligible/absent, and instead the crack-growth on the matrix surface was seen to be spanned by nanotubes bridging across the cracks, assisted by a network of fibrils; this was evident from the SEM images (Figure 7.9d). However, the formation of a rheologically percolated network above 1 wt% MWNT completely changes the matrix behavior; in fact, the crack toughness of the nanocomposite with 1.5 wt% MWNT becomes a dynamic interplay of percolation and nanotube bridging, leading to a semi-ductile behavior. Thus, the drastic hindrance of ductile yielding in the polymer appears to be controlled by the state of dispersion of the MWNT, the interface, and the MWNT-induced structural reorganization (nucleation).

The kinetics of crack propagation aspects has been evaluated using a single-specimen technique, the aim being to qualitatively identify the characteristics of crack growth phenomenon in these PP-MWNT-based nanocomposites. The plot of CTOD versus time in Figure 7.9a shows a linear time-dependence of CTOD, although a significant difference in the slope between the PP matrix and the nanocomposites is also observable. This effect was also reported to be well in agreement with the strain field analysis [102]. Typically, in the case of semicrystalline polymers, the yielding behavior can also be characterized qualitatively by creating plots of maximum local stresses versus time, and on a comparative scale the slope of such plots differs from composition to composition. It was reported that in the case of PP the slope of the local strain versus time plot undergoes a transition at ~40 s, whereas in case of the composite with 0.5 wt% MWNT content such a transition would occur at ~60 s; in other words, the ductile yielding is largely delayed by ~20 s. The change in the slope of the maximal local strain as a function of time curves indicates a higher CTOD-rate associated with crack initiation. In other words, the shift from low to high CTOD rate occurs earlier in PP than in the nanocomposites. Hence, such an observation indicates, qualitatively, that the toughening–that is, the enhanced resistance to crack propagation as observed from the magnitude of βw_p–is conceptually attributed to a delayed yielding response of the nanocomposites.

The kinetics of crack propagation, as shown in $d\delta/dt$ (CTOD rate) versus Δa plot (Figure 7.9b), clearly reveals three stages of crack growth based on the CTOD rate, namely Stage-I, Stage-II and Stage-III, corresponding to $d\delta/dt$ in the range of 0–4×10^{-3} mm s^{-1}, 4–12×10^{-3} mm s^{-1}, and $\geq 12 \times 10^{-3}$ mm s^{-1}, respectively. The steady-state $d\delta/dt$ for PP has been observed to be almost twofold higher than the nanocomposite, which indicates that the incorporation of nanotubes into the semicrystalline matrix substantially alters the mode of crack propagation with respect to the CTOD; that is, the inherent resistance of the material through crack blunting. Interestingly, such a change in the mode of crack propagation is in striking contrast to the response of microcomposites under similar situations, where normally the dδ/dt remains appreciably on a similar level to that of the polymer matrix.

To further characterize the nature of the fracture dynamics and the consequent shift in deformation behavior under biaxial stress in the presence of double-edged notches, the crack extension (Δa) and the CTOD (δ) are plotted in Figure 7.9c as a function of MWNT content at a fixed CTOD rate and fixed crack extension, respectively. The sharp drop in crack-blunting efficiency with the incorporation of 0.5 wt% MWNT in the PP matrix (compared to PP), followed by an increase in the CTOD for a fixed Δa in case of composites with $\geq$1.5 wt% MWNT, indicate a transition in the nature of crack propagation. At $\geq$1.5 wt% MWNT, the ductile yielding is remarkably arrested, accompanied by a decrease in the post-yield blunting efficiency. This in turn leads (qualitatively) to a theoretical possibility of the nanoeffect being controlled by other organizational complications in the polymer nanocomposites at the structural level, such as the dispersion, crystallinity and size of the spherulites. It should be noted that the presence of nanotubes in the PP dramatically alters the deformation mode; this may be attributed first to a physical adsorption of the polymer chains onto the MWNTs (due to a high organophilic surface-affinity of nanotubes through van der Waals interaction), and second to the influence of nanotubes on the morphological dimensions (i.e., the size of the crystallites), since nanotubes potentially act as α-nucleating agents.

Investigations into the micromechanistic attributes prior to failure by fractured surface morphology, and a strain field analysis illustrating the intensification of response to biaxial tensile-stress across the composite surface, have each been carried out. The SEM images of the fractured surfaces show clearly that the network-density of fibrils decreases in the nanocomposite with 0.5 wt% MWNT when compared to pure PP. This indicates a drastic hindrance of the yielding process by MWNT, despite plastic deformation, whilst still partly retaining the ductile nature of PP [101, 102, 104]. At 1.5 wt% of MWNTs the network-density of fibrils was almost absent, whereas cracks were spanned by nanotubes or nanotube bundles. Therefore, the formation of a percolated network above 1 wt% of MWNT completely changes the matrix behavior and thus the crack toughness behavior of nanocomposites with MWNT contents higher than 1.5 wt%. However, the formation of a percolated network above 1 wt% MWNT completely altered the matrix behavior, such that the crack toughness of the nanocomposite with 1.5 wt% MWNT becomes a dynamic interplay of percolation and nanotube bridging, leading to a semiductile behavior.

The fractured surface of nanocomposite with 1 wt% MWNT showed the appearance of nanotubes with higher diameters at their broken/debonded ends, as compared to other parts. This indicated that the polymer chains prior to failure were stretched to their maximum, followed by recoiling and balling-up around the tip of the broken end of the nanotubes. Such a phenomenon has also been observed experimentally in the case of PC-MWNT-based nanocomposites by Ding et al. [16], using an individual nanotube pull-out test with an AFM tip. In this case, the outer polymer sheath becomes contracted and balled-up during the fracture process.

Further (theoretical) support of such thermodynamic feasibily was provided from complying molecular mechanics calculations which suggested that the conformational helices of polymer chains would wrap around the nanotube [17].

The radius of gyration, which scales as $R_g \sim M^{0.5}$, remains appreciably within the same range as the diameter of the MWNT (0.1–10 nm), and has been estimated to be approximately 9.6 nm based on the molecular mass of PP (M_n = 92.1 kg mol^{-1}) as determined by gel-permeation chromatography (GPC) measurements. Thus, the topological feasibility of polymer chains dynamically wrapping-up around the individual nanotubes is further strengthened, albeit on a conceptual basis. However, as the overall crystallinity remained unaffected despite the nanotubes being strongly α-nucleating, and the T_g being lowered in the nanocomposites, the inevitable structural implications may rest upon the fact that nanotubes do not interact appreciably with the crystalline domains. In contrast, the reduction in T_g hints at an easy mobility of the polymer chains, a phenomenon which (in theory) might be possible if the amorphous free volume were to be increased. These observations prompt first, towards an understanding that the polymer-wrapping phenomenon is largely confined to the amorphous regions; and second, towards the fact that the immobilized polymer layer at the polymer–nanotube interface creates a low enthalpy to greatly facilitate mobility of the free chain ends. The i-PP exists in a 3_1-helix conformational arrangement with a radial dimension of 2.096 nm about its mean chain axis (c-axis). These implications are further strengthened by the increase in the onset of degradation temperatures in the case of nanocomposites, indicating the excellent adsorption of the PP-chains onto the surfaces of the nanotubes.

7.5
Concluding Remarks

Studies of the fracture behavior of polymer–nanotube nanocomposites, by employing fracture mechanics concepts, are still in their infancy despite widespread reports of improvements in mechanical properties, especially with regards to tensile strength and modulus. To reach an optimal balance of toughness and stiffness remains an unsolved research problem, especially with respect to nanocomposites. Consequently, the problem of tailoring the stiffness and toughness of nanocomposites represents a critical phase in the development of new materials concepts. The future creation of advanced nanomaterials pertaining to polymer–CNT-based nanocomposites will only be determined if certain fundamental questions can be answered, including:

- What possible routes may lead to the quantification and mitigation of matrix sensitivity effects of fracture behavior?
- How critical is the role of added polarity or interfacial compatibilization between the polymer and carbon nanotube in controlling the mechanical properties?

- What level of toughness of these nanocomposites should be optimal for an objective application?
- Is the kinetics of fracture behavior of these nanocomposites matrix-controlled, morphology-controlled, or interface effects-controlled?
- What is the nature of strain field development?

And last, but not least:

- In what possible ways might these new insights help in the design of materials for practical applications?

Acknowledgments

The authors gratefully acknowledge their fruitful collaborations with Dr D. Jehnichen (IPF Dresden, Germany) for his support and helpful discussions with the WAXD measurements. They also thank Mr A. Janke and Ms U. Reuter for preparing the TEM and AFM images (IPF Dresden, Germany), M. Thunga from FSU-Jena (Jena, Germany) for his support in the dynamic-mechanical analysis of the composites, and Dr K. Schneider (IPF-Dresden, Germany) and Dr R. Lach (PSM Merseburg, Germany) for helpful discussions regarding the fracture mechanics and strain field analysis.

References

1. Ijima, S. (1991) *Nature*, **354**, 56–58.
2. Tjong, S.C. (2006) *Mater. Sci. Eng. R*, **53**, 73–197.
3. Thostenson, E.T. and Chou, T.W. (2003) *J. Phys. D.: Appl. Phys.*, **36**, 573–582.
4. Coleman, J.N., Khan, U., Blau, W.J., and Gunko, Y.K. (2006) *Carbon*, **44**, 1624–1652.
5. Thostenson, E.T., Li, C., and Chou, T.W. (2005) *Compos. Sci. Technol.*, **65**, 491–516.
6. Seidel, G.D. and Lagoudas, D.C. (2006) *Mech. Mater.*, **38**, 884–907.
7. Coleman, J.N., Khan, U., and Gunko, Y.K. (2006) *Adv. Mater.*, **18**, 689–706.
8. Xie, X.L., Mai, Y.W., and Zhou, X.P. (2005) *Mater. Sci. Eng. R*, **49**, 89–112.
9. Lau, K.T., Gu, C., and Hui, D. (2006) *Compos., Part B*, **37**, 425–436.
10. Liu, P. (2005) *Eur. Polym. J.*, **41**, 2693–2703.
11. Chang, T.E., Jensen, L.R., Kisliuk, A., Pipes, R.B., Pyrz, R., and Sokolov, A.P. (2005) *Polymer*, **46**, 439–444.
12. Wagner, H.D., Lourie, O., Feldman, Y., and Tenne, R. (1998) *Appl. Phys. Lett.*, **72**, 188–190.
13. Cooper, C.A., Young, R.J., and Halsall, M. (2001) *Composites, Part A*, **32**, 401–411.
14. McCarthy, B., Coleman, J.N., Czerw, R., Dalton, A.B., In het Panhuis, M., Maiti, A., Drury, A., Bernier, P., Nagy, J.B., Byrne, H.J., Caroll, D.L., and Blau, W.J. (2002) *J. Phys. Chem. B*, **106**, 2210–2216.
15. Zalamea, L., Kim, H., and Pipes, R.B. (2007) *Compos. Sci. Technol.*, **67**, 3425–3433.
16. Ding, W., Eitan, A., Fisher, F.T., Chen, X., Dikin, D.A., Andrews, R., Brinson, L.C., Schadler, L.S., and Ruoff, R.S. (2003) *Nano Lett.*, **3**, 1593–1587.

17 Lordi, V. and Yao, N. (2000) *J. Mater. Res.*, **15**, 2770–2779.
18 Singh, S., Pei, Y., Miller, R., and Sundararajan, P.R. (2003) *Adv. Funct. Mater.*, **13**, 868–872.
19 Wan, H., Delale, F., and Shen, L. (2005) *Mech. Res. Commun.*, **32**, 481–489.
20 Frankland, S.J.V., Caglar, A., Brenner, D.W., and Griebel, M. (2002) *J. Phys. Chem. B*, **106**, 3046–3048.
21 Haque, A. and Ramasetty, A. (2005) *Compos. Struct.*, **71**, 68–77.
22 Guz, A.N. and Rushchitskii, Y. (2003) *Int. Appl. Mech.*, **39**, 1271–1293.
23 Guz, I.A., Rodger, A.A., Guz, A.N., and Rushchitsky, J.J. (2007) *Composites, Part A*, **38**, 1234–1250.
24 Wagner, H.D. (2002) *Chem. Phys. Lett.*, **361**, 57–61.
25 Odegard, G.M., Gates, T.S., Wise, K.E., Park, C., and Siochi, E.J. (2003) *Compos. Sci. Technol.*, **67**, 1671–1687.
26 Namilae, S., and Chandra, N. (2006) *Compos. Sci. Technol.*, **66**, 2030–2038.
27 Bower, C., Rosen, R., Jin, L., Han, J., and Zhou, O. (1999) *Appl. Phys. Lett.*, **74**, 3317–3319.
28 Waters, J.F., Guduru, P.R., and Xu, J.M. (2006) *Compos. Sci. Technol.*, **66**, 1138–1147.
29 Li, C. and Chou, T.-W. (2006) *Compos. Sci. Technol.*, **66**, 2409–2414.
30 Andrews, R., Jacques, D., Minot, M., and Rantell, T. (2002) *Macromol. Mater. Eng.*, **287**, 395–403.
31 Wei, C., Srivastava, D., and Cho, K. (2002) *Nano Lett.*, **2**, 647–650.
32 Wei, C. (2006) *Appl. Phys. Lett.*, **88**, 093108.
33 Brown, D., Clarke, H.R., Okuda, M., and Yamazaki, T. (1994) *J. Chem. Phys.*, **100**, 1684–1692.
34 Wichmann, M.H.G., Schulte, K., and Wagner, H.D. (2008) *Compos. Sci. Technol.*, **68**, 329–331.
35 Liu, L. and Wagner, H.D. (2005) *Compos. Sci. Technol.*, **65**, 1861–1868.
36 Vaia, R. and Wagner, H.D. (2004) *Mater. Today*, **7**, 32–37.
37 Wagner, H.D. and Vaia, R.A. (2004) *Mater. Today*, **7**, 38–42.
38 Jiang, K.R. and Penn, L.S. (1992) *Compos. Sci. Technol.*, **45**, 89–103.
39 Schadler, L.S., Giannaris, S.C., and Ajayan, P.M. (1998) *Appl. Phys. Lett.*, **73**, 3842.
40 Gao, H. and Ji, B. (2003) *Eng. Fract. Mech.*, **70**, 1777–1791.
41 Magonov, S.N., Elings, V., and Whangbo, M.-H. (1997) *Surf. Sci.*, **375**, 385–391.
42 Satapathy, B.K., Lach, R., Weidisch, R., Schneider, K., Janke, A., and Knoll, K. (2006) *Eng. Fract. Mech.*, **73**, 2399–2412.
43 Cooper, C.A., Ravich, D., Lips, D., Mayer, J., and Wagner, H.D. (2002) *Compos. Sci. Technol.*, **62**, 1105–1112.
44 Zhu, J., Peng, H., Rodriguez-Macias, F., Margrave, J.L., Khabashesku, V.N., Imam, A.M., Lozano, K., and Barrera, E.V. (2004) *Adv. Funct. Mater.*, **14**, 643–648.
45 Qin, S., Qin, D., Ford, W.T., Resasco, D.E., and Herrera, J.E. (2004) *Macromolecules*, **37**, 752–757.
46 Safadi, B., Andrews, R., and Grulke, E.A. (2002) *J. Appl. Polym. Sci.*, **84**, 2660–2669.
47 Qian, D., Dickey, E.C., Andrews, R., and Rantell, T. (2000) *Appl. Phys. Lett.*, **76**, 2868–2870.
48 Kong, H., Gao, C., and Yan, D. (2004) *J. Am. Chem. Soc.*, **126**, 412–413.
49 Jin, Z., Pramoda, K.P., Xu, G., and Goh, S.H. (2001) *Chem. Phys. Lett.*, **337**, 43–47.
50 Christopher, A.D. and Tour, J.M. (2004) *J. Phys. Chem. A*, **108**, 11151–11159.
51 Blake, R., Gun'ko, Y.K., Coleman, J., Cadek, M., Fonseca, A., Nagy, J.B., and Blau, W.J. (2004) *J. Am. Chem. Soc.*, **126**, 10226–10227.
52 Paiva, M.C., Zhou, B., Fernando, K.A.S., Lin, Y., Kennedy, J.M., and Sun, Y.-P. (2004) *Carbon*, **42**, 2849–2854.
53 Liu, L., Barber, A.H., Nuriel, S., and Wagner, H.D. (2005) *Adv. Funct. Mater.*, **15**, 975–980.
54 Ramasubramaniam, R., Chen, J., and Liu, H. (2003) *Appl. Phys. Lett.*, **83**, 2928–2930.
55 Sandler, K.W., Kirk, J.E., Kinloch, I.A., Shaffer, M.S.P., and Windle, A.H. (2003) *Polymer*, **44**, 5893–5899.

56 Pötschke, P., Abdel-Goad, M., Alig, I., Dudkin, S., and Lellinger, D. (2004) *Polymer*, **45**, 8863–8870.

57 Kota, A.K., Cipriano, B.H., Duesterberg, M.K., Gershon, A.L., Powell, D., Raghavan, S.R., and Bruck, H.A. (2007) *Macromolecules*, **40**, 7400–7406.

58 Nogales, A., Broza, G., Roslaniec, Z., Schulte, K., Šics, I., Hsiao, B.S., Sanz, A., García-Gutiérrez, M.C., Rueda, D.R., Domingo, C., and Ezquerra, T.A. (2004) *Macromolecules*, **37**, 7669–7672.

59 Chen, J., Liu, H., Weimar, W.A., Halls, M.D., Waldeck, D.H., and Walker, G.C. (2002) *J. Am. Chem. Soc.*, **124**, 9034–9035.

60 Wong, M., Paramsothy, M., Xu, X.J., Ren, Y., Li, S., and Liao, K. (2003) *Polymer*, **44**, 7757–7764.

61 Ajayan, P.M., Schadler, L.S., Giannaris, C., and Rubio, A. (2000) *Adv. Mater.*, **12**, 750–753.

62 Andrews, R., Jacques, D., Qian, D., and Rantell, T. (2002) *Acc. Chem. Res.*, **35**, 1008–1017.

63 Zhang, X., Liu, T., Sreekumar, T.V., Kumar, S., Moore, V.C., Hauge, R.H., and Smalley, R.E. (2003) *Nano Lett.*, **3**, 1285–1288.

64 Xia, H., Wang, Q., and Qiu, G. (2003) *Chem. Mater.*, **15**, 3879–3886.

65 Sabba, Y. and Thomas, E.L. (2004) *Macromolecules*, **37**, 4815–4820.

66 Chen, J., Rajagopal, R., Xue, C.H., and Liu, H.Y. (2006) *Adv. Funct. Mater.*, **16**, 114–119.

67 Wenseleers, W., Vlasov, I., Goovaerts, E., Obraztsova, E.D., Lobach, A.S., and Bouwen, A. (2004) *Adv. Funct. Mater.*, **14**, 1105–1112.

68 Kashiwagi, T., Fagan, J., Douglas, J.F., Yamamoto, K., Heckert, A.N., Leigh, S.D., Obrzut, J., Du, F., Lin-Gibson, S., Mu, M., Winey, K.I., and Haggenmueller, R. (2007) *Polymer*, **48**, 4855–4866.

69 Koerner, H., Liu, W., Alexander, M., Mirau, P., Dowty, H., and Vaia, R.A. (2005) *Polymer*, **46**, 4405–4420.

70 Pötschke, P., Fornes, T.D., and Paul, D.R. (2002) *Polymer*, **43**, 3247–3255.

71 Chang, T.E., Kisliuk, A., Rhodes, M., Brittain, W.J., and Sokolov, A.P. (2006) *Polymer*, **47**, 7740–7746.

72 Deng, F., Ogasawara, T., and Takeda, N. (2007) *Compos. Sci. Tech.*, **67**, 2959–2964.

73 Guzman de Villoria, A. and Miravete, R. (2007) *Acta Mater.*, **55**, 3025–3031.

74 Du, F., Scogna, R.C., Zhou, W., Brand, S., Fischer, J.E., and Winey, K.I. (2004) *Macromolecules*, **37**, 9048–9055.

75 Kashiwagi, T., Grulke, E., Hilding, J., Groth, K., Harris, R., Butler, K., Shields, J., Kharchenko, S., and Douglas, J. (2004) *Polymer*, **45**, 4227–4239.

76 Assouline, E., Lustiger, A., Barber, A.H., Cooper, C.A., Klein, E., Wachtel, E., and Wagner, H.D. (2003) *J. Polym. Sci., Part B: Polym. Phys.*, **41**, 520–527.

77 Yang, J., Lin, Y., Wang, J., Lai, M., Li, J., Liu, J., Tong, X., and Cheng, H. (2005) *J. Appl. Polym. Sci.*, **98**, 1087–1091.

78 Seo, M.K., Lee, J.R., and Park, S.J. (2005) *Mater. Sci. Eng. A*, **404**, 79–84.

79 Leelapornpisit, W., Ton-That, M., Perrin-Sarazin, F., Cole, K.C., Denault, J., and Simard, B. (2005) *J. Polym. Sci., Part B: Polym. Phys.*, **43**, 2445–2453.

80 Dondero, W.E., and Gorga, R.E. (2006) *J. Polym. Sci., Part B: Polym. Phys.*, **44**, 864–878.

81 Moore, E.M., Ortiz, D.L., Marla, V.T., Shambaugh, R.L., and Grady, B.P. (2004) *J. Appl. Polym. Sci.*, **93**, 2926–2933.

82 (a) Bhattacharya, A.R., Sreekumar, T.V., Liu, T., Kumar, S., Ericson, L.M., Hauge, R.H., and Smalley, R.E. (2003) *Polymer*, **44**, 2373–2377.
(b) Ash, B.J., Siegel, R.W., and Schadler, L.S. (2004) *J. Polym. Sci., Part B: Polym. Phys.*, **42**, 4371–4383.

83 Cadek, M., Coleman, J.N., Ryan, K.P., Nicolosi, V., Bister, G., Fonseca, A., Nagy, J.B., Szostak, K., Béguin, F., and Blau, W.J. (2004) *Nano Lett.*, **4**, 353–356.

84 Chow, T.S. (1980) *J. Mater. Sci.*, **15**, 1873–1888.

85 Bliznakov, E.D., White, C.C., and Shaw, M.T. (2000) *J. Appl. Polym. Sci.*, **77**, 3220–3227.

86 Szazdi, L., Pukanszky, B., Jr., Vancso, G.J., and Pukanszky, B. (2006) *Polymer*, **47**, 4638–4648.
87 Satapathy, B.K., Weidisch, R., Pötschke, P., and Janke, A. (2007) *Compos. Sci. Technol.*, **67**, 867–879.
88 Mai, Y.W. and Powell, P. (1991) *J. Polym. Sci., Part B: Polym. Phys.*, **29**, 785–793.
89 Wu, J., Mai, Y.-W., and Cotterell, B. (1993) *J. Mater. Sci.*, **28**, 3373–3384.
90 Chen, Y.H., Mai, Y.-W., Tong, P., and Zhang, L.C. (2000) *Fracture of Polymers, Composites and Adhesives, ESIS Publication* (eds J.G. Williams and A. Pavan), Elsevier, Amsterdam.
91 Fisher, F.T., and Brinson, L.C. (2003) 44th AIAA/ASME/ASCE/AHS/ASC Structures, Structural Dynamics, and Materials Conference, Norfolk, VA.
92 Schadler, L.S. (2003) *Nanocomposite Science and Technology* (eds P.M. Ajayan, L.S. Schadler, and P.V. Braun), Wiley-VCH Verlag GmbH.
93 Reincke, K., Grellmann, W., Lach, R., and Heinrich, G. (2003) *Macromol. Mater. Eng.*, **288**, 181–189.
94 (a) Grellmann, W., Heinrich, G., and Cäsar, T. (2001) *Deformation and Fracture of Polymers* (eds W. Grellmann and S. Seidler), Springer, Heidelberg, Berlin, p. 479.
(b) Reincke, K., Lach, R., Grellmann, W., and Heinrich, G. (2001) *Deformation and Fracture of Polymers* (eds W. Grellmann and S. Seidler), Springer, Heidelberg, Berlin, p. 493.
95 Lach, R., Kim, G.M., Michler, G.H., Grellmann, W., and Albrecht, K. (2006) *Macromol. Mater. Eng.*, **291**, 263–271.
96 Adhikari, R., Lach, R., Michler, G.H., Weidisch, R., and Knoll, K. (2003) *Macromol. Mater. Eng.*, **288**, 432–439.
97 Grellmann, W. (2001) *Deformation and Fracture of Polymers* (eds W. Grellmann and S. Seidler), Springer, Heidelberg, Berlin, p.3.
98 Lach, R., Weidisch, R., Janke, A., and Knoll, K. (2004) *Macromol. Rapid Commun.*, **25**, 2019–2024.
99 Cooper, C.A., Cohen, S.R., Barber, A.H., and Wagner, H.D. (2002) *Appl. Phys. Lett.*, **81**, 3873–3875.
100 Xia, Z., Riester, L., Curtin, W.A., Li, H., Sheldon, B.W., Liang, J., Chang, B., and Xu, J.M. (2004) *Acta Mater.*, **52**, 931–944.
101 Satapathy, B.K., Ganß, M., Weidisch, R., Pötschke, P., Jehnichen, D., Keller, T., and Jandt, K.D. (2007) *Macromol. Rapid. Commun.*, **28**, 834–841.
102 Ganß, M., Satapathy, B.K., Thunga, M., Weidisch, R., Pötschke, P., and Jehnichen, D. (2008) *Acta Mater.*, **56**, 2247–2261.
103 Argon, A.S. and Cohen, R.E. (2003) *Polymer*, **44**, 6013–6032.
104 Ganß, M., Satapathy, B.K., Thunga, M., Weidisch, R., Pötschke, P., and Janke, A. (2007) *Macromol. Rapid Commun.*, **28**, 1624–1633.

8
Barrier Resistance Generation in Polymer Nanocomposites[1]
Vikas Mittal

8.1
Introduction

The development of polymer nanocomposites during the past decade has revolutionized research into composites technology [1–4]. The incorporation of inorganic fillers was achieved long ago to enhance the properties of neat polymers, and nanocomposites are specialty products where these inorganic fillers are dispersed in the nanoscale, and with at least one dimension of the filler phase being less than 100 nm. The fillers used in such composites are high aspect ratio layered silicates, mostly montmorillonite (MMT), which are surface-modified with the long alkyl chain ammonium cations exchanged on their surfaces and with sodium and potassium cations naturally present. The filler platelets in natural form are electrostatically held, leading to interlayers in between the platelets, thus requiring their surface modification for easy delamination in the polymer matrix. This nanoscale dispersion of the filler leads to increased interfacial contacts with the polymer matrix, and subsequently to a significant enhancement of the composite's properties at very low filler volume fractions. This helps the composite materials to retain their transparency and low density.

The food and drug industries employ vast amounts of polymers as packaging materials. Moreover, trends of their ever-increasing use in such applications are apparent in terms of their rapidly increasing substitution for traditional materials such as glass and metal. In such applications, transparency and gas barrier are necessary criteria. In operation, these materials are required to demonstrate a high permeation resistance to humidity and oxygen, and must also withstand high temperatures during sterilization stages. The inclusion of high aspect ratio layered silicate particles can be expected to bring about the permeation barrier in polymers by hindering the penetration of gas molecules and increasing their average path length [5–7]. However, this is not straightforward, as the barrier performance

---

1) These studies were carried out at the Institute of Chemical and Bioengineering, Department of Chemistry and Applied Biosciences, ETH Zurich, 8093 Zurich, Switzerland.

Optimization of Polymer Nanocomposite Properties. Edited by Vikas Mittal
Copyright © 2010 WILEY-VCH Verlag GmbH & Co. KGaA, Weinheim
ISBN: 978-3-527-32521-4

of the layered silicate filled polymer nanocomposites may be affected by a number of factors, including the chemical architecture of the polymers, the chemical nature of the surface modification present on the clay surface and its compatibility with the polymer matrix, along with the interactions of the permeant molecules with the polymer matrices and organic–inorganic interfaces. Thus, it is very important to understand these factors in order to optimize the barrier properties of the nanocomposite materials. For this, the behavior of polar and nonpolar polymers for the nanocomposite synthesis and filler delamination or exfoliation at the nanoscale must be understood. Similarly, different polar and nonpolar surface modifications of the layered silicates should be investigated. Finally, the permeation behavior of the interacting and noninteracting permeant molecules, such as water vapor and oxygen, respectively, should be analyzed through these nanocomposites.

8.2
Theory of Permeation

According to Fick's law of diffusion, the flux J through a polymer membrane is described by the following equation:

$$J = \frac{DS(p_1 - p_2)}{d} \tag{8.1}$$

where D is the diffusion coefficient, S is the solubility coefficient, and p_1 and p_2 are ambient pressures on the two sides of a film of thickness d. *Permeability* is defined in terms of a thermodynamic factor related to Henry's law, and a kinetic term which is a function of several factors including molecular packing (free volume), temperature, orientation, and crystallinity [8]. Thus, Equation 8.1 can be rewritten as:

$$J = \frac{P(p_1 - p_2)}{d} \tag{8.2}$$

where P is the permeability coefficient.

Nielsen predicted the permeation of gases through the composites containing inorganic fillers, and the reduction in the permeation through the composite, P_c, was related to the permeation of the polymer matrix, P_m, as a function of filler volume fraction, ϕ_f and filler aspect ratio, α by the following equation [9]:

$$\frac{P_c}{P_m} = \frac{\phi_m}{1 + \frac{\alpha \cdot \phi_f}{2}} \tag{8.3}$$

There are a number of other modified finite element models predicting the permeation reduction through the composites after the incorporation of inorganic fillers, such as the Cussler–Aris model [10], the Frederickson–Bicerano model [11], and the Lusti–Gusev model [12], which are respectively described in Equations 8.4–8.6.

$$P_1/P_m = 1/(1+\mu x^2) \tag{8.4}$$

Here, the geometric function μ is a function of aspect ratio of the filler platelets, while x is defined as the product of aspect ratio and filler volume fraction:

$$P_1/P_m = [1/(2+a_1\kappa x)+1/(2+a_2\kappa x)]^2 \tag{8.5}$$

where a_1 and a_2 are constants and κ is a function of aspect ratio of filler platelets:

$$\frac{P_1}{P_m} = \exp\left[-\left(\frac{x}{x_0}\right)^\beta\right] \tag{8.6}$$

where x_0 has a value of 3.47 and β is equal to 0.71.

The theoretical models described above assume the perfect alignment of the filler inclusions for the prediction of composite properties, as shown in Figure 8.1a. The more real case scenario is the complete misalignment of the filler platelets or partial alignment, depending on the shear in the flow direction during the composite synthesis. One such example of complete misalignment is shown in the case of epoxy nanocomposites in Figure 8.1b. The corresponding hypothetical path of permeant molecules through the composites with aligned and misaligned platelets is compared in Figure 8.1a and c. It is clear that the misaligned platelets are less effective as permeation barriers when compared to aligned platelets, and therefore the models must be modified by incorporating misalignment considerations along with other conditions, such as exfoliation of the platelets in varying extents, instead of the complete exfoliation which is generally assumed. Several

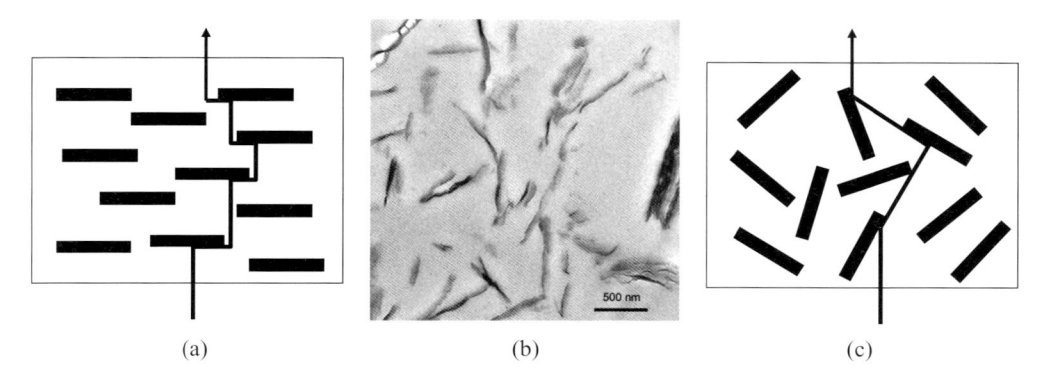

Figure 8.1 (a) Representation of the completely aligned platelets and path of a permeant molecule in such composite; (b) Transmission electron microscopy image of an epoxy composite, depicting the misaligned morphology; (c) A representation of a composite with misaligned platelets, and the path of a permeant molecule from one such composite.

modified models have incorporated these effects, and these are discussed later in the chapter.

8.3
Barrier Generation in Polar Nanocomposites

Polymer nanocomposites with a wide variety of polymer matrices have been synthesized. These have been more successful in terms of filler exfoliation, owing to the better match of polarity of the polar polymer with the partially polar alkyl ammonium ion-modified layered silicates. Although the layered silicates are organically modified, owing to the lower area of cross-section of the ammonium ions in comparison to the area per charge available on the surface of the MMT platelets, there remain residual forces of interaction present between the filler platelets, and this leads to a partial polarity of the filler interlayers. The polar polymer chains can thus intercalate in the filler interlayers, during which process these platelets are pushed larger distances, which leads to their exfoliation. Many studies involving such nanocomposites have reported significant enhancements in the mechanical properties of the composites [13–18]. For example, Wang *et al.* reported an increase of more than 100% in tensile strength, tensile modulus and strain-at-break of the polyurethane (PU) nanocomposites at a loading of 10 wt% organically modified clay modified with protonated octadecyl amine [13]. Messersmith *et al.* reported the synthesis of exfoliated epoxy nanocomposites in which the storage modulus was increased by 450% in the rubbery region, and by 58% in the glassy region, at a filler volume fraction of 4% as compared to the pure epoxy resin [16]. However, the barrier properties of these composites were generally neglected, most likely because it is generally assumed that improvements in mechanical properties automatically improve the other composite properties. However, as barrier properties are more sensitive to the microstructure of the composites along with the interface between the organic and inorganic phases, it is possible that improvements in mechanical performance do not automatically lead to improvements in the barrier resistance of the composites.

The importance of interactions between the surface modification on the filler surface and the polymer in the polar polymer nanocomposites was confirmed by Osman *et al.* [19]. Here, epoxy nanocomposites were synthesized by using MMTs modified with several different polar and nonpolar surface modifications. Consequently, it was observed that, when the clay was modified with a benzylhexadecyldimethylammonium (BzC16) modification, a basal plane spacing of 1.87 nm was observed, as compared to the spacing of 1.25 nm for pristine, unmodified MMT. When the modified clay was suspended in a solvent, the solvent dispersion did not show any diffraction peak in the X-ray diffractograms, which indicated that the filler was totally dispersed in the solvent and that the periodicity had been completely lost. However, when this dispersion was added with an epoxy prepolymer and the polymer was allowed to intercalate well around the dispersed clay platelets, a basal plane spacing at 3.55 nm was observed in the diffractograms. The epoxy prepolymer had a low molecular weight, and was expected to dissolve around the layered platelets, but to keep the platelets dispersed at sufficiently long distances as to have no peak in the diffractograms. However, the presence of a basal plane spacing peak in the diffractograms following the addition of a prepolymer indicated that there might be a certain mismatch between the prepolymer molecules and the dispersed clay platelets. On the other hand, when the clay was modified with a benzyldibutyl(2-hydroxyethyl)ammonium (Bz10H) modification, a basal plane spacing of 1.52 nm was observed in the modified clay. During the dispersion of this clay in the solvent, a basal plane spacing of 1.9 nm was observed, which indicated that these platelets were not completely dispersed in the solvent, most likely because of a higher polarity of the surface modification, and there a certain amount of periodicity was present in the clay platelets. However, when the epoxy prepolymer was added to the system, the basal plane spacing was not negatively impacted and remained the same, at 1.9 nm. This clearly indicated that, in this case, the prepolymer possibly had no negative interaction with the modified clay. As the modified clay in both the cases was different only in terms of the surface modification, it is clear that the observed changes in basal plane spacing are the result of interactions related to this surface modification. In the case of the benzylhexadecyldimethylammonium modification, the polar prepolymer molecules may not have been compatible with the modification molecules, whilst in the case of the more polar modification molecules of benzyldibutyl(2-hydroxyethyl)ammonium, a much better polarity match with the prepolymer molecules can be expected. The result of these interactions at the interface, owing to the compatibility between the clay modification and the polymer chains, results in the corresponding oxygen barrier performance of the composites in these epoxy nanocomposites. A decrease of 20% was observed at 5 vol% of the benzylhexadecyldimethylammonium-modified clay, whereas a much higher decrease in oxygen permeation (75%) was observed at 5 vol% benzyldibutyl(2-hydroxyethyl)ammonium-modified clay. This clearly indicates that, for polar polymer systems, it is not the initial high basal plane spacing of the modified clay which helps in improved barrier performance, but rather more the interactions between the organic and inorganic phases at the interface which dictate the barrier performance of the composites. Similarly, when the MMTs were modified with dioctadecyldimethylammonium (2C18) modifications, the prepolymer chains were not compatible with the surface modification, owing to a polarity mismatch which resulted in a similar decrease in the basal plane spacing of the clay dispersion in solvent as when the prepolymer was added to it. As a result, the permeation behavior of these composites was also negatively impacted, and an increase in oxygen permeation was observed in the composites. Another point to consider is that, although the initial basal plane spacing of the filler was highest in this case (2.51), the final oxygen barrier performance was worst, thus again confirming the importance of the interfacial interactions. A similar increase in oxygen permeation was also observed in the case of PU nanocomposites, where the similar dioctadecyldimethylammonium-modified MMT was added [20]. The permeation was also observed to increase as a function of the filler volume fraction, which clearly indicated that incompatibility at the interface was increased when the filler amount was increased, and this subsequently led an increase in oxygen permeation through the nanocomposites.

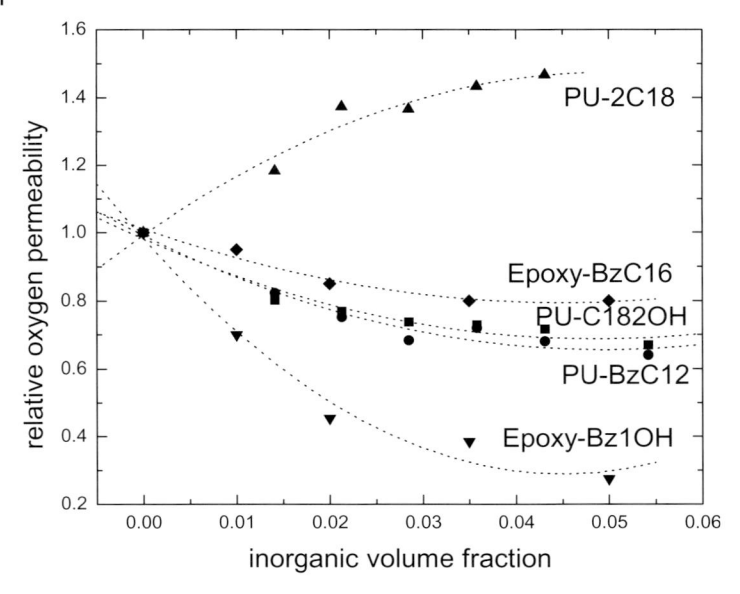

Figure 8.2 Oxygen permeation through the polyurethane and epoxy composites as a function of different filler surface modifications and filler volume fraction.

The other modifications, which were more polar in nature, resulted in a decrease in oxygen permeation as a function of the respective interactions with the polymer matrix. Figure 8.2 shows the oxygen permeation behavior of these nanocomposites as a function of filler volume fraction and different filler modifications.

It was also shown quantitatively [19] that the basal plane spacing of the filler in the composite, or the increase in the basal plane spacing of the filler in the composite as compared to the modified clay, was not the cause of the barrier resistance generation in the polar nanocomposites. Rather, the exfoliated composites, as a result of positive interactions at the interface, were responsible for this barrier performance. Normally, investigations with X-ray diffractograms are used to classify the composite morphology as intercalated and exfoliated. However, owing to the qualitative nature of the X-ray signals, it may not be the true representative of the real composite morphology. In reality, the nanocomposites generally have a mix of intercalated and exfoliated platelets, the quantification of which is not easy when using only X-ray signals, as these tend to be affected by the sample preparation, the orientation of the clay platelets, and also by any defects in the crystal structure. Therefore, even though a signal is present in the diffractograms showing the presence of some intercalated platelets, it might still be possible that the majority of the platelets would be exfoliated. As an example, the X-ray diffractograms of epoxy composites with both benzylhexadecyldimethylammonium- and benzyldibutyl(2-hydroxyethyl)ammonium-modified MMTs are shown in Figure 8.3. The presence of diffraction signals in these cases leads to information that

8.3 Barrier Generation in Polar Nanocomposites | 179

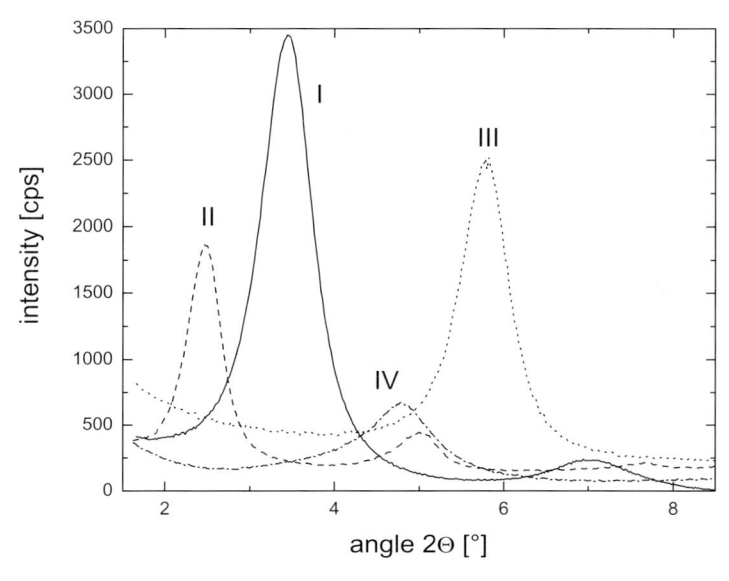

Figure 8.3 X-ray diffractograms of the benzylhexadecyldimethylammonium-(I) and benzyldibutyl(2-hydroxyethyl)ammonium-(III) modified montmorillonites, and their 3 vol% epoxy composites II and IV, respectively.

both are only intercalated, but in reality the composites with benzyldibutyl (2-hydroxyethyl)ammonium-modified MMTs are much more exfoliated, as confirmed by the much better barrier properties of these composites. The transmission electron microscopy images of these benzyldibutyl(2-hydroxyethyl)ammonium and benzylhexadecyldimethylammonium systems are shown in Figure 8.4, which also confirms the permeation findings. The platelets in the first case were visible to be extensively exfoliated (Figure 8.4a), while in the second case the platelets were more intercalated (Figure 8.4b). In the case of PU nanocomposites with modifications of a different chemical nature, the X-ray diffractograms have a similar nature, as shown in Figure 8.5. Irrespective of the extent of exfoliation of platelets in the composite, the intercalated signals were visible in the diffractograms. Likewise, in the PU nanocomposites reported elsewhere, similar dioctadecyldimethylammonium-modified MMTs led to significant improvements in the mechanical properties of the composites. However, as noted above, the oxygen barrier properties were negatively impacted, thus confirming the notion that it is absolutely important to optimize the barrier properties of nanocomposites separately from any mechanical considerations.

There has been always a need to relate the increase in basal plane spacing with the improvement in barrier properties, so that the barrier properties of nanocomposites can be optimally designed. However, due to the above-mentioned reasons, this is not always easy to achieve. As the benzyldibutyl(2-hydroxyethyl)ammonium-modified MMT led to a significant reduction in oxygen permeation, it was of

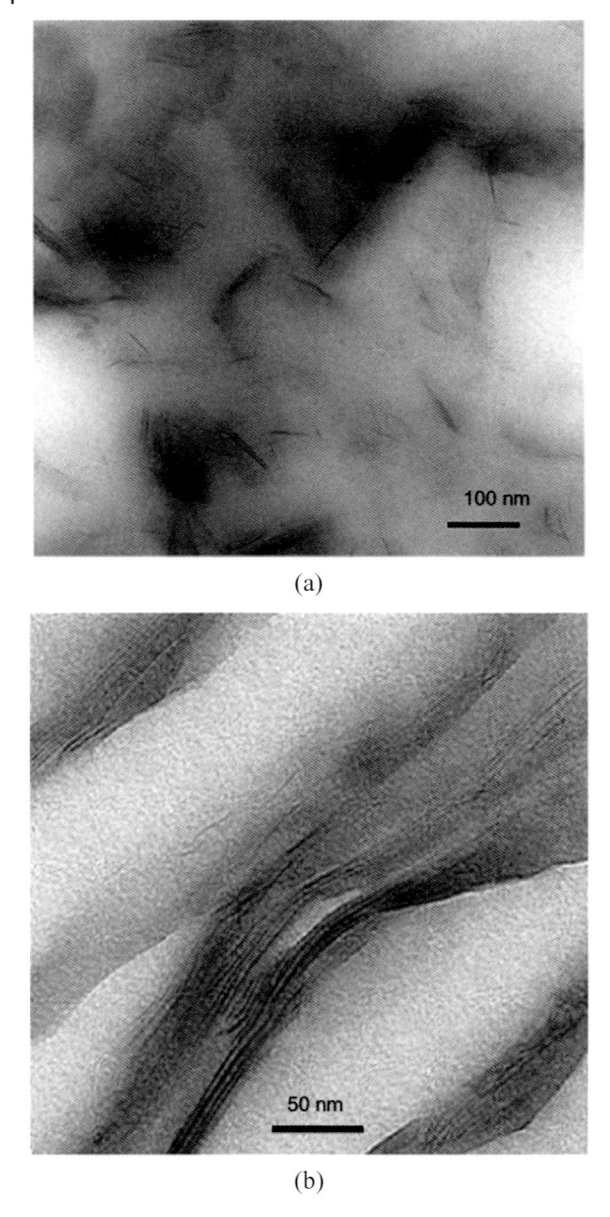

Figure 8.4 Transmission electron microscopy images of the 3 vol% epoxy nanocomposites consisting of (a) benzyldibutyl(2-hydroxyethyl)ammonium- and (b) benzylhexadecyldimethylammonium-modified montmorillonites.

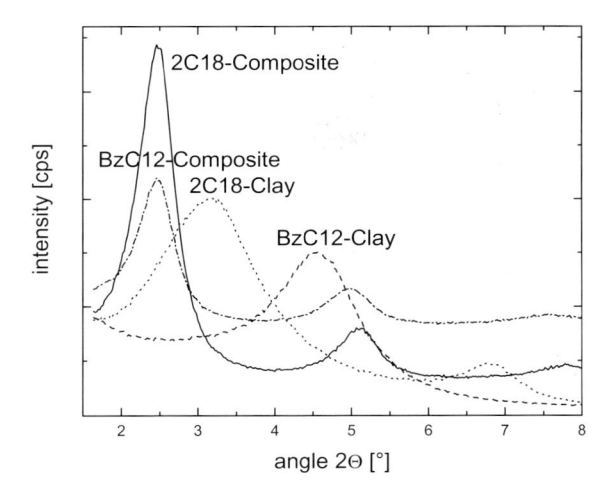

Figure 8.5 X-ray diffractograms of benzyldodecyldimethylammonium (BzC12) and dioctadecyldimethylammonium (2C18)-modified montmorillonites and their 3 vol% polyurethane nanocomposites.

interest to study the possible presence of higher basal plane spacing filler platelets apart from the signals seen in wide angle diffractograms. As shown in Figure 8.6a, the composite films had a strong anisotropic scattering signal, and a q-value of 0.0463 Å^{-1} was observed, which meant a basal plane spacing of 13.6 nm. However, owing to the similar signal observed from the substrate polypropylene foil and the qualitative nature of the X-ray signal, it is still the best to quantitatively analyze the composite properties (e.g., the gas barrier) in order to analyze the effect of interactions of the surface modifications with the polymer, and the resultant microstructure developments in the nanocomposites.

The nature of the permeant molecules and the subsequent effect of interfacial interactions on their diffusion behavior can be very different. Oxygen permeation in the above-mentioned examples of PU and epoxy nanocomposites decreased when the interface between the organic and inorganic phases was compatible. Yet, the oxygen permeation decreased to a smaller extent, or even increased, when the interfacial compatibility was not optimum. However, water vapor permeation was not affected by these factors. For example, water vapor permeation in PU nanocomposites with dioctadecyldimethylammonium-modified clay was decreased by more than 50% at 5 vol% filler fraction. This was the same system where oxygen permeation was increased on increasing the filler volume fraction. A similar water vapor resistance was also observed in the epoxy nanocomposites. The reasons for these different behaviors of oxygen and water vapor permeation within the same nanocomposite systems may lie in the nature of the permeant molecules themselves. Oxygen molecules are noble in nature, and do not interact with each other, nor with the polymer matrix. Therefore, whenever there is a slight incompatibility at the interface which leads to an increased free volume or micro voids, the

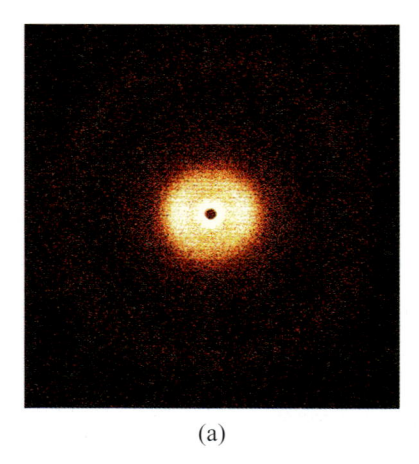

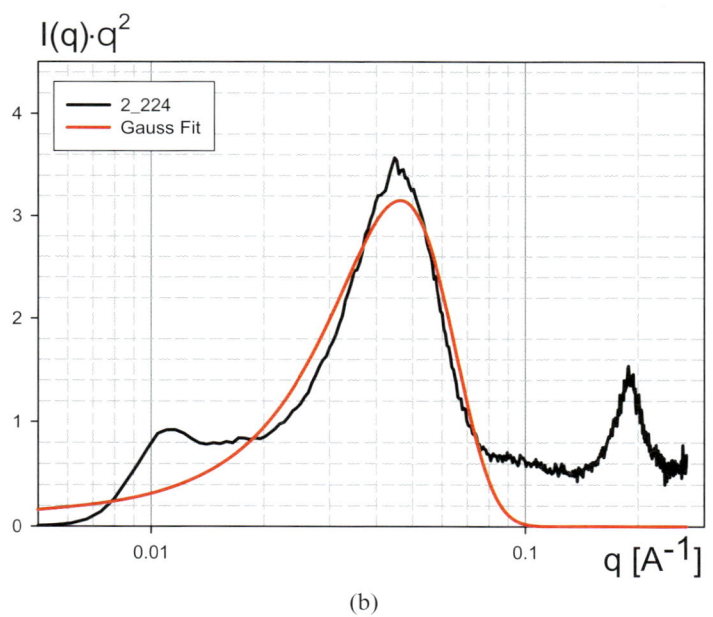

Figure 8.6 Small-angle X-ray diffraction studies on the epoxy nanocomposites containing benzyldibutyl(2-hydroxyethyl)ammonium-modified montmorillonite. (a) Scattering signal; (b) Diffractogram with $I(q) \cdot q^2$ as a function of q.

diffusion of oxygen and water through these areas is increased. However, water vapor molecules are very interactive in nature, and will interact not only with themselves but also with the polymer matrices by forming hydrogen bonds, and subsequently large clusters. Therefore it is quite possible that, even if there is a presence of small areas of high-void fractions at the organic–inorganic interface, the very large water molecule clusters are still unable to pass through. As the water vapor reduction through the nanocomposites was found to be more of a function of the hydrophobicity of the surface modification, the dioctadecyldimethylammonium modification was found to be superior in this case; unfortunately, however, it was of absolutely no use for oxygen permeation.

8.4
Barrier Generation in Nonpolar Nanocomposites

It has been observed in the case of polar nanocomposites, that the interfacial interactions are the key to achieving higher degrees of filler exfoliation, and subsequently barrier properties. However, the behavior of nonpolar polymer matrices towards barrier resistance generation differs from that of polar polymer systems, owing to the absence of any positive interfacial interactions between the organic and inorganic phases. In order for the mixtures of long-chain homopolymers with organically modified clays to be thermodynamically stable, the Flory–Huggins interactions parameter must be less than zero. Even then, such structures were predicted to exhibit an intercalated morphology, without exfoliation for polyolefins. Increasing the length of the tethered surfactants was predicted to improve the thermodynamic state of the system, as a greater distance generated among the clay layers would help to reduce the effective interactions between the clay sheets. For a given density of alkyl chains on the surface, long chains were predicted to form a more homogeneous phase than short chains [21]. Thus, even in the absence of any attractive interaction between the long polymer chains and the surfactant molecules (i.e., Flory–Huggins interactions parameter equal to zero) – that is, at theta conditions – the increase in the d-spacing caused by incorporating longer surfactant chains may help to produce a greater degree of delamination. The surface modification of a similar chemical nature to the polyolefin matrix itself is often assumed to be sufficient for exfoliation of the platelets; however, this does not occur owing to the absence of any interfacial interactions.

One method which is frequently used to overcome the problem of hydrophobicity of the polyolefin matrices is to partially polarize the matrix by adding low-molecular-weight compatibilizers which, owing to their amphiphilic nature, help to compatibilize the organic and inorganic phases [22–31]. Although the compatibilizer helps to improve the mechanical properties to a certain extent, it also plasticizes the matrix, and this leads to a reduction in the modulus. The addition of a compatibilizer may also negatively impact upon the barrier properties of the materials; consequently, it is not always true that improvements in the mechanical properties will lead automatically to improvements in the barrier properties of the nanocomposites, because such properties depend on different microstructural factors. For example, the barrier properties are much more sensitive to the interface between the components; hence, the opposite trends in properties could be observed if there was a presence of a certain incompatibility at the interface. When only 2 wt% of a compatibilizer with a different molecular weight and polarity was added to polypropylene nanocomposites with 3 vol% of the filler and with a dioctadecyldimethylammonium modification, the barrier performance either remained the same or permeation was increased through the composite films [32], whereas the tensile modulus was seen to increase. It was considered that, a mismatch between the polar compatibilizer chains and nonpolar surface modification chains at the interface might lead to the generation of an increased free volume or micro voids, which would help to increase the flow of permeant through the composites. Thus, an increase in permeation caused by interfacial incompatibility and a decrease in permeation caused by filler exfoliation would compete with each other. Compatibilizers with different molecular weights, as well as different polarities, affected the barrier properties in different ways, which indicated that the interfacial interactions also differed in each case. A maximum increase in permeation or a deterioration in barrier performance was observed with the most polar compatibilizer. A similar behavior was also seen in the PU and epoxy nanocomposites (as mentioned above), where polar polymer matrices were seen to show incompatibility with the nonpolar surface modifications; subsequently, permeation through the nanocomposite films was increased significantly as a function of the filler volume fraction [19, 20]. An exchange of polar modifications on the filler surface may represent a solution to achieve compatibility with the polar compatibilizer chains, although in such a scenario it cannot be guaranteed that the polypropylene chains would intercalate with the clay interlayers at all, owing to a too-polar environment within the interlayers.

A second approach to improving the barrier and mechanical properties of the nanocomposites is to optimize the filler modification. As noted above, the theoretical predictions suggest that, in case of nonpolar polymer composites, an increased basal plane spacing can lead to an improved delamination of the filler in the matrix, and this in turn would lead to better composite properties. When a comparison of the barrier properties of composites with octadecyltrimethylammonium- and docosyltriethylammonium-modified MMTs was conducted [33], a small increase in the chain length of the surface modification led to the docosyltriethylammonium system being almost twice as effective as the octadecyltrimethylammonium system. A 30% decrease in oxygen permeation, as compared to pure polypropylene, was observed when the long chains were exchanged on the surface, compared to a 17% decrease achieved with the short-chain counterparts. An increased chain length effectively reduces the interfacial interactions between the clay platelets holding them, and thus increases their potential of exfoliation in the polymer matrix under the action of shearing when compounded with a polymer at high temperature. Trials were also undertaken to exchange much longer modifications on the surface; however, owing to the polar ester groups present in the modification chains the permeation barrier was not significantly enhanced, for the above-mentioned reasons.

As it is difficult to exchange preformed long modification chains on the filler surface due to problems not only of solubility but also steric hindrance, another method of achieving a higher basal plane spacing in the modified MMTs would be to increase the chain density in the surface modifications. By doing this, the chains would lie straighter on the surface, thus increasing their angle with the surface and causing an increase in the basal plane spacing. The modified clay, with its reduced electrostatic forces of interaction between the platelets, would then be more dispersible than the MMT, with a lower chain density modification. The cation-exchange capacity (CEC) also has a significant effect on the composite properties, as a higher CEC leads to a greater number of charges per unit area (i.e., a lower area available per charge in the high-CEC clay), which in turn means that more organic modifications can be exchanged on the surface. Thus for the same cation, the MMT with a higher CEC would have a higher basal plane spacing. However, care must be exercised when selecting the MMTs, as the high-CEC MMTs would be more difficult to disperse in water and their surfaces would subsequently become modified. The basal plane spacing values of the octadecyltrimethylammonium (C18), dioctadecyldimethylammonium (2C18), and trioctadecylmethylammonium (3C18)-modified MMTs, and the corresponding oxygen permeation values, are listed in Table 8.1 [33]. The basal plane spacing was increased by increasing the chain density in the surface modification, and subsequently the oxygen permeation behavior of the composites was also improved. The rate of oxygen permeation through the composites without any filler was $89\,cm^3\,\mu m^{-1}\,m^{-2}\,d^{-1}\,mmHg^{-1}$, and at 3 vol% trioctadecylmethylammonium (3C18)-modified MMT, reductions of 40% and 35%, respectively, were observed for the modified MMTs with CECs of 880 and $680\,\mu Eq\,g^{-1}$. These values, in comparison with the polyethylene nanocomposites synthesized with the MMTs of CECs of 680 and $1000\,\mu Eq\,g^{-1}$, are shown in Figure 8.7. It is clear that, for both the polyethylene and polypropylene systems, the increases in both CEC and density of the chains in the surface modification would help to generate a better barrier performance in the composites. The TEM image of the polypropylene nanocomposite with 3 vol% dioctadecyldimethylammonium-modified MMT is shown in Figure 8.8. It was observed that that a partial exfoliation of the filler could still be achieved without using conventional compatibilizers, thus underlining the importance of filler surface optimization to improve the composite properties.

Table 8.1 Comparison of basal plane spacing and oxygen permeation values though the 3 vol% filler–polypropylene composites as a function of modification chain density [33].

Modification	M680		M880	
	Spacing (nm)	Oxygen permeation ($cm^3.\mu m.m^{-2}.d^{-1}.mmHg^{-1}$)	Spacing (nm)	Oxygen permeation ($cm^3.\mu m.m^{-2}.d^{-1}.mmHg^{-1}$)
C18	1.82	74	1.84	74
2C18	2.45	61	2.51	59
3C18	3.29	58	3.42	53

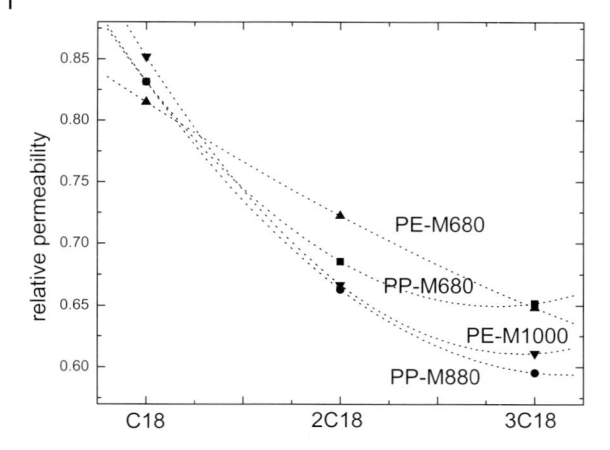

Figure 8.7 Relative gas permeability of polypropylene and polyethylene nanocomposites as a function of cation-exchange capacity, as well as density of the chains in the surface modifications [33, 34].

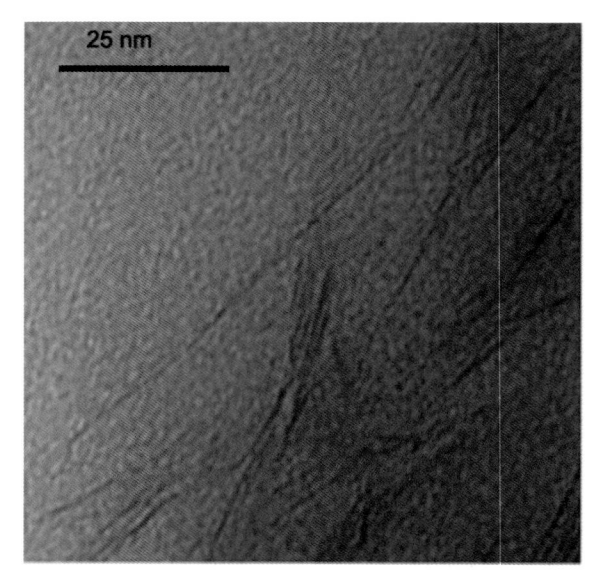

Figure 8.8 Transmission electron microscopy image of the 3 vol% 2C18-polypropylene nanocomposite. The black lines represent the cross-sections of silicate platelets.

Although the basal plane spacing values in surface-modified MMTs can be increased by the exchange of longer ammonium chain modifications, or with modifications of higher chain density, this is still insufficient to achieve complete exfoliation of the platelets in the polymer matrix. Thus, several other approaches to achieve optimal surface modifications have been developed.

One approach is to synthesize polymer chains that can bind ionically to the surface; these filler-bound chains are produced by first immobilizing either a monomer or initiator on the filler surface; this can subsequently be used to generate polymer chains attached to the clay surface by adding an external monomer [35–38]. This system can be controlled by altering the amount of platelet surface area covered with the initiator or monomer molecules, while the remainder can be exchanged with nonreactive alkyl ammonium chains. *Living polymerization techniques* can also be used to further control the polymerization reactions. These polymer-modified clays can then be compounded along with polyolefins, and represent very high-potential materials where the platelet surfaces are much more efficiently organophilized.

A second approach for filler organophilization involves the use of surface reactions on the filler surface; an example is surface esterification [39]. In this approach, the filler can be first modified with a reactive surface modification (e.g., modification with terminal hydroxy groups), and these then can be reacted with acid groups, for example, to generate long chains attached onto the filler surface. This system is much more simple than the surface polymerization reactions, and can lead to much higher basal plane spacing, owing to the problem of early termination in the surface-polymerization system. The X-ray diffractograms of the 2C18- and 3C18-modified MMTs, compared with MMTs modified by surface polymerization and esterification reactions, are shown in Figure 8.9. The basal

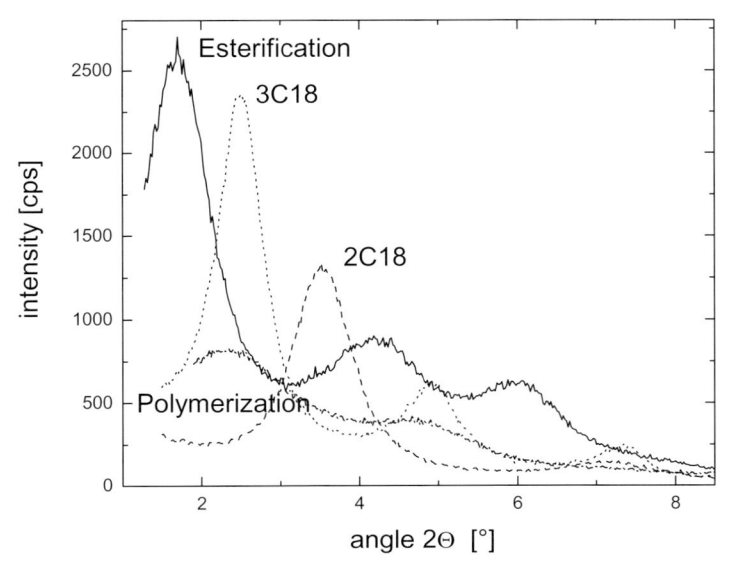

Figure 8.9 X-ray diffractograms of the 2C18- and 3C18-modified montmorillonites, and their comparison with the montmorillonites modified on the surface by surface polymerization and esterification reactions.

spacing values in these systems were observed, respectively, as 2.51, 3.42, 3.80, and 5.27 nm. Thus, higher basal plane spacing values can be achieved by using such surface-reaction methods.

A third method of achieving a better organic modification of the filler surface is to use *physical adsorption*, where organic molecules such as alkyl alcohols or polymer chains are adsorbed onto patches left uncovered on the filler surface following the surface modification [40]. The area of the ammonium modification head group attached onto the clay surface is generally smaller than the area available per charge on the filler surface. Therefore, the filler surface will remain polar, even after the ion exchange, and these voids can be filled by the adsorption of organic molecules from solution. The thermogravimetric analysis (TGA) of various surface-modified clays, namely dioctadecyldimethylammonium-modified MMT (2C18), trioctadecylmethylammonium-modified MMT (3C18), MMT after a surface esterification reaction, and MMT after the physical adsorption of poly(vinylpyrrolidone) (PVP) are shown in Figure 8.10. A comparison of the 2C18 and 3C18 methods with other methods shows clearly that the amount of organic matter bound to the clay surface is significantly increased after surface reaction or physical adsorption. This in turn would lead to a much higher basal plane spacing that would not be possible by exchanging the preformed chains.

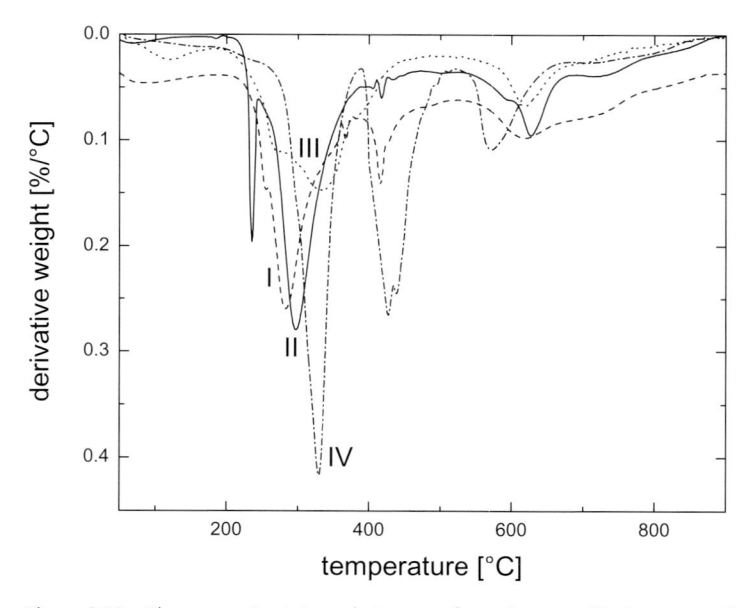

Figure 8.10 Thermogravimetric analysis curve for various modified montmorillonites. I = 2C18; II = 3C18; III = esterification on the clay surface; IV = physical adsorption on the clay surface.

8.5
Modeling of Barrier Properties of Composites

As noted above, most of the models used to describe or predict the barrier properties of nanocomposites are based on assumptions such as perfect alignment and/or complete exfoliation of the platelets. However, in reality this is generally not the case. In order to better predict the composite's properties, several different approaches have been developed. The results generated from a finite element model to predict the barrier properties of nanocomposites, and which take into account both misalignment and any incomplete exfoliation effects of the platelets, is shown in Figure 8.11 [41, 42]. It is possible, by using such models, to compare experimental values with predictions and also to predict the nanocomposite's properties. It is also possible to gain insights into the degree of randomness of the platelets in the composite, as a function of the reduction in the composite's properties with increasing filler volume fraction. Moreover, the average aspect ratio of platelets in the composite can also be predicted, as estimated from experimental values of the reduction in permeation through the composite. Most importantly, this method provides a much better evaluation of the average aspect ratio of the composites, which otherwise could not be estimates from TEM images, owing to the bending and folding of the platelets. The permeation reduction of different polymer matrices, as a function of the different surface modifications, may then be compared and the amount and nature of components or modifications for optimum improvements in the barrier properties predicted (see Figure 8.11).

One other method of predicting the barrier properties of nanocomposites involves experiments where factorial and mixture designs are generated for the purpose of predicting or modeling the nanocomposite's properties [43]. Parameters studied for such factorial design have included the inorganic volume fraction, the CEC of the MMT, and the number of octadecyl chains in the ammonium modification which are exchanged ionically on the clay surface. As shown in Figure 8.12, by holding a particular set of values from these factors, a complete set of permeation properties can be predicted that would assist in selecting the system components so as to achieve an optimum reduction in permeation through the nanocomposites.

In the mixtures design, the oxygen permeation of nanocomposites was analyzed in the light of components such as the amount of polymer, the amount of organic modification on the clay surface, and the amount of inorganic filler. The dotted line in Figure 8.13 shows the area of operation as the mixture was set with constraints of the amounts of each component. As none of the mixture components can be utilized over the ranges of 0 to 100% of the total weight, a constrained mixture was generated with constraints set on the polymer from 84 to 100%, inorganic from 0 to 11%, and the corresponding filler modification from 0 to 5% of the total weight of composite. From this, a contour plot can be generated that can be used to predict the permeation properties of a wide spectrum of values for the various amounts of components in the mixture.

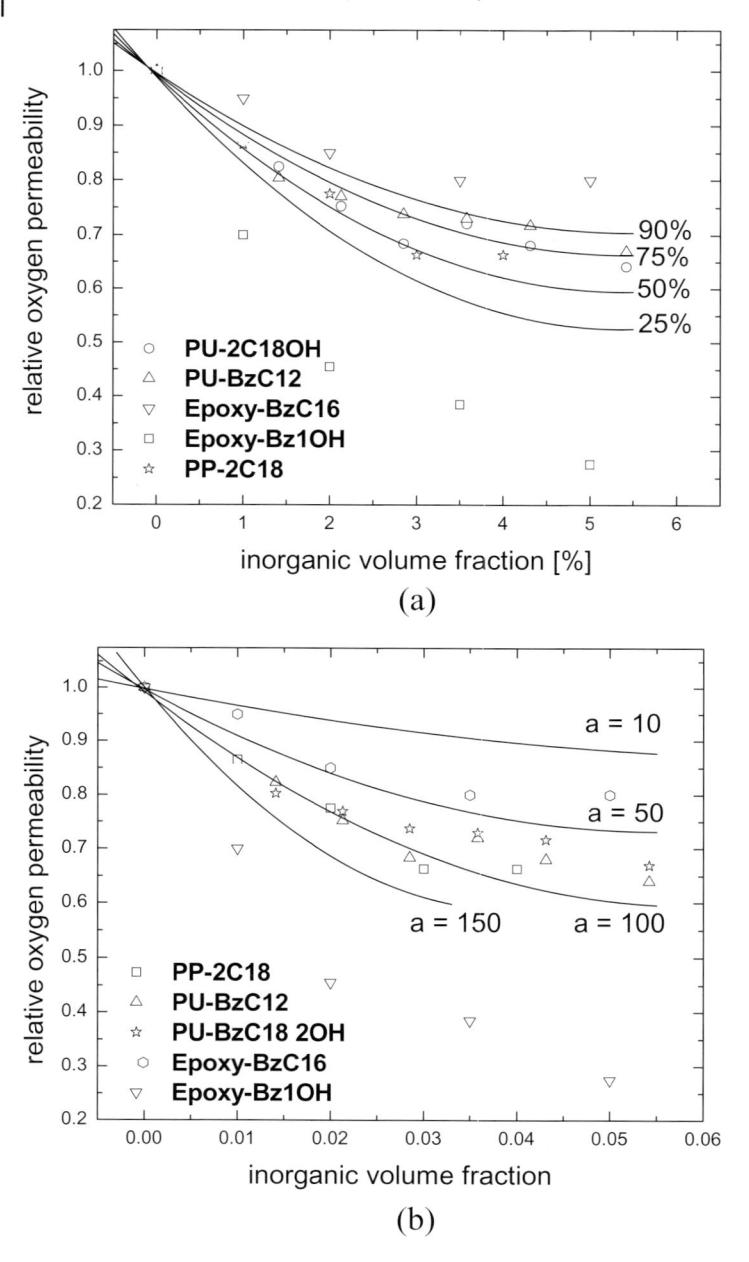

Figure 8.11 Modeling of the barrier properties of the nanocomposites as a function of (a) degree of randomness of platelets and (b) aspect ratio of the platelets. The various surface modifications as well polymer matrices have been incorporated into the comparison.

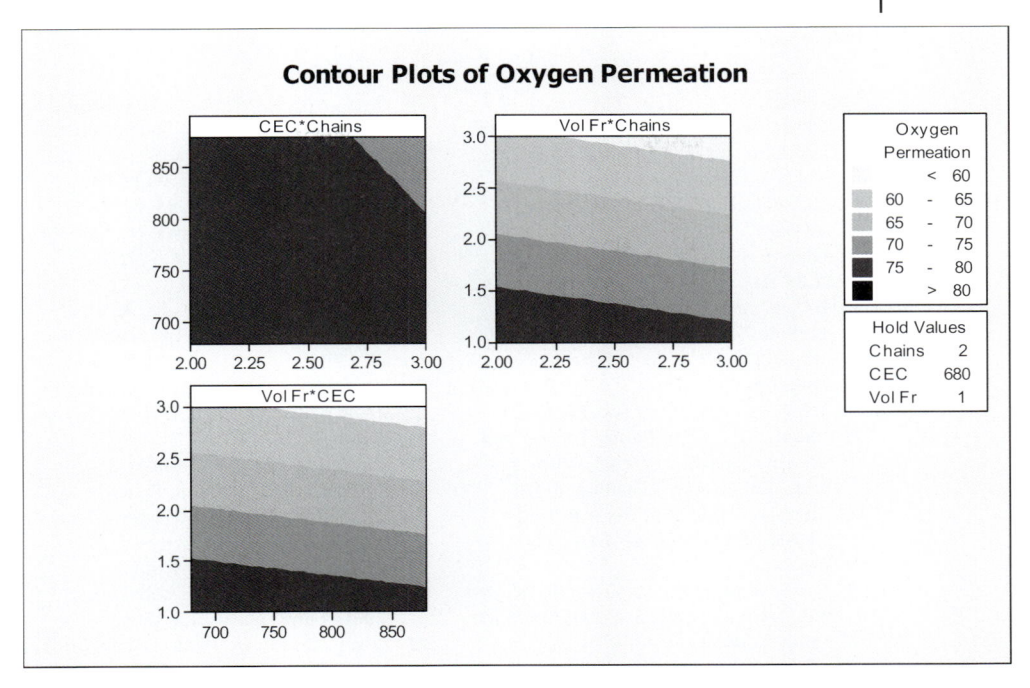

8.5 Modeling of Barrier Properties of Composites

Figure 8.12 Design of experiments methodology to predict the barrier properties of polymer nanocomposites by using the factorial design method. CEC = cation-exchange capacity.

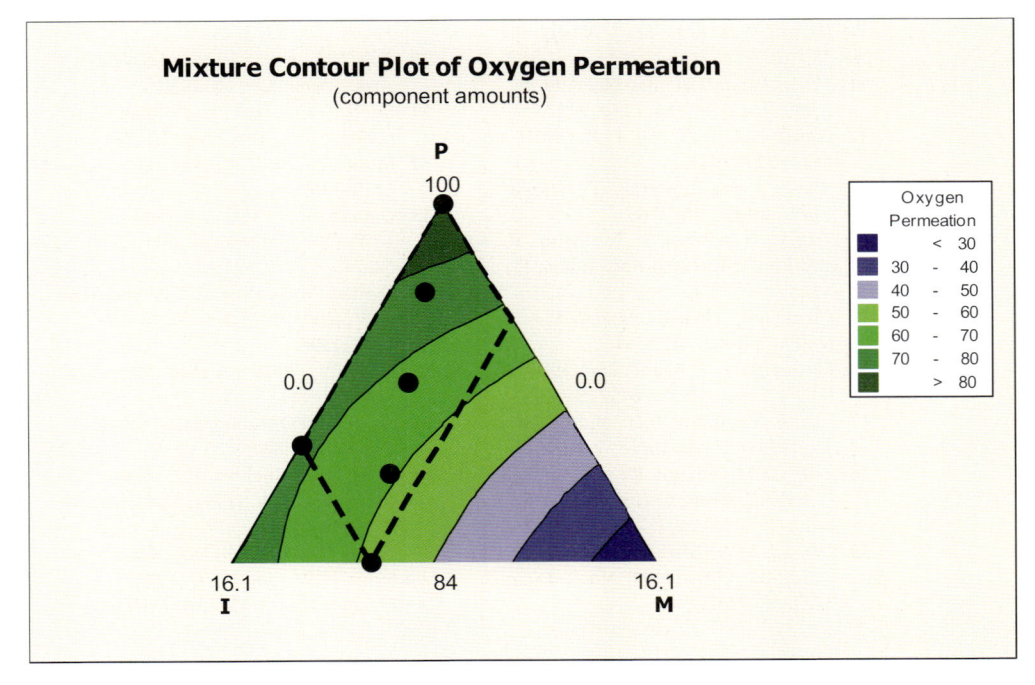

Figure 8.13 Design of experiments methodology to predict the barrier properties of polymer nanocomposites by using the mixture design method. P is the amount of polymer; M is the amount of organic modification on the clay surface; and I is the amount of inorganic filler.

References

1 Usuki, A., Kojima, Y., Kawasumi, M., Okada, A., Fukushima, Y., Kurauchi, T., and Kamigaito, O. (1993) *J. Mater. Res.*, **8**, 1179–1184.
2 Giannelis, E.P. (1996) *Adv. Mater.*, **8**, 29–35.
3 Gilman, J.W., Jackson, C.L., Morgan, A.B., Harris, R., Manias, E., Giannelis, E.P., Wuthenow, M., Hilton, D., and Phillips, S.H. (2000) *Chem. Mater.*, **12**, 1866–1873.
4 LeBaron, P.C., Wang, Z., and Pinnavaia, T.J. (1999) *Appl. Clay Sci.*, **15**, 11–29.
5 Bissot, T.C. (1990) *Barrier Polymer and Structures* (ed. W.J. Koros), American Chemical Society, Washington, DC.
6 Eitzman, D.M., Melkote, R.R., and Cussler, E.L. (1996) *AIChE J.*, **42**, 2–9.
7 Barrer, R.M. (1968) *Diffusion in Polymers* (eds J. Crank and G.S. Park), Academic Press, London.
8 Silvis, H.C. (1997) *Trends Polym. Sci.*, **5**, 75–79.
9 Nielsen, L.E. (1967) *J. Macromol. Sci.*, **A1**, 929.
10 Cussler, E.L., Hughes, S.E., Ward, W.J., III, and Aris, R. (1988) *J. Membr. Sci.*, **38**, 161.
11 Fredrickson, G.H. and Bicerano, J. (1999) *J. Chem. Phys.*, **110**, 2181.
12 Gusev, A.A. and Lusti, H.R. (2001) *Adv. Mater.*, **13**, 1641.
13 Wang, Z. and Pinnavaia, T.J. (1998) *Chem. Mater.*, **10**, 3769.
14 Zilg, C., Thomann, R., Mulhaupt, R., and Finter, J. (1999) *Adv. Mater.*, **11**, 49.
15 Chen, T.K., Tien, Y.I., and Wei, K.H. (2000) *Polymer*, **41**, 1345.
16 Messersmith, P.B. and Giannelis, E.P. (1994) *Chem. Mater.*, **6**, 1719.
17 Lan, T., Kaviratna, P.D., and Pinnavaia, T.J. (1995) *Chem. Mater.*, **7**, 2144.
18 Zilg, C., Muelhaupt, R., and Finter, J. (1999) *Macromol. Chem. Phys.*, **200**, 661.
19 Osman, M.A., Mittal, V., Mobridelli, M., and Suter, U.W. (2004) *Macromolecules*, **37**, 7250–7257.
20 Osman, M.A., Mittal, V., Mobridelli, M., and Suter, U.W. (2003) *Macromolecules*, **36**, 9851–9858.
21 Heinz, H. and Suter, U.W. (2004) *Angew. Chem., Int. Ed.*, **43**, 2239.
22 Kawasumi, M., Hasegawa, N., Kato, M., Usuki, A., and Okada, A. (1997) *Macromolecules*, **30**, 6333.
23 Usuki, A., Kato, M., Okada, A., and Kurauchi, T. (1997) *J. Appl. Polym. Sci.*, **63**, 137.
24 Hasegawa, N., Kawasumi, M., Kato, M., Usuki, A., and Okada, A. (1998) *J. Appl. Polym. Sci.*, **67**, 87.
25 Reichert, P., Nitz, H., Klinke, S., Brandsch, R., Thomann, R., and Muelhaupt, R. (2000) *Macromol. Mater. Eng.*, **275**, 8.
26 Manias, E., Touny, A., Wu, L., Strawhecker, K., Lu, B., and Chung, T.C. (2001) *Chem. Mater.*, **13**, 3516.
27 Zhang, Q., Fu, Q., Jiang, L., and Lei, Y. (2000) *Polym. Int.*, **49**, 1561.
28 Oya, A., Kurokawa, Y., and Yasuda, H. (2000) *J. Mater. Sci.*, **35**, 1045.
29 Xu, W., Liang, G., Wang, W., Tang, S., He, P., and Pan, W.P. (2003) *J. Appl. Polym. Sci.*, **88**, 3225.
30 Ellis, T.S. and D'Angelo, J.S. (2003) *J. Appl. Polym. Sci.*, **90**, 1639.
31 Su, S., Jiang, D.D., and Wilkie, C.A. (2004) *Polym. Deg. Stab.*, **83**, 321.
32 Mittal, V. (2008) *J. Appl. Polym. Sci.*, **107**, 1350–1361.
33 Osman, M.A., Mittal, V., and Suter, U.W. (2007) *Macromol. Chem. Phys.*, **208**, 68–75.
34 Osman, M.A., Rupp, J.E.P., and Suter, U.W. (2005) *J. Mater. Chem.*, **15**, 1298–1304.
35 Mittal, V. (2007) *J. Colloid Interface Sci.*, **314**, 141–151.
36 Fu, X. and Qutubuddin, S. (2001) *Polymer*, **42**, 807.
37 Meier, L.P., Shelden, R.A., Caseri, W.R., and Suter, U.W. (1994) *Macromolecules*, **27**, 1637.
38 Velten, U., Shelden, R.A., Caseri, W.R., Suter, U.W., and Li, Y. (1999) *Macromolecules*, **32**, 3590.

39 Mittal, V. (2007) *J. Colloid Interface Sci.*, **315**, 135–141.
40 Mittal, V. and Herle, V. (2008) *J. Colloid Interface Sci.*, **327**, 295–301.
41 Osman, M.A., Mittal, V., and Lusti, H.R. (2004) *Macromol. Rapid Commun.*, **25**, 1145.
42 Lusti, H.R. (2003) Property predictions for short fiber and platelet filled materials by finite element calculations. PhD thesis, ETH, Zurich.
43 Mittal, V. (2008) *J. Thermoplastic Comp. Mater.*, **21**, 9.

9
Mechanisms of Thermal Stability Enhancement in Polymer Nanocomposites
Krzysztof Pielichowski, Agnieszka Leszczyńska, and James Njuguna

9.1
Introduction

An appropriate understanding of the thermal decomposition process of a particular polymer is an important stage in the thermal stabilization of that polymer. The polymer decomposition process itself is a complex procedure which consists of numerous chemical reactions in the solid material, as well as physical processes that occur simultaneously upon heating and lead to the formation of gaseous and solid products [1]. The degradation of a pure polymer begins at weak sites, such as unstable end or side chain groups, and is followed by processes such as main chain random scission, β-scission, hydrogen transfer, and chain stripping at higher temperatures. In addition to these decomposition reactions, successive chemical reactions such as radical recombination, crosslinking, condensation, and cyclization occur simultaneously. A good understanding of thermal stability of polymers is therefore highly important to:

- develop a rational technology for polymer processing;
- synthesize fire-safe polymeric materials;
- ensure polymers usage, especially at elevated temperatures;
- develop new technologies for efficient energy management; and
- conduct the thermal recycling of waste plastics.

Current research investigations have indicated that the introduction of a nanoparticulate filler into a polymer matrix will result in the versatile improvement of a variety of properties, including an increase of thermal stability [2, 3]. Nanofillers have high specific surface and will develop a very large interfacial area if correctly dispersed in a polymer matrix. In comparison with classical microcomposites, nanofillers exhibit a greater ability to interact with a polymer matrix due to the nanoscopic dimensions of the filler particles, the closeness to molecular dimensions, and an ability to modify polymer properties in the interfacial area. As a result, a higher fraction of polymer with modified properties will be gained at a relatively low filler content, and this will result in significant changes of the macroscopic properties of the material.

Optimization of Polymer Nanocomposite Properties. Edited by Vikas Mittal
Copyright © 2010 WILEY-VCH Verlag GmbH & Co. KGaA, Weinheim
ISBN: 978-3-527-32521-4

In this chapter, attention is focused on ongoing research studies representing the main issues in the area of thermal stability of polymeric nanocomposite materials. Selected examples are provided that illustrate the mechanisms of thermal stability enhancement in polymeric nanomaterials, in addition to factors which influence thermal stability, including composition, structure, and the role of interfacial interactions.

9.2
The Mechanisms of Thermal Stability Improvement by Different Nanofillers

To date, the mechanisms relating to the improvement of thermal stability in polymer nanocomposites are not fully understood [4, 5]. Current research studies have not (yet) indicated a specific mechanism, and it is likely that the eventual mechanism, when elucidated, will consist of both physical and chemical processes. The change in the chemical reactions during the thermal degradation of a polymer is a specific effect that is characteristic for each polymer/nanofiller system. However, it may also bring about significant changes in thermal stability that are greater than would be caused by physical shielding alone.

9.2.1
Clay Minerals

Clay minerals such as montmorillonite (MMT) and other clays are of major interest due to their natural abundance, their nontoxic properties, and their versatile improvement of polymer properties, including mechanical strength, thermal stability, barrier and flame performance related to the lamellar shape of nanoparticles, characteristic structure of layers in polymer matrix and nanoscopic dimensions of filler particles [6, 7]. Several effects have been observed that help to explain these changes in thermal properties. The *barrier effect* is, for example, a physical mechanism, whereas changes in molecular dynamics, catalytic effects for charring reactions and radical trapping, are of a chemical nature.

9.2.1.1 Barrier Effect
It is fundamentally believed that nonpermeable nanolayers which are distributed homogeneously in a polymer matrix disturb the diffusion of small molecules by creating a maze or "tortuous path" that retards the progress of the gas molecules through the matrix resin. In relation to the thermal degradation process, the barrier effect may slow down a rate of mass loss by retarding the escape of volatile products of thermal degradation outside the degrading material, whilst also hindering the diffusion of oxygen into the polymer bulk during the course of thermo-oxidative degradation. The primary observations that support the barrier mechanism of lamellar nanofillers show a slowing down in the rate of mass loss [8], and a more pronounced effect of stabilization in an oxidative atmosphere as compared to pyrolytic conditions [9, 10]. For instance, in the recent studies on

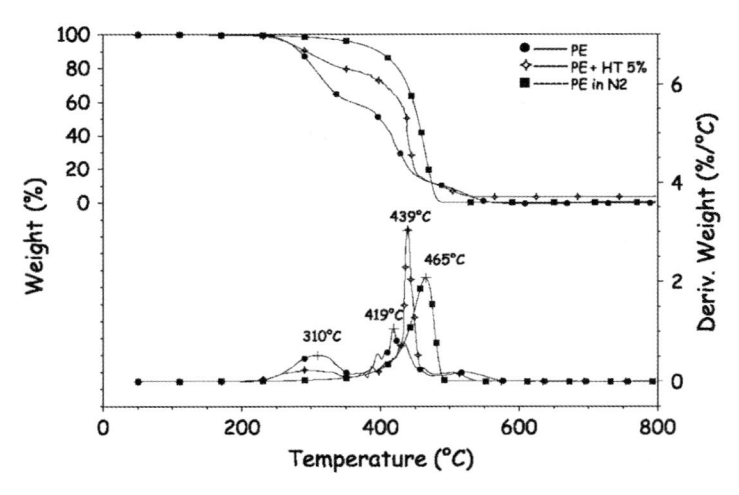

Figure 9.1 Thermogravimetric analysis of polyethylene (PE) and its nanocomposite in air; comparison with thermal degradation of pure PE in an inert atmosphere. Reprinted from Ref. [10], with permission from Elsevier.

polyoxymethylene/montmorillonite (POM/MMT) nanocomposites the barrier effect of nanolayers towards oxygen permeation was evidenced by the quantitative changes of volatile compounds that evolved during oxidative degradation [11]. Specifically, during the course of oxidative degradation of POM-based nanocomposites in comparison to neat POM, the amount of oxidation products was significantly reduced, while the concentration of volatile products typical for pyrolytic degradation was increased. Similarly, in the case of polyethylene (PE), the lamellae of hydrotalcite prevent the polymer from oxidative degradation, shifting the thermogravimetric (TG) curve towards the temperature region of pyrolytic degradation (Figure 9.1) [10].

Elsewhere, the kinetic analysis of degradation indicates that the rate-limiting step in the thermo-oxidative degradation of linear low-density PE (LLDPE) nanocomposites changes from a peroxide radical decomposition to a random scission decomposition [12]. In studies conducted by Cai et al., TG analysis enabled the observation of a significant mass increase due to the oxidation process of the polymer, which was almost completely inhibited in the nanocomposites [13]. These observations show clearly the barrier action of clay towards oxygen diffusion.

Apart from the mass transport barrier mechanism, another phenomenon which plays a role in improving the thermal stability of nanomaterials is the *heat insulation* effect [14]. For instance, in the case of a styrene-acrylonitrile/MMT system that does not form an enhanced amount of char, the insulative effect of MMT nanoparticles is also evidenced by a reduction in the backside temperature of a sample in cone calorimetry measurements [15].

9.2.1.2 Restricted Thermal Motions

One of the very first mechanisms to be proposed for thermal stability improvement in polymer–clay nanocomposites was the restriction of molecular motions and related changes in the kinetics and mechanisms of reactions and physical processes in the condensed phase [16]. More recent studies have confirmed changes in the molecular dynamics of polymer chains anchored to the filler surface that have resulted in an increase in the glass transition temperature (T_g). For example, Chen et al. showed that the glass transition in a polystyrene (PS) nanocomposite was characterized by a significantly higher activation energy than that in virgin PS [17]. Further, on the basis of the heat capacity data, the volume of cooperatively rearranging regions for nanocomposites was found to be approximately 1.8-fold larger than that for virgin PS. In other words, translational motions in the polymer/clay system require a larger degree of cooperativity [18, 19]. Because molecular mobility is the major factor contributing to the transport of reactive species within the polymer, the nanocomposites are likely to have a lower reactivity and, therefore, a greater chemical and thermal stability than the virgin polymer. Regardless of the increased activation energy of glass transition, measured in lower-temperature experiments, the increase in melt viscosity appears to be commonly observed for various polymer/clay nanocomposites, and has a major impact on the kinetics of chemical reactions that occur in viscous media [20].

Recent rheological studies of polymer/clay nanocomposites have introduced new insights into the structure formation, and have indicated the formation of a spatial network of clay nanoparticles in a polymer matrix that can restrain the mobility of the polymer chains [21].

9.2.1.3 Char Forming and Catalytic Effects

Current reports of the thermal stability of polymer/clay nanocomposites have provided contradictory conclusions with regards to the efficiency of clay in improving thermal stability. For example, Vaia et al. revealed that refractory char resulted in an order-of-magnitude decrease in the mass loss rate relative to the neat polymer, even for as little as 2 wt% exfoliated layered silicate [22]. On the other hand, Tidjani et al. [9], while investigating poly(propylene-graft-maleic anhydride)/layered silicate nanocomposites, suggested that the presence of the clay exerted only a minor effect on the thermal degradation, while the nanocomposites showed hardly any additional residue. Thus, it was concluded that the clay layers functioned rather as diffusion barriers in the exfoliated systems [23]. However, if differences in the methodology were excluded, then the many divergences in the results indicated the occurrence of effects of a chemical nature which, unlike the physical effects, were different for each polymer-filler system. Thus, the role of the dispersed nanofiller in the thermal decomposition of a polymer would depend critically on the specific mechanisms associated with the polymer degradation reaction, as well as on the chemical properties of the nanofiller. Subsequent systematic studies involving the thermal stability of a number of polymers have led to the proposal that the clay affects the polymer degradation not only quantitatively but also qualitatively.

Large differences in the efficiency of MMT in improving the thermal stability of various polymers have been considered in terms of the complexity of degradation pathways, or in terms of radical stability [24–27]. When there is more than one degradation pathway – as in the case of PS, where both monomer and oligomers are produced – the presence of the clay can promote one degradation pathway at the expense of another. If the pathway which is promoted leads to a higher-molecular-weight material, then the polymer will be degraded more slowly than it would be in the absence of the clay. However, if there is only a single degradation pathway – or, more theoretically, probable ways but leading to production of the same products, as for instance in the case of poly(methyl methacrylate) (PMMA) – the clay cannot promote an evolution of different degradation products. In the case of radical stability, if the stability of radical species produced during the thermal decomposition of polymer is high, and they exhibit longer lifetimes, then the probability that they will undergo secondary intermolecular reactions (especially radical recombination reactions) is also high. The role of the clay is then to prevent mass transport from the bulk, and to permit radical recombination reactions, thus exerting a stabilization effect in the polymer/layered silicate nanocomposite. This mechanism seems to be supported by many investigations that have confirmed the formation of new products of degradation. The change in the total heat of pyrolytic degradation of PS-based nanocomposites also suggests a possible alternation of the degradation mechanism that may be related to changes in the branching ratios of the individual paths, as well as to the formation of new degradation products [18]. It should be stressed that barrier and radical trapping models have been used successfully to explain why the degradation of a polymer/clay system is slower; however, these models do not offer any straightforward explanation of the changes in the thermal effect, nor of predicting changes in the degradation mechanism.

The typical clays used in polymer-based nanocomposites have a twofold catalytic effect on the thermal degradation of organic molecules in close contact with the clay layers (organic modifier). Carbon–carbon bond scission is accelerated by the clay when the temperature is increased in an inert atmosphere (thermal degradation), and this competes with the acceleration of carbon–hydrogen bond scission in the presence of oxygen (thermal oxidation). Whilst thermal degradation often leads to volatile materials, oxidative dehydrogenation leads to unsaturated moieties evolving thermally stable charred material [28].

An accelerated char formation due to the presence of dispersed MMT layers has been indicated as a mechanism of thermal stability improvement in many studies, including those of Agag and Takeichi [29], Becker et al. [30], and Gilman et al. [31, 32]. Different specific mechanisms have been proposed for enhanced charring in nanocomposites. For example, the catalytic activity of clay towards the hydrogen abstraction reaction seems to prevail in polyolefins and their olefin-containing copolymers [33]. Yet, under oxidative conditions the exfoliated MMT nanoparticles would most likely catalyze the hydroperoxides decomposition (Schemes 9.1 and 9.2) [34].

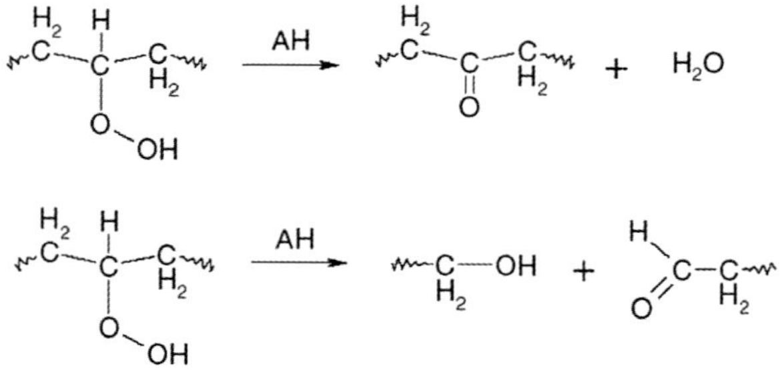

Scheme 9.1 Acid-catalyzed decomposition of polyethylene (PE) hydroperoxides. Reprinted from Ref. [34], with permission from Springer.

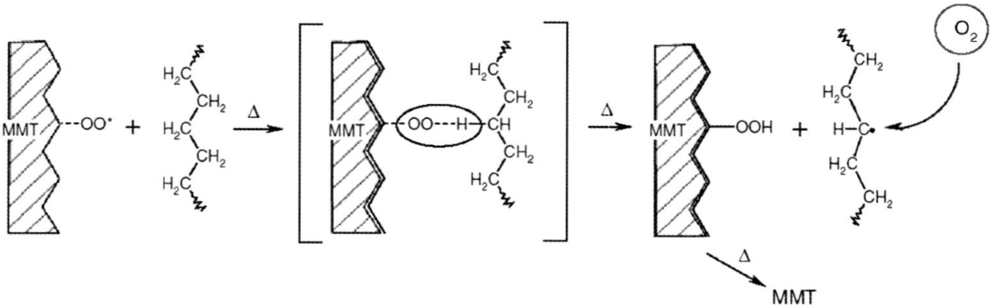

Scheme 9.2 The earliest stages of the PE thermo-oxidative degradation process in the presence of exfoliated MMT nanoparticles. Reprinted from Ref. [34], with permission from Springer.

Metals present in the clay and in acidic sites inherently present on the clay surface, or formed as a result of the decomposition of alkylammonium salts, have shown catalytic activity towards degradation reactions [35–37]. The possible scheme of subsequent events is an unsaturated and conjugated bond formation, cyclization, aromatization, fusion of aromatic rings, turbostratic char formation and, finally, graphitization, that comprise a complex charring process in polymer nanocomposites.

For numerous nanocomposite systems, the nanofiller induces degradation reactions before the onset of thermal decomposition of the pure polymer, thus promoting the charring process and, finally, slowing down the rate of mass loss in a subsequent stage of degradation. This is the case, for example, with ethylene-propylene-diene terpolymers (EPDM) [38], polypropylene (PP) [34, 39], and PMMA [35].

The catalytic activity of clay towards the degradation of polymer has also been observed during nanocomposite compounding and aging in a thermal oxidative environment [40].

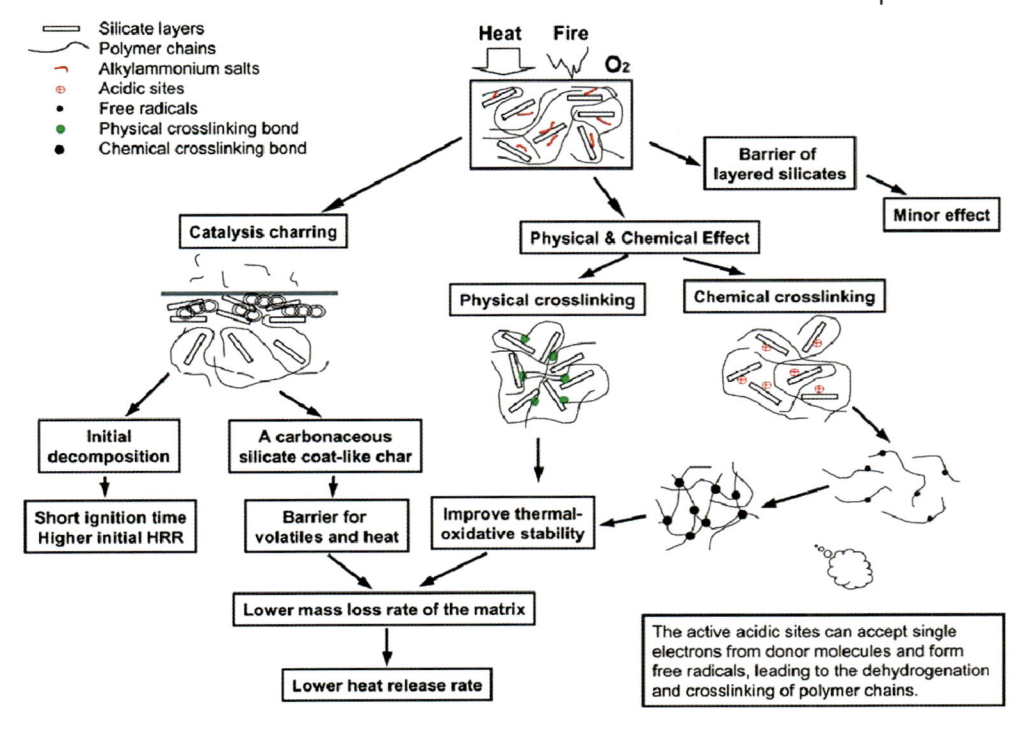

Figure 9.2 Schematic representation of decomposition mechanisms of PP/clay nanocomposites during combustion. Reprinted from Ref. [42], with permission from Elsevier.

Char formation during the thermal degradation of a nanocomposite is an indication of the enhanced flame resistance of the material [41]. Therefore, the thermal stability of nanocomposites, and especially the thermo-oxidative stability, is often investigated in relation to their fire retardancy. Conversely, observation made during burning experiments might add into an understanding of the clay action onto a polymer material in the presence of heat and oxygen. As an example, Qin et al. [42] showed that the shorter ignition time and higher initial heat release rate (HRR) of a nanocomposite was not due to volatilization of the organic modifier in the organoclay, but rather to decomposition of the polymer matrix, catalyzed by the clay itself. Qin and colleagues later proposed a block scheme that visualized a multiplicity of processes, both chemical and physical in nature, to which the nanofiller would contribute, making the overall degradation mechanism much more complex (Figure 9.2).

9.2.1.4 Radical Trapping and Sorption Mechanisms

It has been found that Fe^{3+} cations present in Mg/Fe-layered double hydroxide prevent nanocomposites from decomposing by radical trapping at the onset temperature region [43]. Similarly, for clay-containing systems it has been suggested that any improvement in the thermal stability of the nanocomposite might be associated with the reactivity of different metals with polymeric radicals ("radical trapping model"), and with the different thermal stabilities of the resultant intermediate products [44]. For instance, although both iron and aluminum contributed to the improvement of thermal stability of PMMA nanocomposites, iron was found more effective.

In addition to radical trapping, the physical adsorption of small molecules onto the filler surface can also influence the degradation process. For example, HCl produced during the thermal degradation of poly(vinylchloride) (PVC) would be adsorbed and reacted with hydrotalcite [45].

Among the commercially available polymeric compositions and masterbatches, different additives are present which contribute to the complex thermal behavior. When the nanofiller is introduced into such a complex system, it is not only polymer–nanofiller interactions that should be considered but also nanofiller interactions with polymer additives, as these may exert profound influences on the material's performance [46].

9.2.2
Carbonaceous Nanofillers

9.2.2.1 Carbon Nanotubes and Carbon Nanofibers

Until now, several mechanisms of thermal stability enhancement have been considered for carbon nanotubes (CNTs) and carbon nanofibers (CNFs). Today, it is acknowledged that the retardation of mass loss in inert atmosphere is likely to be the result of a physical adsorption effect of the polymer molecules onto the nanotube surfaces, or the absorption of free radicals generated during polymer decomposition by the activated carbon surface [48]. Another effect associated with carbonaceous nanofillers is a restriction of the mobility of the polymer molecules. This reduces the tension induced by the thermal excitation of the C–O–C bond, and leads to an enhancement of the thermal stability and, notably, to an increase in the degree of crystallinity around the CNFs [49]. Although CNTs do not offer an intercalation phenomenon, their stabilizing effects may be rationalized similar to the effect of layered nanofillers, in terms of a barrier effect. Thus, CNTs and their aggregates will hinder diffusion of the degradation products from the bulk of the polymer into the gas phase [49, 50], an effect which is especially pronounced in an inert atmosphere and is characterized by a shift in the degradation onset temperature [51].

Marosfői et al. showed that the thermal stabilization effect of CNTs could be attributed to the increased interfacial interactions between the nanoadditive and the polymer (PP), which in turn would lead to an increase in the activation energy of decomposition [52]. Marosfői and coworkers subsequently demonstrated that the presence of multiwalled nanotubes (MWNTs) would slightly delay thermal volatilization (by 15–20 °C), without any modification of the thermal degradation mechanism [53], whereas thermal oxidative degradation in air would be delayed by about 100 °C, independent of the MWNT concentration (0.5–3.0 wt%). This stabilizing effect was ascribed to the formation of a thin (1–2 μm-thick) protective film of MWNT/polyaromatic carbon char that was generated on the surface of the nanocomposites. The char comprised a network of homogeneously dispersed MWNTs which became entangled to form a nonwoven, math-like structure, as revealed by scanning electron microscopy (SEM) analysis. The mechanism of formation of the protective surface char was likely to involve charring by oxidative dehydrogenation, catalyzed by the MWNTs. Any further oxidation of the polymer would be prevented by an oxygen-screening action of the composite film, which was thermally stable to about 400 °C. Under fire conditions, the protective layer of nanotubes has been shown physically to shield a polymer and to reduce the external heat flux by 50% [50]. Interestingly, a PP/carbon black sample did not form a network layer, and its heat-release rate did not differ significantly from that of PP. These results indicated that the flame-retardant effectiveness of PP/MWNT nanocomposites was due mainly to the extended shape of the MWNTs [54]. Recent reports have indicated that the formation of an MWNT interconnected network structure during burning would increase the melt viscosity of a nanocomposite, and prevent its dripping and flowing [55].

Of special interest has been the suggestion that thermal shielding might also explain the fire-retardancy contribution of CNTs. However, this fails to account for the reported delay in polymer/CNT thermal oxidation, the rate of which was controlled by oxygen radical initiation [50].

Carbon nanoparticles may also play an accelerating role at the start of polymer decomposition, especially under oxidative conditions [56], although most current explanations have attributed this effect to the remaining traces of inorganic elements, such as Fe, Al and Co [57, 58]. Interestingly, although this effect has been observed for carbon nanoparticles, the presence of metal elements has not been detected [59].

The degradation behavior of polymer/CNT nanocomposites depend very heavily on the type of modification to which CNTs are subjected [51]. In this respect, two phenomena may be considered: (i) the change in degradation route by reactive chemical groups, as introduced by the surface modification of CNTs; and (ii) the increase in thermal stability due to a finer dispersion. An excellent example here is that of CNTs which have been plasma-modified with maleic anhydride (mCNT). These have been shown to improve the thermal stability of polyimide nanocomposites by enhancing their dispersion, thus enabling the mCNT to either bond chemically or interact physically with the polyimide matrix [60]. It has also been shown that CNTs functionalized with amine groups can cause significant improvements in thermal stability when bound covalently to the epoxy matrix [61].

Chipara et al. have shown that the thermal stability of polymer/CNF nanocomposites is strictly related to the filler content [62]. Here, an important role has been ascribed to the interactions between macromolecular chains and CNFs, which contribute to the formation of a polymer–filler interface with an enhanced thermal stability. A two-layer structure has been proposed for a PP/CNF interface

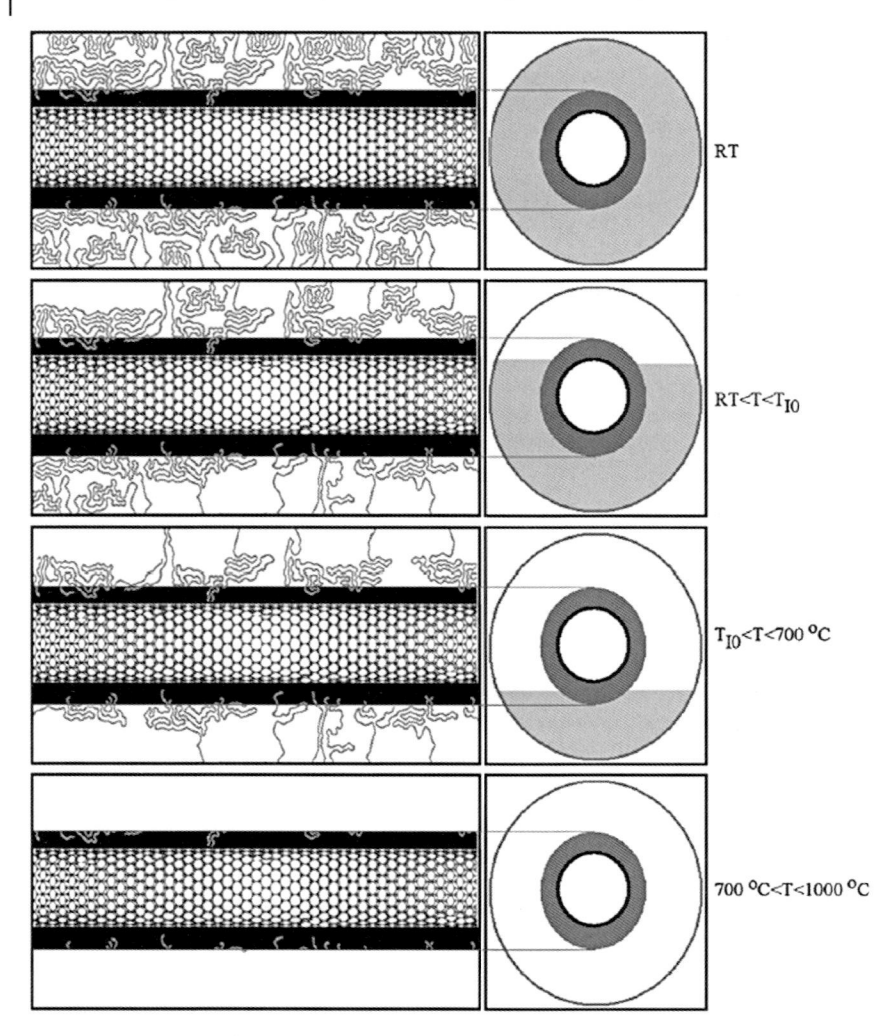

Figure 9.3 A model of the adhesion of polypropylene chains to carbon nanofibers (CNFs) and morphological changes in the course of degradation (RT = room temperature; T_{10} = temperature of maximum mass loss). Reprinted from Ref. [62], with permission from Elsevier.

(Figure 9.3) that undergoes morphological changes during the course of degradation. The external layer is soft, has a thickness of about 100 nm, and is indicated by light-grey color, while the core layer is rigid, has a thickness on the order of few nanometers, and is indicated by dark-grey color. Over a temperature range of 700 to 1000 °C, the polymer is fully vaporized, and only those molecules which are trapped by strong van der Waals interactions will be left on the CNFs. For these molecules, the energy of the van der Waals interaction (with the nearest molecules of the CNFs) is stronger than the energy of the C–C bond.

9.2.2.2 Carbon Black

Although reports of the thermal properties of polymer composites with nanostructured carbon black (CB) are rare, it has been shown that CB can enhance the thermal stability of polymers by increasing the activation energy of the degradation process [63]. Observations from comparative studies, however, have shown that this enhancement is inferior to that of a clay nanofiller [64].

9.2.2.3 Fullerenes

Fullerenes are widely considered as stabilizing modifiers for polymers as they are known to be active radical scavengers; one C_{60} molecule can trap between one and 34 free radicals [65]. For example, in PP the initial decomposition temperature is significantly delayed in an inert atmosphere, and a slight enhancement has been observed under oxidative conditions [66]. Functionalized fullerenes can act as crosslinking agents, causing an increase in the thermal stability of natural rubber, for example, as a result of additional strong physical junctions of the rubber network [67]. Thus, a physical mechanism of stabilization is also feasible.

9.2.2.4 Graphite

Due to the highly anisometric shape of the elemental particles of expanded graphite, and to its intrinsic heat resistance, physical barriers for heat and diffusing gases are likely to be produced, thereby delaying the mass loss of polymer during the course of degradation, in similar fashion to clay nanofillers. The physical mode of action is supported by observations that graphite does not interfere with chemical routes of polymer degradation, but rather shifts the mass loss towards higher temperatures [68, 69]. A clear dependence of the thermal properties of nanocomposites on the content and dispersion degree of the filler has also been observed [70]. The commonly performed oxidation of graphite via chemical oxidation increases the formation of polar groups on the graphite. Analogously, a mechanism of thermal stabilization of polymers involving the formation of strong interfacial interactions has been proposed [71]. Recent studies have shown the use of oxidized graphite to be superior in terms of structure formation, and to result in improvements in thermal properties, in comparison with unmodified graphite [72].

9.2.3 Silica-Based Nanofillers

9.2.3.1 Silica Oxide

The proposed mechanisms of thermal stability enhancement of polymers by silica oxide are similar to those of clay layers. Typically, changes in the molecular dynamics [73], physical crosslinking and strong interactions due to high interface areas are specified. Experiments with nanocomposites of polymaleimide (PMI), containing both covalently bonded and physically mixed colloidal silica nanoparticles, have shown that the inorganic networks chemically bound to polymaleimides function to improve the thermal resistance of the nanocomposites [74]. In contrast, the incorporation of colloidal silica does not make any positive contribution to the thermal stability, although the weight fraction of the inorganic part of the composite is increased significantly. The introduction of nanoparticles into a polymer causes physical or chemical crosslinking; thus, additional bonding forces and residual crystallinity must be surmounted by thermal energy over the melting temperature.

9.2.3.2 Polyhedral Oligomeric Silsesquioxane

Today, it is recognized that polyhedral oligomeric silsesquioxanes (POSS)-containing nanomaterials may be applied to a wide variety of processes, ranging from surface modifiers and coatings to catalysts and membrane materials. These inorganic–organic hybrid polymers possess several attractive properties, including an increased thermal stability, a higher T_g value, better flame and heat resistance, and enhancements in the modulus and melt strengths. Most importantly, these property enhancements occur at low POSS contents (<10 mol.%) [75].

The aniline groups of octa(aminophenyl)silsesquioxane (OAPS) offer versatility, both as reaction sites to which other nanobuilding blocks can be added, and as starting points for generating other functional groups, thereby providing access to diverse and novel nanocomposites. The key problems with almost all silsesquioxane materials investigated to date have been that the aliphatic components limit the thermal stability of the resultant nanocomposites, they have a strong (lowering) influence on T_g-values, and they decrease mechanical properties potentials. However, when reacted with diepoxides or dianhydrides, OAPS nanocomposites have been proposed to provide high-crosslink density materials that have good thermal stability and good-to-excellent tensile and compressive strengths [76]. The reaction of OAPS with pyromellitic anhydride (PMA) at >300 °C yields a material that exhibits a 5% mass loss at a temperature of 540 °C (air and N_2) and a 75 wt% char yield at >1000 °C (N_2). Such a robust nature is exemplary and offers potential for many diverse applications.

9.2.4
Metals and Metal Oxides

Increases in the thermal stability of polymer/metal or metal oxide nanocomposites may be explained by the formation of a polymer–nanoparticle network by the physical crosslinking of a polymer via metal particles; this causes the whole system to be stabilized due to the thermal motions of the polymer chains being restricted. Examples include polyurethane/TiO_2 [77] and polyacrylate/ZnO systems [78].

In contrast, a decrease in thermal stability, which is frequently reported for hybrid films, can most likely be attributed to metal-catalyzed oxidative decomposition pathways in the nanocomposites [79, 80]. Recently, several studies have suggested that gold nanoparticles promote the thermal decomposition of the polymer due to its catalytic ability [81–83]. It is likely that metal-containing nanocomposites are more resistive to pyrolytic degradation, although enhanced mass losses are to be expected in nanocomposites under an oxidative atmosphere due to the catalytic effects towards oxidation reactions [84, 85] (Schemes 9.3 and 9.4).

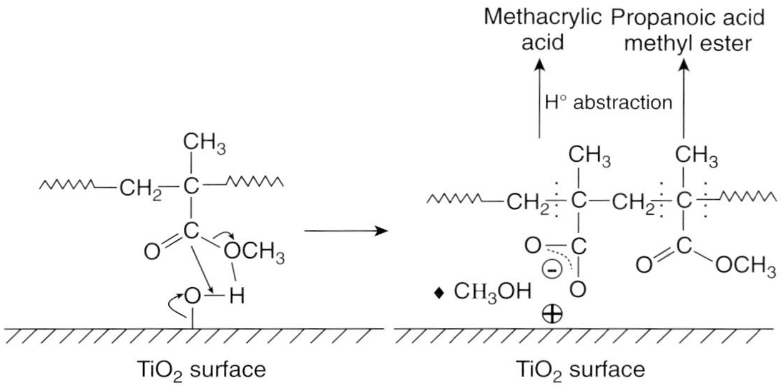

Scheme 9.3 Scheme of the degradation mechanism of PMMA–TiO$_2$ nanocomposites. Reprinted from Ref. [84], with permission from Elsevier.

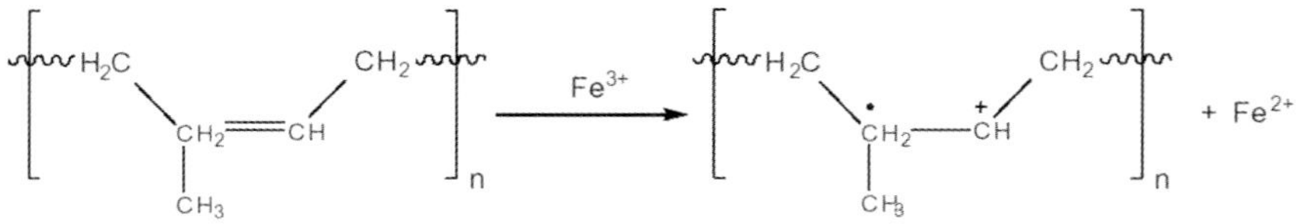

Scheme 9.4 Mechanism of the thermal degradation of natural rubber (NR) in the presence of ferric ion. Reprinted from Ref. [85], with permission from Elsevier.

9.2.5
Other Fillers

Among the many nanofillers used in modern polymer nanotechnology, halloysite nanotubes (HNTs) – a type of aluminosilicate clay with a hollow nanotubular structure – exhibit stabilizing effects on PP decomposition due their ability to entrap the decomposition products [86]. It is believed that the products become entrapped within the HNT lumens at the initial stage of degradation, but are then freed when the temperature is raised.

9.3
Concluding Remarks

The mechanisms of thermal stabilization of polymer nanocomposites are related to the nanocomposite structure, the morphological features of polymer, and to the interfacial interactions of the components. The structure itself is affected by many factors, including the filler content, the fabrication method, and the miscibility of the components.

The changes in thermal stability of polymer matrices upon nanofiller modification may be divided basically into three groups:

- An improvement of thermal stability in terms of initial temperature of degradation (T_d^i), temperature of maximum rate of degradation (T_{max}), char residue and rate of volatilization. This effect is mostly observed for inert nanofillers.

- A diminishment of T_d^i, with simultaneous decrease of degradation rate, which results in a lower conversion of nanocomposites at temperatures corresponding to the temperature of the maximum rate of degradation for the neat polymer (catalytic effect toward the charring reactions).

- A deterioration of thermal stability in terms of T_d^i, T_{max} and volatilization rate over the whole temperature range.

However, complex chemical reactions and physical processes which take place in polymer nanocomposites under degradative conditions, may cause different effects that are not easily categorized, as compared to classical composite materials.

As the thermal stability of nanofillers is usually much better than that of the majority of polymers used as composite matrices, the thermal stability of the resultant composite will depend mainly on the thermal properties of the polymer matrix, the organic modifying agent (usually ammonium salt), and the interactions that occur in the interfacial region.

Barrier formation, restricted thermal motions, radical trapping, sorption and catalytic effects are each considered to be important features that contribute to the mechanism of thermal stability enhancement in polymeric nanocomposites.

Most importantly, the presence or absence of oxygen in the atmosphere in which degradation is carried out will very often determine whether the nanoparticles have an advantageous influence on the polymer.

References

1 Pielichowski, K. and Njuguna, J. (2005) *Thermal Degradation of Polymeric Materials*, Rapra, Shawbury.
2 Utracki, L.A. (2004) *Clay-Containing Polymeric Nanocomposites*, Rapra, Shawbury.
3 Pavlidou, S. and Papaspyrides, C.D. (2008) *Prog. Polym. Sci.*, **33**, 1119–1198.
4 Pandey, J.K., Raghunatha, R.K., Pratheep, K.A., and Singh, R.P. (2005) *Polym. Degrad. Stab.*, **88**, 234–250.
5 Morgan, A.B. and Wilkie, C.A. (eds) (2007) *Flame Retardant Polymer Nanocomposites*, John Wiley & Sons, Inc., New York.
6 Leszczyńska, A., Njuguna, J., Pielichowski, K., and Banerjee, J.R. (2007) *Thermochim. Acta*, **453**, 75–96.
7 Leszczyńska, A., Njuguna, J., Pielichowski, K., and Banerjee, J.R. (2007) *Thermochim. Acta*, **454**, 1–22.
8 Shi, X. and Gan, Z. (2007) *Eur. Polym. J.*, **43**, 4852–4858.
9 Tidjani, A., Wald, O., Pohl, M.-M., Hentschel, M.P., and Schartel, B. (2003) *Polym. Degrad. Stab.*, **82**, 133–140.
10 Costantino, U., Gallipoli, A., Nocchetti, M., Camino, G., Bellucci, F., and Frache, A. (2005) *Polym. Degrad. Stab.*, **90**, 586–590.
11 Pielichowski, K. and Leszczyńska, A. (2008) *Mod. Polym. Mater. Environ. Appl.*, **3**, 97–102.

12 Qiu, L., Chen, W., and Qu, B. (2006) *Polymer*, **47**, 922–930.
13 Cai, J., Yu, Q., Han, Y., Zhang, X., and Jiang, L. (2007) *Eur. Polym. J.*, **43**, 2866–2881.
14 Chang, J.-H., Kim, S.J., Joo, Y.L., and Im, S. (2004) *Polymer*, **45**, 919–926.
15 Stretz, H.A., Wootan, M.W., Cassidy, P.E., and Koo, J.H. (2005) *Polym. Adv. Technol.*, **16**, 239–248.
16 Blumstein, A. (1965) *J. Polym. Sci., Part A: General Papers*, **3**, 2665–2672.
17 Chen, K., Wilkie, C.A., and Vyazovkin, S. (2007) *J. Phys. Chem. B*, **111**, 12685–12692.
18 Chen, K., Susner, M.A., and Vyazovkin, S. (2005) *Macromol. Rapid Commun.*, **26**, 690–695.
19 Vyazovkin, S. and Dranca, I. (2004) *J. Phys. Chem. B*, **108**, 11981–11987.
20 Vyazovkin, S., Dranca, I., Fan, X., and Advincula, R. (2004) *J. Phys. Chem. B*, **108**, 11672–11679.
21 Wang, K., Liang, S., Deng, J., Yang, H., Zhang, Q., and Fu, Q. (2006) *Polymer*, **47**, 7131–7144.
22 Vaia, R.A., Price, G., Ruth, P.N., Nguyen, H.T., and Lichtenhan, J. (1999) *Appl. Clay Sci.*, **15**, 67–92.
23 Levchik, S. and Wilkie, C. (2000) *Fire Retardancy*, Marcel Dekker, New York.
24 Jang, B.N., Costache, M., and Wilkie, C.A. (2005) *Polymer*, **46**, 10678–10687.
25 Jang, B.N. and Wilkie, C.A. (2005) *Polymer*, **46**, 2933–3000.
26 Jang, B.N. and Wilkie, C.A. (2005) *Polymer*, **46**, 9702–9713.
27 Costache, M.C., Wang, D., Heidecker, M.J., Manias, E., and Wilkie, C.A. (2006) *Polym. Adv. Technol.*, **17**, 272–280.
28 Bellucci, F., Camino, G., Frache, A., and Sarra, A. (2007) *Polym. Degrad. Stab.*, **92**, 425–436.
29 Agag, T. and Takeichi, T. (2000) *Polymer*, **41**, 7083–7090.
30 Becker, O., Varley, R.J., and Simon, G.P. (2004) *Eur. Polym. J.*, **40**, 187–195.
31 Gilman, J.W. (1999) *Appl. Clay Sci.*, **15**, 31–49.
32 Gilman, J.W., Kashiwagi, T., Brown, J., Lomakin, S., and Giannelis, E. (1998) Proceedings of the 43rd International SAMPE Symposium, Anaheim, USA, May 31–June 4, 1998.
33 Zanetti, M., Bracco, P., and Costa, L. (2004) *Polym. Degrad. Stab.*, **85**, 657–665.
34 Lomakin, S.M., Novokshonova, L.A., Brevnov, P.N., and Shchegolikhin, A.N. (2008) *J. Mater. Sci.*, **43**, 1340–1353.
35 Kong, Q., Hu, Y., Yang, L., Fan, W., and Chen, Z. (2006) *Polym. Compos.*, **27**, 49–54.
36 Zanetti, M., Kashiwagi, T., Falqui, L., and Camino, G. (2002) *Chem. Mater.*, **14**, 881–887.
37 Wan, C., Qiao, X., Zhang, Y., and Zhang, Y. (2003) *Polym. Testing*, **22**, 453–461.
38 Chen, W., Feng, L., and Qu, B. (2004) *Chem. Mater.*, **16**, 368–370.
39 Tang, Y., Hu, Y., Song, L., Zong, R., Gui, Z., Chen, Z., and Fan, W. (2003) *Polym. Degrad. Stab.*, **82**, 127–137.
40 Ramos Filho, F.G., Mélo, T.J.A., Rabello, M.S., and Silva, S.M.L. (2005) *Polym. Degrad. Stab.*, **89**, 383–392.
41 Costache, M.C., Heidecker, M.J., Manias, E., Camino, G., Frache, A., Beyer, G., Gupta, R.K., and Wilkie, C.A. (2007) *Polymer*, **48**, 6532–6545.
42 Qin, H., Zhang, S., Zhao, C., Hu, G., and Yang, M. (2005) *Polymer*, **46**, 8386–8395.
43 Ding, Y., Gui, Z., Zhu, J., Hu, Y., and Wang, Z. (2008) *Mater. Res. Bull.*, **43**, 3212–3220.
44 Zhu, J., Uhl, F.M., Morgan, A.B., and Wilkie, C.A. (2001) *Chem. Mater.*, **13**, 4649–4654.
45 Bao, Y.-Z., Huang, Z.-M., Li, S.-X., and Weng, Z.-X. (2008) *Polym. Degrad. Stab.*, **93**, 448–455.
46 Bocchini, S., Morlat-Therias, S., Gardette, J.L., and Camino, G. (2008) *Eur. Polym. J.*, **44**, 3473–3481.
47 Yang, J., Lin, Y., Wang, J., Lai, M., Li, J., Liu, J., Tong, X., and Cheng, H. (2005) *J. Appl. Polym. Sci.*, **98**, 1087–1091.
48 Shaffer, S.P. and Windle, A.H. (1999) *Adv. Mater.*, **11**, 937–941.
49 Chatterjee, A. and Deopura, B.L. (2006) *J. Appl. Polym. Sci.*, **100**, 3574–3578.
50 Kashiwagi, T., Grulke, E., Hilding, J., Harris, R., Awad, W., and Douglas, J.F. (2002) *Macromol. Rapid Commun.*, **13**, 761.
51 Bikiaris, D., Vassiliou, A., Chrissafis, K., Paraskevopoulos, K.M., Jannakoudakis,

A., and Docoslis, A. (2008) *Polym. Degrad. Stab.*, **93**, 952–967.

52 Marosfői, B.B., Szabó, A., Marosi, G., Tabuani, D., Camino, G., and Pagliari, S. (2006) *J. Therm. Anal. Calorim.*, **86**, 669–673.

53 Bocchini, S., Frache, A., Camino, G., and Claes, M. (2007) *Eur. Polym. J.*, **43**, 3222–3235.

54 Kashiwagi, T., Grulke, E., Hilding, J., Groth, K., Harris, R., Butler, K., Shields, J., Kharchenko, S., and Douglas, J. (2004) *Polymer*, **45**, 4227–4239.

55 Schartel, B., Potschke, P., Knoll, U., and Abdel-Goad, M. (2005) *Eur. Polym. J.*, **41**, 1061–1070.

56 Sarno, M., Gorrasi, G., Sannino, D., Sorrentino, A., Ciambelli, P., and Vittoria, V. (2004) *Macromol. Rapid Commun.*, **25**, 1963–1967.

57 Hirschler, M.M. (1984) *Polymer*, **25**, 405.

58 Carty, P. and White, S. (1994) *Fire Saf.*, **23**, 67.

59 Vassiliou, A., Bikiaris, D., Chrissafis, K., Paraskevopoulos, K.M., Stavrev, S.Y., and Docoslis, A. (2008) *Compos. Sci. Technol.*, **68**, 933–943.

60 Chou, W.-J., Wang, C.-C., and Chen, C.-Y. (2008) *Polym. Degrad. Stab.*, **93**, 745–752.

61 Chen, X., Wang, J., Lin, M., Zhong, W., Feng, T., Chen, X., Chen, J., and Xue, F. (2008) *Mater. Sci. Eng. A*, **492**, 236–242.

62 Chipara, M., Lozano, K., Hernandez, A., and Chipara, M. (2008) *Polym. Degrad. Stab.*, **93**, 871–876.

63 Chrissafis, K., Paraskevopoulos, K.M., Stavrev, S.Y., Docoslis, A., Vassiliou, A., and Bikiaris, D.N. (2007) *Thermochim. Acta*, **465**, 6–17.

64 Kim, E.-S., Kim, H.-S., Jung, S.-H., and Yoon, J.-S. (2007) *J. Appl. Polym. Sci.*, **103**, 2782–2787.

65 Krusic, P.J., Wasserman, E., Keizer, P.N., Morton, J.R., and Perton, K.F. (1991) *Science*, **254**, 1183–1185.

66 Fang, Z., Song, P., Tong, L., and Guo, Z. (2008) *Thermochim. Acta*, **473**, 106–108.

67 Jurkowska, B., Jurkowski, B., Kamrowski, P., Pesetskii, S.S., Koval, V.N., Pinchuk, L.S., and Olkhov, Y.A. (2006) *J. Appl. Polym. Sci.*, **100**, 390–398.

68 Duquesne, S., Bras, M.L., Bourbigot, S., Delobel, R., Camino, G., Eling, B., Lindsayd, C., and Roels, T. (2001) *Polym. Degrad. Stab.*, **74**, 493–499.

69 Cerezo, F.T., Preston, C.M.L., and Shanks, R.A. (2007) *Compos. Sci. Technol.*, **67**, 79–91.

70 George, J.J. and Bhowmick, A.K. (2008) *J. Mater. Sci.*, **43**, 702–708.

71 Xiao, M., Sun, L., Liu, J., Li, Y., and Gong, K. (2002) *Polymer*, **43**, 2245–2248.

72 George, J.J., Dyopadhyay, A.B., and Bhowmick, A.K. (2008) *J. Appl. Polym. Sci.*, **108**, 1603–1616.

73 Bershtein, V.A., Egorova, L.M., Yakushev, P.N., Pissis, P., Sysel, P., and Brozova, L. (2002) *J. Polym. Sci., Part B: Polym. Phys.*, **40**, 1056–1069.

74 Lu, G.T. and Huang, Y. (2002) *J. Mater. Sci.*, **37**, 2305–2309.

75 Pielichowski, K., Njuguna, J., Janowski, B., and Pielichowski, J. (2006) *Adv. Polym. Sci.*, **201**, 225–303.

76 Tamaki, R., Tanaka, Y., Asuncion, M.Z., Choi, J., and Laine, R.M. (2001) *J. Am. Chem. Soc.*, **123**, 12416–12417.

77 Chen, J., Zhou, Y., Nan, Q., Ye, X., Sun, Y., Zhang, F., and Wang, Z. (2007) *Eur. Polym. J.*, **43**, 4151–4159.

78 Liufu, S.C., Xiao, H.N., and Li, Y.P. (2005) *Polym. Degrad. Stab.*, **87**, 103–110.

79 Sawada, T. and Ando, S. (1998) *Chem. Mater.*, **10**, 3368–3378.

80 Rancourt, J.D. and Taylor, L.T. (1987) *Macromolecules*, **20**, 790–795.

81 Walker, C.H., John, J.V.S., and Wisian-Neilson, P. (2001) *J. Am. Chem. Soc.*, **123**, 3846–3847.

82 Chang, C.M. and Chang, C.C. (2007) *Polym. Degrad. Stab.*, **93**, 109–116.

83 Huang, H.M., Chang, C.Y., Liu, I., Tsai, H., Lai, M., and Tsiang, R.C. (2005) *J. Polym. Sci., Part A: Polym. Chem.*, **43**, 4710–4720.

84 Laachachi, A., Ferriol, M., Cochez, M., Ruch, D., and Lopez-Cuesta, J.M. (2008) *Polym. Degrad. Stab.*, **93**, 1131–1137.

85 Pojanavaraphan, T. and Magaraphan, R. (2008) *Eur. Polymer. J.*, **44**, 1968–1977.

86 Du, M., Guo, B., and Jia, D. (2006) *Eur. Polymer. J.*, **42**, 1362–1369.

10
Mechanisms of Tribological Performance Improvement in Polymer Nanocomposites

Ga Zhang and Alois K. Schlarb

10.1
Introduction

The field of tribological systems extends over an area from "living systems" such as knee-joints, to technical applications such as slide bearings, gear wheels, or sliding elements. Every tribological system has in common that, at minimum, two components are interacting in motion, whereas generally the components consist of two different materials. One component consists generally of steel, while the second will consist increasingly of plastics or polymer-based composites. Today, these materials are being used increasingly as tribo-materials due to their light weights and excellent tribological performances, especially under dry sliding conditions [1–4].

The principles of friction were first studied systematically by Leonardo da Vinci [5], and even today tribological test equipments are based essentially on da Vinci's principles. The many different tests methods used in practice are all based on corresponding types of tribological loading, such as sliding, rolling sliding with rolling, or oscillating sliding. Sliding wear test methods, such as pin on disc and block on ring tests, are used widely in the wear testing of plastics [6]. The performance of the tested materials is effectively described by two characteristics – the *coefficient of friction* and the *wear rate*. Both characteristics are time-dependent, and depend additionally on the load on the specimen, the test speed, and the surrounding temperature.

The tribological behaviors of polymers can be improved with the addition of appropriate reinforcing and/or lubricating fillers. Conventional reinforcing fillers include microsized rigid particles, carbon fibers, and glass fibers. Whilst these reinforcing fillers generally increase the stiffness and strength of the matrix, they may accelerate the abrasiveness if they are torn from the matrix and act as a third-body on the contact surface. Short carbon fibers (SCFs) are used extensively as conventional reinforcing fillers in polymeric tribo-materials, due to their high mechanical properties and high reinforcing effects. In addition, because of the layered structure of carbon fibers, a progressive thinning process of carbon fibers is possible, and this can reduce the risk of being torn out. With high pv factors,

Optimization of Polymer Nanocomposite Properties. Edited by Vikas Mittal
Copyright © 2010 WILEY-VCH Verlag GmbH & Co. KGaA, Weinheim
ISBN: 978-3-527-32521-4

however, fiber failures and the abrasiveness by removed fibers will result in an increased wear rate [7].

Lubricating fillers, which include graphite, poly(tetrafluoroethylene) (PTFE), graphite, and MoS_2, reduce the frictional force between sliding pairs. In addition, *internal lubricants* contribute to the formation of a homogeneous transfer film on the counterface. This reduces the friction and the wear by avoiding direct contact between the sliding pairs. The addition of an internal lubricant to an SCF-filled polymeric composite will result in further improvements of the wear resistance [1, 8]. Combined roles of the conventional fillers have been reported.

Today, nanofiller-reinforced polymers are attracting increasing attention due to their unique properties that result from their nanoscale structures. The most common nanofillers include inorganic nanoparticles, carbon nanotubes (CNTs), carbon nanofibers (CNFs), and nanoclay. The extremely high specific surface area represents one of the most attractive characteristics of nanofillers, because it facilitates the creation of a large amount of interphases in a composite and, thereby, a strong interaction between the fillers and the matrix, even at a rather low nanofiller loading [9]. The nanoscopic confinement of matrix polymer chains exerted by the nanofillers is important to improve their interfacial interaction with the matrix polymer [10]. The incorporation of nanofillers into a polymer matrix has been proved to be an effective way of improving the mechanical properties of the matrix. Due to restrictions on polymer molecules by nanofillers, the stiffness and creep resistance [11, 12] of the host polymer are generally improved. Numerous studies have reported considerable improvements in the toughness of polymer nanocomposites via different mechanisms, such as an increase in fracture area owing to crack deflection [13–15], cavitation [16, 17], the formation of nanovoids [15, 18] and the development of thin dilatation zones [19]. Compared to microscale particulate fillers, nanofillers are generally more effective at improving the toughness of the rigid matrix, because their small dimensions reduce local stress concentrations and create large numbers of subcritical microcracks, microvoids, or nanovoids.

Based on the results of numerous studies, the size of the reinforcing particles has been shown to play a significant role in the tribological performance of particulate-reinforced polymers. Although different results have been reported [20], in general fine particles appear to lead to a better tribology profile under dry sliding conditions. However, when the particles size is reduced to nanoscale level (<100 nm), the wear resistance of these nanocomposites is significantly different compared to that of microparticles-filled systems. From a tribological point of view, the major advantages of polymer nanocomposites relative to microfiller-filled composites may be summarized as follows [1]:

- Generally, a lower abrasiveness due to the reduced angularity and size.
- Reinforcing effects – that is, enhanced stiffness, strength, and toughness.
- Generally, a high effectiveness at very low contents.
- A more proactive influence on the formation of homogeneous and strong transfer films, due to their high reactivity.

The rapid development of nanomaterials has provided a high potential for developing highly wear-resistant polymeric nanocomposites. Since the 1990s, CNTs have generated vast activity in many areas of science and technology, due to their excellent mechanical and physical properties. This novel nanomaterial has also been used during the past 10 years to improve the tribological performance of polymers [21–27]. Nanoparticles and CNTs are two most widely employed nanofillers in thermosetting and thermoplastic materials, to improve wear resistance. Notably, in recent years nanoparticles and sub-microparticles have been incorporated into conventional filler-reinforced composites so as to further improve the tribological performance of the latter materials. Indeed, the results of several studies have indicated that a combination of nanofillers (or sub-microparticles) and microfillers can provide a better tribological output than can composites filled uniquely with nanofillers or microfillers [1, 28–30].

In the following sections, the roles of nanoparticles and CNTs, and especially the synthetic role of nanoparticles with conventional fillers in hybrid composites, will be reviewed. However, it is important first to note that the frictional process of a polymer material is very complex, and the tribological behavior is essentially governed by the process that occur in the surface layer involved in friction. Many factors can simultaneously influence the tribological mechanisms, notably the adhesion between rubbing pairs, the repeated stressing and unstressing, the strain status of the frictional layer, and their repetitive effects determine the friction dissipation and material failures such as fatigue, abrasion, and delamination. In addition, the tribo-chemical reaction and formation of a transfer film will influence the interaction between rubbing pairs.

10.2
Nanoparticle Reinforcements

Until now, many nanoparticles have been incorporated into thermoplastic and thermoset materials in order to improve their tribological performances. The improvements in tribological performances were mainly related to an optimization of the transfer film, and to structural modifications of the materials. A thin, uniform and tenacious transfer film will improve the tribological performance in general, and indeed the incorporation of nanoparticles often improves the quality of the transfer film. In addition, as noted above, nanoparticles can generate stiffening, strengthening, and toughening effects on the host polymer. Although a direct correlation between mechanical improvements and tribological improvements is very often invalid [31, 32], the matrix/nanofillers stress transfer that occurs in the frictional layer can lead to a different loading behavior (normal and tangential) of the layer. This is believed to influence the friction dissipation and material failures. Although the dispersion state of nanoparticles in a matrix has rarely been indicated, a good dispersion of nanoparticles is believed to contribute better to tribological performance improvements. Clearly, agglomerated nanoparticles will take on a similar function as micron particles [33, 34]. In particular, a good matrix/

nanoparticles interfacial adhesion is helpful for the matrix/fillers stress transfer. In fact, it has been proven that a good interfacial adhesion is beneficial to the wear resistance of nanocomposites [4, 35–37].

10.2.1
Improvement of Wear Performance by Using Nanoparticles

Various inorganic nanoparticles, including Si_3N_4, SiO_2, SiC, ZrO_2, Al_2O_3, TiO_2, ZnO, CuO, and $CaCO_3$, have been incorporated into poly(etheretherketone) (PEEK) [38–43], poly(phenylene sulfide) (PPS) [44, 45], poly(methyl methacrylate) (PMMA) [46], epoxy [4, 36, 37, 46–49], and PTFE [50, 51] matrices, in order to improve their wear performances. The significantly enhanced wear behavior at very low volume contents of the nanoparticles has been summarized [1]. Depending on the polymeric matrix, the nanoparticle and its size in particular, the optimum content of the additive is in the range of 1–4 vol% for most systems. It should be clear that a high filler content will lead to a deterioration of the wear properties due to a tendency for nanoparticle agglomeration. Only in the case of a PTFE matrix was a maximum reduction in wear rate achieved at an elevated nanoparticle content of 10–15 vol% [50, 51], this being most likely related to the peculiar ribbon-like structure of PTFE. In this case, the improvement in material stiffness would be the primarily important factor to achieve a good tribological profile. It should be noted that the optimum content of a nanoadditive may be related to its dispersion states, with better-dispersed nanoparticles generally providing more effective reinforcing effects.

10.2.2
Roles of Nanoparticles on Transfer Film Formation

Although an in-depth understanding of the formation of transfer films and the chemical, mechanical interactions of transfer films has yet to be clearly identified, many studies have confirmed that the improvement in tribological performance effected by nanoparticles is accompanied by improvements in the quality or homogeneity of the transfer film [38–44]. In fact, the transfer film act as a spacer between the sliding pairs, whilst its morphology, bonding strength to the counterpart and mechanical properties can each influence the tribological behavior of the system. Thus, the roles of nanoparticles on tribo-chemical reactions, and also on the formation and detachment, and the mechanical properties of transfer films, should each be considered. Some interesting studies highlighting the importance of the transfer film are summarize in the following sections.

When Xue *et al.* studied the tribological performance of Si_3N_4 [40], SiO_2 [41], SiC [38], ZrO_2 [43] nanoparticles-filled PEEK, they reported that the inorganic nanoparticles with optimum contents contributed to reductions in the frictional coefficient and wear rate of PEEK. It was found that a good tribological performance corresponded to a homogeneous transfer film being formed on the counterbody surface. Xue also showed that when PTFE and nano-SiC particles were jointly incorporated into PEEK, the friction reduction and wear resistance capacities of the nanocomposites became worse [52]. This was attributed to the tribo-reaction between nano-SiC and PTFE during the sliding process, yielding SiF_x, which was assumed to deteriorate the formation of transfer films on the slider contact surface.

Later, Schwartz *et al.* [53] investigated the tribological performance of nano-Al_2O_3 (33 nm, 1–10 vol%)-reinforced PPS. For this, wear tests were performed using a pin-on-disc rig with tool steel counterparts having a roughness which varied from 27 nm to 100 nm. By measuring the bonding strength between the transfer film and the counterface, the lowest wear rate obtained for the composite with 2 vol% filler was found to occur when the counterpart roughness (Ra) was higher than the diameter of the nanoparticles. Moreover, the tribological performances of the composites were found to be closely related to the bonding strength between the transfer film and the counterpart. Based on these results, it was assumed that the anchoring effect of the nanofillers was crucial for the transfer film formation, and thereby for the tribological performances of the composites. Elsewhere, Bahadur *et al.* [44] reported that TiO_2 and CuO particles were able to reduce the wear resistance of PPS, whereas ZnO and SiC exhibited an opposite effect. The optimum wear resistance was obtained with 2 vol% CuO or TiO_2. The wear behavior of the composites could be explained in terms of the topography of the transfer film and its adhesion to the counterface. A good correlation between the transfer film, counterface bond strength and wear resistance was also noted in these studies. Although, for all materials transfer films were observed on the counterpart surface, when CuO and TiO_2 were used as fillers the transfer films were thinner and more homogeneous than ZnO and SiC. However, with pure PPS the film was thick and grainy, but the coverage was good. A further increase in filler content from 2 vol% led to a less-uniform transfer film and a consequent decrease in the wear resistance of the composites. X-ray photoelectron spectroscopy (XPS) analyses conducted on the transfer films proved that, for the CuO- and TiO_2-filled PPS, chemical reactions had taken place between the PPS and the counterpart which may have improved the adhesion strength of the transfer film to the counterface. However, for ZnO- and SiC-reinforced composites, no chemical reaction was shown to occur.

As noted above, those factors which have the most influence on transfer film formation and interactions between the transfer film and worn surfaces (especially the tribo-chemical reactions) remain unclear. Consequently, further efforts to understand these relationships in greater detail will form the basis of many current and future studies [54, 55].

10.2.3
Structure–Tribological Property Relationships

As a good nanoparticles/matrix interfacial adhesion will benefit the stress transfer between nanoparticles and matrix, a good interfacial adhesion will generally improve the tribological performance. To achieve this, several approaches have been taken, including both physical and chemical methods such as the surface treatment of nanoparticles, the use of coupling agents and compatibilizers, and grafting agents. Compared to physical treatment methods, chemical methods are more effective as they produce a covalent boding between the fillers and the matrix [10]. Among chemical methods, the addition of coupling agents such as silanes is one of the most common. In particular, in the case of epoxy nanocomposites, a grafting treatment of nanoparticles led to a significant reduction in the wear rate [56–58].

Although a direct correlation between the tribological characteristics and the mechanical properties of polymers is often invalid, it is considered that for glassy polymers such as epoxy (and also some brittle thermoplastics), a low fracture toughness and a poor fatigue resistance are the main factors that lead to high wear rates. In this case, the improved fracture toughness and fatigue resistance achieved by incorporating nanoparticles may represent an important contribution to any tribological improvement. For ductile thermoplastics however – and especially for those under a high contact pressure – both two-body abrasion and the plastic flow of the surface layer will determine the tribological behavior. The worn surface of pure PEEK produced under 4 MPa and at $0.6\,m\,s^{-1}$ using a block-on-ring rig method is shown in Figure 10.1. Here, the sliding direction of the ring is indicated by a black arrow on the surface. Besides the plows parallel to the sliding direction, periodic ripple-like deformations are also observed. These deformations, which are vertical to the sliding direction, are assumed to be indicative of a stick-slip process of the surface layer due to its plastic flow. The consequence of this plastic flow is that the matrix material is progressively pushed to one side of the specimen along the sliding direction such that, finally, a large flake-like debris is formed [59]. In this case,

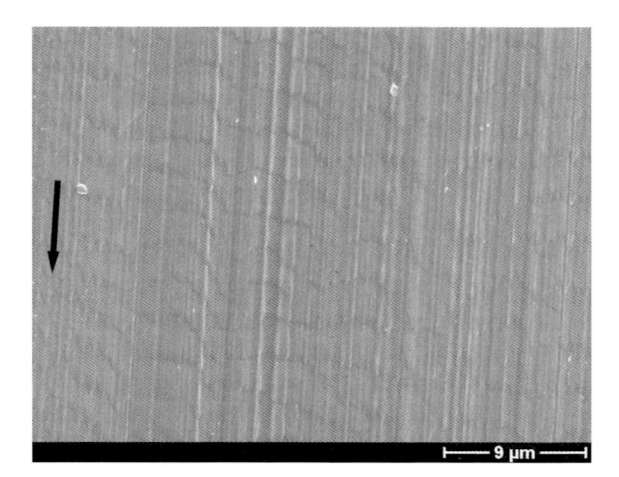

Figure 10.1 Worn surface of pure PEEK produced under 4 MPa and at $0.6\,m\,s^{-1}$. Besides plows parallel to the sliding direction, periodical ripple-like deformations vertical to sliding direction are also noted. The arrow indicates the sliding direction of the counterpart.

it is valid to assume that the stiffening effect of nanoparticles on ductile thermoplastics may be an important factor in the improvement of tribological performance. In the present authors' current studies [60], the roles of nanoparticles on the mechanical and the surface behaviors (e.g., scratch and tribological properties) of PEEK were studied. The aim of these studies was to acquire an understanding of the structure–tribology relationship of nanoparticles-filled PEEK composites. As noted below, the results obtained support the argument that the stiffening effect of nanoparticles on the PEEK matrix is an important factor with regards to tribological improvement.

10.2.3.1 Effect of Grafting Treatment of Nanoparticles on Tribological Improvement of Epoxy Nanocomposites

Compared with thermoplastic materials, and due to the brittle nature and low fatigue resistance of epoxy, in addition to two-body abrasions exerted by the hard protruding regions on the counterface, fatigue and delamination of the surface material are also important mechanisms that contribute to material loss. Consequently, cracks which were vertical to the sliding direction were observed on the worn surface of pure epoxy [48, 56, 57]. Following the addition of nanoparticles, the main wear mechanism of the epoxy is changed to mild abrasion, while cracks vertical to the sliding direction that had originated from fatigue or delamination were either reduced or had disappeared [48, 56, 57]. Cracks on the worn surface of pure epoxy are thought to be related to the repeated stressing and unstressing process of the friction layer. The matrix/nanoparticles stress transfer that occurs in this zone can alter the dynamic stress field in the friction layer, which in turn may change the fracture mode of the layer. In this case, a good matrix/nanoparticles interfacial adhesion and a good nanoparticle dispersion state are believed to contribute more to property improvements.

Zhang et al. found that, by grafting appropriate polymer coatings onto nanoparticle fillers it was possible to improve the tribological performance of the epoxy [4, 37, 46, 56, 57]. These grafting surface treatments helped to build a three-dimensional (3-D) network structure; in other words, the grafted polymer reacted with the epoxy, thus forming a chemical bonding between it and the nanoparticles. A grafting treatment of the nanoparticles was proven to be superior to a physical treatment and the more commonly used chemical treatment, because the low molecular weight of the monomers helps their penetration into the agglomerated nanoparticles [56]. In another study [57], polyglycidyl methacrylate (PGMA) was grafted onto nano-SiC particles as a measure of surface pretreatment. Here, a grafted polymer was selected because the epoxide groups on the PGMA would take part in the curing reaction of the epoxy resin and covalently connect the nanoparticles with the matrix. In comparison to composites filled with untreated nano-SiC particles, the composites with grafted nano-SiC exhibited an improved sliding wear resistance and a reduced frictional coefficient, owing to the chemical bonding at the filler/matrix interface. The friction coefficients and wear rates of the epoxy, when reinforced with pristine and PGMA-grafted SiC nanoparticles, are plotted as a function of nanoparticle content

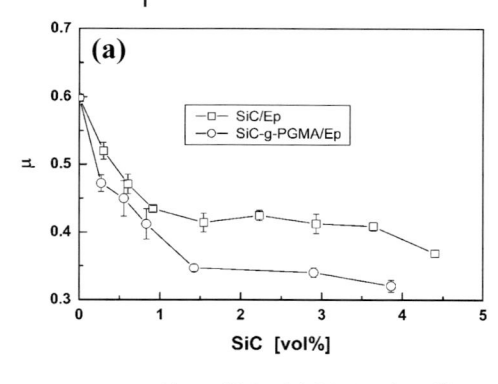

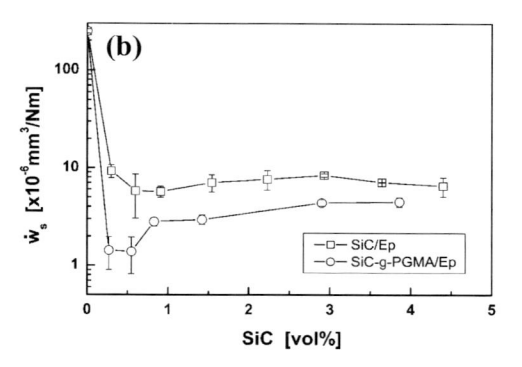

Figure 10.2 (a) Frictional coefficient and (b) specific wear rate of neat epoxy and its composites filled with nano-SiC before and after surface modification. Test rig: block-on-ring, pressure: 3 MPa, sliding velocity: 0.42 m s^{-1} [57]. Reprinted with permission from John Wiley & Sons Ltd.

in Figure 10.2. Here, the designation SiC-g-PGMA/Ep represents the nanocomposite reinforced with the grafted nanoparticles, while SiC/Ep indicates nanocomposites filled with pristine nanoparticles.

In Figure 10.3, the worn surfaces of neat epoxy, epoxy nanocomposites filled with untreated nano-SiC, and nano-SiC grafted with PGMA are compared. The severe wear in the case of neat epoxy, caused by delamination, was reduced when untreated nanoparticles were used. Furthermore, the worn surface of the PGMA-grafted SiC filled composites showed only traces of mild wear. However, these phenomena must be correlated with the role of the grafting polymers; that is, the grafted polymer enhances the filler/matrix interfacial interaction and strengthens nanoparticle agglomerates, thus improving the efficiency of the nanoparticles. The grafting treatment can also improve the dispersing state of the nanoparticles, while the reduced nanoparticle agglomerates and improved filler/matrix interfacial bonding may reduce the third-body abrasions that are exerted by the agglomerates and which are torn out and entrapped on the contact surface. When the same authors also grafted SiO_2, SiC, Al_2O_3, Si_3N_4, nanoparticles with PAAM (polyacrylamide), PS (polystyrene), a copolymer of glycidyl methacrylate, and styrene [35, 54–56], similar improvements were obtained.

From the above-described investigations, it is clear that the interaction between the filler and the matrix is very important for tribological improvements. Likewise, the dispersion state and interfacial adhesion are key factors that determine the reinforcing efficiency of nanoparticles.

10.2.3.2 Role of Nano-SiO$_2$ Particles on the Mechanical and Tribological Behaviors of PEEK

In a recent study [60], PEEK was compounded with nano-SiO$_2$ particles via a ball-milling technique. Here, fumed silica (Aerosil® 7200) with a methacrylsilane

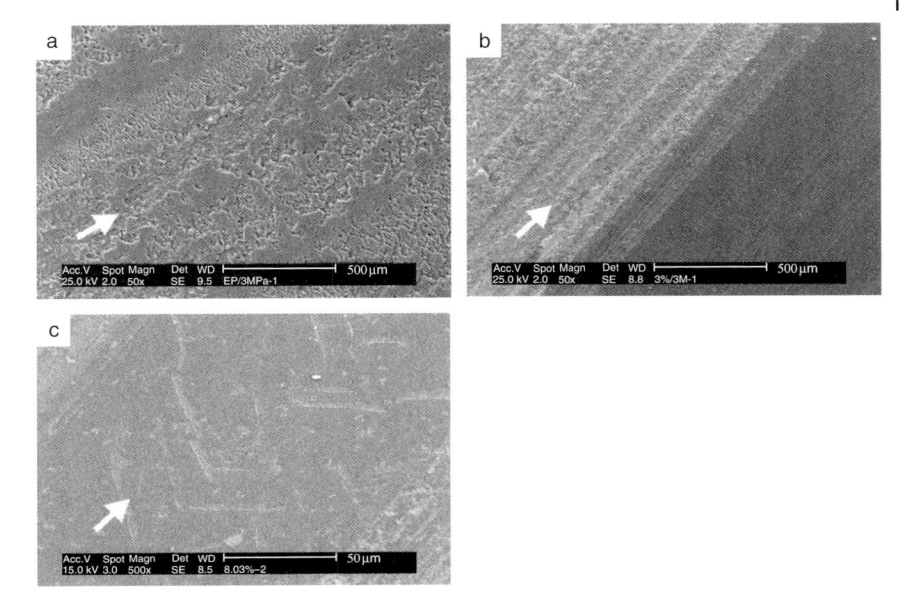

Figure 10.3 Scanning electron microscopy images of the worn surface of (a) neat epoxy, (b) SiC/Ep, and (c) SiC-c-PGMA/Ep. Test rig: block-on-ring, pressure: 3 MPa, sliding velocity: 0.42 m s^{-1}. The contents of the nanoparticles in SiC/Ep and SiC-c-PGMA/Ep were 0.91 vol% and 0.83 vol%, respectively [57]. The arrows indicate the sliding direction of the counterpart. Reprinted with permission from John Wiley & Sons Ltd.

surface treatment was used as nanofiller, whilst the nanoparticles had an average size of 13 nm and a specific area of 125–175 m^2 g^{-1}. The long-term (24 h) repetitive deformation of PEEK powder (average 10 μm diameter), which was the result of collisions with the milling balls, guaranteed a relatively homogeneous dispersion of the nanoparticles in the matrix. Subsequent differential scanning calorimetry (DSC) analyses showed that the nanoparticles had a minor influence on the crystallization temperature, but a negligible influence on the final crystallinity of the matrix. As shown in Figure 10.4, the incorporated nanoparticles remarkably improved the tensile modulus and maintained the tensile strength, but significantly decreased the elongation at break of the matrix. As the content of nano-SiO$_2$ was increased from 0.5 vol% to 2.0 vol%, so too was the tensile modulus of PEEK. However, a further increase in nano-SiO$_2$ content, from 2.0 vol% to 4.0 vol%, caused a slight decrease in the tensile modulus, possibly due to nanoparticle agglomeration.

Figure 10.5 shows the fracture initiation zones on the fracture surfaces of the PEEK and PEEK/nano-SiO$_2$ composites. Fractures of pure PEEK initiate from randomly distributed impurities, and the drawn nature of the deformation in this region suggests ductile deformation rather than stable crack growth (c.f. Figure 10.5a). From Figure 10.5b, it can be seen that for 1 vol% SiO$_2$-reinforced PEEK,

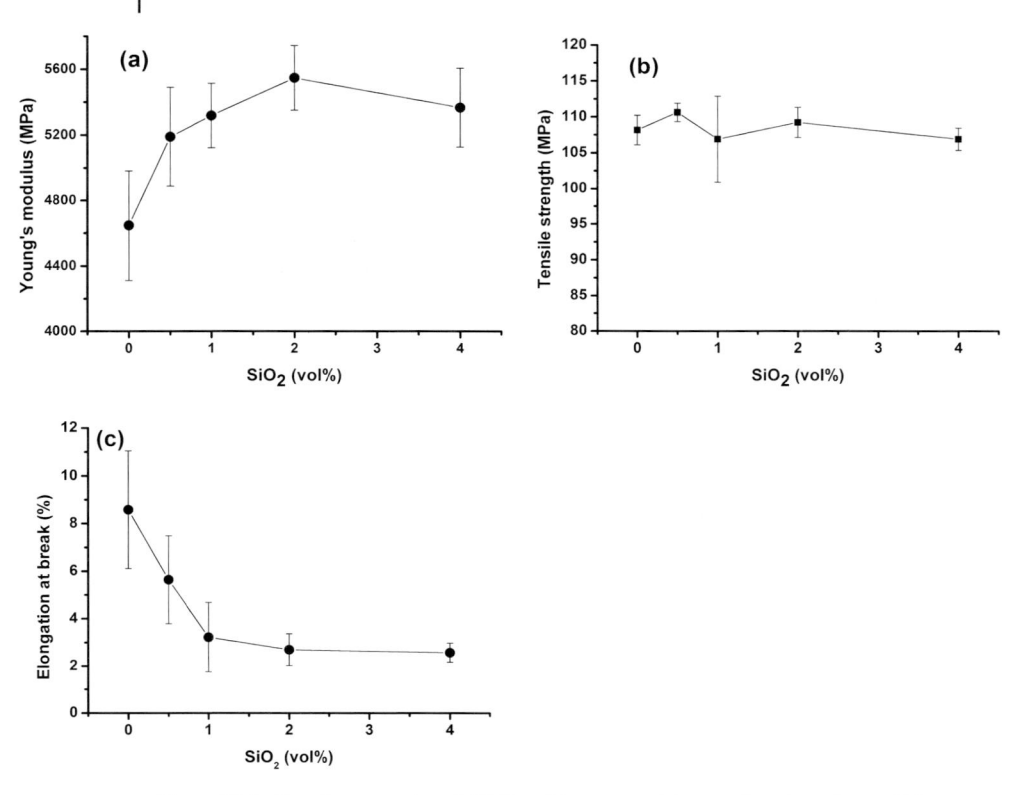

Figure 10.4 Tensile properties of PEEK and its composites as a function of nano-SiO$_2$ volume fraction. (a) Young's modulus; (b) Tensile strength; (c) Elongation at break [60]. Reprinted with permission from Elsevier.

the drawn fibration process is very limited compared to that of pure PEEK; this may explain the decrease in matrix ductility following incorporation of the nano-SiO$_2$ particles. Numerous dimples are found in this region due to secondary cracks, their formation being promoted by the nano-SiO$_2$ particles which serve as stress concentration sites. Figure 10.5c shows the ductile region on the fracture surface of 4 vol% nano-SiO$_2$-filled PEEK. With increasing nano-SiO$_2$ content, the typical size of the dimples is decreased due to larger amounts of nano-SiO$_2$ particles per unit area. Based on the above discussions, the molecular immobility of PEEK caused by the incorporation of nanoparticles is clearly responsible for the improvement in modulus and decrease in ductility.

Figure 10.6a and b show the scratched surfaces of pure PEEK and PEEK/nano-SiO$_2$ (1 vol%). The scratches were produced using a diamond conical indenter with a tip radius of 100 μm and an apex angle of 120°, with a load of 4 N and scratch speed of 50 μm s^{-1}. The scratch direction is indicated on the figures, with zig-zag scratch marks clearly visible on the surface of the pure PEEK (c.f. Figure 10.6a).

10.2 Nanoparticle Reinforcements | 221

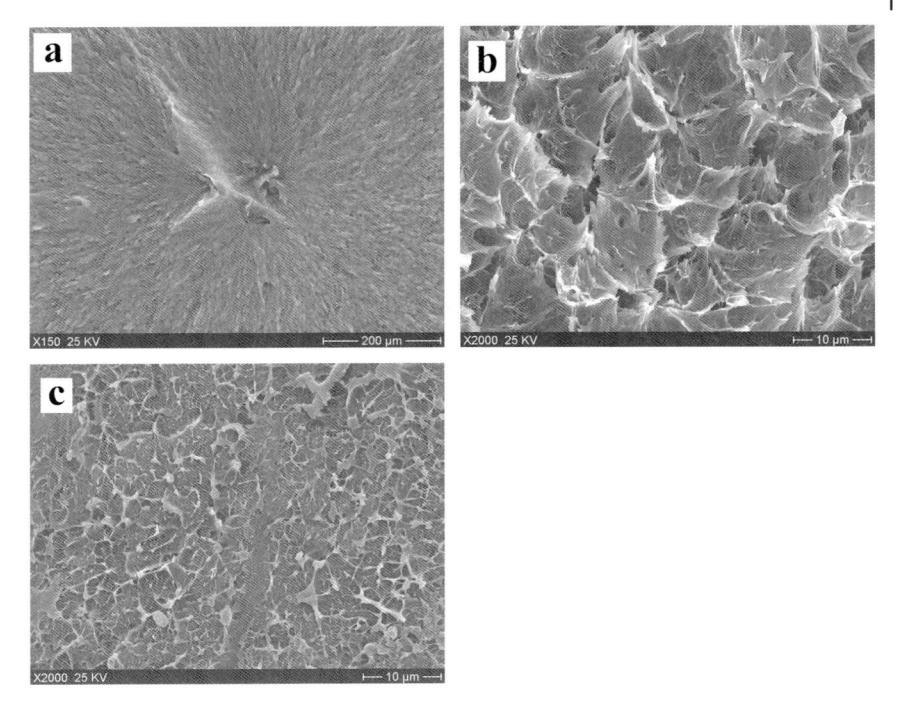

Figure 10.5 Morphologies of fracture surfaces. (a) Pure PEEK; (b) PEEK/1 vol% SiO$_2$; (c) PEEK/4 vol% SiO$_2$ [60]. Reprinted with permission from Elsevier.

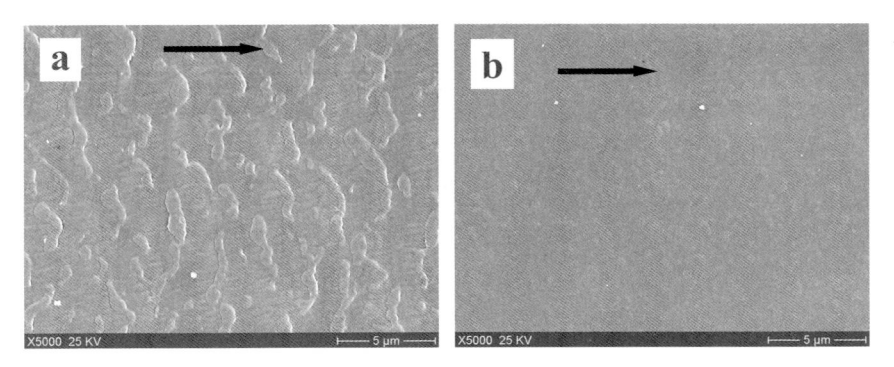

Figure 10.6 Scratch surfaces of pure (a) PEEK and (b) PEEK/nano-SiO$_2$ [30]. Reprinted with permission from Elsevier.

This morphology is indicative of a "stick-slip" phenomenon of the PEEK surface layer. Stick-slip behavior is the result of a plastic flow and a strain-hardening effect of the material stack ahead of the indenter, and occurs due to the ductile features of the material [61]. The plastic flow of PEEK matrix was reduced when nano-SiO$_2$

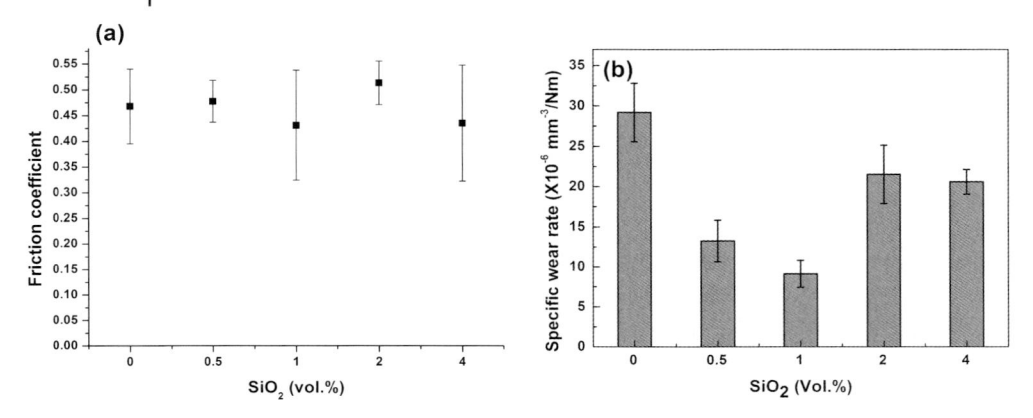

Figure 10.7 (a) Friction coefficients and (b) wear rates (b) of nano-SiO$_2$-filled PEEK as a function of nano-SiO$_2$ content. Test rig: block-on-ring, apparent pressure: 4 MPa, sliding velocity: 1 m s^{-1} [60]. Reprinted with permission from Elsevier.

was incorporated into the composite; therefore, regardless of the higher complexity in true frictional tests, the plastic flow can be reduced by the molecular immobility that is caused by the incorporated nanoparticles.

The friction coefficients and wear rates of pure PEEK, and of the nanocomposites, plotted against the nanoparticle loading, is shown graphically in Figure 10.7. The incorporation of nano-SiO$_2$ did not induce any significant change in the friction coefficient (Figure 10.7a), although the wear resistance of PEEK was significantly improved (Figure 10.7b). Although, initially, the incorporation of nano-SiO$_2$ led to a ~70% reduction in wear rate, further increases in nano-SiO$_2$ content led to an escalation in wear rate. However, with a loading of 4% SiO$_2$ the wear rate of the composite was still almost 30% less than that of pure PEEK.

The worn surfaces of the composites present distinctly different morphologies compared to pure PEEK. For example, whilst plows were clearly visible on the pure PEEK surface (Figure 10.8a), the worn surface of 1 vol% SiO$_2$-filled PEEK was very smooth (Figure 10.8b). However, a high-loading of SiO$_2$ caused obvious abrasive damage (as indicated by an arrow in Figure 10.8c). This phenomenon can be ascribed to the unavoidable agglomeration of nanoparticles, and may explain the reduction in wear resistance with further increases in nanoparticle concentration above 1 vol%, since higher nanoparticle contents increase the tendency of the nanoparticles to agglomerate. The increased stiffness of PEEK, which is achieved by incorporating nano-SiO$_2$ particles, reduces the normal deformation of PEEK, which in turn alleviates the plowing severity of the worn surface. Moreover, as noted above, the reduced plastic flow of the frictional surface layer may represent an additional factor for the reduced wear rate.

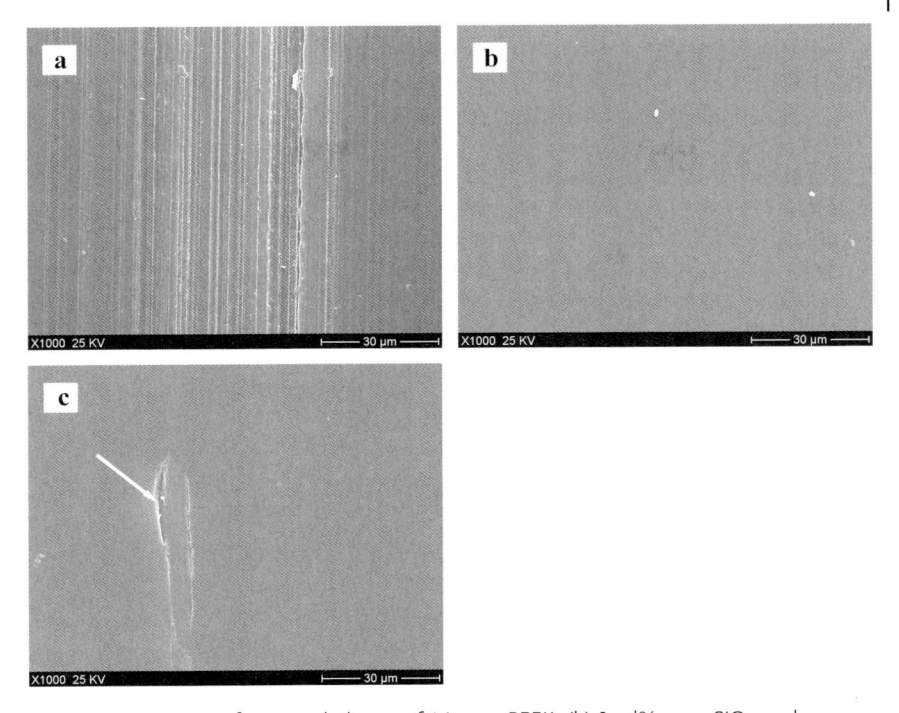

Figure 10.8 Worn surface morphologies of (a) pure PEEK; (b) 1 vol% nano-SiO$_2$- and (c) 4 vol% nano-SiO$_2$-reinforced PEEK [60]. The sliding direction of the counterpart is from up to down. Reprinted with permission from Elsevier.

10.3
Carbon Nanotubes

Carbon nanotubes constitute a rather novel class of nanomaterials, with their exceptionally high axial strengths, axial moduli of the order of terra Pa, and high electrical and thermal conductivities [62]. The exact magnitude of these properties depends on the diameter and chirality of the CNT, and whether they are in single-wall or multiwall form. Because of their outstanding physical properties and their high aspect ratio, CNTs have high application potentials in polymer composites. A low-loading of CNTs in a polymer matrix significantly improves the mechanical properties [62–64]. It should be noted that, as with other types of nanofiller, a good dispersion state of CNTs in a matrix and a good adhesion with the matrix are necessary requirements in order to achieve superior reinforcing effects. The dispersal of CNTs is very difficult, however, as they tend to agglomerate and bundle and entangle together easily, thus limiting their reinforcement efficiency on polymers [65]. Consequently, much effort has been expended in attempts to achieve a homogeneous distribution of CNTs into a host polymer, with dispersing techniques that include mechanical stirring [66], surfactant-assisted processing [67], and solution-evaporation methods with sonication [62].

The exceptional properties of CNTs have also attracted the attention of tribologists, and promising results have been obtained in terms of tribological performance improvements when CNTs have been used [21–27]. Besides their excellent mechanical properties, the high thermal conductivity of CNTs is of major interest for the reinforcement of tribomaterials, as frictional heat can be transferred more rapidly to the bulk material, such that the interface temperature can be reduced. Furthermore, when compared to nanoparticles, the large aspect ratios of CNTs can lead to different load actions, both vertically and tangentially. Unfortunately, systematic theories and knowledge revealing the tribological mechanisms of CNT-filled polymer nanocomposites are still lacking; however, some interesting results from a variety of studies are reviewed and evaluated in the following sections.

Chen and coworkers [21] studied the effect of CNTs fraction, varying from 2.5 vol% to 30 vol%, on the tribological behavior of CNTs-filled PTFE composites. Tribological tests were performed on a plate-on-ring rig with a load of 200 N. The study results indicated that the friction coefficients and wear rates decreased with increasing CNT content, with the improvement being ascribed to the reinforcing effect of CNTs on PTFE. Due to the low stiffness and strength of PTFE, the stiffening effect of CNTs should be significant, and this may be the reason why a high filler fraction is required. However, the dispersion state of CNTs was not characterized in these studies. Especially with a high CNT fraction, the agglomeration and entanglement of CNTs should be significant, and would undoubtedly influence the tribological performance of the composites.

Jacobs et al. [27] investigated the roles of CNTs on the tribological performance of epoxy using a ball-on-prism tribometer. In particular, they investigated the effects of CNTs pretreatment and stirring methods on the tribological performances of nanocomposites. Notably, their results indicated that, with appropriate CNT pretreatment and mixing, the addition of a small amount of CNTs to the epoxy matrix could significantly enhance its wear resistance. For example, whilst the simple stirring of untreated CNTs into an epoxy resin did not cause any noticeable improvement, activation of the CNT surface by boiling in nitric acid for a short period (2 h) improved the wettability of the CNTs by the epoxy matrix and yielded a better wear resistance. However, the application of silane as a coupling agent caused no further improvement in wear behavior of the compounds. Regardless of the pretreatment and dispersal techniques used, the epoxy composites with 1 wt% CNTs showed an optimal wear resistance, but further increases in CNT content caused a slight decrease in wear resistance. The results of these studies also indicated that the pretreatment and dispersion of CNTs should be carried out very carefully in order to prevent their destruction and reduce their reinforcing effects.

When polyamide 6 (PA6)/CNTs nanocomposites containing 1 wt% CNTs were compounded, using a twin-screw extruder, by Meng et al. [26], the addition of a small fraction of CNTs remarkably improved the tensile modulus and tensile strength of the host polymer. In addition, due to the nucleating action of CNTs, the crystallinity of PA6 was increased from 30.8% to 36.7% and, as a result, the nanocomposites exhibited a lower water absorption ratio than did the neat material. Under both dry and water lubrication conditions, CNTs lowered the friction coefficient and wear rate of PA6. With water lubrication, both PA6 and its composite showed higher wear rates than under dry conditions, possibly due to the softening and plasticization of the material surface layer after it had absorbed water. This improved tribological performance was related to the enhanced mechanical properties. The self-lubricating effect of CNTs was proposed as another beneficial factor; according to this suggestion, CNTs can be transferred to a counterface and serve as a solid lubricant, although this has not yet been proven.

Zoo et al. [24] studied the effect of CNTs on the tribological behavior of ultra high-molecular-weight polyethylene (UHMWPE) using a ball-on-disc test rig with a load of 5 N at $0.3\,\mathrm{m\,s^{-1}}$, while the CNT contents ranged from 0.1 wt% to 0.5 wt%. The study results indicated that small amounts of CNTs would raise the friction coefficient but significantly improve the wear resistance. Improvements in wear performance were attributed to the increased hardness of the composite following the addition of CNTs. Such a proposal was feasible, as the vertical deformation of the material would be reduced due to stiffening effects of the CNTs, such that the micro-cutting effect would also be reduced.

Cai et al. [23], in their studies on the tribological properties of *in situ*-fabricated polyimide (PI)/CNTs, found that CNT addition led to remarkable decreases in both friction coefficient and wear rate. Typically, reduced adhesions and scuffing were identified on the worn surfaces of CNTs-reinforced PI compared to pure PI, and related this better tribological performance to the better quality of the transfer film. Nevertheless, the manner in which CNTs improved the quality of the transfer film was unclear. Tanaka et al. [22] used single-wall carbon nanohorns (SWNHs, a member of the CNT family) and multi-wall carbon nanotubes (MWCNTs) to improve the tribological performance of PI. The MWCNTs were produced by the CO_2 laser vaporization of graphite at room temperature, and always aggregated to form a spherical particle with multiple horns. It was expected that the PI/CNHs would provide a better tribological profile than the PI/CNTs, considering the fact that the SWNH aggregates had a spherical shape with sub-micron diameter. Both, SWNHs and CNTs significantly lowered the wear rate of PI, but more interestingly the SWNHs played a slightly superior role on wear reduction than did the CNTs. The two fillers also lowered the friction coefficient of PI. As indicated elsewhere, the fillers only exist between the boundaries of the PI powder (<50 μm), since no melt-stirring or solution-assisted stirring was employed and consequently severe agglomeration occurred. Thus, the potential of SWNHs and CNTs was not fully realized.

It is clear that CNTs have demonstrated a great potential for formulating polymeric tribomaterials. However, based on the above-described studies, research into the tribology of CNTs/polymer composites are at a very early stage, and further efforts are required to obtain an understanding of the tribological mechanisms of these composites. In particular, fundamental studies of the interaction between CNTs and the matrix in the frictional layer should assist in identifying the role(s) of CNTs in this situation.

10.4
Synthetic Roles of Nanoparticles with Traditional Fillers

During recent years, nano and sub-micro inorganic particles have been incorporated into traditional SCFs and solid (micro-sized) lubricants such as graphite and PTFE, in order to improve the tribological performances of these micro-sized filler-reinforced composites [1, 28–30, 68]. It has been proved that a combination of traditional fillers and nano or sub-micro fillers can result in the best tribological profile. That is, the friction coefficient and wear rate of multiple fillers-filled hybrid composites may be lower than those of composites filled uniquely with traditional fillers or nanofillers. Based on this fact, the synthetic roles of both nanofillers and traditional fillers are highlighted in the following sections.

10.4.1
Tribological Behavior of Traditional and Nanofillers (or Sub-Micro)-Filled Epoxy

Zhang et al. [1, 28] studied the effects of sub-micro TiO_2 (300 nm) on the tribological performance of SCF-filled epoxy. Using a pin-on-disc rig under different pv conditions, the addition of 5 vol% TiO_2 fine particles reduced the friction coefficient, contact temperature and wear rate of SCF/PTFE/graphite-filled epoxy. This positive effect of TiO_2 was especially effective under high pv factors. As noted above, for the SCF/PTFE/graphite-filled composites, the SCFs play a key role. As shown in Figure 10.9, the thinning of fibers, breakage, fiber pulverization and removal determines the tribological behavior of the traditional fillers-reinforced composite. Under a high pv factor, the failure of fibers was more severe, and thus the composites presented higher wear rates. As shown in Figure 10.10, for a hybrid composite filled both with traditional fillers and sub-micron fillers, even under a

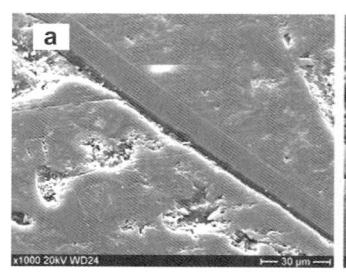

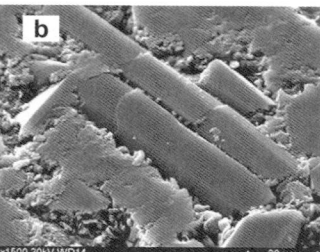

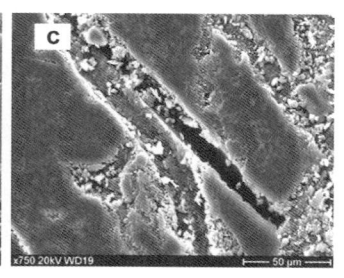

Figure 10.9 Wear process of short carbon fiber (SCF)-reinforced epoxy composite (epoxy + 5 vol% graphite + 5 vol% PTFE + 15 vol% SCF). (a) Fiber thinning; (b) Fiber breakage; (c) Fiber pulverization and interfacial removal. Wear conditions: normal pressure = 1 MPa; sliding velocity = 1 m s^{-1}; pin-on-disk apparatus [1]. Reprinted with permission from Elsevier.

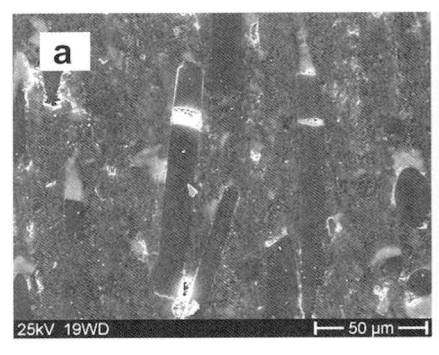

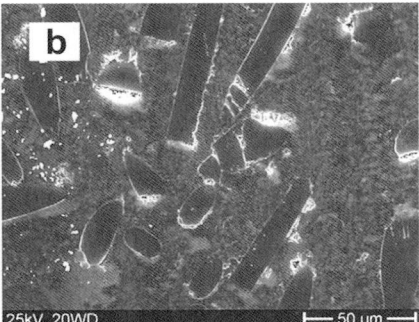

Figure 10.10 Scanning electron microscopy images of the worn surfaces of the nanocomposite (epoxy + 5 vol% graphite + 5 vol% nano-TiO$_2$ + 15 vol% SCF) sliding at various contact pressures and a constant velocity. (a) 1 MPa; (b) 12 MPa. Wear conditions: sliding velocity = 1 m s^{-1}; duration = 20 h; pin-on-disk apparatus [1]. Reprinted with permission from Elsevier.

high pressure the worn surface became much smoother. Basing on the above findings, these authors suggested the existence of a "rolling effect" of the fine particles; that is, the sub-micro fillers roll between the sliding pairs and thereby protect the SCF/matrix interface.

Guo et al. [29] also reported the synthetic roles of nano-SiO$_2$ (12 nm) and SCFs on the tribological performance of epoxy. A combination of nanoparticles together with traditional SCFs provides a much higher wear resistance than would any composite filled only with SCFs or nanoparticles. Moreover, grafting of the copolymers of styrene and maleic anhydride onto the nanoparticles, by which the nanoparticles are covalently connected to the matrix, can further reduce the wear rate of the hybrid composites. From the worn surface, the nanoparticles were found not only to effectively protect the SCFs but also to improve the homogeneity of the transfer films.

10.4.2
Roles of Nanoparticles on the Tribological Behavior of SCF/PTFE/Graphite-Filled PEEK

The incorporation of sub-micro ZnS (300 nm) and TiO$_2$ (300 nm) particles into SCF/graphite-filled PEEK significantly improved the high-temperature wear resistance of the conventional composite [69]. However, under temperatures less than 70 °C the fine particles did not show any positive effects on the wear resistance [69]. Although the wear resistance of a conventional composite of PEEK was already very high at room temperature, it might be improved by reducing the particle size to the nanoscale.

In the present authors' current investigations [30], the roles of low-loading (1 vol%) nano-SiO$_2$ particles (13 nm) on the tribological behavior of SCF/PTFE/

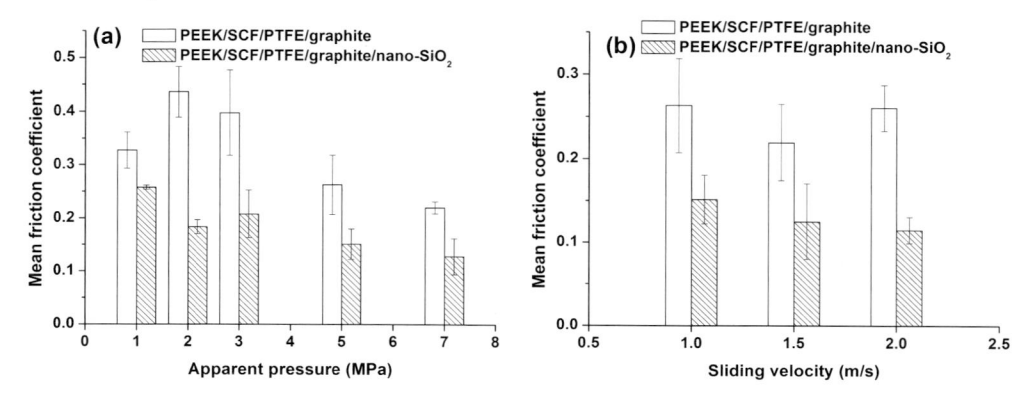

Figure 10.11 Effects of (a) apparent pressure and (b) sliding velocity on the mean friction coefficients of PEEK/SCF/PTFE/graphite and PEEK/SCF/PTFE/graphite/nano-SiO$_2$. The sliding velocity and apparent pressure in (a) and (b) were respectively 1 m s^{-1} and 5 MPa [30]. Reprinted with permission from Elsevier.

graphite (micro-sized, 10 wt% for each)-filled PEEK were examined. For this, commercially available SCF/PTFE/graphite (micro-sized)-filled PEEK granules (Evonik, Germany) were used, while the nanoparticles (Aerosil® 7200) were the same as used in Section 10.2.3.2. The compounding of nano-SiO$_2$ with the composite granules was carried out using a laboratory-scale melting mixer (Brabenbder® kneader), such that the content of nano-SiO$_2$ was 1 vol% (1.51 wt%). For simplification, the two composites are referred to here according to their components, as PEEK/SCF/PTFE/graphite and PEEK/SCF/PTFE/graphite/nano-SiO$_2$, respectively. Tribological tests were carried out on a block-on-ring tribometer at room temperature over a wide pressure range (from 1 to 7 MPa) and a wide sliding velocity range (from 1 to 2 m s^{-1}).

Figure 10.11a shows the effect of apparent pressure on the mean friction coefficients of the composites, with and without nano-SiO$_2$, and with a sliding velocity of 1 m s^{-1}. It was clear that, within the studied range, the incorporation of nano-SiO$_2$ caused a significant reduction in the friction coefficients. The effect of sliding velocity on the friction coefficient under 5 MPa is shown in Figure 10.11b. Notably, across the entire studied range (i.e., from 1 m s^{-1} to 2 m s^{-1}) the nanoparticles always caused significant reductions in the friction coefficients.

Figure 10.12a shows the effect of apparent pressures on the specific wear rates of the two composites, where the sliding velocity is 1 m s^{-1}. At a pressure below 1 MPa the nanoparticles had a negative effect on the wear resistance of the traditional fillers-reinforced composite, but at pressures >2 MPa the nanocomposite presented remarkably lower wear rates than the composite containing only with traditional fillers. The effect of sliding velocity on specific wear rates measured at pressures below 5 MPa is illustrated graphically in Figure 10.12b. The increase in sliding velocity, from 1 m s^{-1} to 2 m s^{-1}, gives rise to the specific wear rate of the composite without nano-SiO$_2$. With regards to the nanoparticles-filled composite,

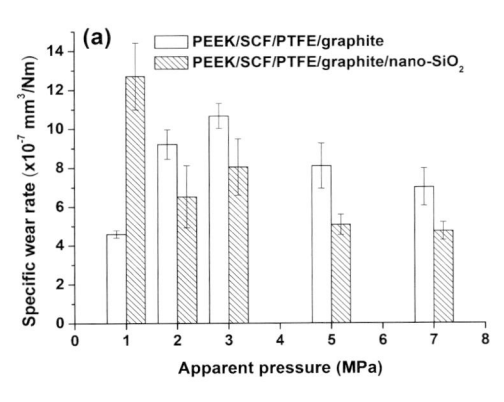

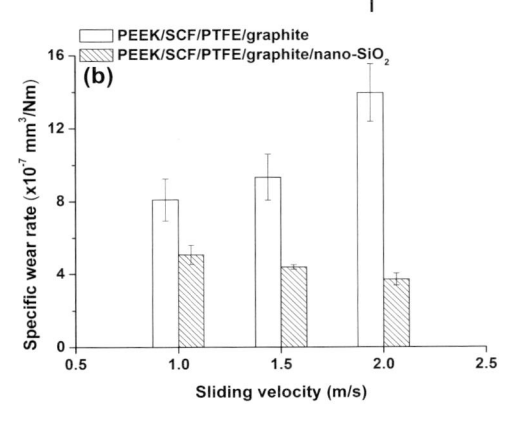

Figure 10.12 Effect of (a) apparent pressure and (b) sliding velocity on the specific wear rates of PEEK/SCF/PTFE/graphite and PEEK/SCF/PTFE/graphite/nano-SiO$_2$. The sliding velocities in (a) and (b) were 1 m s^{-1}; the apparent pressure in (c) was 5 MPa [30]. Reprinted with permission from Elsevier.

however, the increase in sliding velocity led to a slight decrease in the specific wear rate. Thus, the positive effect of nano-SiO$_2$ on wear resistance was more pronounced at high velocities.

Figure 10.13a shows the worn surface of PEEK/SCF/PTFE/graphite produced under a pressure of 1 MPa and a sliding velocity of 1 m s^{-1}; the sliding direction is indicated by an arrow. The SCF surface is clearly smooth, with no obvious failure of the SCFs being noted, whilst the thinning of fibers mainly determines the tribological behavior. Figure 10.13b shows the worn surface of PEEK/SCF/PTFE/graphite/nano-SiO$_2$ produced under the same conditions, but here the worn surface presents a distinctly different morphology from that of the conventional composite, with the fibers and SCF/PEEK interface appearing to be destroyed. Figure 10.13c shows the worn surface of a fiber, observed at high magnification. Here, the fiber outline is indicated by a dashed oval, and the cracks that have occurred in the fiber are noticed. The abrasiveness exerted by possible nano-SiO$_2$ agglomerates serving as a "third-body" seems to have severely destroyed the fibers. This may explain why, under low pressure, the incorporation of nano-SiO$_2$ reduces the wear resistance of the composite.

Under high pressure, and especially with a high sliding velocity, the failures of SCFs dominate the wear of the material. Figure 10.14a shows the worn surface of PEEK/SCF/PTFE/graphite produced under a pressure of 5 MPa and a sliding velocity of 1 m s^{-1}. Here, deep scratch traces generated by the broken fibers are noted in some zones on the worn surface, whilst in some regions separation of the SCFs from the PEEK matrix has occurred (see Figure 10.14b). It is interesting to note that the separation of SCF/PEEK is always initiated from the side along the sliding direction, as indicated by white arrows. The separation of SCF/PEEK is assumed to be the result of interfacial fatigue occurring in the zones where the fibers are most highly loaded. Under high pressures, the fibers bear high tension

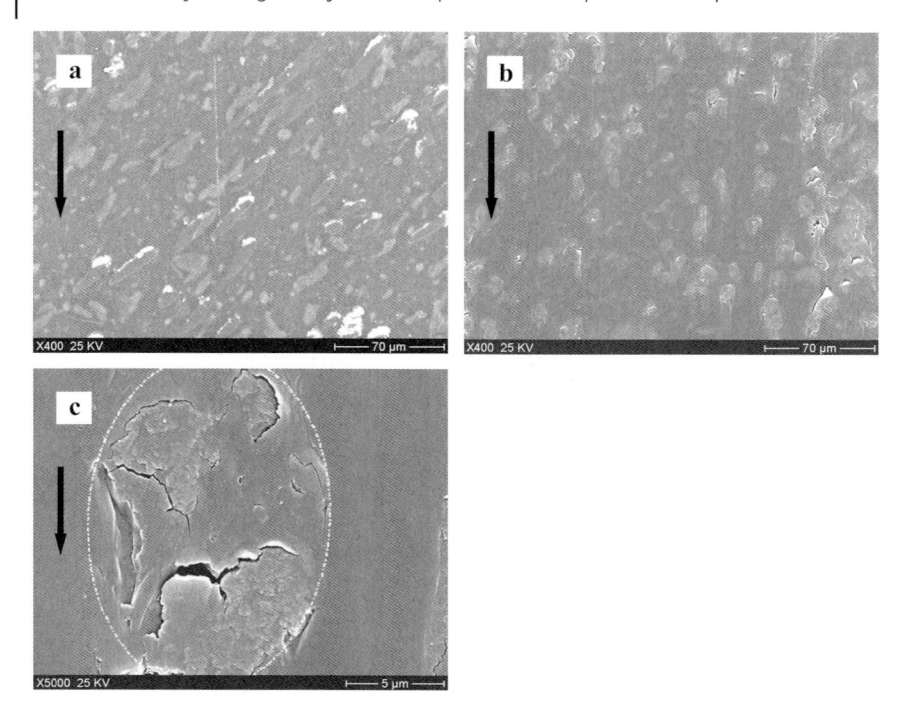

Figure 10.13 Worn surfaces (produced under 1 MPa pressure, sliding velocity 1 m s^{-1}) of PEEK/SCF/PTFE/graphite (a) and PEEK/SCF/PTFE/graphite/nano-SiO$_2$ at low (b) and high (c) magnification [30]. Reprinted with permission from Elsevier.

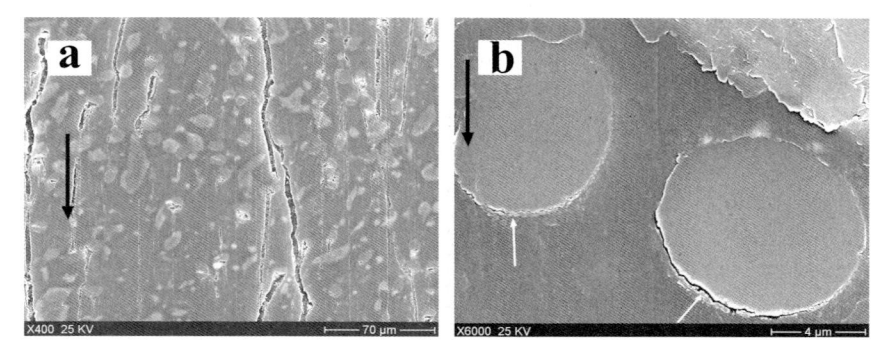

Figure 10.14 The worn surface of PEEK/SCF/PTFE/graphite produced under 5 MPa pressure, 1 m s^{-1} sliding velocity, at low (a) and high (b) magnification [30]. Reprinted with permission from Elsevier.

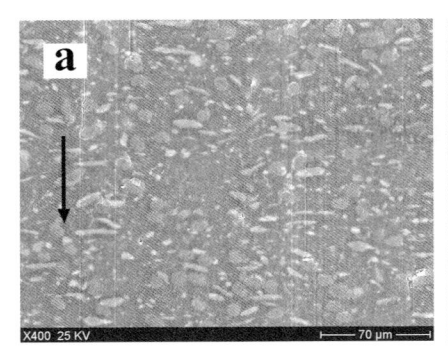

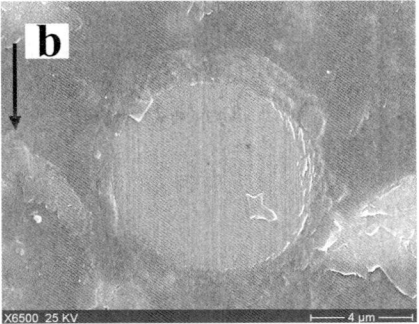

Figure 10.15 The worn surface of PEEK/SCF/PTFE/graphite/nano-SiO$_2$ produced under 5 MPa pressure, 1 m s^{-1} sliding velocity, at low (a) and high (b) magnification [30]. Reprinted with permission from Elsevier.

and shear stresses [70, 71]. In addition, due to the distinctly different moduli of the SCFs and the matrix, stress concentration occurs on SCFs in the frictional layer, and this may lead to a significant deformation of SCFs along the sliding direction. With cyclic effects, the high stress and strain of SCFs can cause failures of SCF/PEEK interfaces in the zones where the fibers are highly loaded. Further stressing on partially debonded SCFs will cause SCF failure, such as interfacial removal and breakage [70].

Figure 10.15a shows the worn surface of PEEK/SCF/PTFE/graphite/nano-SiO$_2$ produced under a pressure of 5 MPa and a sliding velocity of 1 m s^{-1}, while Figure 10.15b shows the typical SCF/PEEK interface under high magnification. On comparing the worn surfaces of the two composites, two different aspects can be addressed. First, the incorporated nano-SiO$_2$ significantly protects the SCF/PEEK bonding during the sliding process, and therefore no obvious SCF separation from the matrix is noted (this may be the main reason why wear resistance is improved after incorporating nano-SiO$_2$). Second, slight plow traces are observed on the fiber surfaces, which can be ascribed to mild abrasions caused by crushed nanoparticle agglomerates. Under high pressure, the large nanoparticle agglomerates may be crushed under an elevated compressive stress, and this is the reason why the agglomerates do not result in such severe SCF destruction as under a low pressure. With regards to the wear rates of the two composites (see Figure 10.12) it can be concluded that, for the studied nanocomposite, a pressure of 2 MPa is critical since, under a higher pressure large nanoparticle agglomerates would be crushed to form smaller units that would no longer exert any severe abrasive force on the SCFs. However, it should be noted that a different velocity or a different dispersion state of the nanoparticles can change the critical pressure. For the nanocomposite, an increase in sliding velocity from 1 m s^{-1} to 2 m s^{-1} will slightly reduce the wear rate, but at a higher velocity the nanoparticle agglomerates can be crushed into smaller particles than are produced at a lower velocity, due to the higher dynamics of the counterpart. Moreover, as the sliding velocity is increased the crushed

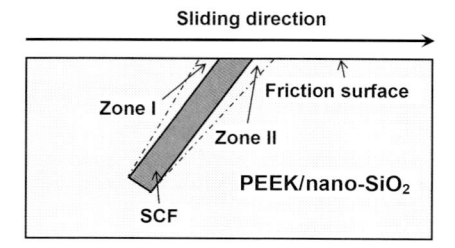

Figure 10.16 Schematic illustration of the effect of nanoparticles on the SCF/matrix interface [30]. Reprinted with permission from Elsevier.

agglomerates will tend to change their movement patterns, from sliding to rolling [72]. Together, these two factors can reduce the plowing and cutting effects on SCFs, and thus also reduce the wear rate of the composite.

With regards to the mechanism of how nanoparticles protect the carbon fibers, it is assumed that the stiffness enhancement of the matrix is an important factor. As shown schematically in Figure 10.16, in zone I (in front of the SCFs) the increased stiffness of PEEK (due to nanoparticles incorporation) can reduce the stress concentration on SCFs. However, during the sliding process, stress transfer occurs from the matrix to the nanoparticles in the frictional layer, such that the stress concentration on the fibers can be reduced. In zone II (behind the SCFs), the increased stiffness of PEEK can reduce the deformation of SCFs under a tension stress. Taken together, these two factors are considered to reduce PEEK/SCF interface fatigue and thus reduce SCF failures in the subsequent sliding process. In order to verify this mechanism, investigations should be made into the fatigue resistances of the PEEK/SCF and PEEK/SCF/nano-SiO$_2$ composites.

To conclude, the synthetic roles of both nanofillers and traditional fillers on the tribological performance of polymer composites have been confirmed, especially under high-pv conditions. It appears that the addition of nanoparticles enhances the critical values of the pv factor in conventional composites, while it can be reasonably expected that a better dispersion state of the nanoparticles may improve the tribological performance, even with a low pv factor. Taken together, these conditions mean that this formulation route becomes increasingly attractive. The combination of multiple-scale fillers-reinforced composites is of particular interest in the development of wear-resistant materials that will be subjected to extreme conditions, such as high pv and high environmental temperatures. Consequently it is believed that, in the future, this formulation route will attract the attention of materials scientists and tribologists alike. With regards to the tribological mechanism, it has been verified by many research groups that the protection of SCFs by nanoparticles or sub-microparticles is responsible for the improvements in their properties. However, further effort is required to acquire an understanding of the mechanism by which nanoparticles protect SCFs.

References

1 Friedrich, K., Zhang, Z., and Schlarb, A.K. (2005) *Compos. Sci. Technol.*, **65**, 2329–2343.
2 Voort, J. and Bahadur, S. (1995) *Wear*, **181–183**, 212–221.
3 Briscoe, B.J., Yao, L.H., and Stolarski, T.A. (1986) *Wear*, **108**, 357–374.
4 Zhang, M.Q., Rong, M.Z., Yu, S.L., Wetzel, B., and Friedrich, K. (2002) *Wear*, **253**, 1086–1093.
5 Persson, B.N.J. (2000) *Sliding Friction: Physical Principles and Applications*, Springer.
6 Grellmann, W. and Seidler, S. (2007) *Polymer Testing*, Carl Hanser.
7 Zhang, G. and Schlarb, A.K. (2009) *Wear*, **266**, 337–344.
8 Voss, H. and Friedrich, K. (1987) *Wear*, **116**, 1–18.
9 Rong, M.Z., Zhang, M.Q., Zheng, Y.X., Zeng, H.M., Walter, R., and Friedrich, K. (2001) *Polymer*, **42**, 167–183.
10 Dasari, A., Yu, Z.Z., and Mai, Y.W. (2009) *Mater. Sci. Eng. R*, **63**, 31–80.
11 Zhang, Z., Yang, J.L., Schlarb, A.K., and Friedrich, K. (2006) *Polymer*, **47**, 2791–2801.
12 Zhang, Z., Yang, J.L., and Friedrich, K. (2004) *Polymer*, **45**, 3481–3485.
13 Faber, K.T. and Evans, A.G. (1983) *Acta Metall.*, **31** (4), 565–576.
14 Faber, K.T. and Evans, A.G. (1983) *Acta Metall.*, **31** (4), 577–584.
15 Wetzel, B., Rosso, P., Haupert, F., and Friedrich, K. (2006) *Eng. Fract. Mech.*, **73**, 2375–2398.
16 Zhang, H., Zhang, Z., Friedrich, K., and Eger, C. (2006) *Acta Mater.*, **54**, 1833–1842.
17 Yang, J.L., Zhang, Z., and Zhang, H. (2005) *Compos. Sci. Technol.*, **65**, 2374–2379.
18 Cotterell, B., Chia, J.Y.H., and Hbaieb, K. (2007) *Eng. Fract. Mech.*, **74**, 1054–1078.
19 Ma, J., Mo, M.S., Du, X.S., Rosso, P., Friedrich, K., and Kuan, H.C. (2009) *Polymer*, **49**, 3510–3523.
20 Durand, J.M., Vardavoulias, M., and Jeandin, M. (1995) *Wear*, **181-183**, 833–839.
21 Chen, W.X., Li, F., Han, G., Xia, J.B., Wang, L.Y., Tu, J.P., and Xu, Z.D. (2003) *Tribol. Lett.*, **15**, 275–278.
22 Tanaka, A., Umeda, K., Yudasaka, M., Suzuki, M., Ohana, T., Yumura, M., and Iijima, S. (2005) *Tribol. Lett.*, **19**, 135–142.
23 Cai, H., Yan, F., and Xue, Q. (2004) *Mater. Sci. Eng. A*, **364**, 94–100.
24 Zoo, Y.S., An, J.W., Lim, D.P., and Lim, D.S. (2004) *Tribol. Lett.*, **16**, 305–309.
25 Chen, H., Jacobs, O., Wu, H., Ruediger, G., and Schaedel, B. (2007) *Polym. Test.*, **26**, 351–360.
26 Meng, H., Sui, G.X., Xie, G.Y., and Yang, R. (2009) *Compos. Sci. Technol.*, **69**, 606–611.
27 Jacobs, O. and Schaedel, B. (2008) *Tribology of Polymeric Nanocomposites* (eds K. Friedrich and A.K. Schlarb), Elsevier, Amsterdam.
28 Chang, L., Zhang, Z., Breidt, C., and Friedrich, K. (2005) *Wear*, **258**, 141–148.
29 Guo, Q.B., Rong, M.Z., Jia, G.L., Lau, K.T., and Zhang, M.Q. (2009) *Wear*, **266**, 658–665.
30 Zhang, G., Chang, L., and Schlarb, A.K. (2009) *Compos. Sci. Technol.*, **69**, 1029–1035.
31 Shi, G., Zhang, M.Q., Rong, M.Z., Wetzel, B., and Friedrich, K. (2004) *Wear*, **256**, 1072–1081.
32 Schwartz, C. and Bahadur, S. (2000) *Wear*, **237**, 261–273.
33 Tanaka, K. (1986) *Friction and Wear of Polymer Composites* (ed. K. Friedrich), Elsevier, Amsterdam.
34 Bahadur, S. and Tabor, D. (1984) *Wear*, **98**, 1–13.
35 Shi, G., Zhang, M.Q., Rong, M.Z., Wetzel, B., and Friedrich, K. (2004) *Wear*, **256**, 1072–1081.
36 Ji, Q.L., Zhang, M.Q., Rong, M.Z., Wetzel, B., and Friedrich, K. (2005) *Tribol. Lett.*, **20**, 115–123.
37 Zhang, M.Q., Rong, M.Z., Yu, S.L., Wetzel, B., and Friedrich, K. (2002) *Macromol. Mater. Eng.*, **287**, 111–115.
38 Xue, Q. and Wang, Q. (1997) *Wear*, **213**, 54–58.

39 Wang, Q., Xue, Q., Liu, H., Shen, W., and Xu, J. (1996) *Wear*, **198**, 216–219.
40 Wang, Q., Xu, J., Shen, Q., and Liu, W. (1996) *Wear*, **196**, 82–86.
41 Wang, Q., Xue, Q., and Shen, W. (1997) *Tribol. Int.*, **30**, 193–197.
42 Wang, Q., Xu, J., Shen, W., and Xue, Q. (1997) *Wear*, **209**, 316–321.
43 Wang, Q., Xue, Q., Shen, W., and Zhang, J. (1998) *J. Appl. Polym. Sci.*, **69**, 135–141.
44 Bahadur, S. and Sunkara, C. (2005) *Wear*, **258**, 1411–1421.
45 Avella, M., Errica, M.E., and Martuscelli, E. (2001) *Nano Lett.*, **1**, 213–217.
46 Rong, M., Zhang, M., Liu, H., Zeng, H., Wetzel, B., and Friedrich, K. (2001) *Ind. Lubr. Tribol.*, **53**, 72–77.
47 Wetzel, B., Haupert, F., Friedrich, K., Zhang, M.Q., and Rong, M.Z. (2002) *Polym. Eng. Sci.*, **42**, 1919–1927.
48 Wetzel, B., Haupert, F., and Zhang, M.Q. (2003) *Compos. Sci. Technol.*, **63**, 2055–2067.
49 Shi, G., Zhang, M.Q., Rong, M.Z., Wetzel, B., and Friedrich, K. (2003) *Wear*, **254**, 784–796.
50 Li, F., Hu, K., Li, J., and Zhao, B. (2002) *Wear*, **249**, 877–882.
51 Sawyer, W.G., Freudenberg, K.D., Bhimaraj, P., and Schadler, L.S. (2003) *Wear*, **254**, 573–580.
52 Wang, Q.H., Xue, Q.J., Liu, W.M., and Chen, J.M. (2000) *Wear*, **243**, 140–146.
53 Schwartz, C.J. and Bahadur, S. (2000) *Wear*, **237**, 261–273.
54 Bahadur, S. (2000) *Wear*, **245**, 92–99.
55 Gao, J. (2000) *Wear*, **245**, 100–106.
56 Rong, M.Z., Zhang, M.Q., Shi, G., Ji, Q.L., Wetzel, B., and Friedrich, K. (2003) *Tribol. Int.*, **36**, 697–707.
57 Luo, Y., Rong, M.Z., and Zhang, M.Q. (2007) *J. Appl. Polym. Sci.*, **104**, 2608–2619.
58 Luo, Y., Rong, M.Z., and Zhang, M.Q. (2005) *Polym. Compos.*, **13**, 245–252.
59 Zhang, G. and Schlarb, A.K. (2009) *Wear*, **266**, 745–752.
60 Zhang, G., Schlarb, A.K., Tria, S., and Elkedim, O. (2008) *Compos. Sci. Technol.*, **68**, 3073–3080.
61 Misra, R.D.K., Hadal, R.S., and Duncan, S.J. (2004) *Acta Mater.*, **52**, 4363–4376.
62 Qian, D., Dickey, E.C., Andrews, R., and Rantell, T. (2000) *Appl. Phys. Lett.*, **76**, 2868–2870.
63 Kanagaraj, S., Varanda, F.R., Zhiltsova, T.V., Oliveira, M.S.A., and Simoes, J.A.O. (2007) *Compos. Sci. Technol.*, **67**, 3071–3077.
64 Kim, J.Y., Han, S.I., Kim, D.K., and Kim, S.H. (2009) *Compos., Part A*, **40**, 45–53.
65 Xie, X.L., Mai, Y.W., and Zhou, X.P. (2005) *Mater. Sci. Eng. R*, **49**, 89–112.
66 Sandler, J., Shaffer, M.S.P., Prasse, T., Bauhofer, W., Schulte, K., and Windle, A.H. (1999) *Polymer*, **40**, 5967–5971.
67 Shaffer, M.S.P. and Windle, A.H. (1999) *Adv. Mater.*, **11**, 937–941.
68 Jiang, Z.Y., Gyurova, L.A., Schlarb, A.K., Friedrich, K., and Zhang, Z. (2008) *Compos. Sci. Technol.*, **68**, 734–742.
69 Chang, L., Zhang, Z., Ye, L., and Friedrich, K. (2007) *Tribol. Int.*, **40**, 1170–1178.
70 Friedrich, K., Váradi, K., Goda, T., and Giertzsch, H. (2002) *J. Mater. Sci.*, **37**, 3497–3507.
71 Goda, T., Váradi, K., Wetzel, B., and Friedrich, K. (2004) *J. Compos. Mater.*, **38**, 1607–1618.
72 Adachi, K. and Hutchings, I.M. (2003) *Wear*, **255**, 23–29.

11
Mechanisms of Biodegradability Generation in Polymer Nanocomposites

Mitsuhiro Shibata

11.1
Introduction

Biodegradable plastics such as aliphatic polyesters, cellulose-based thermoplastics and other polysaccharide-based plastics have been extensively investigated, since the 1970s, in order to reduce the environmental pollution caused by plastic wastes [1]. With regards to biodegradable polyesters, polylactide (PLA), poly(butylene succinate) (PBS), poly(ε-caprolactone) (PCL), poly(butylene adipate-co-terephthalate) (PBAT), and poly(3-hydroxybutyrate-co-3-hydroxyvalerate) (PHBV) have been already commercialized. Although the market for biodegradable polyesters is gradually increasing, these environmentally benign polyesters have not yet sufficiently replaced the conventional nonbiodegradable polymers such as poly(ethylene terephthalate) (PET), because of their high expense and inferior mechanical and thermal properties. Recently, the reinforcement of such biodegradable polyesters with layered silicates as environmentally benign materials has generated significant research attention, because these materials often exhibit greatly improved mechanical, thermal, barrier, and fire-retardant properties at low clay contents, in comparison with more conventional microcomposites. When naturally occurring clay is used as a filler of the biodegradable polyester, it is to be expected that a small amount of the residual clay will return to nature after biodegradation. In recent years, many reports have been made regarding the preparation, morphology, and properties of biodegradable polyester/layered silicate nanocomposites, including PLA [2–21], PBS [22–28], PCL [29–44], PBAT [45–48], and PHBV [49–51] /layered silicate nanocomposites. Far fewer reports have been made, however, on their biodegradability, and consistent results have not yet been obtained. Lee *et al.* reported that the biodegradability of an aliphatic polyester/organomodified montmorillonite (MMT), as evaluated from its weight loss in activated soil, decreased with the increasing clay content [52]. Likewise, Wang *et al.* reported that the biodegradability of a PHBV/organomodified MMT, as evaluated by its weight loss in a soil suspension, decreased with the increasing clay content [49]. Ray *et al.* reported that PLA nanocomposites with a trimethyloctadecylammonium-modified MMT had a higher biodegradability than control PLA, when biodegradability was evaluated by compost treatment at 58 °C for 60 days [19]. It should be noted that the biodegradation of PLA by compost treatment is rapid, whereas biodegradation in activated sludge at room temperature is much slower than that for other aliphatic polyesters such as PHBV, PBS, and PCL. The essential biodegradation mechanisms of the polyester/layered silicate nanocomposites should be the same as those of the control polyesters. In general, the aerobic biodegradation of aliphatic polyesters proceeds via a two-step process:

- The molecular weight of polyesters is reduced by extracellular enzymatic hydrolysis and/or nonenzymatic hydrolysis to produce metabolizable oligomers.

- The intermediates are then metabolized by microorganisms and finally decomposed to carbon dioxide and water.

Importantly, the layered silicates in the polyester nanocomposites physically and chemically affect these biodegradation processes.

In this chapter, the morphologies, thermal, mechanical, and biodegradable properties of PBAT/layered silicate [45, 48] and PBS/layered silicate [26] nanocomposites, prepared using a melt-intercalation method, are described (Figure 11.1). Attention is focused especially on the influence of the dispersion and organomodification of layered silicates on the biodegradability of PBAT and PBS nanocomposites, as evaluated using aerobic biodegradability tests both in soil and in aqueous media using activated sludge.

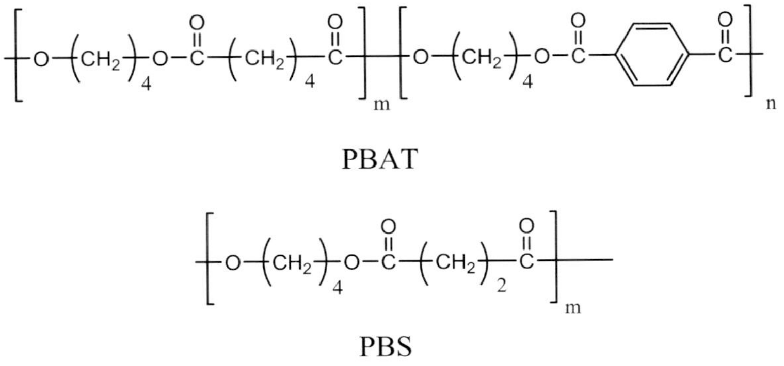

PBAT

PBS

Figure 11.1 The structures of poly(butylene adipate-co-terephthalate) (PBAT) and poly(butylene succinate) (PBS).

11.2
PBAT Nanocomposites

11.2.1
Preparation and Morphology of PBAT Nanocomposites

PBAT is a biodegradable aliphatic–aromatic copolyester, which is a flexible thermoplastic designed for film extrusion and extrusion coating marketed by BASF under the tradename Ecoflex® [melt flow rate (190 °C, 2.16 kg): 3.3–6.6 g per 10 min; specific gravity 1.26; glass transition temperature (T_g): −30 °C by differential scanning calorimetry (DSC), melting temperature (T_m) 115 °C (by DSC); butylene terephthalate/adipate composition ca. 4/6]. Sodium MMT, with the tradename Kunipia® F (cation exchange capacity 115 mEq per 100 g) was supplied from Kunimine Industries Co. Ltd. (Japan). Three types of amine compound, namely dodecylamine (DA), octadecylamine (ODA), and N-dodecyldiethanolamine (DEA) (as shown in Figure 11.2) were used for the preparation of organomodified MMT. The DA-, ODA-, and DEA-modified MMTs (DA-M, ODA-M, DEA-M) were prepared by the ion-exchange reaction of MMT and the protonated ammonium cations of DA, ODA and DEA, prepared *in situ* using aqueous hydrochloric acid. The organic fractions measured by thermogravimetric analysis (TGA) for DA-M, ODA-M, and DEA-M were 20.4, 27.4, and 23.9 wt%, respectively. The cation-exchange rates calculated from the organic fractions were 116, 117, and 97%, respectively. Melt mixing of PBAT with MMTs was performed on a Laboplasto-mill with a twin rotary roller-mixer (Toyo Seiki Co. Ltd, Japan). The mixing was carried out for 5 min with a rotary speed 50 rpm at 140 °C. The inorganic content of the blends varied from 3.0 to 10.0 wt%; for example, the abbreviation PBAT/ODA-M3 represents the PBAT/ODA-M composite with an inorganic content of 3 wt%. Mixing was carried out for 5 min with a rotary speed 50 rpm at 140 °C, after which the obtained compounds were injection-molded at a cylinder temperature of 200 °C and a molding temperature of 60 °C, using a desk injection-molding machine (Little-Ace I Type; Tsubaco KI Co., Ltd, Japan).

Figure 11.3 shows the X-ray diffraction (XRD) patterns of MMT and organoclays, where the peaks correspond to the [001] basal reflection of the MMT aluminosilicate. The interlayer spacing (d_1) of each of the clays was determined from the angular location of the peaks and the Bragg condition. The calculated values are summarized in Table 11.1. The difference in the interlayer spacing between the natural and organoclay is due to intercalation of the ammonium surfactant which,

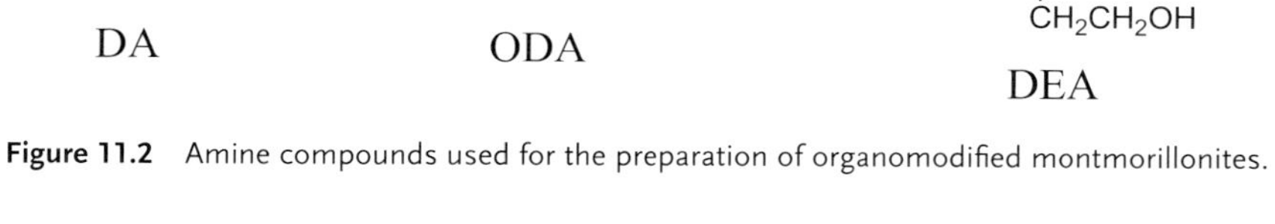

$$CH_3\text{-}(CH_2)_{11}\text{-}NH_2 \qquad CH_3\text{-}(CH_2)_{17}\text{-}NH_2 \qquad CH_3\text{-}(CH_2)_{11}\text{-}N(CH_2CH_2OH)_2$$

DA ODA DEA

Figure 11.2 Amine compounds used for the preparation of organomodified montmorillonites.

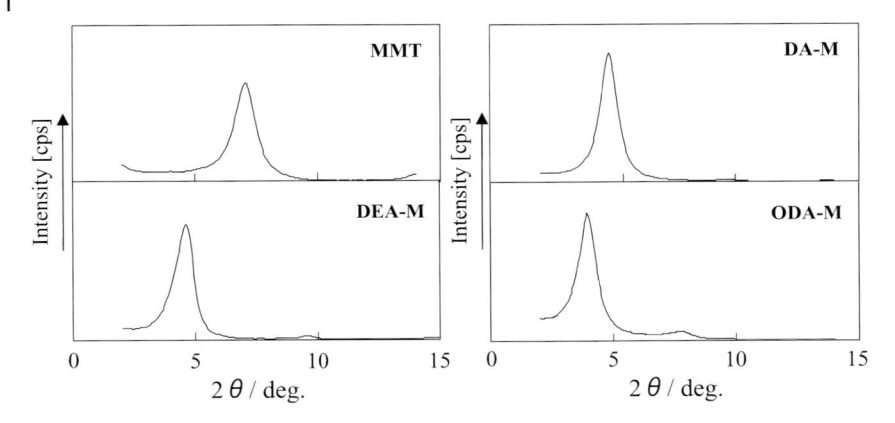

Figure 11.3 X-ray diffraction patterns of MMT and organomodified MMTs. Reprinted from Ref. [26] with permission from John Wiley & Sons Ltd.

Table 11.1 Interlayer spacing as determined by X-ray diffraction (XRD) analysis for clays and PBAT nanocomposites with an inorganic content of 3 wt%.

Clay	XRD peak position (2θ)		Interlayer spacing [001] (nm)		Δd (nm)[a]
	In clay	In composite	In clay (d_1)	In composite (d_2)	
MMT	7.02	6.76	1.26	1.31	0.05
DA-M	5.00	3.22, 6.46[b]	1.77	2.74	0.97
ODA-M	4.84	–	1.82	–	–
DEA-M	4.86	2.76, 5.62[b]	1.82	3.20	1.38

a) $\Delta d = d_2 - d_1$ (nm).
b) The weak peak with a higher 2θ value is due to [002] plane of the silicate layers.
Reprinted from Ref. [48] with permission from John Wiley & Sons Ltd.

having a bulkier substituent on the nitrogen atom, shows a higher interlayer spacing. Figure 11.4 shows the XRD patterns of PBAT and PBAT composites with inorganic content 3 wt%. The difference (Δd) in interlayer spacing between clay and PBAT/clay relates to the degree of intercalation. The Δd of PBAT/MMT was almost zero, indicating that little or no intercalation has occurred. The Δd of PBAT/DA-M and PBAT/DEA-M were 0.97 and 1.38 nm, respectively, which indicated that PBAT had been intercalated into the gallery of silicate layers in these composites (see Table 11.1). The PBAT/ODA-M composite showed only a broad peak at around 2θ = 3°, although this peak was thought mainly to relate to the PBAT component, because neat PBAT and PBAT/MMT composites show a similar peak at around 3°. Even if the peak of the PBAT/ODA-M composite contained the [001] basal reflection of the silicate plates, the contributing proportion would be small. Also, although clear peaks corresponding to the [002] plane were observed

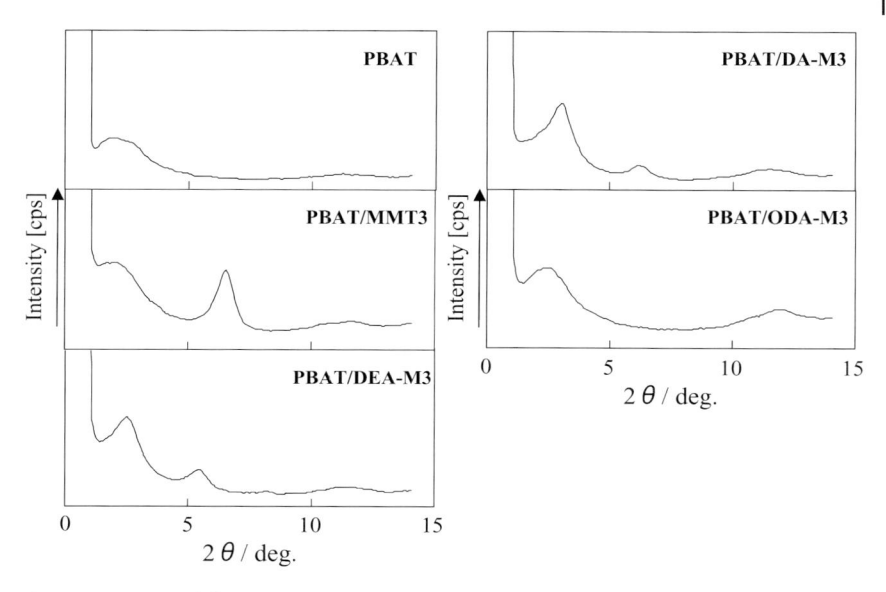

Figure 11.4 X-ray diffraction patterns of PBAT and PBAT/clay composites with an inorganic content of 3 wt%. Reprinted from Ref. [48] with permission from John Wiley & Sons Ltd.

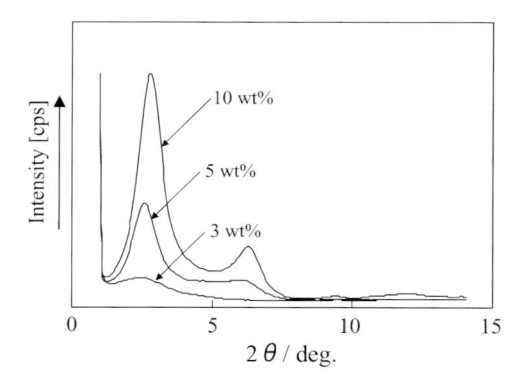

Figure 11.5 X-ray diffraction patterns of PBAT/ODA-M composites with various inorganic contents. Reprinted from Ref. [48] with permission from John Wiley & Sons Ltd.

at 5.6 and 6.4° for PBAT/DEA-M and PBAT/DA-M composites, respectively, there was no peak corresponding to the [002] plane for the PBAT/ODA-M composite. It is supposed that the main peak corresponding to the silicate layer would be hidden in a smaller angle region than 1° of the XRD curve of PBAT/ODA-M3. In that case, there would be a possibility of the formation of a partially exfoliated nanocomposite. For the PBAT/ODA-M composites with inorganic contents of 5 and 10 wt%, two clear diffraction peaks corresponding to the [001] and [002] planes were observed around 3° and 6° (Figure 11.5). The Δd of PBAT/ODA-M5 was calculated to be 1.20 nm, indicating that the increase of ODA-M caused formation of the intercalated nanocomposite.

Figure 11.6 shows the transmission electron microscopy (TEM) images with a lower magnification for PBAT composites with an inorganic content of 3 wt%. It is clear that ODA-M is more finely dispersed in the matrix than either DA-M or DEA-M. The PBAT/DA-M and PBAT/DEA-M composites also contained some clusters of clays. In the higher-magnification TEM images shown in Figure 11.7, it is clear that both the PBAT/DA-M and PBAT/DEA-M composites

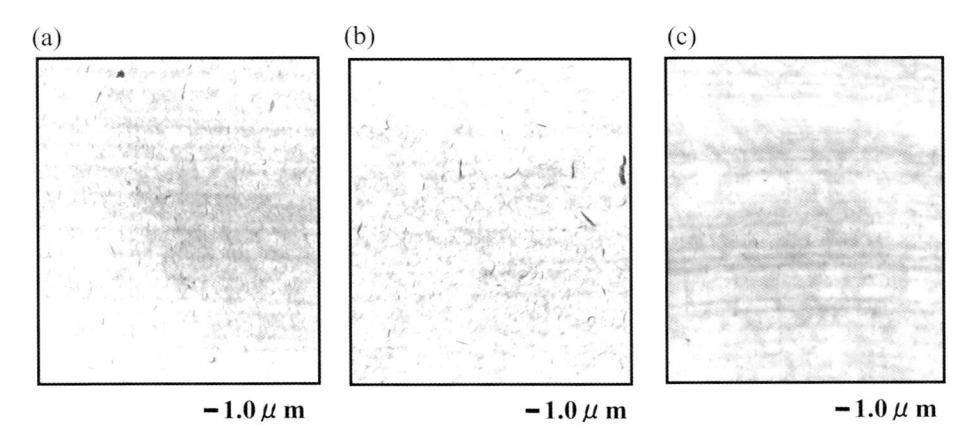

Figure 11.6 Transmission electron microscopy images with a lower magnification for PBAT composites. (a) PBAT/DA-M3; (b) PBAT/DEA-M3; (c) PBAT/ODA-M3. Reprinted from Ref. [48] with permission from John Wiley & Sons Ltd.

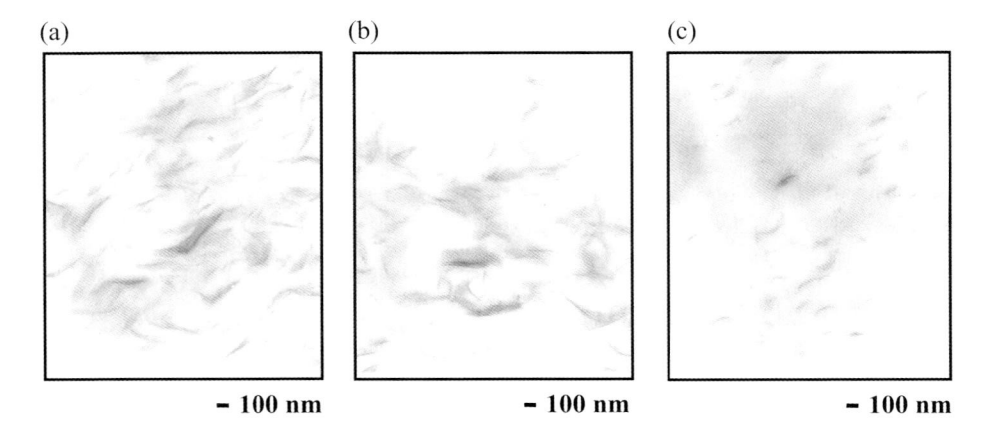

Figure 11.7 Transmission electron microscopy images with a higher magnification for PBAT composites. (a) PBAT/DA-M3; (b) PBAT/DEA-M3; (c) PBAT/ODA-M3. Reprinted from Ref. [48] with permission from John Wiley & Sons Ltd.

are composed mainly of a stacked intercalated structure, although in the case of PBAT/ODA-M a smaller amount of stacked plates appeared in the broad- and obscure-shaded region. Based on this result and the XRD findings it was concluded that exfoliation had occurred mainly in the broad-shade region.

11.2.2
Mechanical Properties of PBAT Nanocomposites

Figure 11.8 shows the tensile properties of the PBAT composites. Overall, the tensile modulus was increased by the addition of clays, whereas the tensile strength

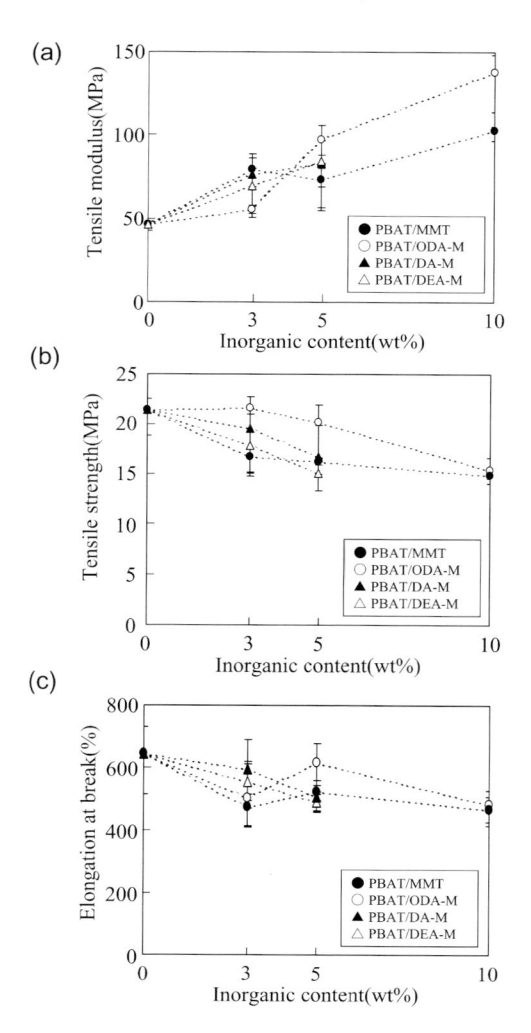

Figure 11.8 Tensile properties of PBAT composites. (a) Tensile modulus; (b) Tensile strength; (c) Elongation at break. Reprinted from Ref. [48] with permission from John Wiley & Sons Ltd.

and elongation at break were rather lowered. Among the PBAT/organoclay composites with an inorganic content of 3 wt%, the intercalated DA-M composite showed the highest modulus, and the exfoliated ODA-M composite the lowest. (The reason for this difference is discussed in Section 11.2.3.) When the inorganic content was 5 wt%, the ODA-M composite showed the highest modulus, while the PBAT/MMT microcomposite showed the lowest modulus. The tensile modulus of the ODA-M composites was increased considerably in line with the inorganic content. Among the PBAT/organoclay composites with an inorganic content of 3 wt%, the exfoliated ODA-M composite showed the highest strength, this being slightly higher than that of neat PBAT; consequently, exfoliation was thought to be effective for maintaining tensile strength. In the case of the intercalated DA-M and DEA-M composites, the strength was seen to decrease in line with inorganic content. For ODA-M composites, the tensile strength decreased with increasing inorganic content, a result which can be attributed to the occurrence of intercalation rather than exfoliation by the increase of ODA-M. It appears that elongation at break is somewhat lowered by the addition of clays. For example, when the PBAT composites with an inorganic content of 5 wt% were compared, the PBAT/ODA-M composite showed the highest tensile modulus, strength, and elongation.

Figure 11.9 shows dynamic viscoelastic curves of PBAT and PBAT composites with inorganic content 3 wt%. All samples showed a primary dispersion around $-20\,°C$, and a secondary dispersion around $60\,°C$ in the $\tan\delta$ curves. The former dispersion corresponds to the motion of the poly(butylene adipate)

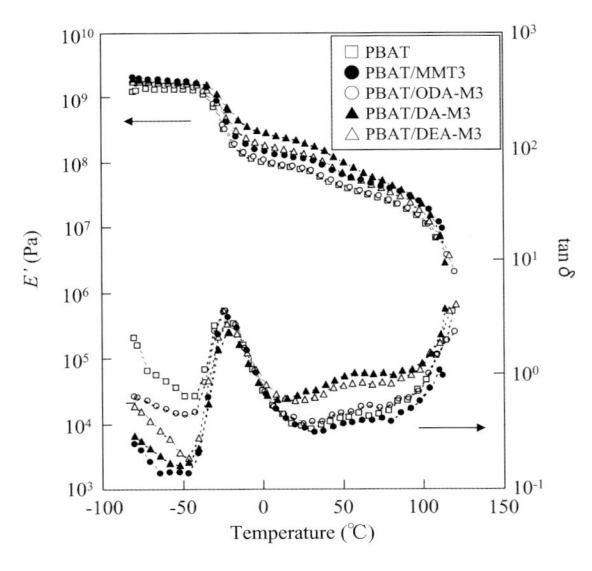

Figure 11.9 Dynamic viscoelastic curves of PBAT and PBAT/clay composites with an inorganic content of 3 wt%. Reprinted from Ref. [48] with permission from John Wiley & Sons Ltd.

unit, and the latter dispersion to the motion of poly(butylene terephthalate). The tan δ peak temperature corresponding to the T_g of the poly(butylene adipate) unit of all the PBAT/clay composites was slightly higher than that of neat PBAT. In agreement with the result of tensile modulus, the intercalated DA-M or DEA-M composites showed a higher storage modulus (E') than the exfoliated ODA-M composite and neat PBAT over the temperature range of −30 to 70 °C. When materials become soft above T_g, the reinforcement effect of the clay particles becomes prominent, due to a restricted movement of the polymer chains, and hence a strong enhancement of modulus appears. There was no significant improvement of E' in the glassy state, where the temperature was below −30 °C.

11.2.3
Thermal Properties of PBAT Nanocomposites

The thermal properties of the PBAT composites, as determined by DSC measurements, are summarized in Table 11.2. When discussing the mechanical properties of PBAT nanocomposites in connection with the degree of intercalation/exfoliation, the influence of degree of crystallinity (χ_c) of PBAT component in the composites should be considered because χ_c may be changed by influence of the organoclay. The original degree of crystallinity of the injection-molded composite can be evaluated from the value of $(\Delta H_m - \Delta H_{g,c})$, where ΔH_m and $\Delta H_{g,c}$ are the heat of melting and crystallization from the glassy state of the injection-molded sample in the first heating DSC scan, respectively. When the composites with inorganic content 3 wt% were compared, the $(\Delta H_m - \Delta H_{g,c})$ value of the PBAT/ODA-M composite was considerably lower than that of other composites, which indicated that the injection-molded PBAT/ODA-M composite had a lower χ_c. The reason for a lower tensile modulus of the exfoliated PBAT/ODA-M3 composite

Table 11.2 Thermal properties of PBAT and PBAT composites.

Sample	$T_{g,c}$ (°C)	$\Delta H_{g,c}$ (J g^{-1})	T_m (°C)	ΔH_m (J g^{-1})	$H\Delta_m - H\Delta_{g,c}$ (J g^{-1})
PBAT	64.3	0.36	115.3	13.3	12.9
PBAT/MMT3	60.6	1.40	115.1	12.8	11.4
PBAT/DA-M3	57.3	0.74	115.4	12.5	11.8
PBAT/DEA-M3	59.3	0.62	114.5	14.2	13.6
PBAT/ODA-M3	57.3	0.99	114.4	9.9	8.9
PBAT/ODA-M5	60.1	0.83	116.7	11.1	10.3
PBAT/ODA-M10	60.5	0.69	117.5	10.2	9.5

$T_{g,c}$ is the crystallization temperature from the glassy state; T_m is melting temperature; $\Delta H_{g,c}$ and ΔH_m are the heat of crystallization from the glassy state and heat of melting per 1 g of PBAT contained in the composite, respectively.
Reprinted from Ref. [48] with permission from John Wiley & Sons Ltd.

should be attributed to the lower χ_c. Exfoliation may influence the retardation of the crystallization of the PBAT component during the injection-molding process. The $(\Delta H_m - \Delta H_{g,c})$ values of PBAT/ODA-M5 and PBAT/ODA-M10 were higher than that of the PBAT/ODA-M3. The fact that the PBAT/ODA-M composites with inorganic contents of 5 and 10 wt% showed considerably higher tensile moduli than the composite with inorganic content 3 wt% may be attributed to the fact that the former composites had a higher χ_c and a higher inorganic content than the latter composite.

11.2.4
Biodegradability of PBAT Nanocomposites

Biodegradability was determined by measuring weight loss of the composite films buried in soil. The soil used in the test was 1 : 1 mixture of black soil and leaf mold for gardening. Each specimen was buried in the soil in planters and incubated at room temperature (25–30 °C), with water being supplied at 2-day intervals to keep the soil wet. Each specimen was removed from the soil after having been buried for 30, 60, 90, 120, 150, and 180 days, respectively, washed with water and then dried to constant weight at 40 °C in a vacuum oven.

Biodegradability was also determined according to JIS K6950 (ISO14851) by measuring the biochemical oxygen demand (BOD) in an aerobic aqueous medium containing activated sludge. Phosphate buffer (pH 7.4, 200 ml) containing 0.25 mM $CaCl_2$, 0.09 mM $MgSO_4$, 0.09 mM NH_4Cl, and 0.9 µM $FeCl_3$ was mixed in a glass bottle. To the mixture was added 4.35 ml of activated sludge containing 30 mg insoluble fraction, which had been obtained from sewerage facilities of the Chiba Institute of Technology. The film samples were then added and the dispersion was steadily stirred using a magnetic stirring bar.

The BOD was measured at 25 °C by using a BOD tester (Model 200F; Taitec Corporation, Japan). The carbon dioxide produced was absorbed into a 50% sodium hydroxide aqueous concentrate, located in a cup within the glass bottle. The volume of the consumed oxygen was measured directly using a scaled cylinder.

Figure 11.10 shows the change in weight loss of control PBAT, PBAT/MMT, and PBAT/ODA-M during burial in soil. For the PBAT/MMT10 composite the weight loss after 240 days was about 4% (the highest value of all samples containing the control PBAT), whereas the weight loss for the PBAT/ODA-M10 composite was about 1% (the lowest value). Since, in soil burial tests the extent of weight loss was generally small, it was necessary to confirm the data obtained with subsequent surface observations and BOD testing. Figure 11.11 shows a series of optical images of the surfaces of samples buried in the soil for 240 days. Clearly, coloring and roughing is promoted by the addition of MMT, although it is interesting to note that the composites with ODA-M show a much lesser degree of change than do those composites with MMT.

Figure 11.12 shows the aerobic biodegradability determined by measuring the BOD of thin film samples composed of PBAT and PBAT composites in an

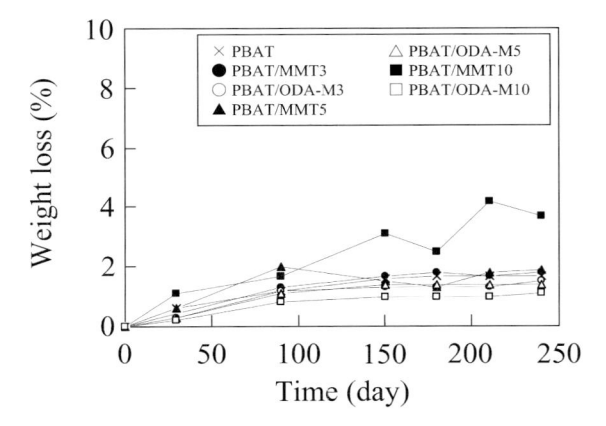

Figure 11.10 Weight loss of PBAT, PBAT/MMT and PBAT/ODA-M after burial in soil for specified periods. Reprinted from Ref. [45] with permission from John Wiley & Sons Ltd.

aqueous medium containing activated sludge for 70 days. The order of biodegradability observed, namely PBAT/MMT10 > PBAT ≈ PBAT/ODA-M10, was in agreement with results obtained after burial in soil. The promotion of biodegradation due to the addition of MMT may be related its highly hydrophilic character. On the other hand, some lowering of biodegradability for PBAT/ODA-M composites may derive from the finely dispersed silicate layers with a large aspect ratio in the PBAT matrix, which force the enzymes and/or water that is diffusing through the bulk of the film to take a more tortuous path. It is possible that the more hydrophobic character of ODA-M than MMT might also contribute in this respect.

11.3
PBS Nanocomposites

11.3.1
Preparation and Morphology of PBS Nanocomposites

Nanocomposites based on PBS and MMTs were prepared using a melt intercalation method, with Bionolle® #1020 [melt flow rate 25 g per 10 min (190 °C, 2.16 kg), specific gravity 1.26, melting temperature (T_m) 115 °C; supplied by Showa Highpolymer Co. Ltd, Japan] being used as the PBS. The MMTs—MMT, DA-M, ODA-M, and DEA-M—were the same as those used for the PBAT nanocomposites. Melt mixing of PBS with MMTs was carried out on a Laboplasto-mill using a twin rotary roller-mixer (Toyo Seiki Co. Ltd, Japan). The inorganic content of the blends varied from 1 to 10 wt%, and mixing was carried out for 5 min with a rotary speed 50 rpm at 140 °C. The compounds obtained were injection-molded at a cylinder

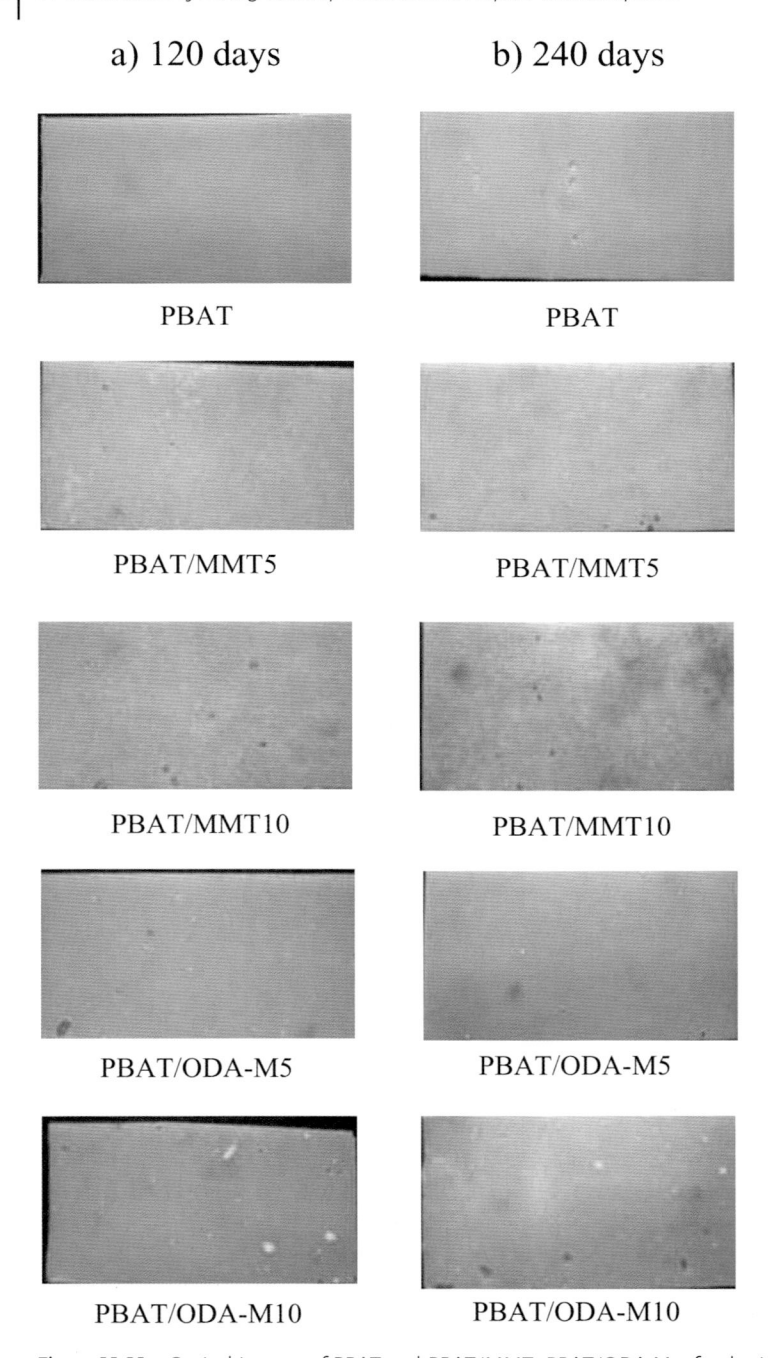

a) 120 days	b) 240 days
PBAT	PBAT
PBAT/MMT5	PBAT/MMT5
PBAT/MMT10	PBAT/MMT10
PBAT/ODA-M5	PBAT/ODA-M5
PBAT/ODA-M10	PBAT/ODA-M10

Figure 11.11 Optical images of PBAT and PBAT/MMT, PBAT/ODA-M, after burial in soil for 240 days. Reprinted from Ref. [45] with permission from John Wiley & Sons Ltd.

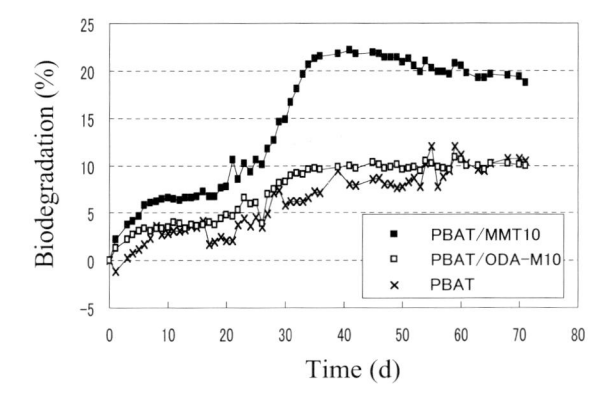

Figure 11.12 Biodegradability of PBAT, PBAT/MMT, and PBAT/ODA-M, as determined by measuring BOD in aqueous medium containing activated sludge. Reprinted from Ref. [45] with permission from John Wiley & Sons Ltd.

Table 11.3 Interlayer spacing as determined by X-ray diffraction (XRD) analysis for PBS nanocomposites with an inorganic content of 3 wt%.

Clay	XRD peak position (2θ)		Interlayer spacing [001] (nm)		Δd (nm)[a]
	In clay	In composite	In clay (d_1)	In composite (d_2)	
MMT	7.02	7.02	1.26	1.26	0.00
DA-M	5.00	3.20, 6.46[b]	1.77	2.76	0.99
ODA-M	4.84	2.70, 5.50[b]	1.82	3.27	1.45
DEA-M	4.86	2.72, 5.70[b]	1.82	3.25	1.43

a) $\Delta d = d_2 - d_1$ (nm).
b) The weak peak with a higher 2θ value is due to [002] plane of the silicate layers.
Reprinted from Ref. [26] with permission from John Wiley & Sons Ltd.

temperature of 140 °C and a molding temperature of 60 °C, using a desk injection-molding machine (Little-Ace I Type; Tsubaco KI Co., Ltd, Japan).

The interlayer spacing of the PBS nanocomposites is summarized in Table 11.3. The differences in interlayer spacing between the natural and organoclay are due to the intercalation of ammonium surfactant. The difference (Δd) in interlayer spacing between clay and PBS/clay relates to the degree of intercalation. For example, the Δd of PBS/MMT is zero, which indicates that little or no intercalation has occurred. All of the PBS/organomodified MMT composites had Δd-values not less than 0.99 nm, which indicated that they were all intercalated nanocomposites. The PBS/ODA-M and PBS/DEA-M composites had a larger Δd than the PBS/DA-M composite, which indicated that a greater amount of polymer had been intercalated between the clay platelets in these cases.

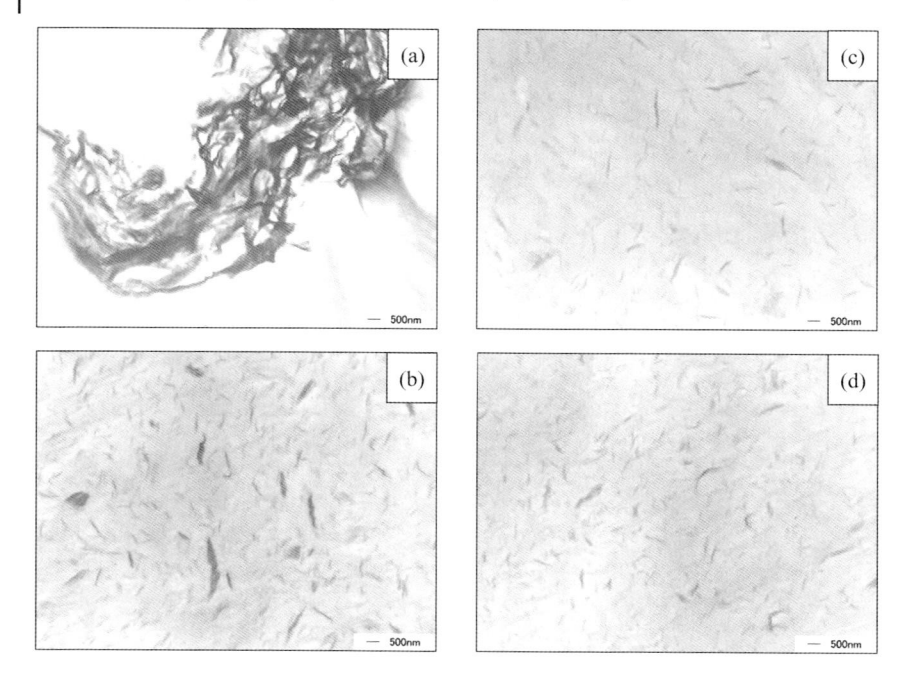

Figure 11.13 Transmission electron microscopy images of PBS nanocomposites. (a) PBS/MMT3; (b) PBS/DA-M3; (c) PBS/ODA-M3; (d) PBS/DEA-M3. Reprinted from Ref. [26] with permission from John Wiley & Sons Ltd.

Figure 11.13 shows some TEM images of PBS nanocomposites with an inorganic content of 3 wt%. From these images, it is clear that the order of finer dispersion of the silicate particles within the PBS matrix is ODA-M ≈ DEA-M > DA-M > MMT. Notably, the PBS/ODA-M and DEA-M composites showed a stacked intercalated structure that was disorderly and uniformly dispersed within the PBS matrix, this being in agreement with the result of XRD measurements. The presence of the exfoliated structure into single layers was not identified. In the case of the PBS/DEA-M composite, some aggregation of the clay platelets was observed.

11.3.2
Mechanical Properties of PBS Nanocomposites

The tensile and flexural properties of PBS composites with an inorganic content of 3 wt% are summarized in Table 11.4. The PBS/DA-M3, PBS/ODA-M3, and PBS/DEA-M3 nanocomposites showed higher tensile and flexural moduli than did the control PBS and PBS/MMT3, although their nanocomposites showed a lower tensile strength and elongation at break. It was considered that such an increase in tensile modulus was due to a better dispersion of the silicate layer

Table 11.4 Mechanical properties of PBS and PBS/clay composites with an inorganic content of 3 wt%.

Sample	Tensile properties			Flexural properties	
	Strength (MPa)	Modulus (GPa)	Elongation at break (%)	Strength (MPa)	Modulus (GPa)
PBS	33.7	0.707	7.60	44.3	0.754
PBS/MMT3	33.3	0.811	7.87	49.9	0.913
PBS/DA-M3	31.3	1.013	6.32	49.6	1.096
PBS/ODA-M3	30.8	1.002	6.34	48.4	1.025
PBS/DEA-M3	29.7	1.044	6.30	48.6	1.036

Reprinted from Ref. [26] with permission from John Wiley & Sons Ltd.

having a high modulus and a high aspect ratio, while the decrease in tensile strength was attributed to a decrease in the elongation at break, which was in turn most likely related to delamination of the polymer–silicate interlayer. On the other hand, these nanocomposites showed only a slightly higher flexural strength than the control PBS.

The reason that a clear decreasing trend in strength was not observed on the flexural mode may be attributed to the fact that the delamination does not significantly affect the strength, because the flexural deformation is composed of both modes of compression and expansion. Figure 11.14 shows, graphically, the relationship between the tensile properties and inorganic content for PBS/MMT and PBS/DEA-M composites. Over the 1 to 10 wt% inorganic content range, the PBS/DEA-M nanocomposite showed a higher tensile modulus and a lower tensile strength and elongation than did the PBS/MMT composite. Moreover, while the tensile modulus of the PBS/DEA-M nanocomposite increased linearly with the clay content, the tensile modulus of the PBS/MMT composite showed no such increase.

Dynamic viscoelastic measurements of the PBS/DEA-M nanocomposite revealed that both E' and T_g were increased with the inorganic content (3–10 wt%). The E'-values and tan δ peak temperature, as measured with dynamic viscoelastic analysis for PBS and all PBS composites, are summarized in Table 11.5. In agreement with the results of tensile modulus measurements, the PBS/DEA-M3 composite showed a higher E'-value over the temperature range of −60 to 90 °C than did the control PBS and PBS/MMT3 composites (Figure 11.15). Furthermore, the tan δ peak temperature corresponding to T_g for the PBS/DEA-M composite was shifted considerably to a higher temperature region in line with the inorganic content (Figure 11.16). In particular, the PBS/DEA-M composite with an inorganic content of 10 wt% showed a significantly higher T_g (−4.2 °C) than did PBS (−18.9 °C). These results indicate that the intercalation of PBS into the galley of silicate layers caused an interference of molecular motion.

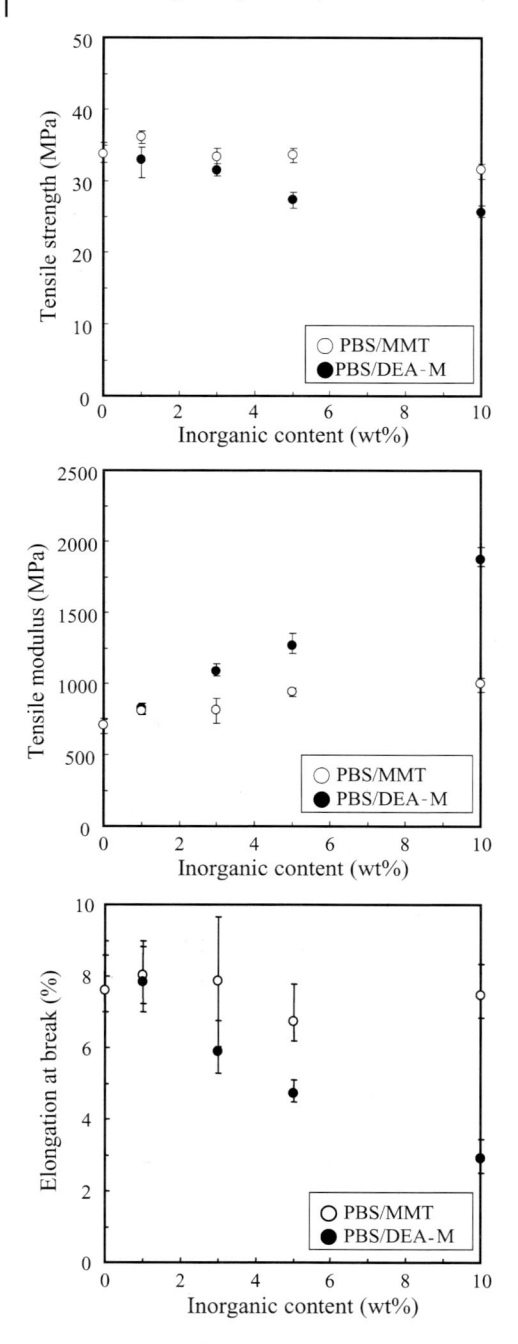

Figure 11.14 Tensile properties versus inorganic content for PBS/MMT and PBS/DEA-M composites. (a) Tensile strength; (b) Tensile modulus; (c) Elongation at break. Reprinted from Ref. [26] with permission from John Wiley & Sons Ltd.

Table 11.5 Dynamic viscoelastic properties of PBS and PBS composites.

Sample	Storage modulus (GPa)			Tan δ peak temperature (°C)
	−60 °C	−20 °C	20 °C	
PBS	2.06	1.09	0.59	−18.9
PBS/MMT1	2.32	1.37	0.67	−15.0
PBS/MMT3	2.13	1.15	0.63	−19.1
PBS/MMT5	2.41	1.49	0.77	−15.0
PBS/MMT10	2.44	1.53	0.79	−16.1
PBS/DA-M3	2.36	1.36	0.72	−18.0
PBS/ODA-M3	2.30	1.42	0.75	−18.9
PBS/DEA-M1	2.12	1.30	0.63	−16.0
PBS/DEA-M3	2.53	1.49	0.81	−20.1
PBS/DEA-M5	2.63	1.78	1.02	−14.0
PBS/DEA-M10	3.06	2.39	1.51	−4.2

Reprinted from Ref. [26] with permission from John Wiley & Sons Ltd.

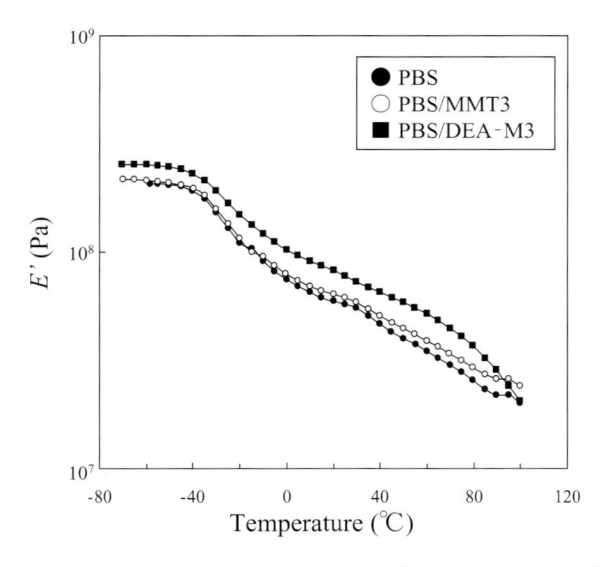

Figure 11.15 Temperature-dependence of storage modulus (E') for PBS, PBS/MMT3, and PBS/DEA-M3. Reprinted from Ref. [26] with permission from John Wiley & Sons Ltd.

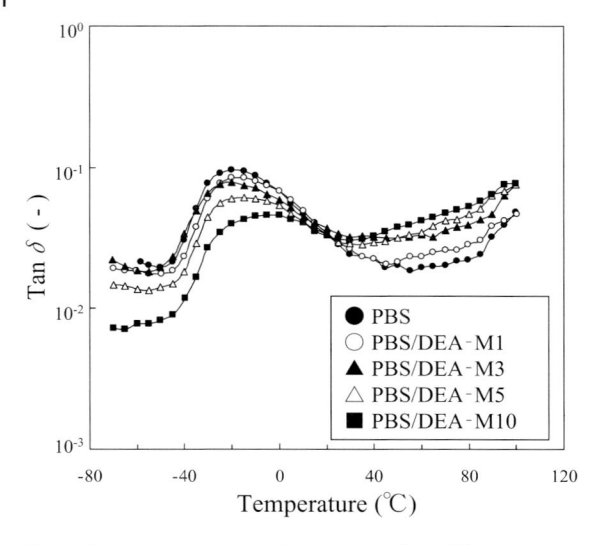

Figure 11.16 Temperature-dependence of tan δ for PBS/DEA-M composites with various inorganic contents. Reprinted from Ref. [26] with permission from John Wiley & Sons Ltd.

However, only a minimal difference was observed in the tan δ peak temperature for PBS/MMT composites.

11.3.3
Thermal Properties of PBS Nanocomposites

The thermal properties of the PBS composites, as determined with DSC measurements, are summarized in Table 11.6. As noted above, when discussing the mechanical properties of PBS nanocomposites in connection with the degree of intercalation, the influence of the degree of crystallinity (χ_c) of the PBS component in the composites should be taken into consideration because it may be influenced by the organoclay. As the original degree of crystallinity of the injection-molded composite can be evaluated from the value ($\Delta H_m - \Delta H_{g,c}$), the degree of crystallinity of the injection-molded PBS/DA-M, ODA-M, and DEA-M samples was evaluated as 93–94% of that of the PBS/MMT composite, based on the ($\Delta H_m - \Delta H_{g,c}$) value. The reason for a lower tensile strength of the PBS/DA-M, ODA-M and DEA-M should not be attributed to a lower degree of crystallinity than for PBS/MMT. Rather, the mechanical properties of the PBS composites were considered not to be seriously affected by changes in the degree of crystallinity. All PBS/clay composites showed higher $T_{m,c}$ and $\Delta H_{m,c}$ values than PBS, which indicated that crystallization from the melt was promoted by the action of the clay as a nucleating agent for PBS crystallization. Likewise, no clear trend was identified between the intercalation and the extent of promotion of crystallization.

Table 11.6 Thermal properties of PBS and PBS composites with an inorganic content of 3 wt%.

Sample	$T_{g,c}$ (°C)	$\Delta H_{g,c}$ (J g^{-1})	T_m (°C)	ΔH_m (J g^{-1})	$\Delta H_m - \Delta H_{g,c}$ (J g^{-1})	$T_{m,c}$ (°C)	$\Delta H_{m,c}$ (J g^{-1})
PBS	93.9	6.3	114.7	81.9	75.7	76.9	63.0
PBS/MMT3	96.7	3.4	114.5	81.1	77.8	81.0	65.2
PBS/DA-M3	98.1	6.9	114.6	79.4	72.5	83.2	66.0
PBS/ODA-M3	93.2	9.0	114.2	82.4	73.4	82.5	64.7
PBS/DEA-M3	98.4	5.4	114.2	77.6	72.2	78.9	63.1

$T_{g,c}$ is the crystallization temperature from the glassy state; T_m is melting temperature; $T_{m,c}$ is the crystallization temperature from the melt; $\Delta H_{g,c}$, ΔH_m and $\Delta H_{m,c}$ are the heat of crystallization from the glassy state, of melting, and crystallization from the melt per 1 g of PBS contained in the composite, respectively.
Reprinted from Ref. [26] with permission from John Wiley & Sons Ltd.

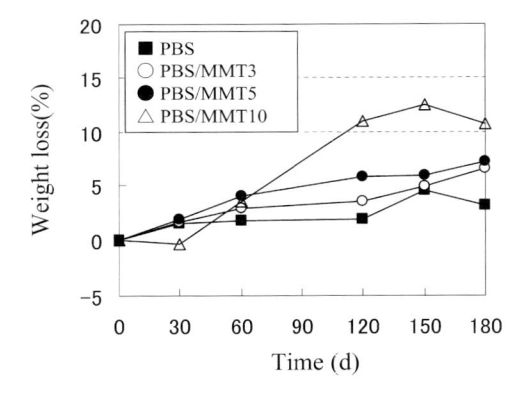

Figure 11.17 Weight loss of PBS and PBS/MMT composites during burial in soil. Reprinted with permission from John Wiley & Sons Ltd.

11.3.4
Biodegradability of PBS Nanocomposites

The biodegradability of PBS nanocomposites was evaluated in a similar manner to that of the PBAT nanocomposites. Figure 11.17 shows the change of weight loss of neat PBS and PBS/MMT composites during burial in soil. Here, the PBS/MMT composites showed a greater degree of weight loss than neat PBS, when compared to the weight loss after 120–180 days. In these studies, the weight loss of the PBS/MMT composites was seen to increase with increasing inorganic content over the range of 3–10 wt%. In contrast, the PBS/DEA-M composites showed a lesser weight loss than did neat PBS (see Figure 11.18). The influence

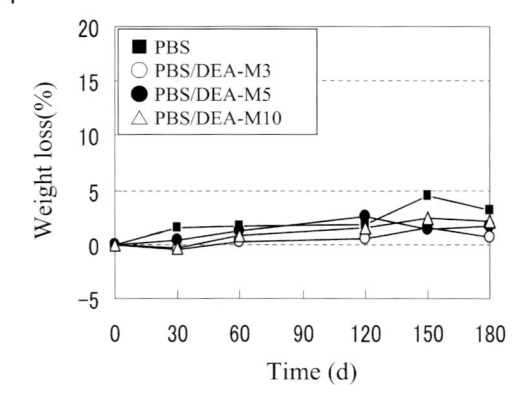

Figure 11.18 Weight loss of PBS and PBS/DEA-M composites during burial in soil. Reprinted with permission from John Wiley & Sons Ltd.

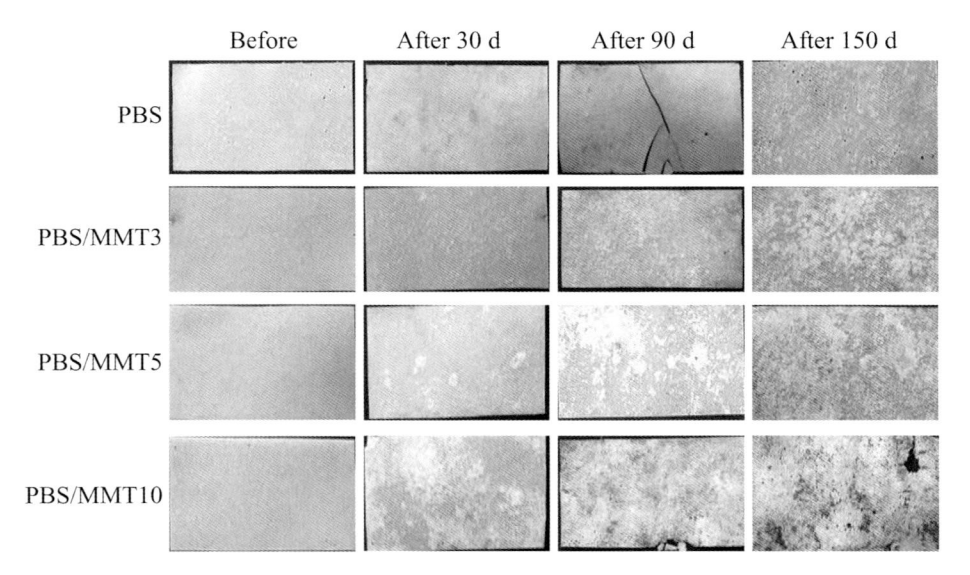

Figure 11.19 Optical images of PBS and PBS/MMT composites after burial in soil for prescribed periods. Reprinted with permission from John Wiley & Sons Ltd.

of DEA-M content on biodegradability could not be elucidated from these weight loss curves.

Figure 11.19 shows the optical images of neat PBS and PBS/MMT composite films after burial in soil. Clearly, the PBS/MMT composites were more degradable than PBS, with the degradability seeming to increase with increasing MMT content. After 150 days, there was a marked difference between the neat PBS film and the PBS/MMT film with an inorganic content of 10 wt%. In contrast to this result, a retardation of biodegradation was observed for PBS/DEA-M films (see Figure 11.20). Although the PBS film was seen to show many "hollows" due to biodegradation, no clear hollows appeared in the PBS/DEA-M10 film after 150 days. Comparative SEM images of the film surfaces of neat PBS and PBS/DEA-M composites are shown in Figure 11.21, where the microcracks can be seen to be first generated at 30 days, and to become gradually larger in the neat PBS film. In contrast, the surface of the PBS/DEA-M composites was almost unchanged after 150 days. These results were in agreement with the trends evaluated by weight loss measurements.

Figure 11.22 shows the aerobic biodegradability of neat PBS and PBS/MMT films, as determined by measuring their BOD for 60 days in an aqueous medium containing activated sludge. Although minimal difference was apparent between neat PBS and PBS/MMT3, the PBS/MMT10 composite was shown to be much more biodegradable than neat PBS. A comparison of the biodegradability of neat PBS and PBS/DEA-M films is shown in Figure 11.23. Here, the PBS/DEA-M3 film was slightly less biodegradable than neat PBS, whereas the biodegradation of PBS/DEA-M10 was retarded to a considerable degree. Such retardation and promotion of biodegradation for DEA-M and MMT, respectively, were in line with the results of weight loss studies after burial in the soil. Whilst the promotion of biodegradability for the PBS/MMT microcomposite may relate to the highly hydrophilic character of the MMT, the reduced biodegradability of the PBS/DEA-M composites may derive from the finely dispersed silicate layers with a large aspect ratio in the PBS matrix, forcing biodegrading microorganisms or water diffusing through the bulk of the film to take a more tortuous path.

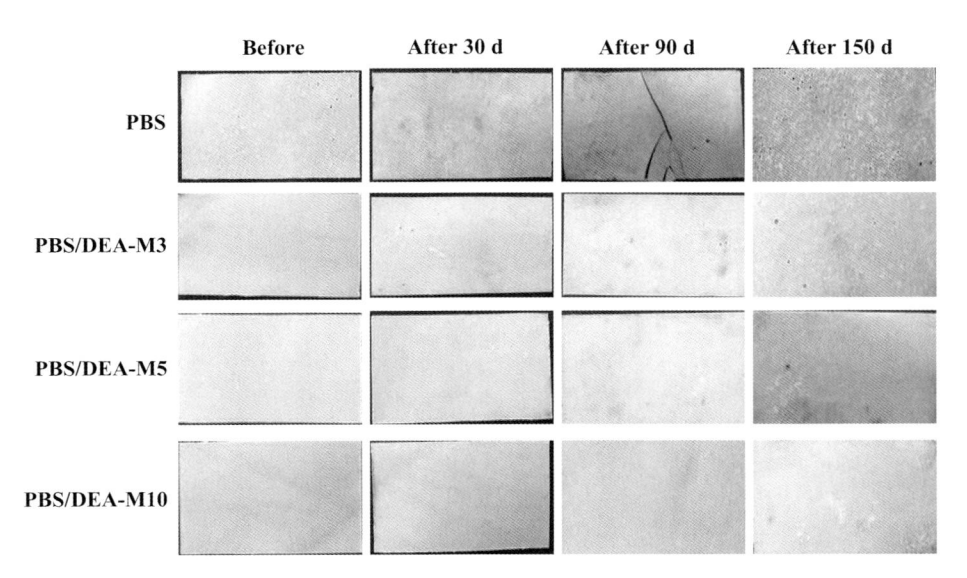

Figure 11.20 Optical images of PBS and PBS/DEA-M composites after burial in soil for prescribed periods. Reprinted with permission from John Wiley & Sons Ltd.

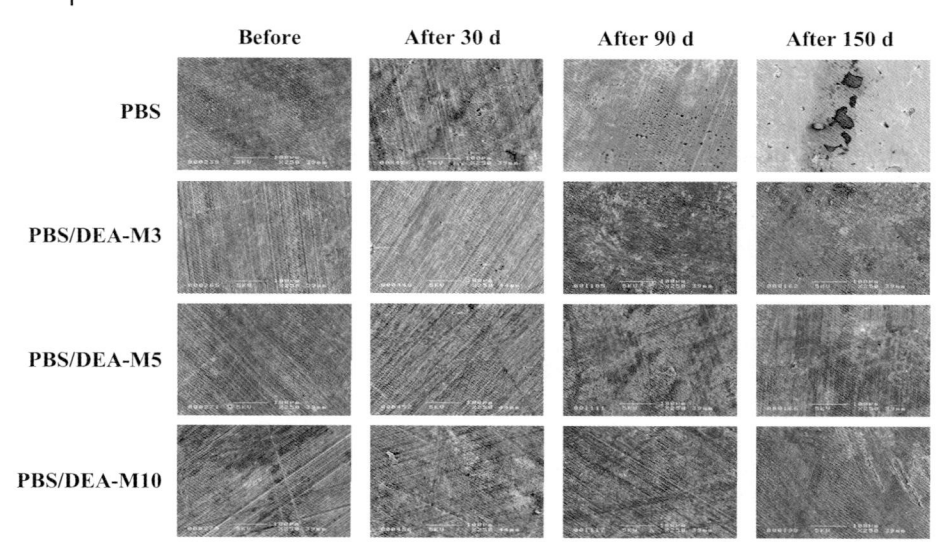

Figure 11.21 Scanning electron microscopy images of the surface of PBS and PBS/DEA-M composites after burial in soil for prescribed periods. Reprinted with permission from John Wiley & Sons Ltd.

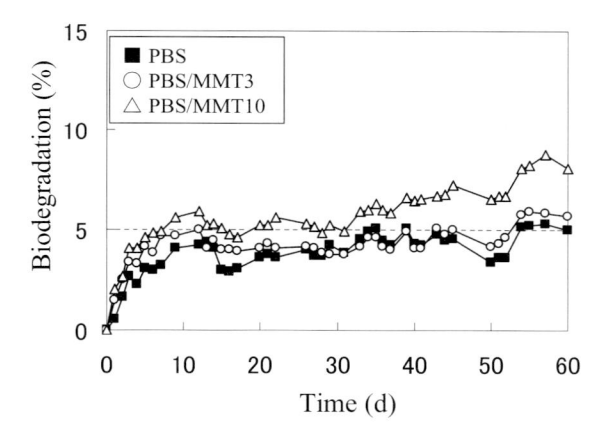

Figure 11.22 Biodegradability of PBS and PBS/MMT composites, as determined by measuring the BOD in an aqueous medium containing activated sludge. Reprinted with permission from John Wiley & Sons Ltd.

11.4
Conclusions

The nanocomposites of biodegradable polyesters (PBAT and PBS) and layered silicates (MMT, DA-M, ODA-M, and DEA-M) have been prepared using melt intercalation method, and their morphology, thermal and mechanical properties, and biodegradability investigated.

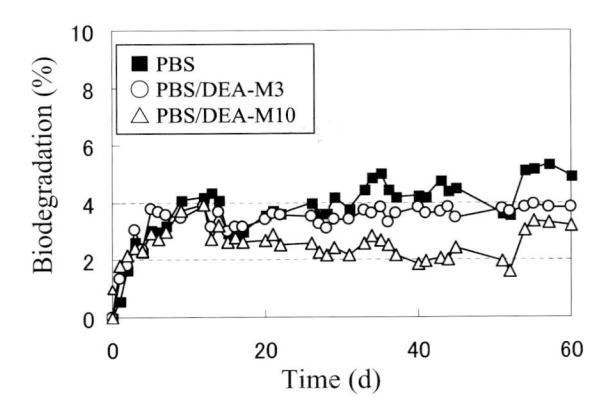

Figure 11.23 Biodegradability of PBS and PBS/DEA-M composites, as determined by measuring the BOD in an aqueous medium containing activated sludge. Reprinted with permission from John Wiley & Sons Ltd.

Based on TEM and XRD observations of PBAT/organoclay composites with an inorganic content of 3 wt%, it was shown that intercalation would occur in the case of DA-M and DEA-M, while exfoliation *and* intercalation would occur in the case of ODA-M. The latter was shown to be more finely dispersed in the matrix than were DA-M and DEA-M and, when the ODA-M content was increased, the XRD peak relating to intercalated clay was also increased. Among the PBAT composites with an inorganic content of 3 wt%, the exfoliated PBAT/ODA-M composite showed the lowest tensile modulus, a finding that was considered to be related to a decrease in crystallinity of the PBAT component in the presence of exfoliated ODA-M. When the tensile properties of the PBAT composites with an inorganic content of 5 wt% were compared, PBAT/ODA-M showed the highest tensile modulus, strength and elongation at break. With regards to biodegradability, following burial in the soil for eight months the PBAT/MMT microcomposite exhibited a greater weight loss than did control PBAT, whilst the PBAT/ODA-M nanocomposite suffered a somewhat lower weight loss. Biodegradability testing by determining BOD-values in aqueous media containing activated sludge showed that the addition of MMT to PBAT would promote biodegradation, whereas the addition of ODA-M had no such effect. It has been suggested thought that hydrophilic character of the MMT may promote the biodegradation of PBAT, and that formation of the nanocomposite in the case of ODA-M depresses the diffusion of those materials that cause the biodegradation.

With regards to PBS/layered silicate nanocomposites, the order of finer dispersion of the silicate particles in the PBS matrix and higher degree of intercalation of PBS molecule into the silicate layers was ODA-M ≈ DEA-M > DA-M > MMT. In addition, the PBS/ODA-M, DEA-M, and DA-M nanocomposites showed a higher tensile modulus and slightly lower tensile strength than for control PBS and PBS/MMT composites. It is possible that this increase in tensile modulus may be due to a better dispersion of the silicate layer having a high modulus and high aspect ratio, while the decrease in tensile strength may be attributed to a decrease of elongation at break which, in turn, may be related to delamination of the polymer–silicate interlayer. The $T_{m,c}$ and ΔH_m of all the PBS/clay composites were higher than those of PBS, which indicated that the clay serves as a nucleating agent and promotes the crystallization of PBS. Soil burial tests of the PBS composites showed the weight loss of PBS/MMT composites to increase with increasing amounts of MMT, whereas that of the PBS/DEA-M composites rather decreased in line with the clay content. Notably, the results of the BOD tests in aqueous media containing activated sludge were in agreement with those of the soil burial tests.

Consequently, in the case of biodegradable PBAT and PBS microcomposites with MMT, the presence of MMT appears to promote biodegradation due to its hydrophilic character. In contrast, for PBAT and PBS nanocomposites with ODA-M and DEA-M, the formation of nanocomposites depressed the diffusion of the materials that caused the biodegradation to be retarded. For biodegradable polyester nanocomposites, it is desirable that characteristics such as strength and heat resistance are retained during any practical usage, and that these materials, when eventually discarded, are completely biodegraded following a suitable period of time. It can be said that the merits of biodegradable polyester/clay nanocomposites are good stability during use, biodegradation of the polyester within a relatively long time after being discarded, and the subsequent return of a small amount of residual clay to nature.

References

1 Stevens, E.S. (2002) *Green Plastics: An Introduction to the New Science of Biodegradable Plastics*, Princeton University Press, Princeton.
2 Zhou, Q. and Xanthos, M. (2008) *Polym. Degrad. Stab.*, **93**, 1450–1459.
3 Shyang, C.W. and Kuen, L.S. (2008) *Polym. Polym. Compos.*, **16**, 263–270.
4 Wu, D., Wu, L., Wu, L., Xu, B., Zhang, Y., and Zhang, M. (2008) *J. Nanosci. Nanotechnol.*, **8**, 1658–1668.
5 Ren, S.-Y., Gu, J., and Dong, B. (2007) *J. Polym. Sci. Part B*, **45**, 3189–3196.
6 Lee, S.-Y., Xu, Y.X., and Hanna, M.A. (2007) *Int. Polym. Process*, **22**, 429–435.
7 Pluta, M., Boiteux, J.K., and Jeszka, G. (2007) *Eur. Polym. J.*, **43**, 2819–2835.
8 Wu, D., Wu, L., Wu, L., Xu, B., Zhang, Y., and Zhang, M. (2007) *J. Polym. Sci. Part B*, **45**, 1100–1113.
9 Yu, L., Petinakis, S., Dean, K., Bilyk, A., and Wu, D. (2007) *Macromol. Symp.*, **249/250**, 535–539.
10 Wu, D., Wu, L., Wu, L., and Zhang, M. (2006) *Polym. Degrad. Stab.*, **91**, 3149–3155.
11 Petersson, L., Oksman, K., and Mathew, A.P. (2006) *J. Appl. Polym. Sci.*, **102**, 1852–1862.
12 Yoshida, O. and Okamoto, M. (2006) *Macromol. Rapid Commun.*, **27**, 751–757.
13 Pluta, M., Paul, M., Alexandre, M., and Dubois, P. (2006) *J. Polym. Sci. Part B*, **44**, 312–325.
14 Pluta, M., Paul, M., Alexandre, M., and Dubois, P. (2006) *J. Polym. Sci. Part B*, **44**, 299–311.
15 Shibata, M., Someya, Y., Orihara, M., and Miyoshi, M. (2006) *J. Appl. Polym. Sci.*, **99**, 2594–2602.
16 Wu, T.-M. and Chiang, M.-F. (2005) *Polym. Eng. Sci.*, **45**, 1615–1621.
17 Paul, M.-A., Delcourt, C., Alexandre, M., Degee, M.P., Monteverde, F., and Dubois, P. (2005) *Polym. Degrad. Stab.*, **87**, 535–542.

18 Ray, S.S. and Okamoto, M. (2003) *J. Nanosci. Nanotechnol.*, **3**, 503–510.
19 Ray, S.S. and Okamoto, M. (2003) *Macromol. Mater. Eng.*, **288**, 203–208.
20 Nam, J.Y., Ray, S.S., and Okamoto, M. (2003) *Macromolecules*, **36**, 7126–7131.
21 Ray, S.S., Maiti, P., Okamoto, M., Yamada, K., and Ueda, K. (2002) *Macromolecules*, **35**, 3104–3110.
22 Makhatha, M.E., Ray, S.S., Hato, J., and Luyt, A.S. (2008) *J. Nanosci. Nanotechnol.*, **8**, 1679–1689.
23 Chen, G.-X., Kim, H.-S., and Yoon, J.-S. (2005) *J. Appl. Polym. Sci.*, **98**, 1727–1732.
24 Chen, G.-X., Kim, H.-S., and Yoon, J.-S. (2007) *Polym. Int.*, **56**, 1159–1165.
25 Chen, G.-X., and Yoon, J.-S. (2005) *J. Polym. Sci. Part B*, **43**, 817–826.
26 Someya, Y., Nakazato, T., Teramoto, N., and Shibata, M. (2004) *J. Appl. Polym. Sci.*, **91**, 1463–1475.
27 Okamoto, K., Ray, S.S., and Okamoto, M. (2003) *J. Polym. Sci. Part B*, **41**, 3160–3172.
28 Ray, S.S., Maiti, P., Okamoto, M., and Okamoto, K. (2002) *J. Nanosci. Nanotechnol.*, **2**, 171–176.
29 Liao, H.-T. (2008) *Polym. Eng. Sci.*, **48**, 1524–1531.
30 Luduena, L.N., Vazquez, A., and Alvarez, V.A. (2008) *J. Appl. Polym. Sci.*, **109**, 3148–3156.
31 Kiersnowski, A., Piglowski, J., Kiersnowski, A., and Gutmann, J.S. (2007) *J. Polym. Sci. Part B*, **45**, 2340–2367.
32 Shibata, M., Teramoto, N., Someya, Y., and Tsukao, R. (2007) *J. Appl. Polym. Sci.*, **104**, 3112–3119.
33 Homminga, D., Goderis, B., Groeninckx, G., and Dolbnya, I. (2006) *Polymer*, **47**, 1620–1629.
34 Chen, B. and Evans, J.R.G. (2006) *Macromolecules*, **39**, 747–754.
35 Di, Y., Iannac, S., and Nicolais, L. (2005) *Macromol. Symp.*, **228**, 115–124.
36 Kiersnowski, A., Dabrowski, P., Piglowski, J., Budde, H., and Kressle, R.J. (2004) *Eur. Polym. J.*, **40**, 2591–2598.
37 Pucciariello, R., Villani, V., Langerame, F., Gorrasi, G., and Vittoria, V. (2004) *J. Polym. Sci. Part B*, **42**, 3907–3919.
38 Kiersnowski, A. and Piglowski, J. (2004) *Eur. Polym. J.*, **40**, 1199–1207.
39 Pucciariello, R., Villani, V., Belviso, S., Gorrasi, G., Tortora, M., and Vittoria, V. (2004) *J. Polym. Sci. Part B*, **42**, 1321–1332.
40 Maiti, P. (2003) *Langmuir*, **19**, 5502–5510.
41 Di, Y., Nicolais, L., Iannace, S., and Maio, E.D. (2003) *J. Polym. Sci. Part B*, **41**, 670–678.
42 Lepoittevin, B., Pantoustier, N., Devalckenaere, M., Alexandre, M., Dubois, P., Calberg, C., Jerome, R., Henrist, C., and Rulmont, A. (2003) *Polymer*, **44**, 2033–2040.
43 Pantoustier, N., Lepoittevin, B., Alexandre, M., Dubois, P., Calberg, C., and Jerome, R. (2002) *Polym. Eng. Sci.*, **42**, 1928–1937.
44 Kubies, D., Jerome, R., Pantoustier, N., Dubois, P., and Rulmont, A. (2002) *Macromolecules*, **35**, 3318–3320.
45 Someya, Y., Kondo, N., and Shibata, M. (2007) *J. Appl. Polym. Sci.*, **106**, 730–736.
46 Chivrac, F., Pollet, E., and Averous, L. (2007) *J. Polym. Sci. Part B*, **45**, 1503–1510.
47 Chivrac, F., Kadlecova, Z., Pollet, E., and Averous, L. (2006) *J. Polym. Environ.*, **14**, 393–401.
48 Someya, Y., Sugahara, Y., and Shibata, M. (2005) *J. Appl. Polym. Sci.*, **95**, 386–392.
49 Wang, S., Song, C., Chen, G., Guo, T., Liu, J., Zhang, B., and Takeuchi, S. (2005) *Polym. Degrad. Stab.*, **87**, 69–76.
50 Chen, G.X., Hao, G.J., Guo, T.Y., Song, M.D., and Zhang, B.H. (2004) *J. Appl. Polym. Sci.*, **93**, 655–661.
51 Chen, G.X., Hao, G.J., Guo, T.Y., Song, M.D., and Zhang, B.H. (2002) *J. Mater. Sci. Lett.*, **21**, 1587–1589.
52 Lee, S.-R., Park, H.-M., Hyuntaek, L., Taekyu, K., Xiucuo, L., Cho, W.-J., and Ha, C.-S. (2002) *Polymer*, **43**, 2495–2500.

12
Self-Healing in Nanoparticle-Reinforced Polymers and other Polymer Systems

Stephen J. Picken, Steven D. Mookhoek, Hartmut R. Fischer, and Sybrand van der Zwaag

12.1
Introduction

The turn of the twenty-first century was marked by two major developments in the field of structural polymer materials, namely the development of nanoparticle-reinforced polymers, and the development of self-healing polymers. Nanoparticle reinforcement forms a logical extension of the successful microparticle reinforcement of engineering polymers aimed at increasing their mechanical properties, and in particular their stiffness and fracture strength, even beyond the current high level. Self-healing polymers form a new route in materials science which is aimed at creating materials with the ability to heal cracks more or less autonomously, and to restore mechanical properties to their original, often lower, level. Consequently, on the one hand the mechanical properties were extended even further, while on the other hand the mechanical properties were sacrificed in order to create a new functionality. In this chapter, the potential of nanoreinforced materials as suitable candidate materials demonstrating self-healing behavior, in combination with acceptable mechanical properties, will be highlighted.

Self-healing materials are loosely defined as materials capable of restoring mechanical damage – that is, a nano-, micro, or sometimes even macroscopic crack, in a more or less autonomous manner. While the mechanism of healing depends on the intrinsic character of the material to heal [1, 2], the common feature in all healing processes is that new material flows into the crack and restores the mechanical traction between both sides of the crack. Thus, in all self-healing materials there is a local flow of "reactive" material – that is, material which transforms from a mobile and hence marginally load-bearing state to a new immobile and load-bearing state.

During the early days of self-healing polymers and other self-healing materials, the functionality of the mobile healing agent and the static matrix material were strictly separated, and healing was obtained by the inclusion of an encapsulated liquid "glue" which would spread itself over the fracture surface as soon as the crack to be healed penetrated the micro-size capsules or sealed hollow

Optimization of Polymer Nanocomposite Properties. Edited by Vikas Mittal
Copyright © 2010 WILEY-VCH Verlag GmbH & Co. KGaA, Weinheim
ISBN: 978-3-527-32521-4

fibers [3, 4]. Provided that the crack faces were not too far apart, and the released healing agent was capable of connecting both sides of the crack, a proper healing agent would lead to a restoration of the load-bearing capabilities of the former crack.

Of course, the presence of a liquid healing agent reduces the initial mechanical properties, in particular the modulus and the tensile strength, as liquids can only transfer loads under hydrostatic loading conditions and are incapable of sustaining tensile or shear forces. Hence (as will be demonstrated later in the chapter), a major challenge is to maximize the availability of a liquid healing agent at the fracture surface at as low a liquid fraction present in the undamaged material. It will be shown that elongated capsules – that is, capsules with a high aspect ratio – represent an important tool to achieve this target.

Since the first demonstration of self-healing behavior via the liquid healing agent route, alternative routes have been identified to induce self-healing behavior; these include:

- Elastic recovery in combination with reversible physical networks as employed in ionomers [5]
- Reversible hydrogen and other chemical bonds, as employed in supramolecular materials [6]
- Crack jamming by the transport of particles under the influence of an electrical field [7]
- Long-distance reptation of dissolved linear molecules in a thermoset matrix [8]
- Atomistic diffusion followed by local formation of strengthening precipitates in aluminum alloys [9]
- Local oxidation of the crack zone, leading to the formation of new phases with a higher specific volume, as in a very special high-temperature ceramic [10]
- Volumetric expansion of a nanostructure clay micro-layer to fill cracks in a multilayered coating [11, 12]
- Local deposits in cracks in bio-concrete grades due to bacterial excrements [13].

In all of the above examples, the matrix is the more rigid phase and the healing agent is the mobile (i.e., less rigid) phase. For such systems, in order to obtain acceptable mechanical properties, the volume fraction of healing agent should be kept relatively small and its configuration optimized so as to achieve the maximum effect at the lowest volume fraction.

In contrast, it could be argued that in nanoreinforced polymers, which consist of a semi-crystalline polymer above its glass transition temperature (T_g), the matrix can be treated as the relatively mobile phase and the nanoparticles as the rigid phase, in order to provide the rigidity and flow stress to the system. Hence, a nanoparticle-reinforced polymer could be designed to become a self-healing system, by tuning the mobility of the matrix and compensating the lower mechanical properties by nanoreinforcements. In this case, the volume fraction of the rigid material would be relatively small and, again, the challenge would be to optimize its configuration in order to obtain the maximum effect at the lowest volume fraction. Again, the particle aspect ratio is the key parameter to increase the efficiency of the reinforcement. Of course, one key difference between liquid-based self-healing polymer systems and self-healing nanoparticle-reinforced polymers is that, in this "first class" of systems, local healing can occur only once, and is lost when the healing agent is used to fill and heal a crack. In contrast, in the "second class" systems, multiple healing can occur locally, as molecular mobility is guaranteed at temperatures above T_g.

Although this chapter focuses on the mechanical properties of bulk polymers and the healing of mechanical damage, it should be pointed out that nanoparticles are also now being used for the restoration of corrosive protection after scratching of the protective coating. Foe example, nanoparticle-filled silane films have been named as a possible replacement for chromate-containing anti-corrosion coatings for aluminum alloys [14]. In this case, a small amount of silica particles (5 ppm) in the silane film will alter the mechanism of cathodic reactions on the alloy surface, such that the film will behave as a cathodic barrier (as is also observed when Ce-compounds are used) [15]. The silica suppresses the cathodic reaction by reacting with the cathodically generated OH^- ions, under the formation of water and SiO_3^{2-} ions. The as-formed SiO_3^{2-} ions subsequently react with Al^{3+} ions at the anodes, forming a passive silicate film. With an increase in silica content (above 15 ppm), such a desirable cathodic inhibitive behavior will disappear again, most likely due to an enhanced porosity.

Nanoreservoirs, when distributed homogeneously in the film matrix, may possess a controlled and corrosion-stimulated inhibitor release to corrosion defects, without any negative effects on the stability of the passive corrosion protective coating matrix. The development of active corrosion protection systems for metallic systems is of prime importance for almost all applications. Oxide nanoparticles containing cerium ions have been investigated as isolated, inhibitor-impregnated reservoirs to be incorporated inside the protective coatings, providing a prolonged release of the inhibitor and self-healing ability [16]. A relatively new contribution to the development of a new protective intelligent system with self-healing ability composed of hybrid sol–gel films doped with nanocontainers that release an entrapped corrosion inhibitor by regulating the permeability of the encapsulation shell in response to pH changes caused by corrosion process, has recently been described [17]. For this, silica nanoparticles of approximately 70 nm diameter, covered layer-by-layer with polyelectrolyte layers and layers of the inhibitor (benzotriazole), were introduced randomly into a silica–zirconia-based hybrid film used to serve as an anticorrosion coating deposited on AA2024 aluminum alloy. The hybrid film with the nanocontainers demonstrated an enhanced long-term corrosion protection compared to the undoped hybrid film, the protective effect being due to a regulated release of the corrosion inhibitor that was triggered by the start of the corrosion process. Here, the released benzotriazole formed a thin adsorption layer on the damaged metallic surface, which not only hindered the anodic and cathodic corrosion processes but also passivated the alloy by replacing the damaged Al_2O_3 film.

12.2
Microstructured Self-Healing Polymer Structures

12.2.1
Liquid-Based Self-Healing Thermosetting Polymers

The development of self-healing materials began with the landmark report of White et al., who showed that a thermosetting epoxy material with embedment of 60–100 μm-diameter spherical urea–formaldehyde capsules filled with dicyclopentadiene, in combination with a dispersed (Grubbs) catalyst, could show substantial self-repair upon fracture [4]. Upon cracking, the released liquid monomer would start off a metathesis polymerization with the dispersed catalyst, which led to a rebonding of the crack surfaces. The data obtained showed that the material would regain about 75% of its original fracture load, provided that a sufficient time for healing was allowed and the crack surfaces were kept in contact during the healing period. Later, Pang et al. [18] reported details of the production of a self-healing fiber composite based on a similar self-healing principle, but more in line with the concept proposed by Dry [3, 18]. In this case, hollow glass capillaries filled with either of the two components of a two liquid-component epoxy-system were embedded. Subsequent compression tests showed that the bleeding composites could re-establish approximately 97% of their original strength after impact.

The original choice of a thermosetting polymer as the matrix material for a self-healing, liquid-based system was considered logical, and stemmed from the initial low viscosity of the thermosetting resin, though this was combined with the low temperature imposing the most favorable conditions for mixing the liquid-filled microcapsules in the matrix material, without excessive premature rupture. Unfortunately, one unavoidable consequence of the crosslinked nature of the matrix polymer was that the matrix itself was highly immobile, and all healing action was related to the formation of a chemical bond between the healing agent (which itself was also a crosslinking system) and the matrix material present at the crack faces. Thus, the matrix material played no role itself in the healing process, other than to provide an adequate anchoring for the healing agent. An additional potential problem was that the crack would be healed by a new material which had different (mechanical) properties from the matrix material, and this might lead to undesirable stress concentrations upon reloading. More recently, Caruso et al. [19] demonstrated self-healing in a thermosetting epoxy resin by using a solvent as the encapsulated phase. This resulted in a significant and surprisingly high healing efficiency of about 80% that could be attributed to the presence of substantial amounts of uncured epoxy moieties within the matrix. The relatively poorly cured state of the epoxy was responsible for the weak mechanical properties of the undamaged material. For a fully cured thermosetting polymer, healing via an encapsulated solvent would not lead to any measurable healing efficiency.

12.2.2
Liquid-Based Self-Healing Thermoplastic Polymers

In contrast to thermosets, thermoplastic polymers can show substantial molecular mobility, and hence self-healing potential, provided that the temperature is well above the T_g (for amorphous polymers), or is above its melting temperature (for semi-crystalline polymers). Molecular mobility can also be enhanced by bringing the polymers into contact with a strong solvent; such enhanced mobility has been used successfully for solvent welding [20, 21].

Very recently, the concepts of solvent welding and liquid encapsulation for thermosets were merged so as to create liquid-based self-healing thermoplastics [22]. These new systems are based on spherical urea–formaldehyde capsules that contain o-dichlorobenzene (DCB) embedded in a mixture of polymethyl methacrylate (PMMA) and polystyrene (PS). This approach was used because a PMMA/PS blend could be polymerized very quickly from a 2 : 1 (by weight) MMA/PS solution resin, and would have a low initial viscosity. The volume fraction of the capsules was 10–15%, with a typical capsule diameter of 60 μm. The encapsulated solvent is known to be an excellent solvent for PS, and has a Hansen solubility sphere radius which is less than the solubility sphere radii of the polymer [23]. Additionally, the DCB has a relative low vapor pressure at room temperature, which prevents rapid evaporation of the solvent when it is released from the capsules. A low vapor pressure allows for an extensive action of the solvent inside the polymer matrix, thus providing a better healing. Unfortunately, using a solvent with a too-low vapor pressure can lead to a weakened interface because, although the crack will appear to be sealed, the solvent cannot escape from the material in time and will act as a plasticizer.

Whilst the mechanical healing efficiencies of this system are yet to be reported, a very elegant proof of the healing process was provided via a study of the healed fracture surface of such a system, using a special version of scanning electron microscopy (SEM)-based X-ray microtomography that provided a spatial resolution of approximately 10 μm. A typical reconstructed image of the volume around the fracture surface showed a depletion of the capsules, and also of the solid contact between the former fracture surfaces (Figure 12.1). While the route of liquid-based thermoplastic polymers is still in its infancy, it represents an interesting and attractive approach, as the use of a solvent as a healing agent has many advantages over the use of a liquid resin to heal a thermoset. Another advantage is that the total market for thermoplastic polymers is much larger than that for thermosets.

12.2.3
Geometric Aspects in Encapsulation

As indicated in Section 12.1, one clear requirement for adequate healing via the liquid encapsulation route is the need to fill the crack gap more or less completely with the healing agent, and preferably without the occurrence of any massive

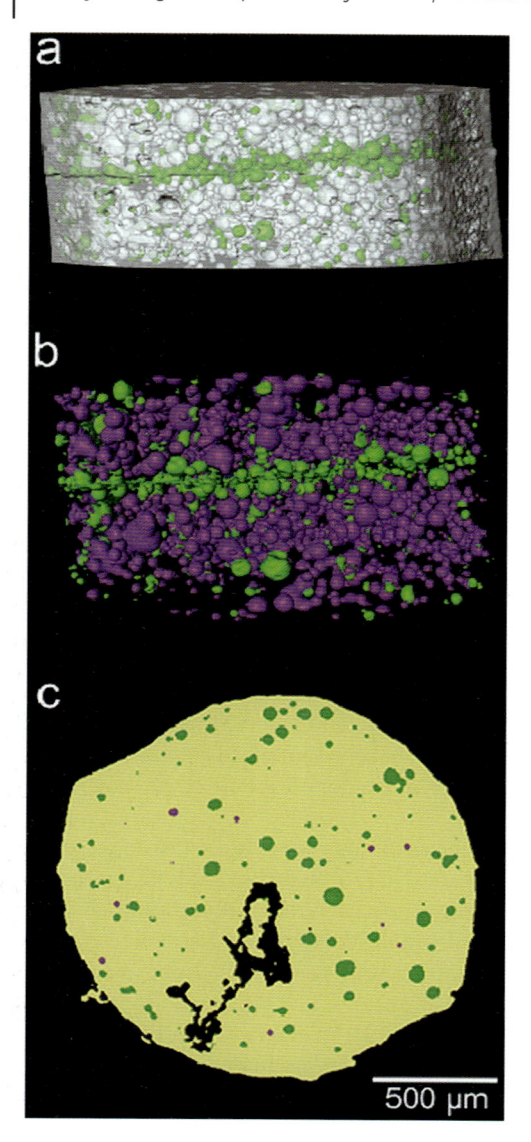

Figure 12.1 Reconstructed tomographic views of the segmented data, showing full capsules (purple) and empty capsules (green) in the vicinity of the crack. (a) Displaying the 3-D rendered dataset of 200 slices including the healed crack and the segmented empty capsules. (b) Showing only the segmented microcapsules, omitting the matrix material. (c) Showing a 2-D segmented slice of the data at the healed crack. The dark area in the lower image corresponds to a fraction of the crack area which did not heal completely. The indicated scale bar applies to all three images.

bleeding [18]. It is intuitively clear that the amount of healing agent entering the crack is determined by the product of the average volume of healing agent stored per capsule and the areal probability of the crack intersecting a capsule. This product depends on the particle size (nanosized capsules inherently contain very small volumes of healing agent indeed), their volume fraction and, in the case of nonspherical particles, their aspect ratio and average orientation of their director.

As a first step in the geometric optimization of liquid-encapsulated self-healing materials, a representative volume element (RVE) is defined in which spherical and cylindrical three-dimensional (3-D) geometric elements are spatially dispersed (Figure 12.2a and b respectively).

The matrix/capsule system can be adjusted by varying the volume of the embedded microparticles (Vcaps), their volume concentration (φ) and the aspect ratio (AR). The theoretical ratio of the volumes of healing agent released for both geometries – that is, the ratio of the released volume for cylindrical and spherical capsules – is denoted as the release improve factor (RIF); it is a function of the cylindrical capsule aspect ratio, and is independent of the capsule size and capsule volume concentration. The RIF, as a function of the AR, is shown graphically in Figure 12.3, where the RIF demonstrates a nonlinear behavior with increasing AR. This increase of RIF as a function of AR demonstrates the ability to enhance the liquid-based healing mechanism by applying cylindrical, rather than spherical, microcapsules.

The benefits of using an elongated capsules are illustrated graphically in Figure 12.4, where the average released liquid volume is plotted as a function of the capsule volume concentration for different ARs, assuming a random orientation of the elongated particles. Based on the data in Figure 12.4, it can be concluded

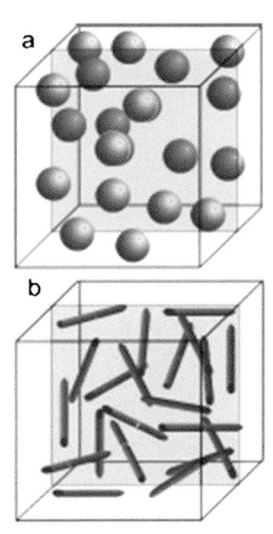

Figure 12.2 Schematic of capsule distribution in representative volume element (RVE) with intersecting fracture-plane. (a) Spherical capsules; (b) Cylindrical capsules.

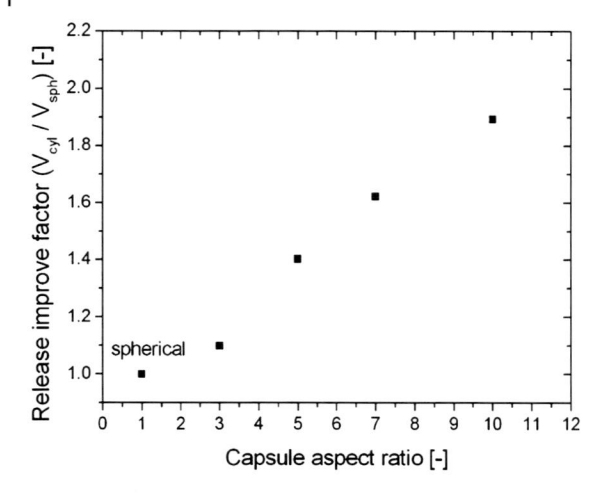

Figure 12.3 Release improve factor (RIF) as a function of the capsule aspect ratio.

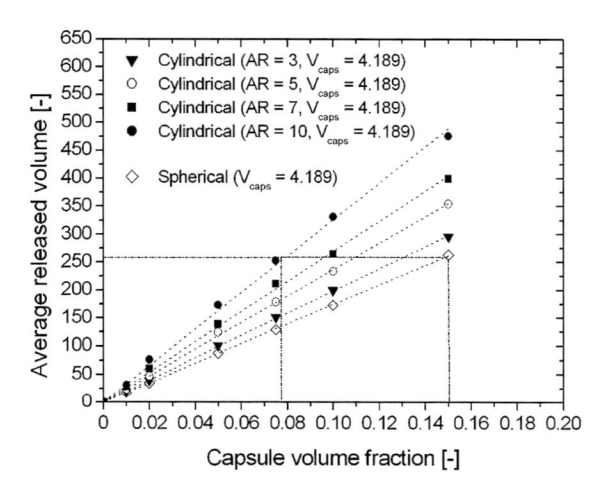

Figure 12.4 Average released liquid volume as function of capsule loading for different capsule aspect ratios.

that similar amounts of healing fluid per fracture plane area can be more easily obtained at a lower volume fraction for elongated particles than for spherical particles. In addition, the smaller volume fraction of the liquid present will have a positive effect on the other mechanical properties of the system.

The average orientation of the cylindrically shaped capsules inside the material can also influence the release of the healing agent. Typically, the orientational order parameter $<P_2>$ for dispersed cylindrical rod-like structures is given by:

$$\langle P_2 \rangle = \frac{\langle 3\cos^2 \theta \rangle - 1}{2} \tag{12.1}$$

where θ is the angle between the cylinder axis and the normal axis (X), perpendicular to the fracture orientation, as shown in Figure 12.1. Now, by varying $<P_2>$ from −0.5 (parallel to the fracture plane) to 0 (random) and to 1 (perpendicular to the fracture plane), the effect of the capsule orientational order can be determined. The RIF, as a function of the orientational order parameter, is shown in Figure 12.5, where the RIF was determined at a given capsule AR of 5. The positive effect of a preferred orientation of the cylinders perpendicular to the fracture plane also offered the possibility of lowering the capsule loading even further with respect to a system based on spherical capsules.

In order to illustrate the magnitude of the effect caused by the orientational order in combination with the AR, the average released volume as function of the capsule concentration can be plotted for the different $<P_2>$ values for dispersed cylinders and with AR = 5, taking the case of homogeneously dispersed spheres as the reference state (Figure 12.6).

In conclusion, irrespective of the nature of the matrix, healing efficiencies in liquid-based self-healing materials can be increased significantly over current levels by using elongated rather than spherical capsules. A decrease in the capsule size to approximately 1 μm may be possible without losing the self-healing properties [24]. However, a further decrease to nanoscale dimensions will limit not only the healing potential (due to an increase in the capillary action of the capsules themselves) but also the minimal amount of healing agent released per capsule.

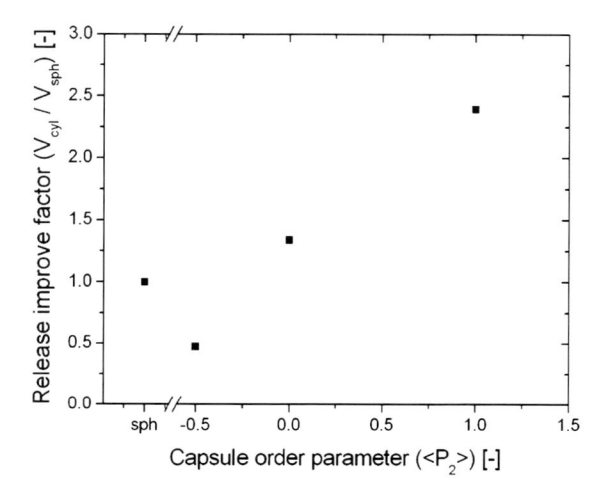

Figure 12.5 Release improve factor as a function of capsule orientation.

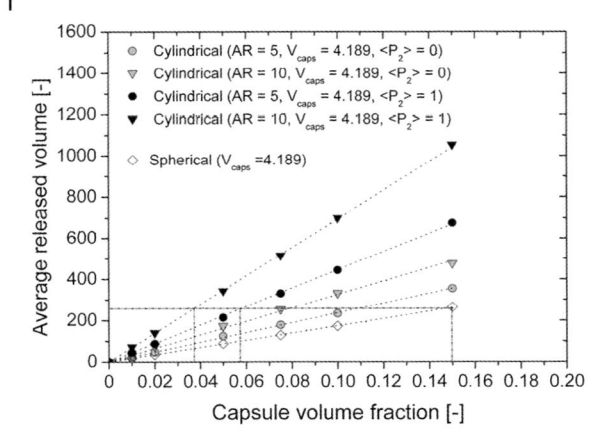

Figure 12.6 Average released liquid volume as a function of capsule loading for different orientational parameters.

12.3
Nanoparticle-Reinforced Self-Healing Polymer Systems

12.3.1
Modeling the Modulus of Nanoparticle-Filled Polymers

A variety of possible approaches exist to assess the mechanical properties of polymer nanocomposites. In fact, one aspect that still remains unclear is the extent to which the presence of nanoparticles gives rise to a change in the local polymer dynamics. For instance, if there is a strong polymer–particle interaction, then a local decrease in dynamics and a corresponding increase in the local T_g may be anticipated. Similarly, a poor interaction may give rise to a faster polymer mobility and local T_g decrease. At present, it is preferable that these issues are not addressed directly; rather, attention should be focused on what might happen with regards to the mechanical properties, purely on the basis of the particle AR and concentration. So, the question is: What might happen if changes in the matrix polymer dynamics are not considered? In fact, it is becoming increasingly clear that interesting phenomena may occur, especially at higher particle ARs.

In order to understand the mechanical properties of these nanocomposites, the decision was taken to use the Halpin–Tsai model (Equation 12.2), which was originally developed to describe the mechanical properties of semi-crystalline polymers [25, 26], using the shape factors derived by Van Es [27]:

$$\frac{E_c}{E_m} = \frac{1+\varsigma\eta\phi_f}{1-\eta\phi_f} \quad \text{in which} \quad \eta = \frac{(E_f/E_m)}{(E_f/E_m)} \tag{12.2}$$

where E_c is the composite Young's modulus, E_f is the filler modulus, E_m is the matrix modulus, ς is the shape factor (this depends on the geometry, AR and orientation), and ϕ_f is the filler volume fraction. Similar expressions can be derived for the other moduli, such as G.

It should be noted that elsewhere, the Halpin–Tsai equations are often described as a semi-empirical approach. Unfortunately, this does not do justice to their original derivation, as they are the result of a close examination of the findings of Kerner, Hill and Hermans [25] in which composite stiffness was analyzed in terms of self-consistent mechanical models. For this, the particles (of certain AR) are embedded in a polymer matrix which, in turn, is embedded in a continuum (consisting of polymer and particles). The self-consistency constraint then gives rise to equations for the various moduli that, under certain assumptions, can be reduced to the form proposed by Halpin and Tsai. Consequently, the Halpin–Tsai equations should be considered as a mechanical mean-field model, and not as an arbitrary choice.

By applying the appropriate shape factors for different particles shapes and orientations, this model can be used to successfully describe the Young's and shear moduli. The specific shape factors can be determined by comparing the model with experimental results, or with more fundamental theories – that is, the Eshelby theory, the Mori–Tanaka theory, and 3-D finite element modeling [27–29].

The shape factors for the tensile moduli of platelet-reinforced composites (a width; b thickness) are [27]:

E11 or E22 $\quad \varsigma = \dfrac{2}{3}\left(\dfrac{a}{b}\right) \quad$ (in the radial direction of the platelets)

E33 $\quad \varsigma = 2 \quad$ (perpendicular to the platelets)

The shape factors for fibers (a length; b width) are [27]:

E11 or E22 $\quad \varsigma = 2 \quad$ (perpendicular to the fiber direction)

E33 $\quad \varsigma = 2\left(\dfrac{a}{b}\right) \quad$ (in the fiber direction)

When the stiffness values of the composite, matrix, and filler are known, the Halpin–Tsai model can also be used to back-calculate the AR of the reinforcing particles.

This will be an effective AR, because the particles can have different shapes, sizes, and thickness, as has been noted elsewhere [28, 30–32]. Instead of using an image analysis of transmission electron microscopy (TEM) images to estimate the AR distribution [28, 30, 31], the effective AR that the model provides based on experimental data is used. This effective AR is at least a reasonable estimate of the average AR [33, 34], and provides a useful parameter to compare different nanocomposite compositions. The Halpin–Tsai equations predict that platelet-shaped particles are hardly effective at very low ARs (<10). The maximum stiffening effect is reached only at ARs above 1000 for platelets, and above 100 for fibers [29] (see Figure 12.7).

When Equation 12.2 is rewritten in one equation (as is shown in Equation 12.3), the physical nature of the Halpin–Tsai model becomes more transparent:

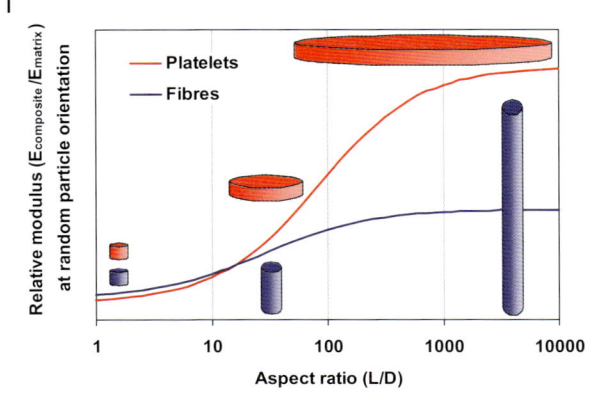

Figure 12.7 Influence of the aspect ratio on the modulus for platelets and fibers [29].

$$E_c = \zeta E_m \frac{(\phi_f + 1/\zeta)E_f + (1-\phi_f)E_m}{(1-\phi_f)E_f + (\phi_f + \zeta)E_m} \qquad (12.3)$$

When the ARs are much smaller than the ratio of the filler and matrix moduli ($\zeta \ll E_f/E_m$) the Halpin–Tsai model gives results close to a series model:

$$E_c = \frac{E_m E_f}{E_f(1-\phi_f) + E_m \phi_f} = \left(\frac{\phi_f}{E_f} + \frac{1-\phi_f}{E_m}\right)^{-1} \qquad (12.4)$$

When the ARs are much larger than the ratio of the filler and matrix moduli ($\zeta \gg E_f/E_m$) the Halpin–Tsai model gives results close to a parallel model:

$$E_c = E_f \phi_f + E_m (1-\phi_f) \qquad (12.5)$$

For nanocomposites, the series model underestimates the modulus. Because the volume fraction of filler is low, the first term in Equation 12.4 dominates, giving values close to the matrix modulus. Similarly the parallel model (Equation 12.5) overestimates the composite modulus at low matrix moduli, since it assumes a continuous reinforcing phase.

Further insight in the Halpin–Tsai model can be obtained if Equation 12.3 is approximated for large ARs and low filler fractions, as shown in Equation 12.6:

$$E_c = \frac{E_m \zeta}{E_f + E_m \zeta}(E_f \phi_f + E_m(1-\phi_f)) \qquad (12.6)$$

Here, it can be observed that if the matrix modulus remains sufficiently high, the pre-factor will reduce to a value close to 1 and the parallel model is obtained. However, the matrix modulus multiplied by the ζ factor is below E_f, the pre-factor will begin to force the composite modulus to lower values, and ultimately the soft matrix will dominate the behavior of the material. This explains why a nanocomposite with large AR particles can perform so well at elevated temperatures when a semi-crystalline polymer matrix. Above the T_g, the Young's modulus of a semi-crystalline polymer is normally too low for any useful application. However, for nanocomposites, the physical network (from the crystalline regions) allows the large AR nanoparticles to contribute to the modulus, roughly in accordance with the parallel model. This means that the filler contribution is additive to the modulus of the soft semi-crystalline matrix, and a useful stiffness results. It should be noted that this situation does not apply to an amorphous polymer, as the modulus in the rubber plateau of an amorphous polymer is far too low for the pre-factor be useful.

12.3.2
Experimental Validation for Non-Self-Healing Systems

To validate the model, nanoparticle-reinforced composites using silicate clays as reinforcement and Nylon 6 as the matrix material were produced and their mechanical properties determined. Special attention was paid to the influence of moisture on the modulus of the composite, as changes in moisture levels allow variation of the modulus of the matrix, while maintaining all other parameters of the composite exactly constant. Some representative data for a variety of dry and humidified PA6 samples are shown in Figure 12.8 where, in all cases, the increase in stiffness is comparable, although the more humid samples will initially have much lower modulus values. If the mechanical model is correct, then the calculated AR of the nanoparticles should be independent of the value of the matrix modulus. This is the case in Figure 12.9, which shows a systematic decrease in the effective particle AR with the increasing weight fraction of silicate nanoparticles, irrespective of the moisture level within the matrix.

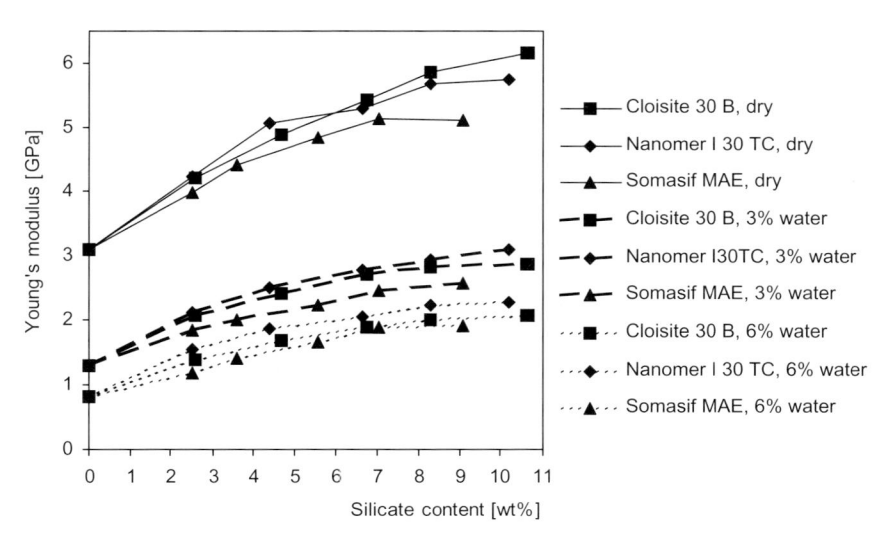

Figure 12.8 Modulus as a function of silicate content for three different moisture contents.

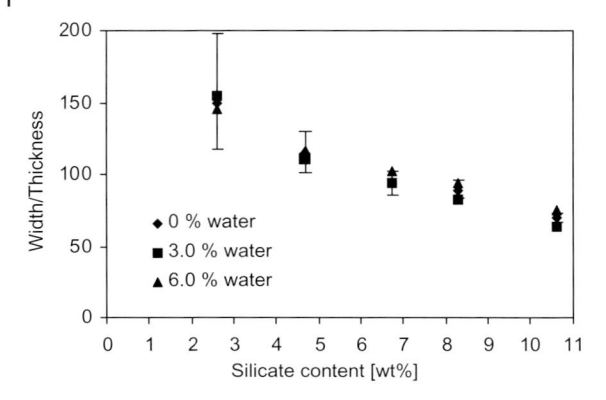

Figure 12.9 Aspect ratios calculated from the moduli as a function of silicate content for three different moisture contents (Cloisite 30B nanocomposites).

The most important point about the data in Figure 12.9 is the fact that the values for three different moisture contents overlap. This shows that the modulus of a nanocomposite can be explained by the combination of a very stiff filler particle with a high AR and the regular matrix modulus, which in this case is varied by altering the moisture content. Apparently, the Halpin–Tsai composite theory can be used to describe the modulus of nanocomposites for a wide variety of matrix moduli, both below and above the T_g of the matrix polymer, with the same AR for all matrix moduli. The fact that the modulus of nanocomposites can be explained by the Halpin–Tsai theory, independent of the modulus of the matrix, proves that the reinforcing mechanism in nanocomposites is similar to that in traditional composites. No additional stiffening of the matrix due to confinement of the polymer has to be assumed to explain the high modulus of nanocomposites.

12.3.3
Design of a Self-Healing Nanoparticle Composite

From the results obtained with PA6-based polymer nanocomposites (PNCs), it can be concluded that the Halpin–Tsai model is able to describe the enhancement of the mechanical properties rather successfully. In addition, it was found that the interaction between the polymer and the nanoparticles could give rise to a (small) yield stress. The use of this insight – that nanoparticles can give a modulus increase as well as a yield stress enhancement – will allow a self-healing PNC to be designed with acceptable mechanical properties.

For this, it is necessary to return to Equation 12.6:

$$E_c = \frac{E_m \zeta}{E_f + E_m \zeta}(E_f \phi_f + E_m(1-\phi_f)) \qquad (12.7)$$

From this expression, it is observed that if the pre-factor is in the order of 1, then the modulus will be dominated by the linear (parallel) addition term $E_f \phi_f$. In the case of PA6, this leads to a good enhancement of the properties above T_g, as the semicrystalline network still provides sufficient stiffness to meet the requirement for the pre-factor to remain close to 1, that is, $E_m\zeta > E_f$.

By extending this argument, it is possible (in principle) to use a soft dynamic matrix polymer (E_m = low) if the shape factor is sufficiently large to prepare materials that combine mechanical stiffness (via the $E_f\phi_f$ term) and local mobility by using a matrix polymer that contains a dynamic network. This class of polymers is known generically as "reversible supramolecular polymers". Examples include the dynamic hydrogen-bonded materials of Sijbesma [6], metal-organic coordination polymers reported by Schmatloch [35], and boric acid-modified polydimethylsiloxane (PDMS), also known as "Silly-Putty." In this class of materials, the polymer chains or crosslinking networks are present only in a dynamic sense. The polymer chains are continuously breaking and forming over time, while the level of interaction is sufficient to have a high dynamic degree of polymerization.

To demonstrate the principle of this idea, Silly-Putty nanocomposites have been prepared by mechanical blending of standard Silly-Putty with about 5% and 10% magnetic anisotropy energy (MAE) nanoparticles; the resultant properties are shown in Figure 12.10.

From these results it was concluded that, although the mechanical stiffness of the matrix living polymer network is still relatively low, the addition of nanoparticles can enhance the dynamics stiffness by a factor 100, or more. Furthermore, at clay loading levels of approximately 5%, the material keeps its shape over extended time periods (months), unlike the unfilled sample, so that the particles also cause a yield stress preventing flow under low-stress conditions.

Additional experiments were performed using Dow Corning 3179 dilatant compound (the professional variant of Silly Putty), with and without nanofiller as the matrix for a glass fiber-reinforced composite (M. Jansen *et al.*, unpublished results). Without the nanoparticles, the resultant material did not show any useful properties, but adding the nanoparticles to the matrix led to a dramatic improvement in

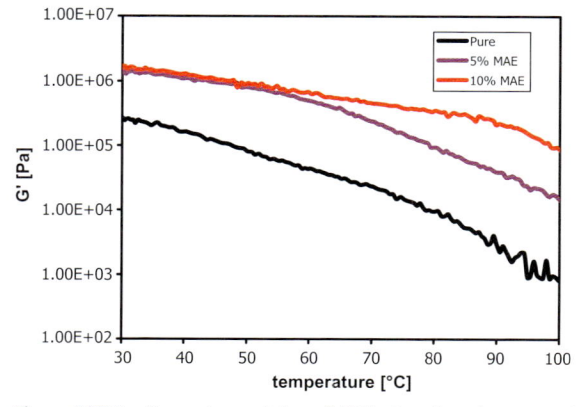

Figure 12.10 Dynamic modulus of "Silly Putty" and nanoparticle-reinforced grades as a function of temperature.

energy absorption, as monitored using a falling weight test. The results obtained are summarised in Table 12.1.

The impact energy absorption capability of this system was still not very high; however, by embedding two layers of glass fabric in the polymer sheet the absorbed impact energy was increased to 55 J per impact. Following the impact test, the samples were left for several days to recover their integrity. The mobility of the nanofilled matrix remained sufficient for rebonding of the impact-damaged part, although clearly the broken glass fibers could not heal together again. The unfilled reference samples allowed imaging by C-scan that clearly demonstrated the self-healing of a damaged zone (Figure 12.11).

The observation (which was mentioned briefly above) that the glass fibers failed under impact was in fact highly relevant to demonstrate the utility of this physical self-healing concept. The occurrence of fiber fracture demonstrated that the limiting properties could be obtained from the properties of the fiber reinforcement, and that the matrix had sufficient dynamic stiffness to efficiently transfer the stress to the reinforcement fibers.

It is, of course, clear that a substantial amount of further development and fine-tuning will be required to fully realize the potential of the physical self-healing of

Table 12.1 Failure energy versus composition of the matrix of glass-fibre reinforced B-PDMS (Dow Corning 3179 dilatant compound) nanocomposites.

Material	MM	Glass fabric	B-PDMS	Failure/J
Reference	0	57	43	3
10%	5	57	38	8
20%	10	57	33	8

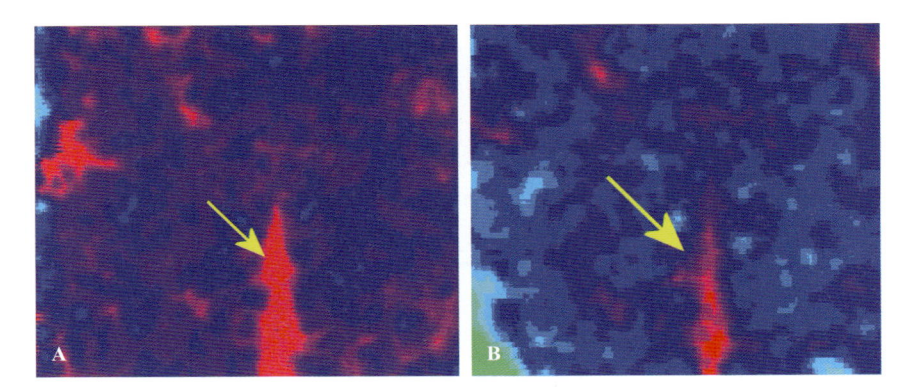

Figure 12.11 Self-healing in (non-nano) polydimethylsiloxane (PDMS)/glass fiber panels. (A) After impact damage; (B) After one week of recovery. The red zones are delaminated. The arrow is a reference position indicator.

living polymer nanocomposites. Nevertheless, the current experimental results have shown clearly that a combination of dynamics and stiffness on the one hand, and repetitive self healing on the other hand, is indeed possible under appropriate conditions and with a judicious design of the material.

12.4
Concluding Remarks

Whilst the development of liquid-based self-healing polymers has been investigated for some time, and the simultaneous control and optimization of mechanical and self-healing properties can now be addressed properly, the creation of self-healing nanoparticle-reinforced composites remains in its infancy. Yet, this concept has great potential, and the development of these materials should be combined with new techniques to quantify the degree of healing, both for single-damage events and for multiple healing. The development of such materials is also greatly facilitated by the existence of validated mechanical models. It should be noted that, although the analysis of polymer dynamics in confined and hybrid systems has been widely discussed, self-healing polymer nanocomposites seem to be one of the first practical applications resulting from fundamental studies of how to best utilize the properties of polymers in complex hybrid systems. With regards to potential applications of these dynamic, self-healing materials, it is possible to envisage personal protection for sports and other outdoor activities, protective clothing (e.g., for motorcyclists) and, indeed, military applications.

Acknowledgments

The authors gratefully acknowledge the contributions by M. Janssen, V. Antonelli, and R. Marissen in developing the nanoreinforced composites.

References

1 van der Zwaag, S. (ed.) (2007) *Self Healing Materials: An Alternative to 20 centuries of materials science*, Springer, Dordrecht.
2 van der Zwaag, S., van Dijk, N.H., Jonkers, H.M., Mookhoek, S.D., and Sloof, W.G. (2009) *Philos. Trans. Roy. Soc. A – Math. Phys. Eng. Sci.*, **367** (1894), 1689–1704.
3 Dry, C. (1996) *Compos. Struct.*, **35**, 263–269.
4 White, S.R., Sottos, N.R., Geubelle, P.H., Moore, J.S., Kessler, M.R., Sriram, S.R., Brown, E.N., and Viswanathan, S. (2001) *Nature*, **409**, 794–797.
5 Varley, R.J. and van der Zwaag, S. (2008) *Acta Mater.*, **56**, 5737–5750.
6 Sijbesma, R.P., Beijer, F.H., Brunsveld, L., Folmer, B.J.B., Hirschberg, J.H.K., Lange, R.F.M., Lowe, J.K.L., and Meijer, E.W. (1997) *Science*, **278** (5343), 1601–1604.
7 Ristenpart, W.D., Jiang, P., Slowik, M.A., Punckt, C., Saville, D.A., and Aksay, I.A. (2008) *Langmuir*, **24** (21), 12172–12180.

8 Hayes, S.A., Jones, F.R., Marshiya, K., and Zhang, W. (2007) *Compos. A*, **38**, 1116–1125.
9 Lumley, R.N., Polmear, I.J., and Morton, A. (2003) *J. Mater. Sci. Technol.*, **19**, 1483–1490.
10 Song, G.M., Pei, Y.T., Sloof, W.G., Li, S.B., De Hosson, J.Th.M., and van der Zwaag, S. (2008) *Scripta Mater.*, **58**, 13–16.
11 Miccichè, F., Fischer, H., Varley, R., and van der Zwaag, S. (2008) *Surf. Coat. Technol.*, **202** (14), 3346–3353.
12 Hikasa, A., Sekino, T., Hayashi, Y., Rajagopalan, R., and Niihara, K. (2004) *Mater. Res. Innovations*, **8** (2), 84–88.
13 Jonkers, H.M. (2007) *Self Healing Materials – An Alternative Approach to 20 Centuries of Materials Science* (ed. S. van der Zwaag), Springer, The Netherlands.
14 Palanivel, V., Zhu, D., and van Ooij, W.J. (2003) *Prog. Org. Coat.*, **477**, 384–392.
15 Parkhill, R.L., Knobbe, E.T., and Donley, M.S. (2001) *Prog. Org. Coat.*, **41**, 261–265.
16 Zheludkevich, M.L., Serra, R., Montemor, M.F., and Ferreira, M.G.S. (2005) *Electrochem. Commun.*, **7**, 836–840.
17 Zheludkevich, M.L., Shchukin, D.G., Yasakau, K.A., Möhwald, H., and Ferreira, M.G.S. (2007) *Chem. Mater.*, **19** (3), 402–411.
18 Pang, J.W.C. and Bond, I.P. (2005) *Compos. A*, **36**, 183–188.
19 Caruso, M.M., Delafuente, D.A., Ho, V., Sottos, N.R., Moore, J.S., and White, S.R. (2007) *Macromolecules*, **40** (25), 8830–8832.
20 Peng-Peng, W. (1994) *J. Polym. Sci.*, **32** (7), 1217–1227.
21 Wu, T. and Lee, S. (1994) *J. Polym. Sci.*, **32** (12), 2055–2064.
22 Mookhoek, S.D. (2010) Novel routes towards liquid based self-healing polymer systems. Ph.D. Thesis, Delft University of Technology.
23 Hansen, C.M. (1999) *Hansen Solubility Parameters: A User's Handbook*, CRC Press.
24 Blaiszik, B.J., Sottos, N.R., and White, S.R. (2008) *Compos. Sci. Technol.*, **68**, 978–986.
25 Halpin, J.C. and Kardos, J.L. (1976) *Polym. Eng. Sci.*, **16** (5), 344–352.
26 Halpin, J.C. (1969) *J. Compos. Mater.*, **3**, 732–734.
27 van Es, M., Xiqiao, F., van Turnhout, J., and van der Giessen, E. (2001) *Specialty Polymer Additives*, Blackwell Science, Oxford.
28 Sheng, N., Boyce, M.C., Parks, D.M., Rutledge, G.C., Abes, J.I., and Cohen, R.E. (2004) *Polymer*, **45** (2), 487–506.
29 van Es, M. (2001) Polymer clay nanocomposites: the importance of particle dimensions. Ph.D. Thesis, Delft University of Technology.
30 Fornes, T.D. and Paul, D.R. (2003) *Polymer*, **44** (17), 4993–5013.
31 Kuelpmann, A., Osman, M.A., Kocher, L., and Suter, U.W. (2005) *Polymer*, **46** (2), 523–530.
32 Wu, Y.P., Jia, Q.X., Yu, D.S., and Zhang, L.Q. (2004) *Polym. Test.*, **23** (8), 903–909.
33 Vlasveld, D.P.N., Groenewold, J., Bersee, H.E.N., Mendes, E., and Picken, S.J. (2005) *Polymer*, **46**, 6102–6113.
34 Vlasveld, D.P.N., Groenewold, J., Bersee, H.E.N., and Picken, S.J. (2005) *Polymer*, **46** (26), 12567–12576.
35 Schmatloch, S., van den Berg, A.M.J., Hofmeier, H., and Schubert, U.S. (2004) *Linear Polym.*, **7**, 191–201.

13
Crystallization in Polymer Nanocomposites
Jyoti Jog

13.1
Introduction

Over the past decade, polymer nanocomposites have been a topic of great interest to research groups in both pure and applied materials science [1–4]. These nanocomposites are two-phase materials wherein a filler with at least one dimension in the range of 1 to 100 nm is dispersed in the polymer matrix. Various nanofillers such as carbon nanotubes (CNTs), inorganic nanoparticles, or layered silicates such as clay are used. These nanocomposites show remarkable property improvements as compared to virgin polymers; moreover, this occurs at very low filler loadings when compared to conventional microcomposites. The properties of these nanocomposites are dependent on the morphology generated during their processing. In the case of semi-crystalline polymers, this is more crucial as the crystallization of polymers takes place during processing. A knowledge of the crystallization process, and the effect of these nanofillers on the crystallization behavior of polymer nanocomposites, is therefore imperative to acquire an understanding of structure–property relationships in polymer nanocomposites.

Crystallization in polymers has been the subject of great academic interest, as polymers are known to exhibit a variety of structures at various length scales, such as the unit-cell, lamella, and spherulites. The crystallization of polymers results in different morphologies, depending on the crystallization conditions and the presence of other components. The presence of a second phase either a polymer or a low-molecular-weight compound or inorganic filler has a profound effect on the crystallization of the polymer. In nanocomposites, the dimensions of the second phase are close to the chain dimensions, and a number of studies have attempted to elucidate the effect of nanosized fillers on the crystallization kinetics, the crystalline morphology, and the crystalline forms of polymers. In this chapter, a brief review is presented of recent investigations into the effects of nanofillers on various aspects of crystallization, such as isothermal crystallization, nonisothermal crystallization, spherulitic growth, and polymorphism.

Optimization of Polymer Nanocomposite Properties. Edited by Vikas Mittal
Copyright © 2010 WILEY-VCH Verlag GmbH & Co. KGaA, Weinheim
ISBN: 978-3-527-32521-4

13.2
Nanofillers

The various types of nanofiller that are commonly used for the preparation of nanocomposites are described in the following sections.

13.2.1
Silicates

The clays used in the formation of organoclays or nanoclays belong to smectite group clays. The following types of silicate are used as nanoclays.

- **Montmorillonite** (MMT) is an aluminum silicate, and is mainly used because of its suitable charge density. Each aggregated particle of MMT consists of stacks of lamella with thickness of about 1 nm. MMT has very high surface area (ca. $750 \, m^2 g^{-1}$). A scanning electron microscopy (SEM) image of an organically modified MMT is shown in Figure 13.1a.

Figure 13.1 Scanning electron microscopy images. (a) Clay; (b) MWCNT; (c) Alumina; (d) Transmission electron microscopy image of BaTiO$_3$.

- **Saponite** is a complex hydrous silicate of aluminum and magnesium. The saponite structure is in the form of ribbons and laths.
- **Attapulgite** is a hydrous magnesium–aluminum silicate mineral. It is present in the form of needles, fibers, or fibrous clusters. Standard attapulgite clays are agglomerated bundles of clay particles between 20 and 100 µm long and less than 1 µm in diameter.
- **Sepiolite** is a family of fibrous hydrated magnesium silicates which can be present in fibrous, fine-particulate, and solid forms. The size of the fibers varies widely, but in most cases they are 10–5000 nm long, 10–30 nm wide, and 5–10 nm thick. The aspect ratio is within the range of 100–300, and the surface area is in the range of 200–300 $m^2 g^{-1}$.
- **Rectorite** is another type of clay mineral, with structure and characteristics similar to those of MMT. The interlayer cations of smectite structure can be replaced easily either by organic or inorganic cations, which makes it possible to separate rectorite layers to prepare polymeric nanocomposites.

The dispersion of these hydrophilic clays in the polymer matrix is a major issue in the development of polymer/clay nanocomposites, as the polymers are hydrophobic. The organic modification of clays is therefore necessary in order to achieve compatibility between the polymer and the clay. The organic modification also facilitates the dispersion of clay layers in the polymer matrix. The organically modified clays can be used to improve the mechanical properties of polymer composites prepared by *in situ* intercalation polymerization, solution-mixing, and melt intercalation.

13.2.2
Carbon Nanotubes

Carbon nanotubes were first reported by Ijima [5] in 1991, while the first polymer nanocomposites using CNTs as filler were reported in 1994 by Ajayan *et al.* [6] CNTs are classified into single-walled carbon nanotubes (SWCNTs), multiwalled carbon nanotubes (MWCNTs) and carbon nanofibers (CNF). The MWCNTs comprise a number of graphene layers coaxially rolled together to form a cylindrical tube. The outer diameter of the MWCNT is approximately 3–10 nm; a SEM image of a MWCNT is shown in Figure 13.1b. CNTs have a unique combination of mechanical, electrical, and thermal properties that make them excellent candidates to either substitute or complement conventional nanofillers in the fabrication of multifunctional polymer nanocomposites.

13.2.3
Exfoliated Graphite

Exfoliated graphite (xGnP) is fabricated from natural graphite, and can be used as a nanoreinforcement for polymers as an alternative to expensive carbon-based nanomaterials. The average thickness of the xGnP is 5–10 nm, and the surface area is more than $100\,m^2\,g^{-1}$. The incorporation xGnP into polymers leads to changes in their electrical conductivity, barrier properties, modulus, and surface toughness.

13.2.4
Other Nanoparticles

Nanosized inorganic particles are used to impart enhanced mechanical properties and decreased thermal expansion coefficients.

- **Alumina** (aluminum oxide) is an electrical insulator with a relatively high thermal conductivity ($40\,W\,m^{-1}\,K$). The most common form of crystalline alumina is α-aluminum oxide; this is of particular interest for health and medical applications, as it is a biocompatible material. A SEM image of alumina nanoparticles is shown in Figure 13.1c.
- **Boehmite** is an aluminum oxide hydroxide (γ-AlO(OH)) mineral. Boehmite powder is composed of micrometer-scaled agglomerates of crystallites of 10–60 nm in size. Boehmite particles have a rod-like shape, with a high degree of anisotropy.
- **Calcium Carbonate** is the most important filler for polymer applications. It is mainly used with polypropylene (PP) and other polymers for various applications such as furniture, in the automotive industry, as packaging, and as fibers.
- **Fumed Silica** is a fluffy white amorphous powder with high surface area. It is a pure form of silicon dioxide, and is prepared by reacting silicon tetrachloride in an oxy-hydrogen flame. The particle size ranges from 0.007 to 0.05 μm, and the surface area from 50 to $600\,m^2\,g^{-1}$.
- **Barium Titanate** ($BaTiO_3$) is a high-dielectric constant material that is used widely in multilayer ceramic capacitors. A transmission electron microscopy (TEM) image of $BaTiO_3$ nanoparticles is shown in Figure 13.1d. The use of $BaTiO_3$ nanopowders can be further extended by combining it with polymers, which offer design and processing flexibility.
- **Silicon Nitride Ceramic** is a widely used material that has superior physical properties; it is also used for various applications such as electronic devices.
- **Zinc Oxide** is a large band gap n-type semiconductor that has attracted great attention because of its specific optical and electronic properties. It is known as a luminescent material.

13.3
Isothermal and Nonisothermal Crystallization in Polymers

Crystallization is widely observed for a large number of polymers with structural regularity. The crystallization process of polymers consists of two steps, namely *nucleation* and *growth*. When nucleation occurs spontaneously, due only to supercooling and where no second phase or other nuclei exist, this type of nucleation is known as *homogeneous nucleation*. When the nucleation is induced in the presence of any other second phase, such as a foreign particle or surface, then the nucleation is referred as *heterogeneous nucleation*. The nucleation, growth, and kinetics of development of these crystalline structures are of both profound fundamental and pragmatic interest. The kinetics of polymer crystallization is controlled by various inherent factors, such as molecular weight, chain flexibility, chain defects, and stereo-regularity. The crystallization process is also affected by experimental conditions such as temperature, pressure, nucleating agents, and stress.

A variety of experimental techniques have been used to investigate the mechanism of crystallization under both isothermal and nonisothermal conditions. The most common techniques used are differential scanning calorimetry (DSC), optical microscopy (OM), wide-angle X-ray diffraction (WAXRD), and Fourier transform-infrared spectroscopy (FTIR). The isothermal crystallization data are generally analyzed using Avarmi's theory of phase transformation [7–9] whereas the nonisothermal crystallization data are generally analyzed using modified Avarmi equations [10–12].

13.3.1
Polypropylene (PP)

13.3.1.1 Crystallization

The crystallization of PP nanocomposites has been studied for a number of nanofillers such as clay [13–15], attapulgite [16, 17], SWCNTs [18–21], MWCNTs [22–25], graphite [26], boehmite [27], fumed silica [28, 29], and $CaCO_3$ [30].

Among the polymer/clay nanocomposites, PP/clay is the most studied system. The nonpolar nature of PP requires the use of a suitable compatibilizer to achieve an adequate intercalation and dispersion of clay in PP matrix [31]. Figure 13.2 depicts the dispersion of nanoclay in a PP matrix, as observed using TEM. The isothermal crystallization behavior of PP/clay nanocomposites indicated that the crystallization process of PP was significantly enhanced [13–15]. The incorporation of nanoclay also results in significant changes in the spherulitic morphology of the nanocomposites. The effect of nanoclay addition on the PP spherulites is illustrated in Figure 13.3.

In polypropylene (PP)/organic-attapulgite (ATP) nanocomposites [16], it was revealed that the ATP nanoparticles were effective nucleating agents and accelerated the crystallization process. In another study by Zhao *et al.* [17], it was reported that the application of ultrasonic oscillation during extrusion further accelerated the crystallization rate.

Bhattacharya *et al.* [18] reported an enhanced nucleation of PP in the presence of SWCNTs, while Valentine *et al.* [19] noted that the crystallization kinetics was strongly affected by the distance between the nanotube bundles in PP/SWCNT composites. Grady *et al.* [20] have reported that the rate of crystallization and the percent crystallinity was higher in PP/SWCNT nanocomposite, thus confirming

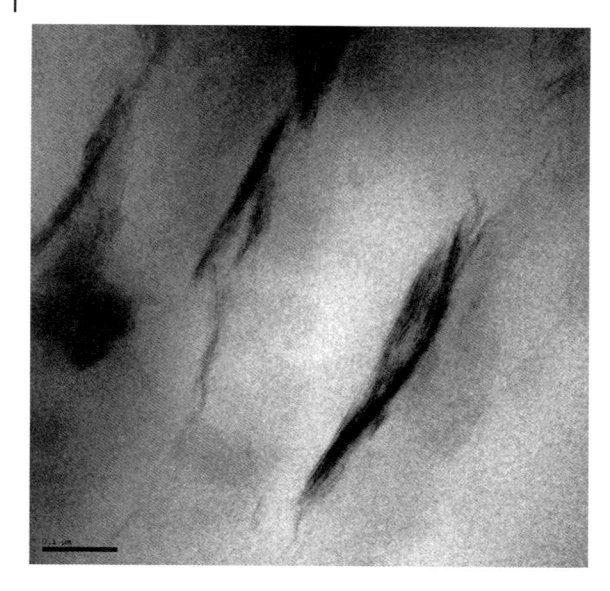

Figure 13.2 Dispersion of clay layers in polypropylene.

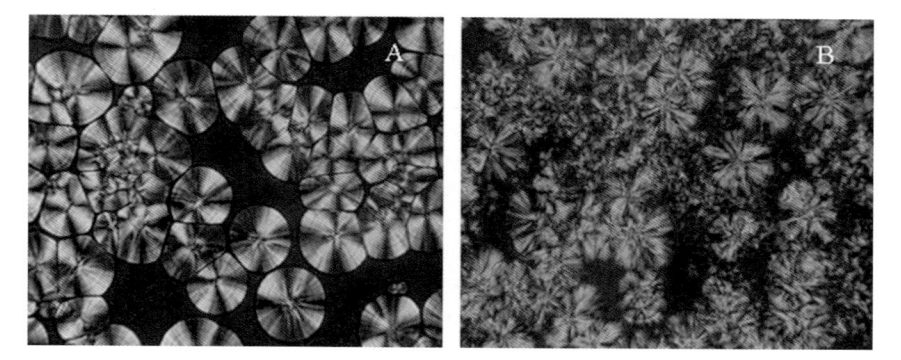

Figure 13.3 Spherulites of polypropylene (PP) and PP/clay nanocomposites.

the enhanced nucleation of PP. Manchado *et al.* [21] have also noted a similar increase in the crystallization rate of PP.

In PP/MWCNT nanocomposites, MWCNTs acted as nucleating agents for PP and syndiotactic PP [22]. A nonisothermal crystallization study of PP/MWCNT showed that the crystallization rate was increased up to 4% MWCNT, this being attributed to heterogeneous nucleation. However, at higher MWCNT loadings, the crystallization rate decreased slightly or remained almost constant. This effect was ascribed to an aggregation of the filler particles at high concentration, which led to a decreased number of heterogeneous nuclei [23]. Bao and Tjong [24] confirmed that the incorporation of MWCNTs into PP resulted in an enhanced nucleation and also a reduced mobility of the polymer chains. A similar enhancement in the nucleation process in PP crystallization was also reported in PP/MWCNT nanocomposites [25].

Page and Gopakumar [26] have shown that the type of nucleation, growth and geometry of PP crystals were markedly changed in the presence of nanosized graphite particles.

Streller *et al.* [27] have noted that nano-boehmite acted as a nucleating agent for PP, and this resulting in a higher density of spherulites with a much smaller size as compared to neat PP.

The crystallization of nanocomposites of PP with fumed silica showed an enhanced nucleation of PP in the presence of SiO_2 nanoparticles [28]. In another study of PP with surface-treated silica, the observed increase was attributed to an enhanced nucleation of PP in the presence of silica nanoparticles [29].

Tao *et al.* [30] have reported that the well-dispersed nano-$CaCO_3$ acted as an effective nucleating agent and accelerated the crystallization of PP.

13.3.1.2 Polymorphism in PP

The crystallization behavior of PP is very complex, as it exhibits four crystal modifications, such as α, β, γ, and smectic [32, 33].

In the case of PP/clay nanocomposites, it was reported that the γ-form was located in the proximity of the silicate layer, whereas the α-phase existed in the bulk [34]. Whilst the existence of the γ-form in PP/clay was also observed by Nam *et al.* [35], it was recently reported that, in melt-compounded PP/organoclay nanocomposites, the low nanoclay concentrations induced formation of the β-form, with such formation being inhibited at higher nanoclay contents [36].

The SWNTs displayed a clear nucleating effect on the PP crystallization, favoring the α-form rather than the β-form [20, 37].

Wan *et al.* [38] noted that the incorporation of surface-modified nanoparticles induced much more β-form crystal in nanocomposites. On the other hand, it was reported that nanocomposites of PP with $CaCO_3$ prepared by a novel process resulted in γ-form PP, whereas only α-form PP was formed during conventional processing [39].

13.3.2
Poly-1-Butene (PB)

Nanocomposites of poly (1-butene) (PB) with organically modified clay were prepared by melt intercalation, the process being facilitated by using organically modified clay. The isothermal crystallization studies clearly indicated an enhancement in the crystallization rate, and resulted in a disturbed spherulitic morphology because of the dispersed clay layers [40, 41].

In nanocomposites of PB with organoclay prepared by melt compounding using a compatibilizer, the presence of organoclay was observed to disrupt the ordered morphology of the polymer [42].

When the effects of MWCNTs on the kinetics of crystallization were investigated [43], the results of isothermal crystallization studies suggested that the crystalliza-

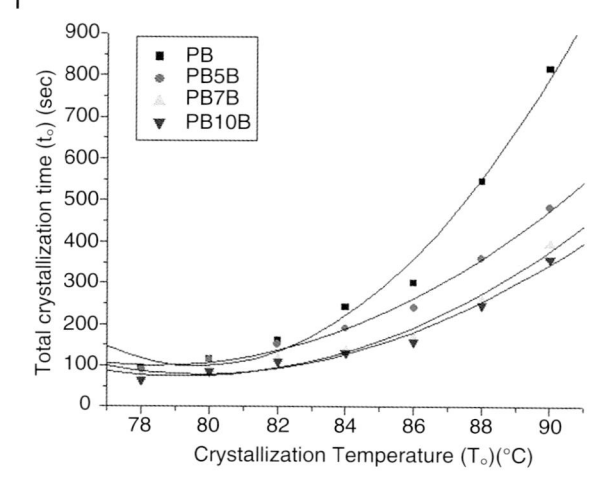

Figure 13.4 Total crystallization time for poly (1-butene) (PB) and PB/BaTiO$_3$ nanocomposites.

tion rate was accelerated in nanocomposites, with the observed changes in crystallization kinetics being ascribed to an enhanced nucleation of PB in the presence of MWCNTs.

In the case of PB/BaTiO$_3$ nanocomposites, lower values of total crystallization time (t_c) were observed (see Figure 13.4). The lowered values suggested an enhanced crystallization of PB in the presence of BaTiO$_3$ nanoparticles, the effect being much more pronounced at higher crystallization temperatures [44].

13.3.3
Polybutylene Terephthalate (PBT)

The crystallization of polybutylene terephthalate (PBT)-based nanocomposites has been reported for nanofillers such as clay [45–47], attapulgite [48], and carbon nanofibers [49].

When the nonisothermal crystallization of a PBT/clay nanocomposite was investigated by Wu et al. [45], it was noted that the incorporation of very small amounts of clay (1%) resulted in an acceleration of the crystallization process, but any further increase in clay content reduced the rate of crystallization. In a recent report, Huang suggested that the clay layers were exfoliated in the PBT matrix in the presence of compatibilizer, and this resulted in an enhanced nucleation of PBT [46, 47].

Chen et al. [48] have reported that PBT/attapulgite nanocomposites exhibited higher crystallization rates than neat PBT. The fold surface free energy of polymer chains in the nanocomposites was less than that in the neat PBT, which suggested that attapulgite acted as a good nucleating agent for the crystallization of PBT.

When the nanocomposites of PBT with carbon nanofibers and organoclay were prepared by melt compounding, isothermal crystallization studies showed that the presence of nanoclays or nanofibers had enhanced the crystallization rate of the PBT [49].

13.3.4
Polyethylene Terephthalate (PET)

The crystallization of PET with nanofillers such as clay [50–53], attapulgite [54], MWCNT [55, 56], silica [57], and alumina [58] has been studied.

The crystallization of PET was accelerated in presence of clay, with optical microscopy studies revealing the presence of irregularly shaped and some small rod-like crystallites which subsequently developed into fan-shaped crystallites, in addition to spherulites [50]. Likewise, Bandopadhay et al. [51] observed an enhanced nucleation and a decreased crystal growth in intercalated PET/clay nanocomposites.

When the effect of the presence of an organic modifier on the crystallization behavior of PET was investigated [52], both nanocomposites were found to have higher degrees of crystallinity and shorter crystallization half-times than neat PET. However, those nanocomposites with organic modifier exhibited a lower degree of crystallinity and a longer half-time of crystallization than that prepared in the absence of an organic modifier. These observed differences in crystallization behavior were ascribed to the presence of an organic modifier, which may act as a crystallization inhibitor. Elsewhere, Hwang et al. [53] noted that, although the dispersion of clay was better for the nanocomposite with a modified clay, a pristine clay exhibited an enhanced nucleation and crystallization rate.

The study of PET/attapulgite nanocomposites indicated that the presence of attapulgite in the nanocomposites caused a composition-dependent increase in the crystallization temperature of PET, and also accelerated the rate of crystallization of PET; these effects were ascribed to the heterogeneous nucleation [54].

The crystallization of PET/MWCNT nanocomposites with very low content of MWCNTs (from 0.01 to 0.2 wt%) displayed a significant nucleating effect on the PET crystallization [55]. However, the presence of MWCNTs also resulted in a decreased chain mobility of PET chains, such that imperfect or smaller/thinner crystallites with a lower melting point were subsequently formed. The crystallization behavior of PET/MWCNT with a weight fraction of MWCNT between 1% and 15% showed that the addition of MWCNTs resulted in an increase in the *trans* conformation and crystallinity of PET at lower MWCNT contents [56].

Tian et al. [57] have studied the crystallization of PET with well-dispersed silica nanoparticles, and concluded that the nanoparticles were acting as an effective nucleating agent, as evidenced by the decreased activation energy.

The nonisothermal crystallization of PET with alumina nanoparticles was studied by Bhimraj et al. [58], who noted that up to 3% alumina content the nanoparticles caused heterogeneous nucleation in PET. Such nucleation was attributed to the curvature of the nanoparticles, which was comparable to the radius of gyration of PET.

13.3.5
Poly Trimethylene Terephthalate (PTT)

The nanocomposites of poly trimethylene terephthalate (PTT) with clay [59–61], MWCNTs [62] and $CaCO_3$ [63] have been investigated for their crystallization behavior.

For example, Ou [59, 60] has reported that presence of clay enhanced the crystallization of PTT in solution-intercalated nanocomposites, and that the enhancement in crystallization was maximum for nanocomposites containing 5% clay.

Hu and Lesser [61] have studied nanocomposites of PTT with various organoclays, and noted that for up to 3% of clay, the nanoscale dispersion of clay in the polymer was effective as a nucleating agent, whereas untreated clay Na^+ dispersed at the micrometer scale did not show much effect on the crystallization of PTT. Similar observations were reported by Liu et al. [62] in melt-blended PTT/clay nanocomposites.

When Xu et al. [63] studied the crystallization behavior of PTT/MWCNT nanocomposites, they found that MWCNT provided more nucleation sites to the crystallizing form, and that this resulted in the formation of smaller spherulites.

Crystallization studies of $PTT/CaCO_3$ nanocomposites, as conducted by Run et al. [64], indicated that the particles of nano-$CaCO_3$ acted as nucleating agents in the composites and accelerated the crystallization rate for up to 2 wt% of the filler.

13.3.6
Polyethylene Naphthalate (PEN)

13.3.6.1 Crystallization
In the case of polyethylene naphthalate (PEN), crystallization was studied in the presence of clay [65, 66], MWCNTs [67], and silica [68].

When the nanocomposites of PEN with organoclay were studied by Wu and Lui [65], they observed that the presence of silicate layers reduced the transportation ability of PEN, as evidenced by an increased activation energy for nanocomposites with increases in clay content. Li et al. also [66] showed that when MMT nanoparticles were dispersed into the PEN matrix they acted as heterogeneous nuclei for PEN and enhanced its crystallization rate. However, a higher loading of MMT resulted in a hindered motion of PEN chains which led in turn to a restrained crystal growth of PEN.

Kim et al. have reported [67] that the incorporation of MWCNTs accelerated the process of nucleation and crystal growth of PEN significantly at a lower MWCNT content.

The crystallization of PEN/silica nanocomposites was investigated by Kim et al. [68], who observed that crystallization was dependent on the silica content, while the addition of silica greatly reduced the work required for chain folding. In a separate study, Kim et al. [69] noted that the silica particles induced an heterogeneous nucleation, as evidenced by the shift of the crystallization peak to a higher temperature.

13.3.6.2 Polymorphism in PEN

It is known that, depending on the crystallization conditions, PEN crystallizes as two different crystal modifications, namely the α and β forms. The annealing of amorphous PEN in the solid state results in the α-form, whereas the β-form can be formed in samples crystallized from melt [70].

Chua et al. [71] have shown that the melt crystallization of PEN in the presence of untreated clay and organoclay enhanced the formation of the β-form at 200 °C, whereas at 180 °C the organoclay facilitated formation of the α-form. These observed changes in crystallization were attributed to the conformational changes that occurred in the modifier at crystallization temperatures.

Kim et al. [72] reported that the incorporation of MWCNTs effectively enhanced the crystallization of the PEN matrix, and facilitated formation of the β-form in the PEN/MWCNT nanocomposites.

13.3.7
Polylactic Acid (PLLA)

The crystallization of nanocomposites of polylactic acid (PLLA) with clay [73, 74], MWCNTs [75], silica [76], and $CaCO_3$ [77] has been reported.

When Krikorian and Pochan [73] investigated the crystallization behaviors of PLLA with two types of MMT they noted that, depending upon the interaction between the polymer and the clay, the nanocomposites were exfoliated and intercalated. The overall crystallization rate in the intercalated system was increased, whilst in the case of the exfoliated system it was found to be retarded. When monitoring the effect of organophilic MMT on the isothermal crystallization of PLLA [74], it was noted that the nucleating effect of clay on the crystallization of PLLA was stronger for natural clay than for the organophilic clay.

The study of nanocomposites of PLLA with surface-modified MWCNTs indicated that the latter would significantly enhance the nonisothermal, melt-crystallization and cold-crystallization rates of PLLA [75].

Nanocomposites of PLLA with silica nanoparticles grafted with L-lactic acid oligomer [76] showed that the oligomer-grafted SiO_2 nanoparticles acted as a nucleating agent for the crystallization of PLLA in the composites.

The incorporation of $CaCO_3$ nanoparticles was reported to result in a decrease in the cold crystallization temperature, this effect being ascribed to the enhanced nucleation of PLLA crystallites during heating [77].

13.3.8
Polyhydroxy Alkonate (PHA)

The nanocomposites of polyhydroxy alkonate (PHA) with clay [78, 79], SWCNTs [80], MWCNTs [81], graphite [82], fumed silica [83] and hydroxyapatite [84], have been studied for their crystallization behavior.

A crystallization study of nanocomposites of poly(3-hydroxybutyrate-co-3-hydroxyvalerate) (PHBV) with organoclay [78] indicated that the organoclay was an effective nucleating agent in the polymer matrix. Nanocomposites of two PHAs, namely poly (3-hydroxybutyrate) (PHB) and PHBV with different types of organoclay, revealed an enhanced nucleation in the nanocomposites, although this effect was much more pronounced for a higher filler content. The extent of the effect (*enhanced nucleation*) was also increased with the increase in polymer/clay affinity [79]. However, for solution-blended nanocomposites of PHBV with organoclay, the crystallization rate was increased [78] only at lower concentrations of clay.

Yun *et al.* [80] have reported the crystallization behavior of PHB and poly(3-hydroxyoctanoate), and shown that 1% SWCNTs effectively nucleated PHB crystallization, whereas higher proportions of SWCNTs were less effective in enhancing nucleation, due to their extensive agglomeration.

Recently, Lai *et al.* [81] studied the nanocomposites of a PHBV with MWCNTs and reported that the crystallization temperature increased as a result of the incorporation of MWCNT.

When the crystallization of PHB in the presence of xGnP nanoplatelets was investigated by Miloaga *et al.* [82], the presence of the latter caused an increase in both the crystallization temperature and crystallization rate for PHB. Subsequent isothermal crystallization studies of nanocomposites of poly (3-hydroxybutyrate-*co*-3-hydroxyhexanoate) with fumed silica showed the crystallization rate to be further reduced, with the extent of such being dependent on the strength of the intermolecular hydrogen bonding [83].

The nonisothermal crystallization of nanocomposites of PHB with hydroxyapatite (HA) revealed that, at low levels of HA, the HA particles served as additional nucleation sites, and that this resulted in a dramatic increase in the crystallization rate. However, at higher HA contents (>20 wt%), the HA particles retarded the growth process due to restricted polymer chain mobility, which in turn caused the crystallization to be retarded [84].

13.3.9
Polyether Ether Ketone (PEEK)

Few reports have been made on the crystallization of polyether ether ketone (PEEK)-based nanocomposites, including nanocomposites of PEEK with alumina and silica [85, 86].

A nonisothermal crystallization study of PEEK filled with alumina nanoparticles, conducted by Kuo *et al.* [85], showed that the presence of alumina nanoparticles would accelerate the nucleation of PEEK due to heterogeneous nucleation, but that the growth rate was reduced due to a retarded polymer chain mobility. Similar results were observed for nanoparticles of alumina and SiO_2 [86].

13.3.10
Nylon 6

13.3.10.1 Crystallization

Nylon 6-based nanocomposites are generally prepared using either an *in situ* polymerization or a melt-compounding technique. The isothermal and nonisothermal crystallization has been investigated for Nylon nanocomposite systems based on various nanofillers, including MMT [87, 88], rectorite [89], MWCNTs [90], xGnP [91], silver nanoparticles[92], and SiO_2 [93].

When the isothermal crystallization of commercially available Nylon 6/clay nanocomposites was investigated by Lincon *et al.* [87], the incorporation of clay into Nylon 6 resulted in an enhanced nucleation and a much weaker temperature dependence of crystallization rate, as compared to the pristine polymer. This lower temperature dependence was attributed to the presence of a strong polymer/clay interaction in Nylon 6/clay nanocomposites. When Fornes and Paul [88] studied the effect of clay concentration and degree of exfoliation on the crystallization kinetics of melt-processed Nylon 6/clay nanocomposites, the addition of clay led to a dramatic increase in the rate of crystallization at clay contents below 3 wt%; however, above 3 wt% the rates were comparable to those of the pristine polymer.

The results of crystallization studies of melt-compounded organic rectorite (OREC) with Nylon 6 indicated that the crystallization rate of the Nylon 6/OREC nanocomposites was faster than that of Nylon 6 at a given cooling rate [89].

The *in situ*-polymerized nanocomposites of Nylon 6 with saponite also revealed that the introduction of saponite into the Nylon matrix resulted in an heterogeneous nucleation and a change in the crystal growth process [90].

A crystallization study of Nylon 6/MWCNT, conducted by Chen and Wu [91], revealed that the addition of 1 wt% f-MWCNT into Nylon 6 induced heterogeneous nucleation. However, a higher MWCNT content resulted in a reduction of the transportation ability of polymer chains during the crystallization process, and thus increased the activation energy.

In a subsequent study, nanocomposites of foliated graphite particles were incorporated into the molten ε-caprolactam monomer, and then polymerized using ultrasonic irradiation. The results obtained indicated that, although foliated graphite nanosheets favored heterogeneous nucleation for the crystallization of Nylon 6, their presence hindered the transport of the polymer chains, and this resulted in a reduction of the crystallization growth rate [92].

When Chae *et al.* [93] studied the effect of silver nanoparticles on the crystallization of Nylon 6, the sharper crystallization peak in the nanocomposites was ascribed to an heterogeneous nucleation of Nylon 6 in the presence of silver particles.

Yang *et al.* [94] have studied the isothermal crystallization of Nylon 6/SiO_2 nanocomposites prepared by *in situ* polymerization. These authors observed that the presence of unmodified silica increased the rate of crystallization, whereas the crystallization rate was slightly decreased in the presence of modified silica. This effect was attributed to the possible better adhesion between modified silica and polymer, which impeded the motion of the polymer chains and thus hampered the crystallization. However, in another study, Rusu and Rusu [95] observed a maximum increase in the crystallization temperature at 2 wt% nanoparticles for both surface-treated and untreated silica.

13.3.10.2 Polymorphism in Nylon 6

Nylon 6 also exhibits polymorphism with two major crystalline forms, namely the α- and γ-forms. The α-form is a monoclinic crystal which consists of a fully extended configuration of the polymer chains, with an anti-parallel orientation. In contrast, the γ-form occurs when the polyamide chains are parallel, with the chain packing of the γ-form resembling that of a hexagonal structure. In general, the α-form is more stable than the γ-form [96, 97].

The incorporation of clay is reported to promote the growth of the γ-form in Nylon 6/clay nanocomposites, whereas the neat polymer tends to crystallize in the α-form [98–102]. This behavior was also observed for clays modified with different swelling agents [103]. The formation of an exclusively γ-form in the nanocomposite was ascribed to an epitaxial crystallization [104].

The effect of annealing on the polymorphic behavior revealed that, in nanocomposites, the γ-form was dominant whereas in Nylon 6 the α-form was predominant, notably when annealed at temperatures between 230 and 250 °C [105, 106].

Xie et al. [107] compared the effect of MMT (plate-like morphology) and sepiolite (needle-like morphology) clays on the structure of Nylon 6. The X-ray diffraction (XRD) analysis showed that sepiolite facilitated formation of the α-form of Nylon 6, while MMT induced the γ-form. This difference in the crystallization behavior of Nylon 6 in the presence of sepiolite and MMT was attributed to a lesser confinement effect in sepiolite as compared to MMT.

In the case of Nylon6/synthetic saponite nanocomposites, Wu et al. have noted that the addition of synthetic saponite increased the crystallization rate of the α-form at a lower saponite content and promoted a heterophase nucleation of the γ-phase at a higher synthetic saponite content [98, 108].

Phang et al. reported that only the α-form crystalline structure was formed upon incorporating MWCNTs into PA6 matrix, and the cooling rate and annealing conditions did not affect the crystal form [109]. However, whilst amino-functionalized MWCNTs induced the γ-form in Nylon 6, MWCNT facilitated formation of the α-form, as observed by Li et al. [110].

The nanocomposites of boehmite treated with a titanate coupling agent and Nylon 6 were investigated by Zdilek et al. [111]. Here, the addition of titanate-modified boehmite favored formation of γ-form, especially at higher concentrations. Such formation of γ-form was attributed to an altered interaction of the boehmite particles with the polymer chains, due to surface modifications.

The polymorphism in Nylon 6/$CaCO_3$ nanocomposites with two types of surface-treated $CaCO_3$, namely calcite and aragonite, was reported by Avella et al. [112]. These authors' results indicated that, although both types of $CaCO_3$ induced the γ-form in Nylon 6, the α-form was dominant in nanocomposites with an aragonite nanofiller.

Zheng et al. [113] have reported that nanosized ZnO particles induced the γ-form in Nylon 6, and that the amount of α-form crystallinity decreased linearly with the increase in ZnO content. The crystallinity of the γ-form increased linearly with the decreasing size of ZnO particles, and was ascribed to a larger interfacial area leading to an enhanced nucleation.

13.3.11
Nylon 66

13.3.11.1 Crystallization

The crystallization study of Nylon 66/clay nanocomposites by Wang et al. [114] showed that the clay platelets played a competing role in the crystallization process of Nylon 66. Although clay acted as a nucleating agent for the Nylon 66 matrix accelerating the crystallization rate, the presence of platelets also resulted in a retardation of the crystal/spherulite growth, especially for nanocomposites with a higher MMT content.

In Nylon 66/clay [115] nanocomposites, the presence of delaminated clay layers resulted in an heterogeneous nucleation, as evidenced by an increased crystallization temperature. Moreover, the spherulitic size was decreased dramatically and the symmetry of spherulites was destroyed due to the presence of MMT.

The isothermal crystallization results for nanocomposites of Nylon 66 with functionalized MWCNTs [116] showed that the crystallization rates were first increased, but then decreased with increasing MWCNT content. This observation was ascribed to the twofold effect of MWCNTs on the crystallization of Nylon 66. The presence of MWCNT provided heterogeneous nucleation sites for Nylon 66 crystallization, while the MWCNT network structure hindered large crystal growth.

Xu et al. [117] observed that silica nanoparticles with a modified surface effectively hindered the diffusion process of polymer chain segments and, at the same time, also acted as a nucleating agent. However, because these two competing processes were acting simultaneously, no distinct changes were observed in the crystallinity of Nylon 66 in the nanocomposites as compared to that of neat Nylon 66.

13.3.11.2 Polymorphism in Nylon 66

Nylon 66 exhibits two crystalline forms, namely the α-form (triclinic) and γ-form (pseudohexagonal) [118]. In the case of nanocomposites of Nylon 66 with organoclay, it was noted that the addition of MMT induced the γ-form of PA66 [115, 119].

13.3.12
Nylon 11

Zhang et al. [120] have observed that the addition of MMT induced a change in the crystalline form of Nylon 11, from the pseudohexagonal δ' form to the more stable hexagonal δ form. This was attributed to the restricted formation of hydrogen-bond sheets, forcing the amide groups of nylon out of the hydrogen-bond sheets in the presence of MMT. In another study, Zhang et al. [121] have confirmed that MMT both induced and stabilized the γ-form of Nylon 11.

13.3.13
Nylon 10,10

The crystallization and melting behavior of Nylon 10,10/MMT composites was reported by Zhang and Yan [122]. Their crystallization studies indicated that, at a low clay content, the crystallization rate was enhanced, whereas at a 10% clay content the physical hindrance of the clay layers resulted in a retardation of crystallization.

Zhang et al. [123] have reported that the degree of crystallinity of Nylon 10,10/clay nanocomposites remained almost constant, while the degree of supercooling decreased with the increase in clay content. This observation was attributed to the heterophase nucleation of Nylon 10,10 due to the presence of clay layers.

Zheng et al. have studied the crystallization of Nylon10,10/MWCNT nanocomposites, and reported that only the α-form of crystals were formed [124].

13.3.14
Polyvinylidene Fluoride (PVDF)

13.3.14.1 Crystallization

Priya and Jog have shown that polyvinylidene fluoride (PVDF)/clay composites exhibited a shift in the crystallization temperature to higher temperatures for all nanocomposites. This indicated that the crystallization process of PVDF in the nanocomposites was accelerated due to the presence of clay [125–127].

13.3.14.2 Polymorphism in PVDF

PVDF exhibits polymorphism, and is known to crystallize in four different forms. The most readily formed crystal structure, the α-form, is obtained when crystallized from the melt, while the β-form can be obtained by orientation of the α-form. The β-form is very important because of its piezoelectric and pyroelectric properties. Transformation of the polymer melt to the α-form is kinetically favored, whereas the β-form is favored for energetic reasons [128–131].

Priya and Jog were the first to report polymorphism in PVDF/clay nanocomposites when it was noted that, in the presence of organoclay, PVDF would crystallize into the β-form, which would remain unaltered even after thermal annealing [128]. The XRD patterns for PVDF and PVDF/clay nanocomposites are shown in Figure 13.5, where a single peak observed at about 20° confirmed the formation of β-form of PVDF in the nanocomposites. The observed β-form was ascribed to the crystallization of PVDF in constrained phase. In the case of PVDF, similar results have been reported by Buckley et al. [132], who studied the effect of clay content on the structure of nanocomposites of PVDF with organoclay. Here, the β-form was

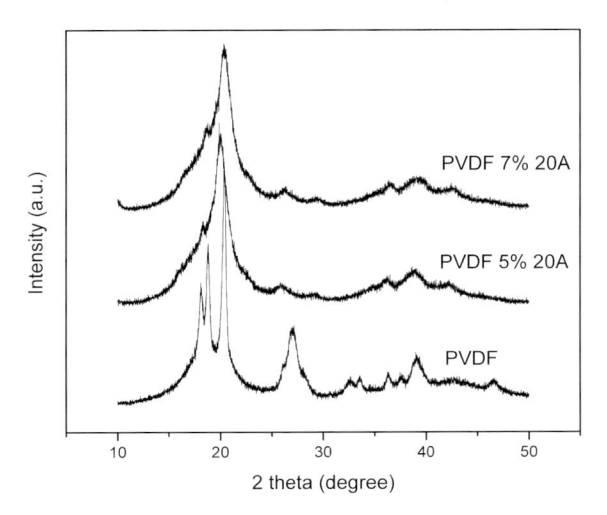

Figure 13.5 X-ray diffraction scans of polyvinylidene fluoride (PVDF) (α-form) and PVDF/clay nanocomposites (β-form).

observed to be dominant when the organoclay content exceeded 0.5%, but for an intermediate clay content (0.025–0.5%) the α- and β-forms coexisted, though the fraction of the β-form was higher than that of the α-form. Yet, for very low content of organoclay (<0.025%) the α-form was dominant. The presence of the organically modified silicate Lucentite STN also resulted in the β-form of PVDF [133]. Even in solution-blended nanocomposites of PVDF with clay, the nucleating effect of clay favored formation of the β-form [134]. In another study, the β-form of PVDF was seen to be present in all of the nanocomposites, regardless of the nanoclay morphology and contents [135].

Manna *et al.* [136] have reported that PVDF/silver nanocomposites exhibited the β-form of PVDF, while the Ag nanoparticles enhanced the nucleation process in PVDF/silver nanocomposites.

13.3.15
Syndiotactic Polystyrene (sPS)

13.3.15.1 Crystallization

Wu *et al.* [137] have reported that the incorporation of clay into syndiotactic polystyrene (sPS) resulted in an heterogeneous nucleation at a lower clay content (0.5%), while the activation energy was increased with increasing clay content; this suggested hindrance to the transport of polymer chains. Similar results were reported by Park *et al.* [138] for nanocomposites with an exfoliated structure. Here, the lower values of the degree of crystallinity for exfoliated nanocomposites were explained on the basis of a hindered crystal growth in the presence of delaminated clay layers.

13.3.15.2 Polymorphism in sPS

Tseng et al. [139] have confirmed that the presence of clay layers facilitated the formation of the thermodynamically more stable β-form when sPS/clay nanocomposites were crystallized from melt, or were cold-crystallized. In a study conducted by Wu et al. [140], it was revealed that formation of the β-form was favored at higher crystallization temperatures, but that the temperature of crystallization and clay influenced the contents of the α- and β-forms. Subsequently, α-crystal formation in sPS was observed in the presence of organoclays, even at high crystallization temperatures, whereas the β-form was formed in nanocomposites with unmodified clay, at all temperatures [141]. Wu et al. [142] reported that the presence of clay resulted in the α-crystalline form of sPS when the samples were quenched from melt. However, melt-crystallized samples showed the presence of a thermodynamically stable β-form of sPS. Similar results were reported by Tseng et al. for melt-blended nanocomposites of sPS and clay [143].

13.4
Conclusions

For most nanocomposites, the addition of nanofillers generally results in an enhanced nucleation at a lower nanofiller content. The few exceptions to this include PHA/fumed silica and Nylon6/modified silica nanocomposites, in which reduced crystallization rates have been observed. Such enhancement is generally observed for low nanofiller contents, while higher filler contents are reported to cause reduce growth rates.

The incorporation of nanofillers also resulted in the γ-form (clay), α-form (SWCNTs), and β-form (surface-treated $CaCO_3$) of PP in nanocomposites. The nanofillers induced the β-form (untreated clay, treated clay and MWCNTs) and α-form (organoclay) in PEN nanocomposites. It was observed that the crystallization of Nylon6/clay nanocomposites resulted in the γ-form, while the presence of sepiolite did not affect the crystal form. In the case of MWCNT, the crystallization form of Nylon 6 remained unchanged, while surface-modified MWCNTs favored formation of the γ-form. The incorporation of nanoparticles such as calcite and ZnO also induced the γ-form in Nylon 6. In the case of PVDF, clay and silver nanoparticles were reported to induce the β-form, whilst for sPS-based nanocomposites the addition of clay resulted in the β-form.

The reported results on the crystallization of polymer nanocomposites can be summarized as follows:

1) The inclusion of nanofillers enhanced nucleation due to a larger surface area for all the nanofillers at lower concentrations.
2) Higher concentrations of nanofillers had adverse effects on crystallization, due to a hindered transportation of the polymer chains.
3) Surface modifications resulted in stronger interactions between the polymer and the filler, which resulted in a retardation of crystallization.
4) Nanofillers induced changes in the crystalline structure.

References

1. LeBaron, P.C., Wang, Z., and Pinnavaia, T.J. (1999) *Appl. Clay Sci.*, **15**, 11–29.
2. Coleman, J.N., Khan, U., Blau, W.J., and Gunko, Y.K. (2006) *Carbon*, **44**, 1624–1652.
3. Zeng, Q.H., Yu, A.B., (Max) Lu, G.Q., and Paul, D.R. (2005) *J. Nanosci. Nanotechnol.*, **5**, 1574–1592.
4. Sinha Ray, S. and Bousmina, M. (2005) *Prog. Mater. Sci.*, **50**, 962–1079.
5. Ijima, S. (1991) *Nature*, **354**, 56–58.
6. Ajayan, P.M., Stephan, O., Colliex, C., and Trauth, D. (1994) *Science*, **265**, 1212–1214.
7. Avrami, M. (1939) *J. Chem. Phys.*, **7**, 1103–1112.
8. Avrami, M. (1940) *J. Chem. Phys.*, **8**, 212–224.
9. Avrami, M. (1941) *J. Chem. Phys.*, **9**, 177–184.
10. Caze, C., Devaux, E., Crespy, A., Carro, J.P., and Carrot, J.P. (1997) *Polymer*, **38**, 497–502.
11. Chuah, K.P., Gan, S.N., and Chee, K.K. (1998) *Polymer*, **40**, 253–259.
12. Di Lorenzo, M.L. and Silvestre, C. (1999) *Prog. Polym. Sci.*, **24**, 917–950.
13. Hambir, S., Bulakh, N., Kodgire, P., Kalgaonkar, R., and Jog, J.P. (2001) *J. Polym. Sci., Part B: Polym. Phys.*, **39**, 446–450.
14. Hambir, S., Bulakh, N., and Jog, J.P. (2002) *Polym. Eng. Sci.*, **42**, 1800–1807.
15. Kodgire, P., Kalgaonkar, R., Hambir, S., Bulakh, N., and Jog, J.P. (2001) *J. Appl. Polym. Sci.*, **81**, 1786–1792.
16. Wanga, L. and Shenga, J. (2005) *Polymer*, **46**, 6243–6249.
17. Zhao, L., Du, Q., Jiang, G., and Guo, S. (2007) *J. Polym. Sci., Part B: Polym. Phys.*, **45**, 2300–2308.
18. Bhattacharya, A.R., Sreekumar, T.V., Liu, T., Kumar, S., Ericson, L.M., Hauge, R.H., and Smalley, R.E. (2003) *Polymer*, **44**, 2373–2377.
19. Valentini, L., Biagiotti, J., Kenny, J.M., and Santucci, S. (2003) *Compos. Sci. Technol.*, **63**, 1149–1153.
20. Grady, B.P., Pompeo, F., Shambaugh, R.L., and Resaco, D.E. (2002) *J. Phys. Chem. B*, **106**, 5852–5858.
21. Lopez Manchado, M.A., Valentini, L., Biagiotti, J., and Kenny, J.M. (2005) *Carbon*, **43**, 1499–1505.
22. Kovalchuk, A.A., Shevchenko, V.G., Shchegolikhin, A.N., Nedorezova, P.M., Klyamkina, A.N., and Aladyshev, A.M. (2008) *J. Mater. Sci.*, **43**, 7132–7140.
23. Peneva, Y., Valcheva, M., Minkova, L., Mi Cusik, M., and Omastov, M. (2008) *J. Macromol. Sci., Part B: Phys.*, **47**, 1197–1210.
24. Bao, S.P. and Tjong, S.C. (2008) *Mater. Sci. Eng. A*, **485**, 508–516.
25. Seo, M.K., Lee, J.R., and Park, S.J. (2005) *Mater. Sci. Eng. A*, **404**, 79–84.
26. Danny, J.Y., Page, S., and Gopakumar, T.G. (2006) *Polym. J.*, **38**, 920–929.
27. Streller, R.C., Thomann, R., Torno, O., and Mulhaupt, R. (2008) *Macromol. Mater. Eng.*, **293**, 218–227.
28. Vladimirov, V., Betchev, C., Vassiliou, A., Papageorgiou, G., and Bikiaris, D. (2006) *Compos. Sci. Technol.*, **66**, 2935–2944.
29. Papageorgiou, G.Z., Achilias, D.S., Bikiaris, D.N., and Karayannidis, G.P. (2005) *Thermochim. Acta*, **427**, 117–128.
30. Tao, R., Liu, Z.Y., Yang, W., and Yang, M.B. (2008) *Polym.-Plastics Technol. Eng.*, **47**, 490–495.
31. Kalgaonkar, R.A., and Jog, J.P. (2007) *Nanofibers and Nanotechnology in Textiles* (eds P.J. Brown and K. Stevens), CRC Press, Woodhead Publishing Ltd, Cambridge, England.
32. Cheng, S.Z.D., Janimak, J.J., and Rodriguez, J. (1995) Crystalline structure of polypropylene homo-and copolymers, in *Polypropylene: Structure, Blends and Composites* (ed. J. Karger-Kocsis), Chapman & Hall, London, pp. 31–40.
33. Turner-Jones, A., Aizlewood, J.M., and Beckett, D.R. (1964) *Makromol. Chem.*, **75**, 134.
34. Lincoln, D.M., Vaia, R.A., Wang, Z.G., Hsiao, B.S., and Krishnamoorthi, R. (2001) *Polymer*, **42**, 9975–9985.
35. Nam, P.H., Maiti, P., Okamoto, M., Kotaka, T., Hasegawa, N., and Usuki, A. (2001) *Polymer*, **42**, 9633–9640.

36 Medellin-Rodriguez, F.J., Mata-Padilla, J.M., Hsiao, B.S., Waldo-Mendoza, M.A., Ramirez-Vargas, E., and Sanchez-Valdes, S. (2007) *Polym. Eng. Sci.*, **47**, 1889–1897.

37 Weawkamol, L., That, M.T., Sarazin, F.P., Cole, K.C., Denault, J., and Simard, B. (2005) *J. Polym. Sci., Part B: Polym. Phys.*, **43**, 2445–2453.

38 Wan, W., Yu, D., Xie, Y., Guo, X., Zhou, W., and Cao, J. (2006) *J. Appl. Polym. Sci.*, **102**, 3480–3488.

39 Xueqin, G., Jing, X., Zhanchun, C., Cong, D., and Kaizhi, S. (2008) *Polym. Int.*, **57**, 23–27.

40 Wanjale, S.D. and Jog, J.P. (2003) *J. Macromol. Sci., Part B: Phys.*, **42**, 1141–1152.

41 Wanjale, S.D. and Jog, J.P. (2003) *J. Polym. Sci., Part B: Polym. Phys.*, **41**, 1014–1021.

42 Causin, V., Marega, C., Marigo, A., Ferrara, G., Idiyatullina, G., and Fantine, F. (2006) *Polymer*, **47**, 4773–4780.

43 Wanjale, S.D. and Jog, J.P. (2006) *Polymer*, **47**, 6414–6421.

44 Wanjale, S.D. and Jog, J.P. (2008) *Polyolefin Composites* (eds D. Nwabunma and T. Kyu), John Wiley & Sons, Inc.

45 Wu, D., Zhou, C., Fan, X., Mao, D., and Bian, Z. (2006) *J. Appl. Polym. Sci.*, **99**, 3257–3265.

46 Huang, J.W. (2008) *J. Appl. Polym. Sci.*, **110**, 2195–2204.

47 Huang, J.W. (2008) *J. Polym. Sci., Part B: Polym. Phys.*, **46**, 564–576.

48 Chen, X., Xu, J., Lu, H., Yang, Y. (2006) *J. Polym. Sci., Part B: Polym. Phys.*, **44**, 2112–2121.

49 Al-Mulla, A., Mathew, J., and Shanks, R. (2007) *J. Polym. Sci., Part B: Polym. Phys.*, **45**, 1344–1353.

50 Wan, T., Chen, L., Chua, Y.C., and Lu, X. (2004) *J. Appl. Polym. Sci.*, **94**, 1381–1388.

51 Bandopadhyay, J., Ray, S.S., and Bousmina, M. (2008) *J. Nanosci. Nanotechnol.*, **8**, 1812–1822.

52 Chung, J.W., Son, S.B., Chun, S.W., Kang, T.J., and Kwak, S.Y. (2008) *J. Polym. Sci., Part B: Polym. Phys.*, **46**, 989–999.

53 Hwang, S.Y., Lee, W.D., Lim, J.S., Park, K.H., and Im, S.S. (2008) *J. Polym. Sci., Part B: Polym. Phys.*, **46**, 1022–1035.

54 Yuan, X., Li, C., Guan, G., Liu, X., Xiao, Y., and Zhang, D. (2007) *J. Appl. Polym. Sci.*, **103**, 1279–1286.

55 Wang, Y., Deng, J., Wang, K., Zhang, Q., and Fu, Q. (2007) *J. Appl. Polym. Sci.*, **104**, 3695–3701.

56 Tzavalas, S., Mouzakis, D.E., Drakonakis, V., and Gregoriou, V.G. (2008) *J. Polym. Sci., Part B: Polym. Phys.*, **46**, 668–676.

57 Tian, X., Ruan, C., Cui, P., Liu, W., Zheng, J., Zhang, X., Yao, X., Zheng, K., and Yong, L. (2006) *J. Macromol. Sci., Part B: Phys.*, **45**, 835–848.

58 Bhimaraj, P., Yang, H., Siegel, R.W., and Schadler, L.S. (2007) *J. Appl. Polym. Sci.*, **106**, 4233–4240.

59 Ou, C.F. (2003) *J. Appl. Polym. Sci.*, **89**, 3315–3322.

60 Ou, C.F. (2003) *J. Polym. Sci., Part B: Polym. Phys.*, **41**, 2902–2910.

61 Hu, X. and Lesser, A.J. (2003) *J. Polym. Sci., Part B: Polym. Phys.*, **41**, 2275–2289.

62 Liu, Z., Chen, K., and Yan, D. (2003) *Eur. Polym. J.*, **39**, 2359–2366.

63 Xu, Y., Jia, H.B., Piao, J.N., Ye, S.R., and Huang, J. (2008) *Mater. Sci.*, **43**, 417–421.

64 Run, M., Yao, C., Wang, Y., and Gao, J. (2007) *J. Appl. Polym. Sci.*, **106**, 1557–1567.

65 Wu, T.M. and Liu, C.Y. (2005) *Polymer*, **46**, 5621–5629.

66 Li, G., Shi, Y., Fan, S., and Yu, X. (2007) *J. Macromol. Sci., Part B: Phys.*, **46**, 1231–1245.

67 Kim, J.Y., Park, H.S., and Kim, S.H. (2006) *Polymer*, **47**, 1379–1389.

68 Kim, J.Y., Kim, S.H., Kang, S.W., and Chang, J.H. (2006) *Macromol. Res.*, **14**, 146–154. Ahn

69 Kim, S.H., Ahn, S.H., and Hirai, T. (2003) *Polymer*, **44**, 5625–5634.

70 Vasanthan, N. and Salem, D.R. (1999) *Macromolecules*, **32**, 6319–6325.

71 Chua, Y.C., and Lu, X., Wan, T. (2006) *J. Polym. Sci., Part B: Polym. Phys.*, **44**, 1040–1049.

72 Kim, J.Y., Sang-Il, H., and Kim, S.H. (2007) *Polym. Eng. Sci.*, **47**, 1715–1723.
73 Krikorian, V. and Pochan, D.J. (2004) *Macromolecules*, **37**, 6480–6491.
74 Fujimori, A., Ninomiya, N., and Masuko, T. (2008) *Polym. Eng. Sci.*, **48**, 1103–1111.
75 Shieh, Y.T. and Liu, G.L. (2007) *J. Polym. Sci., Part B: Polym. Phys.*, **45**, 1870–1881.
76 Yan, S., Yin, J., Yang, Y., Dai, Z., and Ma, J., Chen, X. (2007) *Polymer*, **48**, 1688–1694.
77 Andricic, B., Kovacic, T., Perinovic, S., and Grgic, A. (2008) *Macromol. Symp.*, **263**, 96–101.
78 Chen, G.X., Hao, G.J., Guo, T.Y., Song, M.D., and Zhang, B.H. (2004) *J. Appl. Polym. Sci.*, **93**, 655–661.
79 Bordes, P., Pollet, E., Bourbigot, S., and Averous, L. (2008) *Macromol. Chem. Phys.*, **209**, 1473–1484.
80 Yun, S.I., Gadd, G.E., Latella, B.A., Lo, V., Russell, R.A., and Holden, P.J. (2008) *Polym. Bull.*, **61**, 267–275.
81 Lai, M., Li, J., Yang, J., Liu, J., Tong, X., and Cheng, H. (2004) *Polym. Int.*, **53**, 1479–1484.
82 Miloaga, D.G., Hosein, H.A., Misra, M., and Drzal, L.T. (2007) *J. Appl. Polym. Sci.*, **106**, 2548–2558.
83 Lim, J.S., Noda, I., and Im, S.S. (2007) *Polymer*, **48**, 2745–2754.
84 Tang, C.Y., Chen, D.Z., Tsui, C.P., Uskokovic, P.S., Yu, P.H.F., and Leung, M.C.P. (2006) *J. Appl. Polym. Sci.*, **102**, 5388–5395.
85 Kuo, M.C., Huanga, J.C., and Chen, M. (2006) *Mater. Chem. Phys.*, **99**, 258–268.
86 Kuo, M.C., Tsai, C.M., Huang, J.C., and Chen, M. (2005) *Mater. Chem. Phys.*, **90**, 185–195.
87 Lincoln, D.M., Vaia, R.A., and Krishnamoorti, R. (2004) *Macromolecules*, **37**, 4554–4561.
88 Fornes, T.D. and Paul, D.R. (2003) *Polymer*, **44**, 3945–3961.
89 Yue, S., Gong, W., Qi, N., and Wang, B. (2008) *J. Appl. Polym. Sci.*, **110**, 3149–3155.
90 Wu, T.M., Lien, Y.H., and Hsu, S.F. (2004) *J. Appl. Polym. Sci.*, **94**, 2196–2204.
91 Chen, E.C. and Wu, T.M. (2008) *J. Polym. Sci., Part B: Polym. Phys.*, **46**, 158–169.
92 Weng, W., Chen, G., and Wu, D. (2003) *Polymer*, **44**, 8119–8132.
93 Chae, D.W., Oh, S.G., and Kim, B.C. (2004) *J. Polym. Sci., Part B: Polym. Phys.*, **42**, 790–799.
94 Yang, F., Ou, Y., and Yu, Z. (2006) *High Perform. Polym.*, **18**, 355–375.
95 Rusu, G. and Rusu, E. (2006) *High Perform. Polym.*, **18**, 355–375.
96 Miyasaka, K. and Ishikawa, K. (1968) *J. Polym. Sci., Part B: Polym. Phys.*, **6**, 1317–1329.
97 Li, Y. and Goddard, W.A., III (2002) *Macromolecules*, **35**, 8440–8455.
98 Wu, T.M., and Liao, C.S. (2000) *Macromol. Chem. Phys.*, **201**, 2820–2825.
99 Wu, T.M. and Chen, C.E. (2002) *Polym. Eng. Sci.*, **42**, 1141–1150.
100 Xie, S., Zhang, S., Liu, H., Chen, G., Feng, M., Qin, H., Wang, F., and Yan, M. (2005) *Polymer*, **46**, 5417–5427.
101 Liu, L., Qi, Z., and Zhu, X. (1999) *J. Appl. Polym. Sci.*, **71**, 1133–1138.
102 Li, T.-C., Ma, J., Wang, M., Tjiu, W.C., Liu, T., and Huang, W. (2007) *J. Appl. Polym. Sci.*, **103**, 1191–1199.
103 Ma, C.M., Kuo, C.T., Kuan, H.-C., and Chiang, C.-L. (2003) *J. Appl. Polym. Sci.*, **88**, 1686–1693.
104 Maiti, P. and Okamoto, M. (2003) *Macromol. Mater. Eng.*, **288**, 440–445.
105 Hu, X. and Zhao, X. (2004) *Polymer*, **45**, 3819–3825.
106 Zhao, X.Y. and Wang, M.Z. (2006) *J. Appl. Polym. Sci.*, **100**, 3116–3122.
107 Xie, S., Zhang, S., Wang, F., Yang, M., Seguela, R., and Lefebvre, J.M. (2007) *Compos. Sci. Technol.*, **67**, 2334–2341.
108 Wu, T.M., Chiang, C.E., and Liao, C.S. (2002) *Polym. Eng. Sci.*, **42**, 141–1150.
109 Phang, I.Y., Ma, J., Shen, L., Liu, T., and Zhang, W.D. (2006) *Polym. Int.*, **55**, 71–79.
110 Li, J., Fang, Z., Tong, L., Gu, A., and Liu, F. (2006) *J. Polym. Sci., Part B: Polym. Phys.*, **44**, 1499–1512.
111 Kazimierczak, C.O., Zdilek, K., and Picken, S.J. (2005) *Polymer*, **46**, 6025–6034.

112 Avella, M., Errico, M.E., and Gentile, G. (2006) *Macromol. Symp.*, **234**, 170–175.
113 Zheng, J., Siegel, R.W., and Toney, C.G. (2003) *J. Polym. Sci., Part B: Polym. Phys.*, **41**, 1033–1050.
114 Wang, B.M.Q., Wang, H., and Jian, L. (2007) *J. Polym. Sci., Part B: Polym. Phys.*, **46**, 1093–1104.
115 Zhang, Q.X., Yu, Z.Z., Ma, M.Y.J., and Mai, Y.W. (2003) *J. Polym. Sci., Part B: Polym. Phys.*, **41**, 2861–2869.
116 Li, L., Christopher, Y.L., Ni, C., Rong, L., and Hsiao, B. (2007) *Polymer*, **48**, 3452–3460.
117 Xu, X., Li, B., Lu, H., Zhang, Z., and Wang, H. (2008) *J. Appl. Polym. Sci.*, **107**, 2007–2014.
118 Colclough, M.L. and Baker, R. (1978) *J. Mater. Sci.*, **13**, 2531–2540.
119 Song, L., Hu, Y., He, Q., and You, F. (2008) *J. Fire Sci.*, **26**, 475–492.
120 Zhang, X., Yang, G., and Lin, J. (2006) *J. Appl. Polym. Sci.*, **102**, 5483–5489.
121 Zhang, G., Li, Y., and Yan, D. (2003) *J. Polym. Sci., Part B: Polym. Phys.*, **42**, 253–259.
122 Zhang, G. and Yan, D. (2003) *J. Appl. Polym. Sci.*, **88**, 2181–2188.
123 Zhang, G., Li, Y., Yan, D., and Xu, Y., Zhang, G., Li, Y., Yan, D., and Xu, Y. (2003) *Polym. Eng. Sci.*, **43**, 204–213.
124 Zeng, H., Gao, C., Wang, Y., Watts, P.C.P., Kong, H., Cui, X., and Yan, D. (2006) *Polymer*, **47**, 113–122.
125 Priya, L. and Jog, J.P. (2003) *J. Polym. Sci., Part B: Polym. Phys.*, **41**, 31–38.
126 Priya, L. and Jog, J.P. (2002) *J. Polym. Sci., Part B: Polym. Phys.*, **40**, 1682–1689.
127 Priya, L. and Jog, J.P. (2003) *J. Appl. Polym. Sci.*, **89**, 2036–2040.
128 Gregorio, R., Jr (2006) *J. Appl. Polym. Sci.*, **100**, 3272–3279.
129 Lovinger, A.J. (1982) *Developments in Crystalline Polymers* (ed. D.C. Basset), Applied Science Publishers Ltd.
130 Bachmann, M. and Lando, J. (1981) *Macromolecules*, **14**, 40–46.
131 Gregorio, R., Jr, Cestari, M., Chaves, N., Nociti, P.S., Mendonca, J.A., and Lucas, A.A. (1996) *The Polymeric Materials Encyclopedia*, CRC Press, Boca Raton, FL.
132 Buckley, J., Cebe, P., Cherdack, D., Crawford, J., Ince, B.S., Jenkins, M., Pan, J., Reveley, M., Washington, N., and Wolchover, N. (2006) *Polymer*, **47**, 2411–2422.
133 Ramasundaram, S., Yoon, S., Jin Kim, K., and Park, C. (2008) *J. Polym. Sci., Part B: Polym. Phys.*, **46**, 2173–2187.
134 Causin, V., Carraro, M.L., Marega, C., Saini, R., Campestrini, S., and Marigo, A. (2008) *J. Appl. Polym. Sci.*, **109**, 2354–2361.
135 Dillon, D.R., Tenneti, K.K., Li, C.Y., Ko, F.K., Sics, I., and Hsiao, B.S. (2006) *Polymer*, **47**, 1678–1688.
136 Manna, S., Batabyal, S.K., and Nandi, A.K. (2006) *J. Phys. Chem. B*, **110**, 12318–12326.
137 Wu, T.M., Hsu, S.-F., Chien, C.-F., and Wu, J.-Y. (2004) *Polym. Eng. Sci.*, **44**, 2288–2297.
138 Park, C., Choi, W.M., Kim, M.H., and Park, O. (2004) *J. Polym. Sci., Part B: Polym. Phys.*, **42**, 1685–1693.
139 Tseng, C.R., Wu, J.Y., Lee, H.Y., and Chang, F.C. (2001) *Polymer*, **42**, 10063–10070.
140 Wu, T.M., Hsu, S.F., and Wu, J.Y. (2002) *Polym. Eng. Sci.*, **42**, 2295–2305.
141 Ghosh, A.K. and Woo, E.M. (2004) *Polymer*, **45**, 4749–4759.
142 Wu, T.M., Hsu, S.F., and Wu, J.Y. (2002) *J. Polym. Sci., Part B: Polym. Phys.*, **40**, 736–746.
143 Tseng, C.R., Lee, H.Y., and Chang, F.C. (2001) *J. Polym. Sci., Part B: Polym. Phys.*, **39**, 2097–2107.

14
Prediction of the Mechanical Properties of Nanocomposites
Qinghua Zeng and Aibing Yu

14.1
Introduction

14.1.1
Nanocomposites

During the past two decades, nanocomposites have emerged as a new class of materials that have attracted considerable interest in their scientific development and engineering applications [1–13]. Nanostructures generated through the introduction of discrete nanoscale fillers of different structural, chemical, and physical properties into a continuous solid matrix (e.g., metal, ceramics, or polymer) represent radical alternatives to the structure of conventional particle-reinforced composites. Such structural features, together with the unique properties of nanoparticles, lead to many new phenomena, exceptional properties, and promising applications [14–18]. In the past, many nanocomposites have been developed, some of which are now commercially available [19, 20]. Some characteristic features and potential applications of these nanocomposites, including metal–matrix nanocomposites, ceramic–matrix nanocomposites, and polymer–matrix nanocomposites, are summarized in Table 14.1.

14.1.2
Some Issues in Nanocomposites

Despite such success, further developments in nanocomposites have been hindered because of the poor understanding of some key fundamental issues, such as the uniform dispersion of nanoparticles, the well-controlled and quantified interface, the rationale design, and the prediction of properties.

14.1.2.1 Dispersion of Nanoparticles
The promise of improved and novel properties of nanocomposites depends largely on the degree of nanoparticle dispersion in the matrix. Given a uniform dispersion, the large specific surface area of nanofillers will produce a large interfacial

Table 14.1 Characteristics and potential applications of nanocomposites.

Nanocomposite type	Matrix	Nanofiller	Characteristics	Applications
Metal–matrix nanocomposite	Aluminum, magnesium, copper, nickel, iron	Nanotube, boride, carbide, nitride, oxide, their mixture	High strength to shear and compression High fracture toughness High abrasion High creep resistance High thermal conductivity	High-temperature applications (e.g., aerospace, automotive, structural materials).
Ceramic–matrix nanocomposite	Alumina, magnesia, mullite, cordierite, silicon nitride, silicon carbide	Silicon carbide, silicon nitride, titanium nitride, metallic particles	High strength, high stiffness, resilience to oxidation and to high temperature from ceramic matrix High fracture toughness from ceramic reinforcements Improved high-temperature mechanical, magnetic, electric, and optical properties	Automotive engines Cutting tools Heat engine components Wear-resistant parts Aerospace components Flexible superconducting wire Fiber-optic connector components Ballistic armor
Polymer-matrix nanocomposite	Thermoplastic, thermoset, elastomer	Carbon black, fumed silica, clay, carbon nanotube, graphite	Extraordinary mechanical properties (e.g., tensile strength, Young's modulus, yield stress) Superior gas-barrier properties Improved thermal stability Increased fire retardancy Improved optical properties	Automotive parts (e.g., timing belt, exterior step assist) Packaging materials (e.g., food, drink, beer) Coatings and adhesives Fire retardants Others (e.g., drug delivery, sensors, medical devices)

area per unit volume between the nanofiller and matrix, which fundamentally differentiates nanocomposites from conventional composites. Recently, a variety of approaches have been used to improve the dispersion of nanoparticles in the matrix, including the modification of nanoparticle surfaces, the *in situ* synthesis of nanoparticles, and the functionalization of polymer chains. Yet, it is very difficult to achieve an optimal dispersion of nanoparticles because of their active surface atoms and their tendency to form agglomerates (e.g., nanoparticle aggregates, carbon nanotube (CNT) bundles, and layered clays in the form of platelet stacks), their chemical dissimilarity with the matrix, and the large amount of defects and an uncertainty of the nanoparticles' properties.

14.1.2.2 Interface

Interfaces are important for many properties and applications of multiphase materials. This is particularly true for particle-reinforced polymer composites, where the interfacial characteristics (e.g., adhesion strength) between the particles and the polymer play a crucial role in the load transfer and mechanical properties. In polymer composites, the interface can be defined as the region beginning at the point of a particle at which the properties differ from those of the bulk particle, and ending at that point in the matrix at which the properties become equal to those of the bulk matrix. With the decrease in particle size from micrometer to nanometer range, the volume fraction of interfacial region is significantly increased, which in turn mainly determines the many mechanical properties of the polymer nanocomposites. Thus, it is important to quantify the properties of the interfacial region and its contribution to the overall properties of polymer nanocomposites. Experimentally, some potential approaches have been developed to characterize the interfacial region, including volume strain measurements, infrared microscopy, dielectric measurements, neutron scattering, nuclear magnetic resonance (NMR), differential scanning calorimetry (DSC) measurements, and mechanical measurements [21].

14.1.2.3 Crystallization

It is well known that the behavior of polymer chains around a particle surface differs from that of bulk polymer chains. Indeed, it has been shown in many studies that such a polymer is often semi-crystalline. Depending on the nanofiller–polymer interactions, the polymer in the vicinity of the nanofiller may be developed into new crystal structures (e.g., the γ-crystal phase of polyamide in clay–nylon 6 nanocomposites), amorphized (e.g., clay–polyethylene nanocomposites), or heterogeneously nucleated (e.g., clay–polyethylene terephthalate nanocomposites). This crystallization of the polymer will lead to dramatic effects on the overall properties of the polymer nanocomposites. Consequently, when predicting the properties of a nanocomposite the morphological features and properties of the interfacial polymer must be taken into account, in addition to the properties of individual nanoparticles and bulk polymer.

14.1.3
Property Predictions

The efficient design of new materials for specific properties and applications requires predictions to be made of the properties of the candidate materials, and the use of these predictions to evaluate, screen, and guide the material synthesis. In order to meet this challenge, it is necessary to extend, significantly, the existing prediction methods for polymer composites and to develop new prediction approaches specifically for nanocomposites. The prediction of nanocomposite properties requires some important information, such as the properties of the individual components (i.e., nanofiller and polymer), the processing methods and conditions, the structure and morphology of the nanocomposites and, more importantly, the nature of the interfacial region.

The prediction methods for nanocomposite properties are available at many different levels of complexity and at different length and time scales, such as molecular, microscale, mesoscale, and macroscale. These methods can be divided into two general classes, namely analytical models and numerical simulations:

- **Analytical models** have the advantage of greater simplicity and require minimal computational resources. However, this simplicity and ease of application may compromise the level of detail that can be taken into account, and also the limited accuracy of the predictions. Thus, analytical models such as classical composite theory are mainly used for quick preliminary parametric screening studies, based on an idealized morphology rather than on the detailed morphology of the system concerned.

- **Numerical simulations**, on the other hand, can provide more accurate predictions because their models are based on more realistic physics and detailed morphology of the system of interest. With the increasing need for an understanding and prediction of nanocomposite properties, increasingly sophisticated numerical simulations will play an important role as they are developed further and integrated more seamlessly with each other. It is highly likely that current advances in computer hardware, simulation techniques, and software development will make it possible to perform rapid numerical simulations with enhanced power and accuracy.

In the remainder of this chapter, recent advances towards the fundamental understanding of nanocomposites will be reviewed. In particular, some promising theoretical and numerical techniques, that are capable of addressing such fundamentals, will be described briefly. Their applications in predicting the mechanical properties of nanocomposites will then be outlined.

14.2
Analytical and Numerical Techniques

14.2.1
Analytical Models

Analytical (or theoretical) models are mathematical models that have a closed-form solution to the equations used to describe changes in a system. Some analytical models have been developed for highly specific applications, and others for general applications. In materials science, micromechanical models are normally developed to analyze the composite or heterogeneous materials on the level of individual constituents. Given the properties of the constituents, micromechanical models can used to predict the properties of the composite materials. Thus, a central theme of micromechanical models is the development of a representative volume element (RVE) to statistically represent the local continuum properties. The RVE may then be used in a repeating or periodic nature in the full-scale model. A micromechanical method can account for interfaces between constituents, for discontinuities, and for coupled mechanical and nonmechanical properties. Notably, such methods have been widely used to estimate the mechanical and transport properties of composite materials.

For polymer nanocomposites, these analytical models are still preferred due to their predictive power, their low cost of easy use, and their reasonable accuracy for some simplified structures. Recently, these analytical models have been extended to estimate the mechanical and physical properties of polymer nanocomposites. Despite their success in predicting the macroscopic behavior of fiber-reinforced composites, the direct use of micromechanical models for nanotube-reinforced composites is doubtful due to the significant scale differences between nanotubes and typical carbon fibers. Thus, two methods have recently been proposed for modeling the mechanical behavior of single-walled carbon nanotube (SWCNT) polymer composites, namely an equivalent-continuum approach and a self-similar approach [22, 23]. A brief introduction to the different analytical models, including the rule of mixtures, the Halpin–Tsai and Mori–Tanaka models, the equivalent-continuum approach, and the self-similar approach is provided in the following sections.

14.2.1.1 Rule of Mixtures

The rule of mixtures is the simplest and most intuitive approach to estimate (approximately) the properties of composite materials. It is based on an assumption that the properties of a composite material are a function of the volume fraction of each constituent phase in the composite. Accordingly, the properties of filled composite materials are estimated based on the properties of filler and matrix, as shown in Table 14.2.

14 Prediction of the Mechanical Properties of Nanocomposites

Table 14.2 Property prediction of composite materials by the rule of mixtures.

Property	Equation
Density	$d_c = d_m v_m + d_f v_f$
Coefficient of thermal expansion	$\alpha = v_f \alpha_f + v_m \alpha_m$
Modulus of elasticity	$E = v_f E_f + v_m E_m$
Average strain	$\bar{\sigma} = v_f \bar{\sigma}_f + v_m \bar{\sigma}_m$
Average stress	$\bar{\varepsilon} = v_f \bar{\varepsilon}_f + v_m \bar{\varepsilon}_m$
Poisson's ratio	$\mu = v_f \mu_f + v_m \mu_m$
Thermal conductivity	$K = v_f K_f + v_m K_m$
Electrical conductivity	$s = v_f s_f + v_m s_m$
Gas diffusivity	$D = v_f D_f + v_m D_m$

v, α, $\bar{\sigma}$, $\bar{\varepsilon}$, μ, K, s, D are the volume fractions, coefficient of thermal expansion, average strains, average stresses, Poisson's ratio, thermal conductivity, electrical conductivity, gas diffusivity, respectively. The subscripts f and m refer to the filler and the matrix, respectively.

14.2.1.2 Halpin–Tsai Model

The Halpin–Tsai model is a well-known composite theory for predicting the stiffness of unidirectional composites as a functional of the aspect ratio. In this model, the longitudinal E_{11} and transverse E_{22} engineering moduli are expressed in the following general form:

$$\frac{E}{E_m} = \frac{1 + \xi \eta v_f}{1 - \eta v_f} \tag{14.1}$$

where E and E_m represent the Young's modulus of the composite and matrix, respectively, v_f is the volume fraction of filler, and η is given by

$$\eta = \frac{E_f/E_m - 1}{E_f/E_m + \xi_f} \tag{14.2}$$

where E_f represents the Young's modulus of the filler, and ξ_f is the shape parameter which depends on the filler geometry and loading direction. When calculating the longitudinal modulus E_{11}, ξ_f is equal to l/t, and when calculating the transverse modulus E_{22}, ξ_f is equal to w/t. Here, the parameters of l, w, and t are the length, width, and thickness of the dispersed fillers, respectively. If $\xi_f \to 0$, the Halpin–Tsai theory converges to the inverse rule of mixtures (lower bound). Conversely, if $\xi \to \infty$, then the theory reduces to the rule of mixtures (upper bound).

14.2.1.3 Mori–Tanaka Model

The Mori–Tanaka model is derived based on the principles of Eshelby's inclusion model for predicting an elastic stress field in and around an ellipsoidal filler in an infinite matrix. The complete analytical solutions for longitudinal E_{11} and transverse E_{22} elastic moduli of an isotropic matrix filled with aligned spherical inclusion are:

$$\frac{E_{11}}{E_m} = \frac{A_0}{A_0 + v_f(A_1 + 2v_0 A_2)} \tag{14.3}$$

$$\frac{E_{22}}{E_m} = \frac{2A_0}{2A_0 + v_f(-2A_3 + (1-v_0)A_4 + (1+v_0)A_5 A_0)} \tag{14.4}$$

where v_0 is the Poisson's ratio of the matrix, and the parameters, $A_0, A_1, \ldots, A_5$ are functions of the Eshelby's tensor and the properties of the filler and the matrix, including Young's modulus, Poisson's ratio, filler concentration, and filler aspect ratio [24].

14.2.1.4 Equivalent-Continuum Approach

In the equivalent-continuum approach, which was proposed by Odegard et al. [25], a molecular dynamics method is used to model the molecular interactions between SWCNTs and a polymer. As a consequence, an homogeneous equivalent-continuum reinforcing element (e.g., a nanotube surrounded by a cylindrical volume of polymer) is constructed. Finally, micromechanics are used to determine the effective bulk properties of the equivalent-continuum reinforcing element embedded in a continuous polymer.

14.2.1.5 Self-Similar Approach

The self-similar approach, as proposed by Pipes and Hubert [26], consists of three major steps. First, a helical array of SWCNTs is assembled; this array is termed a nanotube "nanoarray," where 91 nanotubes make up the cross-section of the helical nanoarray. The nanotube nanoarrays are then surrounded by a polymer matrix and assembled into a second twisted array, termed a nanotube "nanowire." Finally, the nanotube nanowires are further impregnated with a polymer matrix and assembled into the final helical array-nanotube "microfiber." The self-similar geometries described in the nanoarray, nanowire, and microfiber allow the same mathematical and geometric model to be used for all three geometries [26].

14.2.2
Numerical Methods

Molecular modeling refers to the theoretical methods and computational techniques to model the behavior of molecules. It has been widely used in the fields of chemistry, biology, and materials science for studying molecular systems ranging from small chemical systems to large biological molecules and material assemblies. The most popular techniques include molecular mechanics, molecular dynamics, and Monte Carlo simulations. Today, these techniques are used routinely to investigate the structure, dynamics, and thermodynamics of inorganic, biological, and polymer systems. For nanocomposites, molecular modeling can be used to predict the thermodynamic and kinetic properties of nanofiller–matrix mixtures, interfacial molecular structure and interactions, molecular dynamic properties, and mechanical properties (e.g., Young's modulus, tensile strength, stress–strain relationship, load transfer, viscoelasticity, and thermoelasticity).

Microscale modeling aims to bridge molecular methods and continuum methods, and to avoid their shortcomings. Specifically, in nanoparticle–polymer systems, the study of structural evolution involves the description of bulk flow and the interactions between nanoparticle and polymer components. It should be noted that hydrodynamic behavior is relatively straightforward to handle by continuum methods, but is very difficult and expensive to treat when using atomistic methods. In contrast, the interactions between components can be examined at an atomistic level but are usually not straightforward to incorporate at the continuum level. Therefore, a variety of microscale methods have been proposed to study the microscopic structure and phase separation of polymer nanocomposites, including Brownian dynamics, dissipative particle dynamics, lattice Boltzmann, time-dependent Ginzburg–Landau theory, and dynamic density functional theory (DFT). These methods usually treat a polymer system with a field description or microscopic particles that incorporate molecular details implicitly; therefore, they are able to simulate the phenomena at length and time scales that are inaccessible by molecular modeling.

Despite the importance of understanding the molecular structure and nature of materials, their behavior can be homogenized with respect to different aspects and at different scales. Typically, the observed macroscopic behavior is usually explained by ignoring the discrete atomic and molecular structure, and assuming that the material is continuously distributed throughout its volume. The continuum material is thus assumed to have an average density and can be subjected to body forces such as gravity and surface forces. Generally speaking, continuum methods at macroscale obey the fundamental laws of continuity, equilibrium, the moment of momentum principle, conservation of energy, and conservation of entropy. These laws provide the basis for the continuum models, and must be coupled with the appropriate constitutive equations and equations of state to provide all the equations necessary for solving a continuum problem. In the following sections, some numerical methods of different scales that have been applied in polymer nanocomposites are briefly introduced.

14.2.2.1 Molecular Dynamics

Molecular dynamics (MD) is a computer simulation technique that allows one to predict the time evolution of a system of interacting particles (e.g., atoms, molecules, granules) and to estimate the relevant physical properties. Specifically, it generates such information as atomic positions, velocities, and forces from which the macroscopic properties (e.g., pressure, energy, heat capacities) can be derived by means of statistical mechanics. MD simulation usually consists of three constituents: (i) a set of initial conditions (e.g., initial positions and velocities of all particles in the system); (ii) the interaction potentials to represent the forces among all the particles; and (iii) the evolution of the system in time by solving a set of classical Newtonian equations of motion for all particles in the system.

A physical simulation involves the proper selection of interaction potentials, numerical integration, periodic boundary conditions, and the controls of pressure and temperature to mimic physically meaningful thermodynamic ensembles of different types such as grand canonical (μVT), microcanonical (NVE), canonical (NVT), and isothermal–isobaric (NPT). The constant temperature and pressure are usually controlled by adding an appropriate thermostat (e.g., Berendsen, Nose, Nose–Hoover and Nose–Poincare) and barostat (e.g., Andersen, Hoover and Berendsen), respectively. Applying MD into polymer composites allows investigations to be made into the effects of fillers on polymer structure and dynamics in the vicinity of the polymer–filler interface, and also to probe the effects of polymer–filler interactions on the material's properties.

14.2.2.2 Monte Carlo

The Monte Carlo (MC) technique is a stochastic method that uses random numbers to generate a sample population of a system from which the properties of interest can be calculated. A MC simulation usually consists of three typical steps. First, a physical problem is translated into an analogous probabilistic or statistical model. Second, the probabilistic model is solved by a numerical stochastic sampling experiment. Third, the obtained data are analyzed by using statistical methods. Thus, MC provides information only on equilibrium properties (e.g., free energy, phase equilibrium); this is different from MD, which provides details of both nonequilibrium and equilibrium properties.

In a typical MC simulation, it is possible to hypothesize a new configuration for a molecular system by arbitrarily or systematically moving one atom from one position to another. Then, due to such atomic movement, the change in system energy can be computed. This new configuration is then evaluated according to the following rules:

- If the energy decreases, the movement is immediately accepted and the displaced atom remains in its new position.
- If the energy increases, the move is accepted only with a certain probability.
- If the new configuration is rejected, the original position is counted as a new one and the process is repeated by using other arbitrarily chosen atoms.

In polymer nanocomposites, MC methods have been used to investigate the molecular structure at the nanoparticle surface and to evaluate the effects of various factors.

14.2.2.3 Brownian Dynamics

Although, Brownian dynamics (BD) simulation is similar to MD simulation, it introduces a few new approximations that allow simulations to be performed on the microsecond timescale, whereas the MD simulation is known to take up to a few nanoseconds. BD uses an implicit continuum solvent description to replace the explicit description of solvent molecules used in MD. Besides, the internal motions of molecules are typically ignored, allowing a much larger time-step than that of MD. Therefore, BD is particularly useful for systems where there is a large gap of time scale governing the motion of different components. For example, in a polymer–solvent mixture a short time-step is required to resolve the fast motion of the solvent molecules, whereas the evolution of the slower modes of the system requires a larger time-step. However, if the detailed motion of the solvent molecules is involved, they may be removed from the simulation and their effects on the polymer represented by dissipative and random forces. Thus, the forces in the MD governing equation are replaced by a Langevin equation.

One consequence of this approximation is that the energy and momentum are no longer conserved, which implies that the macroscopic behavior of the system will not be hydrodynamic. In addition, the effect of one solute molecule on another through the flow of solvent molecules is neglected. Thus, BD can reproduce the diffusion properties, but not the hydrodynamic flow properties, as the simulation does not obey the Navier–Stokes equations.

14.2.2.4 Dissipative Particle Dynamics

Dissipative particle dynamics (DPD), which was originally developed by Hoogerbrugge and Koelman [27], can be used to simulate both Newtonian and non-Newtonian fluids, including polymer melts and blends, on microscopic length and time scales. Although, like MD and BD, DPD is a particle-based method, its basic unit is not a single atom or molecule but rather a molecular assembly (i.e., a particle). DPD particles are defined by their mass, position, and momentum. The interaction force between two DPD particles can be described by a sum of conservative, dissipative, and random forces.

While the interaction potentials in MD are high-order polynomials of the particle separation, the potentials in DPD are softened so as to approximate the effective potential at microscopic length scales. The form of conservative force is chosen to decrease linearly with increasing particle separation. Beyond a certain cut-off separation, the weight functions, and thus the forces, are all zero, and because the forces are pairwise and momentum is conserved, the macroscopic behavior directly incorporates Navier–Stokes hydrodynamics. However, energy is not conserved because of the presence of the dissipative and random force terms, which are similar to those of BD but incorporate the effects of Brownian motion on larger length scales. DPD has several advantages over MD; notably, the hydrodynamic behavior is observed with far fewer particles than is required in a MD simulation because of its larger particle size. In addition, its force forms allow larger time steps to be taken than those in MD.

14.2.2.5 Lattice Boltzmann

Lattice Boltzmann (LB) is another microscale method that is suited for the efficient treatment of polymer solution dynamics. This method has recently been used to investigate the phase separation of binary fluids in the presence of solid particles. The LB method originated from Lattice Gas Automaton, the latter being constructed as a simplified, fictitious molecular dynamic in which space, time, and particle velocities are all discrete.

An important advantage of LB method is that microscopic physical interactions of fluid particles can be conveniently incorporated into the numerical model. Compared with the Navier–Stokes equations, the LB method can handle the interactions among fluid particles and reproduce the microscale mechanism of hydrodynamic behavior. Therefore, it belongs to the MD in nature and bridges the gap between the molecular level and the macroscopic level. However, its main disadvantage is that it is typically not guaranteed to be numerically stable, and may lead to physically unreasonable results, an example being the case of high forcing rate or high interparticle interaction strength.

14.2.2.6 Time-Dependent Ginzburg–Landau Method

The time-dependent Ginzburg-Landau (TDGL) method is a microscale method used to simulate the structural evolution of phase separation in polymer blends and block copolymers. It is based on the Cahn–Hilliard–Cook (CHC) nonlinear diffusion equation for a binary blend, and falls under the more general phase-field and reaction-diffusion models. In the TDGL method, a free energy function is minimized to simulate a temperature quench from the miscible region of the phase diagram to the immiscible region. Thus, the resultant time-dependent structural evolution of the polymer blend can be investigated by solving the TDGL/CHC equation for the time dependence of the local blend concentration.

To simplify TDGL, Oono and coworkers proposed a cell dynamic method (CDM) which was derived from the discretized TDGL equation [28, 29], but where the Laplacian term is replaced by its isotropic discretized counterpart. This model reproduces the growth kinetics of the TDGL model, demonstrating that such quantities are insensitive to the precise form of the double-well potential of the bulk free energy term. The TDGL and CDM methods have recently been used to investigate the phase separation of polymer nanocomposites and polymer blends in the presence of nanoparticles.

14.2.2.7 Dynamic Density Functional Theory

The dynamic DFT method is normally used to model the dynamic behavior of polymer systems, and has been implemented in the software package Mesodyn™ from Accelrys. The DFT models the behavior of polymer fluids by combining Gaussian mean-field statistics with a TDGL model for the time evolution of conserved order parameters. However, in contrast to traditional phenomenological free-energy expansion methods employed in the TDGL approach, the free energy is not truncated at a certain level, but instead retains the full polymer path integral numerically. At the expense of a more challenging computation, this allows detailed information to be obtained about a specific polymer system beyond simply the Flory–Huggins parameter, while mobility can also be included in the simulation. In addition, viscoelasticity, which is not included in TDGL approaches, is included at the level of Gaussian chains. A similar dynamic DFT approach has also been developed, and this forms the basis of the software tool SUSHI (Simulation Utilities for Soft and Hard Interfaces) for the molecular and mesoscale modeling of polymer materials [30].

The essence of the dynamic DFT method is that the instantaneous unique conformation distribution can be obtained from the off-equilibrium density profile by coupling a fictitious external potential to the Hamiltonian. Once such a distribution is known, the free energy is then calculated using standard statistical thermodynamics. The driving force for diffusion is obtained from the spatial gradient of the first functional derivative of the free energy with respect to the density.

14.2.2.8 Finite Element Method

The finite element method (FEM) is a general numerical method for obtaining approximate solutions in space to initial-value and boundary-value problems, including time-dependent processes. It employs preprocessed mesh generation, which enables the model to fully capture the spatial discontinuities of highly inhomogeneous materials. It also allows complex, nonlinear tensile relationships to be incorporated into the analysis. Thus, it has been widely used in mechanical, biological, and geological systems.

In FEM, the energy taken from the theory of linear elasticity – and thus the input parameters – are simply the elastic moduli and the density of the material. Since these parameters are in agreement with the values computed by MD, the simulation is consistent across the scales. The FEM has been incorporated into some commercial software packages and open source codes (e.g., ABAQUS, ANSYS, Palmyra and OOF), and is used widely for evaluating the mechanical properties of polymer composites. Some attempts have recently been made to apply the FEM to nanoparticle-reinforced polymer nanocomposites.

14.2.2.9 Boundary Element Method

The boundary element method (BEM) is a numerical computational method which involves solving boundary integral equations for the evaluation of stress and strain fields. This method uses elements only along the boundary other than the whole elements throughout the volume, as used in FEM. Thus, BEM is often more efficient than other methods (e.g., FEM) in terms of computational resources for problems where there is a small surface-to-volume ratio. In BEM, it is assumed that a material continuum exists, and therefore, the details of molecular structure and atomic interactions are ignored.

However, for many problems BEM is significantly less efficient than volume-discretization methods (e.g., FEM, finite difference method, finite volume method). Boundary element formulations typically give rise to fully populated matrices, which means that the storage requirements and computational time will tend to grow according to the square of the problem size. By contrast, finite element matrices are typically banded, since elements are only locally connected and the storage requirements for the system matrices typically grow quite linearly with the problem size. BEM can be applied in many areas of engineering and science, including fluid mechanics, acoustics, electromagnetic, and fracture mechanics. More recently, it has been applied to CNT–polymer composites for predicting their mechanical properties.

14.2.3
Multiscale Modeling

14.2.3.1 Challenges

A main goal of computational materials science is the rapid and accurate prediction of new materials and their new properties and features. However, with currently available computer power such prediction is very difficult to achieve when using the analytical and/or numerical models that have been developed traditionally for a single length and time scale. Therefore, the expectation would be to use multiscale simulation strategies to bridge the models, and simulation techniques across a broad range of length and time scales so as to address the macroscopic or mesoscopic behaviors of materials. The challenge for multiscale simulations is to move – as seamlessly as possible – from one scale to another so that the calculated parameters, properties, and numerical information can be efficiently transferred across scales.

In particular, the challenge for polymer nanocomposites is to predict accurately their hierarchical structures and behaviors, and to capture the phenomena on length scales that span typically five to six orders of magnitude, and time scales that can span a dozen orders of magnitude. For example, a clay particle with a diameter of 0.5 μm and 100 layers would contain about 85 million atoms, which would be too large for molecular modeling. A complete description of a polymer nanocomposite typically requires a wide range of length scales, from the chemical bond at around 1 Å in length, up to chain aggregates which extend for many hundreds of Angstroms, and beyond. There is also a wide range of time scales, with chemical bond vibrations occurring over tens of femtoseconds and, at the other extreme, the collective motions of many chains taking seconds or much longer. From this point of view, new strategies for multiscale modeling are essential to predict accurately the physical/chemical properties and material behavior. Some efforts have recently been made to develop multiscale models that span from molecular to macroscopic levels, and which can be divided into two types of approach, namely sequential and concurrent [31].

14.2.3.2 Sequential and Concurrent Approaches

In *sequential approaches*, a series of hierarchical computational methods are linked in such a way that the calculated quantities from a computational simulation at one scale are used to define the parameters of the model operative on the adjoining larger scale. Such parameter-passing, sequential approaches have proven effective, especially when material behavior can be parsed into several scales, each with its own distinct characteristics. Kremer and Müller-Plathe [32] described a sequential approach which spans from atomistic or molecular scale (atoms) to microscale (monomers) and then to macroscale (chains). The approach adopted the parameter-passing strategy in which the effective free-energy terms for the coarse-grained model were obtained by statistical averaging of fast degrees of freedom within the more detailed model. Recently, Doi [30] has developed a suite of state-of-the-art simulation tools (i.e., OCTA) to model polymers at the molecular scale and mesoscale. Although each tool performs calculations by using only one method, the output from one method can be used directly as the input for another, allowing an off-line bridging of length and time scales.

In *concurrent approaches,* several computational methods are linked together in a combined model in which different scales of material behavior are considered concurrently, and communicate by using some sort of "handshaking" procedure. Such approaches resemble the multigrid techniques used for partial differential equations or the general renormalization-group idea of statistical physics. Whilst intellectually most appealing, these approaches have not yet been developed into a general scheme. Ortiz *et al.* [33] described a new concurrent multiscale approach which is based on an atomistic–continuum handshake arranged in such a way that the atomic details of the material's microstructure are resolved only where necessary. The rest of the material is treated as a continuum within a finite element framework. By using an adaptive mesh refinement, the interaction between the atomistic and continuum descriptions can be organized in an entirely seamless fashion.

14.2.3.3 Applications in Polymer Nanocomposites

The development of polymer nanocomposites relies largely on understanding the structure–property relationship of the materials, which in turn requires a multiscale model to predict the material's properties from the information of particle properties, molecular structure, molecular interaction, and mesoscale morphology. Recently, much effort has been made to employ multiscale strategies to predict the multiscale level of structure, properties, and processing performance of polymer nanocomposites containing nanoparticles (e.g., nanosphere, nanotube, nanorod, and clay platelet), as shown in Table 14.3.

14.3
Prediction of Nanocomposite Properties

The property of polymer composites is closely related to the properties (i.e., particle size, shape, and aspect ratio) and dispersion of the fillers, and their interactions with the polymer. With particle size reduced from micrometer to nanometer, a dramatic improvement in mechanical behavior can be achieved. Clay platelets are thought to possess better reinforcement effects than those of spherical and rod-like particles. For example, the complete exfoliation would benefit the strength and stiffness to a significant degree, whereas partial exfoliation would result in a reduction of the reinforcement efficiency. An understanding of the structural origins of the exceptional mechanical properties of polymer nanocomposites is of significance in the design and development of nanocomposites for engineering applications.

The mechanical properties and failure modes of polymer nanocomposites can be predicted by using analytical models and numerical simulations over a wide range of length and time scales, for example, from molecular scale (e.g., MD),

Table 14.3 Selected applications of multiscale modeling in polymer nanocomposites.

Multiscale model	Nanocomposite system	Reference(s)
DFT-SCF model	Phase behavior of clay–polymer nanocomposites	[34–40]
MD-DPD model	Morphology of clay–polymer nanocomposites	[41, 42]
MD-TDGL-FEM model	Molecular phenomena, mesoscale structure and macroscale mechanical properties of nanoparticle-filled polymers	[43, 44]
Integrated SCF and Halpin–Tsai model	Mechanical properties and structure–property relationship of a natural hybrid nanocomposite	[45]
Coupled CHC-BD-LSM model and SCF-DFT-LSM model	Mechanical behavior of spherical and rod-like nanoparticle–diblock copolymer composites	[46, 47]
Combined MD with stress-relaxation simulations and material point method	Shear modulus of cylindrical nanoparticle–polymer nanocomposites	[48]
CHC-BD-LSM model	Mechanical and electrical properties of nanocomposites containing nanorods and a binary polymer blend	[49]
Coupled MD and continuum models (e.g., FEM, Mori–Tanaka, Halpin–Tsai)	Stiffness of clay–polymer nanocomposite	[50]
Coupled MD and effective medium theory	Thermal conductivity of nanotube–polymer composites	[51]

microscale (e.g., Halphin–Tsai) to macroscale (e.g., FEM), and their combinations. MD simulation is inevitable for the analysis of local load transfers, interface properties, or failure modes at the nanoscale, whereas micromechanical models and continuum models generally assume simplified geometries for each component (i.e., with regards to its molecular configuration), and thus provide a simple and rapid means of predicting the global mechanical properties of nanocomposites and to correlate them with the key factors (e.g., particle volume fraction, particle geometry and orientation, and property ratio between particle and matrix). Recently, some of these models (e.g., Halpin–Tsai, Mori–Tanaka, lattice spring model, and FEM) have been applied to polymer nanocomposites to predict their thermal-mechanical properties, Young's modulus, reinforcement efficiency, and the dependence of the materials' moduli on the nature of individual nanofillers. This has led to significant progress being made towards achieving a fundamental understanding of the mechanical properties and failure modes of polymer nanocomposites, and these will be discussed in the following sections.

14.3.1
Mechanical Properties

14.3.1.1 Stiffness and Strength

Stiffness is often a critical property of a material, as it determines how much a component will deform in response to a force (e.g., stretching, compression, or bending), or how much elastic energy is required to distort a component into a given shape. In polymer nanocomposites, one of the most important theoretical efforts is to predict the *elastic modulus*.

Molecular modeling has been used to predict the stiffness and strength of CNT–polymer composites. For example, Frankland *et al.* [52] predicted that the stiffness enhancement would depend on the length of the CNT and the direction of applied loading. Specifically, when subjected to longitudinal loading, long nanotube composites demonstrate significant enhancements in stiffness, whereas short nanotube composites have no enhancement relative to the polymer. When subjected to transverse loading, however, both long and short nanotube composites exhibit no enhancement relative to that of the polymer. Zhu *et al.* [53] reported similar results in nanotube-Epon 862 composites, with the predicted stiffness being approximately 10-fold greater than that of Epon 862 matrix for long nanotubes, and only about 20% greater for short nanotubes. In a report based on a molecular mechanics simulation, Mokashi *et al.* [54] noted that both CNT length and polymer configuration played an important role in the tensile response and fracture of the nanocomposites. In particular, when reinforced with long CNTs, crystalline polyethylene demonstrated only moderate improvements in tensile strength and elastic stiffness, whereas an amorphous polyethylene demonstrated a significant increase in overall tensile properties. When reinforced with short CNTs, amorphous polyethylene showed a significant decrease in the Young's modulus and tensile strength, indicating a poor load transfer from the matrix to the nanotubes. Another MD simulation predicted that the modulus of nanotube–polymer composites could be effectively improved by deliberately adding chemical crosslinking [55].

Molecular modeling has also been used to predict the stiffness and strength of polymer nanocomposites containing spherical and platelet nanoparticles. Based on a MD simulation, Gersappe [56] attributed the improvement in the toughness of spherical nanofiller–polymer composites to equivalent time scales of motion of the nanofiller and polymer chain, as a result of similar size. Adnan *et al.* [57] predicted, by MD simulation, that the elastic properties of amorphous polyethylene nanocomposites containing fullerene bucky-ball fillers would be significantly enhanced upon the reduction of fullerene size, this effect being attributed to a densification of the polymer matrix near the fullerene and an attractive fullerene–polymer interaction. In the presence of spherical nanoparticles, Sen *et al.* [58] predicted, from MC simulation, that either increasing the chain length or decreasing the temperature would increase the end-to-end distance of polymer chains and shrink the elastic region during deformation, which in turn would have a significant effect on the elastomeric properties of the nanocomposites. Kalfus and Jancar

[59] used a reptation model and the percolation concept to interpret the modulus recovery times observed in nanocomposites containing poly(vinylacetate) (PVA) and hydroxyapatite of either spherical or platelet nanoparticles. Here, the storage modulus recovery time was seen to be governed by the relaxation processes of the polymer chains adsorbed onto the particle surface, other than the particle network and its rearrangement.

The accurate quantification of the elastic moduli of individual components is critical in order to predict the overall elastic properties of nanofiller–polymer composites. To this end, Suter et al. [60] used a large-scale MD simulation to predict the mechanical properties of individual clay platelets. The bending modulus of the clay platelets was estimated as 1.6×10^{-17} J, which corresponded to an in-plane Young's modulus of 230 GPa. By using a MD simulation, Capaldi et al. [61] predicted the average moduli of a crystalline octacyclopentyl polyhedral oligomeric silsesquioxane, namely a bulk modulus of 7.5 GPa, a Young's modulus of 11.78 GPa, and a shear modulus of 4.75 GPa. Patel et al. [62] used a MD simulation to analyze the silsesquioxane packing around the polymer backbone and its effect on polymer motion, and predicted the elastic moduli by the static deformation method. The above-predicted properties may be applicable to micromechanical models to calculate the overall elastic properties of polymer nanocomposites.

Micromechanical models have been revised and extended to predict the mechanical properties of polymer nanocomposites containing nanotubes, nanoclay, or other nanoparticles. Zheng et al. [63] combined the Mori–Tanaka model with the probability distribution functions of the Doi–Hess hydrodynamic theory, and used the model to predict the effective elastic properties of stationary and flow-induced nanorod-polymer composites. The same group also examined the effects of nanorod volume fraction, flow type, and flow rate on the elasticity tensors. de Villoria and Miravete [64] developed a new micromechanical model – the dilute suspension of clusters – to estimate the influence of nanofiller dispersion on the Young's modulus of nanocomposites. This model led to a significant improvement in the theoretical–experimental correlation for epoxy composites reinforced with the clusters of CNTs. Wagner [65] used a continuum-based method (i.e., a modified Kelly–Tyson approach) to predict the interfacial shear strength of SWCNT–polymer composites by assuming uniform interfacial shear and axial normal stresses.

In addition to nanotube–polymer composites, micromechanical models have also been used to evaluate the mechanical properties of polymer nanocomposites with platelet and ellipsoid nanofillers. Lee and Paul [66] developed a model to predict the mechanical properties of composites containing complex inclusions with no axes of symmetry (e.g., three-dimensional ellipsoids) by using Eshelby's equivalent tensor with the Mori–Tanaka model. When Lee and Paul also examined the influences of the primary and secondary aspect ratios on the effective elastic moduli of nanocomposites containing aligned isotropic inclusions, their model predicted that the longitudinal Young's modulus would increase with the primary and secondary aspect ratios, whereas the transverse Young's modulus and shear modulus would decrease as the secondary aspect ratio increased. Subsequently,

Luo and Daniel [67] developed a three-phase structure (i.e., epoxy matrix, exfoliated clay platelet, and clay cluster) to represent the partially exfoliated and intercalated clay and applied a Mori–Tanaka model to predict the modulus of clay–polymer nanocomposites. For this, they correlated the mechanical properties with various parameters, and found that: (i) a higher degree of nanofiller dispersion led to a higher modulus; (ii) a large intercalation basal spacing was preferred, but its effect on the modulus was marginal unless accompanied by a high cluster aspect ratio; and (iii) the exfoliation ratio and aspect ratio of clay platelet had a distinct effect on stiffness enhancement. Jo et al. [68] developed a constitutive model to predict the elastic modulus and tensile strength of intercalated and agglomerated poly(methylmethacrylate) nanocomposite foams, and found the predicted stress–strain curves to be in agreement with experimental measurements. By using the Halpin–Tsai model, Brune and Bicerano [69] predicted that both partial exfoliation and imperfect platelet alignment would significantly decrease the elastic moduli of clay–polymer nanocomposites. When Wu et al. [70] used composite theories (i.e., Guth, Halpin–Tsai, and modified Halpin–Tsai equation) and a reduced modulus for platelet-like fillers to predict the moduli of rubber–clay nanocomposites, they found that the Guth and modified Halpin–Tsai equations could predict the moduli of rubber–clay nanocomposites over a wide range of clay volume fractions. However, the Halpin–Tsai equation proved to be suitable only for those nanocomposites with a low clay volume fraction.

Continuum mechanics, and in particular FEM, has also been used widely to predict the mechanical properties of polymer nanocomposites. Fisher et al. [71] integrated the FEM with a micromechanical method (i.e., Mori–Tanaka model) to predict the modulus of a polymer that had been reinforced with a distribution of wavy nanotube. They predicted that nanotube waviness could significantly weaken the modulus enhancement of the nanotube-reinforced polymers. Bradshaw et al. [72] used a 3-D FEM of a single infinitely long sinusoidal fiber within an infinite matrix to calculate the dilute strain concentration tensor. This tensor was then utilized in a Mori–Tanaka model to predict the modulus of a polymer reinforced with aligned or randomly oriented nanotubes. Ashrafi and Hubert [73] predicted the elastic properties of nanoarrays of SWCNTs by finite element analysis. Then, the predicted elastic properties were used in micromechanical models to calculate the properties of carbon nanoarray–polymer composites and to estimate the effects of the volume fraction and aspect ratio of the nanoarrays. Song and Youn [74] predicted the elastic properties of CNT–polymer composites by using an asymptotic expansion–homogenization method that was able to transform a heterogeneous medium to an equivalent homogeneous medium, and subsequently performed both localization and homogenization for the heterogeneous medium. The predicted elastic modulus was in good agreement with that obtained using an analytical model.

In addition to nanotube–polymer nanocomposites, FEM has been used to predict the mechanical properties of polymer nanocomposites reinforced with other nanoparticles. For example, the elastic properties of silica nanoparticle–polymer nanocomposites were predicted using various FEM-based computational models which reasonably reproduced the nanocomposite microstructure (e.g., particle size, volume fraction, and spatial arrangement) [75–77]. These studies showed that it is important to treat the interphase layer around particles as a third constituent in predicting the elastic modulus. Later, Saber-Samandari and Khatibi [78] developed a 3-D unit cell model to predict the elastic modulus of polymer nanocomposites by considering three constituent phases, namely the particle, interphase, and matrix. These authors estimated the effects of both matrix and filler stiffness, as well as interphase thickness, on the Young's modulus of clay-based polymer nanocomposites. When Tseng and Wang [79] recently developed a constitutive model to predict the elastic properties of nanoparticle-reinforced polymer composites, they showed the effects of interparticle interaction to be significant and not to be neglected, especially in the case of high particle loading.

Comparisons have recently been made as to the predictive capability of micromechanical models and FEM for the mechanical properties of polymer nanocomposites. When Hbaieb *et al.* [80] assessed the accuracy of the Mori–Tanaka model against FEM in predicting the stiffness of clay–polymer nanocomposites they showed that, unlike the 3-D FEM models, the 2-D FEM models could not predict an accurate elastic modulus. Although the Mori–Tanaka model can predict accurately the stiffness of intercalated clay–polymer nanocomposites with a clay volume fraction up to 5 wt%, it underestimates the stiffness at higher volume fractions and overestimates the stiffness of exfoliated clay–polymer nanocomposites. Lagoudas and coworkers [81, 82] predicted the elastic properties of CNT-reinforced composites by using micromechanics techniques and FEM. For this, the effective properties of the CNTs were calculated by composite cylinders micromechanics, and then used in the self-consistent and Mori–Tanaka methods to predict the elastic properties of polymer composites reinforced with SWCNTS or multiwalled carbon nanotubes (MWCNTs). Such a composite cylinders method could reasonably predicts the elastic constants of the composites in which the volume fraction of nanotubes was below a critical value. However, above such a critical volume fraction, FEM may be more advantageous for handling the complicated fiber–polymer interactions. Recently, when Liu and Chen [83–85] used FEM and BEM to predict the elastic properties of CNT composites, the predicted stress gradient across the nanotube–matrix interface was very high, and thus the number of elements might become prohibitively large for the FEM in the large-scale modeling of nanotube composites. The predicted Young's moduli, by using BEM, were very close to those obtained from a MD-based multiscale approach reported elsewhere. It is to be expected that the MD predictions (i.e., elasticity of nanotube fibers, interfacial conditions) could be incorporated readily into such a fast multipole BEM. Huang *et al.* [86] extended the above method by introducing an equivalent length coefficient to consider the 3-D end effects of SWCNTs in a matrix. For this, they first developed a modified mixture model for the prediction of Young's modulus. Whilst this model could be applied to a CNT loading of between 0% and 5%, that of Liu and Chen was only suitable for a nanotube loading close to 5%. Miysgawa *et al.* [87] proposed a pseudo-inclusion model to describe the behavior of polymer nanocomposites that contained randomly oriented and uniformly dispersed clay platelets. In this case, the Tando–Weng and Halpin–Tsai equations, when combined with the pseudo-inclusion model, were shown to be useful for predicting the elastic modulus and coefficient of thermal expansion of clay–polymer nanocomposites.

The equivalent continuum model developed by Odegard et al. [25, 88–90] predicted that, when the effective particle size was less than 1000 , the Young's modulus and shear modulus would each increase with particle size, albeit less than the corresponding values predicted from the Mori–Tanaka model. However, with a particle size in excess of 1000 , the moduli predicted from both models were consistent. A self-similar approach, developed by Pipes and Hubert [26, 91–93], was also used to predict the mechanical properties of the helical nanoarray, nanowire, and microfiber, such as the axial Young's modulus, shearing modulus, and Poisson's ratio. Odegard et al. [23] applied both an equivalent-continuum model and a self-similar approach to predict the elastic properties of SWCNT–polymer composites, and obtained similar values of Young's modulus in both cases.

14.3.1.2 Stress Transfer

Efforts have recently been made to predict stress/load transfer across the nanoparticle–polymer interface, by using molecular modeling. For example, MD simulations [55] predicted that the load transfer of nanotube–polymer composites could be effectively enhanced by deliberately adding chemical crosslinking. When both Liao et al. [94] and Wong et al. [95] used molecular mechanics to quantify the CNT–polymer interfacial shear stress, the predicted value was about one order of magnitude higher than that of microfiber-reinforced composites. Subsequently, by using molecular mechanics and MD, Qian et al. [96] studied the load transfer of a SWCNT bundle (i.e., closest-packed nanotubes) instead of individual and separated nanotubes. Here, the surface tension and inter-tube corrugation were found to contribute to load transfer in the parallel bundle, while twisting of the nanotubes in the bundle significantly enhanced the load transfer.

Gou et al. [97, 98] used molecular mechanics and MD simulations to study the molecular interaction and load transfer in the presence of SWCNTs and a nanotube rope by pulling out a single nanotube from the rope, or a nanotube/rope from the epoxy matrix. The interfacial shear strength between the nanotube and epoxy resin was predicted to be up to 75 MPa, which indicated an effective stress transfer from the epoxy resin to the nanotube. It was also shown that the load transfer would depends not only on the physical interactions between the nanotube and epoxy matrix, but also on the internal interactions within the nanotube rope. The individual nanotubes have stronger interactions with epoxy resins, and hence provide a better load transfer than that of the nanotube ropes. As discussed above, Frankland et al. [52] reported the different behaviors of loading transfer between short- and long-nanotube-reinforced polymer nanocomposites.

Continuum-based models have also been used to predict the stress transfer of polymer nanocomposites. For example, Li and Chou [99] developed a continuum-based model to calculate the interfacial stress transfer of nanotube-reinforced polymer composites. Specifically, they employed molecular structural mechanics to characterize the nanotube, and FEM to model the polymer matrix. In another study, Gao and Li [100] developed a shear-lag model to predict the interfacial stress of a CNT–polymer composite in which the volume element was represented by a composite cylinder embedded with a capped nanotube. The continuum-based shear-lag analysis was performed using the elasticity theory for axisymmetric problems. These studies shed new light on understanding the load transfer across the nanotube–matrix interface, and also provide some guidance for the development of analytical models as well as novel composite materials. When Tsai and Sun [101] used a shear lag model to investigate the load transfer efficiency of clay platelets in a polymer matrix, they predicted that well-dispersed clay platelets – due to their high aspect ratio – could significantly enhance the load transfer efficiency, whereas clay tactoids (i.e., stacks or clusters of clay platelets) would lead to a much lower enhancement of load transfer. By using micromechanical models, Sheng et al. [50] predicted that particles with a high aspect ratio would shield the matrix from straining, and thus reduce the efficiency of load transfer to neighboring particles.

14.3.1.3 Mechanical Reinforcement

The molecular origins of mechanical reinforcement of polymer nanocomposites have been studied using a variety of analytical and numerical methods, which in turn have led to two different opinions. The first opinion attributes the reinforcement exclusively to nanoparticle aggregation or clustering, while the second opinion attributes it to the formation of an interphase (i.e., a transient polymer network) between the nanoparticle and the bulk polymer. For example, Ganesan and coworkers [102] have predicted that, at high particle loadings, the reinforcement occurs due to particle jamming, whereas at low particle loadings the polymer–particle network is important in mechanical reinforcement. Similarly, Salaniwal et al. [103] reported that a strong polymer–particle interaction results in the formation of a solid-like network, with an almost infinite relaxation time.

In order to clarify these two reinforcement mechanisms, Sen et al. [104] conducted an extensive MD study on amorphous polymers filled with solid nanoparticles. It was predicted that the mechanical reinforcement observed experimentally had resulted from either a particle agglomeration or from a polymer-based network. The relative contributions of these would depend heavily on the particle–polymer interaction energy, the particle volume fraction, and the state of particle dispersion. By using MD, Chauve et al. [105] were able to examine the reinforcing effects of cellulose whisker ethylene-vinyl acetate (EVA) copolymer nanocomposites, and predicted that a strong filler–polymer affinity would lead to greater reinforcing effects.

Micromechanical and continuum models can provide a rapid assessment of the reinforcement of nanofillers, and the effects of various parameters. For instance, Paul and coworkers applied the composite theory of Chow [106], as well as those of Halpin–Tsai and Mori–Tanaka [107], to predict the reinforcement efficiency of layered clays and glass fibers in a nylon 6 matrix. These authors attributed the superior reinforcement in exfoliated nylon 6 nanocomposites to the high modulus and high aspect ratio of clay platelets, and their ability to reinforce in two directions. Specifically, the reinforcing efficiency was dramatically decreased if the number of platelets per stack and the gallery spacing between platelets were increased, which is attributed to reductions in both the aspect ratio and the effective modulus of the clay clusters. Furthermore, it was predicted that the superior reinforcement in clay-based nylon 6 nanocomposites would result primarily from a combination of the high modulus and high aspect ratio of the clay platelets, rather than from any changes in the polymer matrix induced by the latter [106]. This conclusion may be arguable, however, as the polymer–filler interphase is often considered to be a critical factor in the property enhancement of polymer nanocomposites. Brune and Bicerano [69] predicted, based on the Halpin–Tsai model, that a partial exfoliation and an imperfect platelet alignment would each cause significant reductions in reinforcement efficiency, as compared to exfoliated clay–polymer nanocomposites.

Wu *et al.* [70] reported that the modulus reinforcement of rubber–clay nanocomposites could be predicted by using composite theories (e.g., Guth, Halpin–Tsai, modified Halpin–Tsai), with the introduction of a modulus reduction factor for the platelet-like fillers. Li *et al.* [108] predicted, via the Halpin–Tsai and Mori–Tanaka models, that the mechanical reinforcement of SWCNT–epoxy composites would be closely related to factors such as covalent nanotube–polymer bonding and matrix crystallinity, as well as the tensile properties of reinforcement and matrix, and bundle curvature and alignment.

Buxton and Balazs [109] used a lattice spring model to examine the effects of filler geometry (e.g., platelet, sphere, and rod), and the intercalation–exfoliation of clay platelets, on the mechanical behavior of particle-reinforced polymers. It was predicted that polymers filled with platelets would experience the most significant increase in reinforcement efficiency, while those with spheres would show the lowest reinforcement efficiency. Moreover, the incomplete exfoliation of clay platelets would lead to a less effective reinforcement of the polymer matrix, a point which was in agreement with experimental observations [110, 111]. The reinforcement efficiency was attributed to the volume of polymer matrix constrained within the proximity of the particles. In other words, a more homogeneous dispersion of clay platelets in the polymer matrix would lead to a greater volume of polymer being constrained by the platelet particles, thereby increasing the reinforcement efficiency. In another study [46], the effects of the spatial distribution of spherical nanoparticles on the mechanical behavior of nanoparticle-filled diblock copolymer composites was investigated by coupling the CHC-BD-LSM models. Here, it was found that the confinement of nanoparticles within a given domain of a bicontinuous diblock mesophase would cause the nanoparticles to percolate and to form essentially a rigid backbone throughout the material. This continuous nanoparticle distribution significantly increased the reinforcement efficiency of the nanoparticles, and also the Young's modulus of the material. In the case of nanorod fillers [49], the reinforcement efficiency of nanocomposites with a bicontinuous morphology was significantly increased compared to that of composites where the nanorods were randomly dispersed in a homogeneous matrix.

14.3.1.4 Interfacial Bonding

The strength of interfacial bonding will determine the formation and nature of the nanofiller–polymer interface, stress transfer, and reinforcement. Thus, quantifying the interfacial interactions and strength will provide further understanding of the above-discussed issues. Lordi and Yao [112] recently studied the interfacial adhesion mechanisms of nanotube composites by using molecular mechanics to determine the binding energy and sliding frictional stress between CNTs and a range of polymer matrices. It was predicted that, although the binding energies and frictional forces would play only a minor role in determining the strength of the interface, the helical polymer conformations would play a very important role. By using MD simulation, Zheng et al. [113] investigated the molecular interactions between SWCNTs and various polymers (e.g., polyethylene, polypropylene, polystyrene, polyaniline) by simulating two typical interactive process modes: (i) polymer wrapping onto the surface of the CNT; and (ii) polymer filling into the CNT. Subsequently, the interaction was found to depend heavily on the specific monomer structure (e.g., aromatic rings) and the nanotube chirality and diameter. During the wrapping process, polystyrene and polyaniline showed a stronger interaction which could be attributed to the groups of aromatic rings. During the filling process, polyethylene, polypropylene and polystyrene showed stronger interactions and could fill into a (10, 10) nanotube cavity due to the attractive van der Waals interactions.

By using MD, Chauve et al. [105] calculated the interaction energy between cellulose whiskers and ethylene–vinyl acetate (EVA) copolymer nanocomposites, and evaluated the work of adhesion necessary to separate the two interacting cellulose surfaces. It was predicted that a greater energy would be needed to separate the polar EVA copolymers from cellulose than for the nonpolar ethylene homopolymer. Meanwhile, the electrostatic interaction was shown to depend heavily on the polymer polarity, whilst a strong hydrogen bonding was observed between the cellulose and polar EVA copolymers. In another study, Minisini and Tsobnang [114] predicted the interaction energy between the polymer and organoclay to be more negative for functionalized polypropylenes than for pure polypropylene. MD simulations conducted by Aleperstein et al. [115] on ethylene–vinyl alcohol copolymer organoclay nanocomposites, showed that the presence of octadecylamine on the clay modification led to the formation of hydrogen bonding between the amine hydrogen of octadecylamine and the hydroxyl oxygen of the copolymer.

14.3.1.5 Viscoelasticity

Viscoelasticity is a property of materials that exhibit both viscous and elastic characteristics when undergoing deformation. Various theoretical and numerical methods have recently been used to explore the viscoelasticity of polymer nanocomposites. For example, when Chabert et al. [116] proposed a combined self-consistent scheme and discrete model to predict the viscoelasticity of nanocomposites, and examined the effect of nanofiller loading and nanofiller–nanofiller interaction, their predictions agreed well with their experimental results. It was also found that, above the percolation threshold, the nanofiller–nanofiller interaction becomes increasingly important with regards to reinforcement. Raos et al. [117] used a DPD simulation to predict the viscoelasticity of a filled crosslinked polymer (i.e., rubber), and examined the effect of nanofiller–nanofiller interaction through morphological systems of dispersed, moderately aggregated, and fully aggregated nanofillers, respectively. Here, it was found that that the interaction has a distinct effect on the dynamics shear modulus. For a given morphology, an increasing interaction always resulted in a certain increase in the modulus. Later, Liu and Brinson [118] developed a hybrid numerical–analytical modeling method to predict the viscoelastic response of polymer nanocomposites reinforced with nanotubes and nanoplatelets, respectively. This method used a finite element technique to determine the strain concentration tensors of the nanoparticle, and then coupled the numerical result with the Mori–Tanaka model to predict the overall nanocomposite response. The predicted pattern of the interphase influence on the overall performance of the nanocomposites was in agreement with experimental observations. Unfortunately, however, a quantitative comparison between the prediction and experimental data was not possible due to some uncertain properties (e.g., nanofiller, interphase, and nanofiller morphology) and to the limitations of the 2-D description in the model.

14.3.2
Mechanical Failure

14.3.2.1 Buckling
Buckling is a structural instability failure mode, and a major concern for structural design. It is related to both the geometry and material properties of the structure. Recently, a number of theoretical and numerical studies have been reported with regards to the prediction of buckling in polymer nanocomposites.

Guz et al. [119] proposed a compressive failure theory for polymer nanocomposites which was based on the microbuckling mechanism and the 3-D linear theory of stability of deformable bodies. For this, two models were employed: (i) an homogeneous anisotropic model; and (ii) a piecewise-homogeneous model. When the compressive failure theory was applied to laminated and fibrous nanocomposites with different microstructures, it was predicted that the helical (torsional) buckling modes would occur in a SWCNT, but not in nanocomposites [120].

Xia and coworkers [121] predicted, again using MD simulation, that the interwall shear coupling would have a strong influence on load transfer and compressive load-carrying capacity. Consequently, they developed a new continuum shear-coupled-shell model which predicted that MWCNTs could be engineered through the control of interwall coupling to increase the load transfer, buckling strength, and energy dissipation by nanotube pullout, so as to provide a good performance for the nanocomposites. Brune and Bicerano [69] subsequently developed an elastic equation based on the Halpin–Tsai model to examine some important aspects of the micromechanics of clay-based polymer nanocomposites, and predicted that the clay platelet would buckle above a critical strain.

14.3.2.2 Fatigue

Fatigue is one of the primary reasons for failure in structural materials. Recent studies have shown that the addition of nanoparticles can improve the fatigue behavior of polymer composites, without sacrificing the stiffness. It is believed that the mechanisms involved in the enhanced fatigue resistance of nanoparticle–polymer nanocomposites include cracking pinning, crack tip deflection, and particle debonding.

Zhang et al. [122] reported that the addition of CNTs resulted in an order of magnitude reduction in the fatigue crack propagation rate for epoxy systems, and also predicted, by using fractography analysis and fracture mechanics modeling, that crack suppression would be caused by crack bridging. In another study, the same group [123] reported that fatigue crack growth rates could be significantly lowered by reducing the nanotube diameter and length and improving their dispersion. Such a reduction was explained, from a qualitative standpoint, by using a fracture mechanics model based on crack-bridging and pull-out of the nanotubes. Under optimal conditions, the fatigue crack propagation rate of a nanocomposite epoxy would be reduced by more than 20-fold compared to an unfilled epoxy.

Mallick and Zhou [124] predicted, using the Eyring equation, that fatigue failure in both polypropylene nanocomposites and polyamide-6 nanocomposites would be initiated at agglomerated nanoparticles. Moreover, polyamide-6 nanocomposites demonstrated a higher fatigue strength but a lower activation volume and energy, and a lower ratio of maximum fatigue stress to yield strength than that of polypropylene nanocomposites.

14.3.2.3 Fracture

With regards to the engineering applications of a material, toughness is one of the most important properties to be considered. Although the toughness and fracture mechanisms of polymer nanocomposites have not yet been investigated systematically, Boo et al. [125] recently summarized the research efforts towards a fundamental understanding of the fracture behavior of polymer nanocomposites, including the various toughening and fracture mechanisms and the effects of the nanofiller on aspect ratio and dispersion.

When Song and Chen [126] investigated the fracture behavior of exfoliated clay–polymer nanocomposites, by using MD simulation, their results showed that the interfacial interactions and the difference of relaxation time between clay and polymer chains had a significant influence on the fracture strength of polymers. In particular, for polymers with a glass transition temperature (T_g) either below room temperature (e.g., polyurethane) or close to room temperature (e.g., nylon), the addition of clay nanoplatelets could significantly enhance the mechanical properties of the polymers. However, the addition of nanoplatelets to polymers with a T_g above room temperature (e.g., epoxy resins and polystyrene) did not improve the toughness of these materials.

Jeong et al., [127] proposed the failure criteria for the fracture of hollow or filled CNTs under biaxial tensile–torsional loads by using a multiscale approach in which continuum mechanics models were used to describe atomistic predictions of failure from MD simulations. Under this type of loading, the failure strength of CNT systems was found to differ significantly from that occurring under uniaxial tensile loading; the values were predicted by using the failure criteria for macroscopic objects. Seshadri and Saigal [128] developed a nanotube-pullout model to predict the enhancement or augmentation of fracture energy in a polymer nanocomposite. This model captured the effect of interfacial adhesion and the viscoelasticity of the polymer, in quantitative manner. The crack opening rate determined the stiffness of the polymer matrix during pullout; that is, a faster rate of pullout rendered the matrix stiffer, which in turn hastened the process of the pullout force reaching a critical value. Fyta et al. [129] employed a tight-binding MD simulation to study the fracture mechanisms and the ideal strength of tetrahedral amorphous carbon and nanodiamond-filled nanocomposites. Since, under tensile or shear load, the fracture in the nanocomposites occurred inter-grain, their ideal strength was similar to the pure amorphous phase. The nanodiamond inclusions significantly enhanced the elastic moduli to values similar to those for diamond.

When Zimmermann et al. [130] used computer simulations to elucidate the evolution of microcracking in alumina/silicon carbide nanocomposites, they found that the region and the fracture mode of microcracking were determined by the intensity and the length scale of the residual stress fields. Ward et al. [131, 132] predicted, by using MD, that the deformation and fracture mechanisms of Al/Si nanocomposites were different from those of single-phase all-aluminum materials. The plastic deformation of all-aluminum polycrystals was associated with a mix of grain boundary deformation and dislocation activity, while the deformation in Al/Si nanocomposites was associated predominantly with grain boundary sliding/shearing at the Al/Si interfaces and little deformation elsewhere. The failure of aluminum polycrystals occurs by crack initiation at triple junctions and grain boundaries, while failure in Al/Si nanocomposites occurs by void damage accumulation, culminating in crack formation, at an Al/Si interface.

14.3.2.4 Wear

Wear is the erosion of material from a solid surface, induced by the action of repeated rubbing. Many types of wear mechanism have been proposed, such as abrasion, adhesion, surface fatigue, corrosion, erosion, and delamination. In a recent book, Friedrich et al. [133] provided a full description of the tribological behavior (e.g., friction, wear, scratch, lubrication) of polymer nanocomposites and coatings. Mai and coworkers [134] also recently reviewed progress in understanding wear and scratch damage in polymer nanocomposites containing different nanoparticles (e.g., TiO_2, Si_3N_4, ZnO, CNT, and nanoclays). It was concluded that nanofillers do not always improve the wear/scratch properties, and that the material properties (e.g., modulus, hardness, fracture toughness, extent of wear rate, scratch penetration depth) are not the sole indicators for assessing candidate materials. Moreover, the apparent contradictions among many studies on several material systems have been attributed to the poor characterization of polymer nanocomposites, and a lack of quantitative descriptions of the observed phenomena.

14.3.2.5 Creep

Creep is a time-dependent plastic deformation (i.e., strain increase) which occurs as a result of the action of a constant applied load. Creep takes place under a stress which is lower than the yielding stress of materials. The results of many studies have shown that the addition of nanofillers into a neat polymer can significantly improve the creep resistance of the latter material; however, few studies have been reported on the prediction of creep behavior of nanofiller–polymer composites [135–137]. Yang *et al.* [135] investigated the tensile creep resistance of polyamide 66 nanocomposites reinforced with either TiO_2 or clay nanoparticles, by using the constitutive Burgers creep model and the empirical Findley power law. The predicted results from both models agreed quite well with the experimental data, and also confirmed the enhanced creep resistance of nanofillers, even over an extended time scale. Ohji [136] reported a delayed fracture behavior (i.e., the creep and creep rupture behavior) under tensile loading at elevated temperatures for three engineering ceramics, namely silicon nitride, silicon carbide, and alumina/silicon carbide nanocomposites. The results showed that silicon carbide had an excellent creep resistance, even at very high temperature, and its creep lives were well described by a diffusive crack growth model. Moreover, the dispersion of silicon carbide nanoparticles into an alumina matrix led to a dramatic improvement in creep resistance when compared to alumina with an equivalent grain size. This improvement was attributed to the impedance of grain boundary sliding, induced by the presence of intergranular silicon carbide nanoparticles.

14.4 Conclusions

The development of polymer nanocomposites requires a comprehensive understanding of the phenomena, and an accurate prediction of the materials' properties and behaviors over different scales of time and length. During the past decade, this has led to a significant stimulation of theoretical efforts in nanoparticle–polymer nanocomposites. Indeed, many analytical and numerical techniques have been employed to predict nanocomposite properties, including Monte Carlo, MD, Brownian dynamics, lattice Boltzmann, Ginzburg–Landau theory, micromechanics, and the FEM. Other novel techniques have also been developed, such as dissipative particle dynamics, equivalent-continuum, and self-similar approaches, which represent methods over various time and length scales from the molecular scale to microscale to macroscale, with each approach having demonstrated success to various degrees when predicting the properties and behaviors of nanocomposites.

The analytical and numerical methods developed to date have different strengths and weaknesses, depending on the needs of research. For example, molecular modeling has been applied to investigate molecular interactions and structure on the nanoscale. Such information is very useful for understanding interaction strengths at nanoparticle–polymer interfaces and the molecular origin of mechanical improvement. However, molecular modeling is computationally very demanding, and cannot be applied to predict mesoscopic structure and properties defined on the microscale, such as the dispersion of nanoparticles in a polymer matrix and the morphology of polymer nanocomposites. For the exploration of morphology on these scales, mesoscopic simulations are more effective. In contrast, the macroscopic properties of materials can often be estimated by using mesoscale or macroscale techniques, although these may be limited when applied to polymer nanocomposites because of difficulties in dealing with the interfacial interaction between nanoparticles and polymers, as well as the nature of the interface. All of these factors are crucial when attempting to predict the properties of nanocomposites with any degree of accuracy.

Despite much progress having been made over the past few years, there remain many challenges in predicting nanocomposite properties, one of which derives from the complicated multiscale morphology and dynamics. Nanocomposites possess morphological features and dynamic phenomena on length and time scales, respectively, that differ by many orders of magnitude. A second challenge derives from the polydispersion of nanofillers in a polymer matrix as a result of the unique nature of nanoparticles, and the uncertainty of surface defects. For example, even in the same mixture clay particles can take different forms, whether aggregated, intercalated, and partially or fully exfoliated, this situation resulting from the nonuniform electrical charges of platelets within the clay stacks. Thus, it is essential to develop new and improved analytical and numerical techniques, in particular to integrate methods on different time and length scales, and to formulate an efficient multiscale method that is capable of spanning from a quantum mechanical domain to molecular domain, a mesoscopic domain, and/or a macroscopic domain. Whilst the development of multiscale methods presents a major challenge, it also represents the future of polymer nanocomposites, and new concepts, theories and computational tools must be developed such that multiscale modeling becomes reality. These developments are crucial in order to predict and design nanocomposite properties. Moreover, continuing advances in computing speed, theoretical and numerical methods, software development, and algorithm efficiency will surely lead to significant progress in this field in the near future.

Acknowledgments

The authors would like to thank Australian Research Council (ARC) for financial support.

References

1. Okada, A. and Usuki, A. (2006) *Macromol. Mater. Eng.*, **291**, 1449.
2. Pinnavaia, T.J. and Beall, G.W. (2000) *Polymer-Clay Nanocomposites*, John Wiley & Sons, Ltd, London.
3. Krishnamoorti, R. and Vaia, R.A. (2001) *Polymer Nanocomposites: Synthesis, Characterization, and Modeling*, American Chemical Society, Washington.
4. Nalwa, H.S. (2003) *Handbook of Organic-Inorganic Hybrid Materials and Nanocomposites*, American Scientific Publishers, Stevenson Ranch, California.
5. Utracki, L.A. (2004) *Clay-Containing Polymeric Nanocomposites*, Rapra Technology Ltd., Shrewsbury.
6. Ke, Y.C. and Stroeve, P. (2005) *Polymer-Layered Silicate and Silica Nanocomposites*, Elsevier, Boston.
7. Nicolais, L. and Carotenuto, G. (2005) *Metal-Polymer Nanocomposites*, John Wiley & Sons, Inc., Hoboken.
8. Koo, J.H. (2006) *Polymer Nanocomposites: Processing, Characterization, and Applications*, McGraw-Hill, New York.
9. Mai, Y.W. and Yu, Z.Z. (2006) *Polymer Nanocomposites*, CRC Press, Cambridge, England.
10. Ray, S.S. (2006) *Polymer Nanocomposites and Their Applications*, American Scientific Publishers, Stevenson Ranch, California.
11. Advani, S.G. (2007) *Processing and Properties of Nanocomposites*, World Scientific Publishing, Singapore.
12. Morgan, A.B. and Wilkie, C.A. (2007) *Flame Retardant Polymer Nanocomposites*, John Wiley & Sons, Inc., Hoboken, New Jersey.
13. Bhattacharya, S.N., Kamal, M.R., and Gupta, R.K. (2008) *Polymeric Nanocomposites: Theory and Practice*, Carl Hanser Publishers, Cincinnati, Ohio.
14. Komarneni, S. (1992) *J. Mater. Chem.*, **2**, 1219.
15. Giannelis, E.P., Krishnamoorti, R., and Manias, E. (1999) *Adv. Polym. Sci.*, **138**, 107.
16. Ray, S.S. and Okamoto, M. (2003) *Prog. Polym. Sci.*, **28**, 1539.
17. Paul, D.R. and Robeson, L.M. (2008) *Polymer*, **49**, 3187.
18. Zou, H., Wu, S.S., and Shen, J. (2008) *Chem. Rev.*, **108**, 3893.
19. Zeng, Q.H., et al. (2005) *J. Nanosci. Nanotechnol.*, **5**, 1574.
20. Goettler, L.A., Lee, K.Y., and Thakkar, H. (2007) *Polym. Rev.*, **47**, 291.
21. Lach, R., et al. (2006) *Macromol. Mater. Eng.*, **291**, 263.
22. Tucker, C.L., III and Liang, E. (1999) *Compos. Sci. Technol.*, **59**, 655.
23. Odegard, G.M., Pipes, R.B., and Hubert, P. (2004) *Compos. Sci. Technol.*, **64**, 1011.
24. Tandon, G.P. and Weng, G.J. (1984) *Polym. Compos.*, **5**, 327.
25. Odegard, G.M., et al. (2003) *Compos. Sci. Technol.*, **63**, 1671.
26. Pipes, R.B. and Hubert, P. (2002) *Compos. Sci. Technol.*, **62**, 419.
27. Hoogerbrugge, P.J. and Koelman, J.M.V.A. (1992) *Europhys. Lett.*, **19**, 155.
28. Oono, Y. and Puri, S. (1988) *Phys. Rev. A*, **38**, 434.
29. Bahiana, M. and Oono, Y. (1990) *Phys. Rev. A*, **41**, 6763.
30. Doi, M. (2003) *Macromol. Symp.*, **195**, 101.
31. Zeng, Q.H., Yu, A.B., and Lu, G.Q. (2008) *Prog. Polym. Sci.*, **33**, 191.
32. Kremer, K. and Muller-Plathe, F. (2001) *MRS Bull.*, **26**, 205.
33. Ortiz, M., et al. (2001) *MRS Bull.*, **26**, 216.
34. Peng, G.W., et al. (2000) *Science*, **288**, 1802.
35. Thompson, R.B., et al. (2001) *Science*, **292**, 2469.
36. Ginzburg, V.V., Qiu, F., and Balazs, A.C. (2002) *Polymer*, **43**, 461.
37. Lee, J.Y., et al. (2002) *Phys. Rev. Lett.*, **89**, 155503.
38. Thompson, R.B., et al. (2002) *Phys. Rev. E*, **66**, 031801.
39. Lee, J.Y., et al. (2002) *Macromolecules*, **35**, 4855.
40. Bicerano, J., et al. (2004) *J. Macromol. Sci., Polym. Rev. C*, **44**, 53.

41 Scocchi, G., et al. (2007) *J. Phys. Chem. B*, **111**, 2143.
42 Fermeglia, M. and Pricl, S. (2007) *Prog. Org. Coat.*, **58**, 187.
43 Glotzer, S.C. and Starr, F.W. (2001) *Foundations of Molecular Modeling and Simulation: Proceedings of the First International Conference on Molecular Modeling and Simulation* (eds P.T. Cummings, P.R. Westmoreland, and B. Carnahan), American Institute of Chemical Engineers, Keystone.
44 Starr, F.W. and Glotzer, S.C. (2001) *Filled and Nanocomposite Polymer Materials, Materials Research Symposium Proceedings* (eds A.I. Nakatani, et al.), Materials Research Society, Warrendale.
45 Porter, D. (2004) *Mater. Sci. Eng. A*, **365**, 38.
46 Buxton, G.A. and Balazs, A.C. (2003) *Phys. Rev. E*, **67**, 031802.
47 Shou, Z.Y., Buxton, G.A., and Balazs, A.C. (2003) *Compos. Interfaces*, **10**, 343.
48 Borodin, O., et al. (2005) *J. Polym. Sci. Part B: Polym. Phys.*, **43**, 1005.
49 Buxton, G.A. and Balazs, A.C. (2004) *Mol. Simul.*, **30**, 249.
50 Sheng, N., et al. (2004) *Polymer*, **45**, 487.
51 Clancy, T.C. and Gates, T.S. (2006) *Polymer*, **47**, 5990.
52 Frankland, S.J.V., et al. (2003) *Compos. Sci. Technol.*, **63**, 1655.
53 Zhu, R., Pan, E., and Roy, A.K. (2007) *Mater. Sci. Eng. A*, **447**, 51.
54 Mokashi, V.V., Qian, D., and Liu, Y.J. (2007) *Compos. Sci. Technol.*, **67**, 530.
55 Frankland, S.J.V., et al. (2002) *J. Phys. Chem. B*, **106**, 3046.
56 Gersappe, D. (2002) *Phys. Rev. Lett.*, **89**, 058301.
57 Adnan, A., Sun, C.T., and Mahfuz, H. (2007) *Compos. Sci. Technol.*, **67**, 348.
58 Sen, T.Z., et al. (2005) *Polymer*, **46**, 7301.
59 Kalfus, J. and Jancar, J. (2007) *J. Polym. Sci. Part B: Polym. Phys.*, **45**, 1380.
60 Suter, J.L., et al. (2007) *J. Phys. Chem. C*, **111**, 8248.
61 Capaldi, F.M., Boyce, M.C., and Rutledge, G.C. (2006) *J. Chem. Phys.*, 124.
62 Patel, R.R., Mohanraj, R., and Pittman, C.U. (2006) *J. Polym. Sci. Part B: Polym. Phys.*, **44**, 234.
63 Zheng, X.Y., et al. (2007) *Continuum Mech. Therm.*, **18**, 377.
64 de Villoria, R.G. and Miravete, A. (2007) *Acta Mater.*, **55**, 3025.
65 Wagner, H.D. (2002) *Chem. Phys. Lett.*, **361**, 57.
66 Lee, K.Y. and Paul, D.R. (2005) *Polymer*, **46**, 9064.
67 Luo, J.J. and Daniel, I.M. (2003) *Compos. Sci. Technol.*, **63**, 1607.
68 Jo, C., Fu, J., and Naguib, H.E. (2006) *Polym. Eng. Sci.*, **46**, 1787.
69 Brune, D.A. and Bicerano, J. (2002) *Polymer*, **43**, 369.
70 Wu, Y.P., et al. (2004) *Polym. Test.*, **23**, 903.
71 Fisher, F.T., Bradshaw, R.D., and Brinson, L.C. (2003) *Compos. Sci. Technol.*, **63**, 1689.
72 Bradshaw, R.D., Fisher, F.T., and Brinson, L.C. (2003) *Compos. Sci. Technol.*, **63**, 1705.
73 Ashrafi, B. and Hubert, P. (2006) *Compos. Sci. Technol.*, **66**, 387.
74 Song, Y.S. and Youn, J.R. (2006) *Polymer*, **47**, 1741.
75 Avella, M., et al. (2004) *Mater. Sci. Technol.*, **20**, 1340.
76 Bondioli, F., et al. (2005) *J. Appl. Polym. Sci.*, **97**, 2382.
77 Cannillo, V., et al. (2006) *Compos. Sci. Technol.*, **66**, 1030.
78 Saber-Samandari, S., and Khatibi, A.A. (2006) *Fracture of Materials: Moving Forwards* (eds H.Y. Liu, X.Z. Hu, and M. Hoffman), Trans Tech Publications, Switzerland.
79 Tseng, K.K. and Wang, L.S. (2004) *J. Nanopart. Res.*, **6**, 489.
80 Hbaieb, K., et al. (2007) *Polymer*, **48**, 901.
81 Seidel, G.D. and Lagoudas, D.C. (2006) *Mech. Mater.*, **38**, 884.
82 Hammerand, D.C., Seidel, G.D., and Lagoudas, D.C. (2007) *Mech. Adv. Mater. Struct.*, **14**, 277.
83 Liu, Y.J. and Chen, X.L. (2003) *Electron. J. Bound. Elem.*, **1**, 316.
84 Liu, Y.J. and Chen, X.L. (2003) *Mech. Mater.*, **35**, 69.
85 Chen, X.L. and Liu, Y.J. (2004) *Comput. Mater. Sci.*, **29**, 1.
86 Huang, G., et al. (2006) *J. Appl. Mech.*, **73**, 737.

87 Miyagawa, H., Rich, M.J., and Drzal, L.T. (2004) *J. Polym. Sci. Part B: Polym. Phys.*, **42**, 4391.
88 Odegard, G.M., et al. (2002) *Compos. Sci. Technol.*, **62**, 1869.
89 Odegard, G.M., Clancy, T.C., and Gates, T.S. (2005) *Polymer*, **46**, 553.
90 Valavala, P.K. and Odegard, G.M. (2005) *Rev. Adv. Mater. Sci.*, **9**, 34.
91 Pipes, R.B. and Hubert, P. (2003) *Nano Lett.*, **3**, 239.
92 Pipes, R.B., et al. (2003) *Compos. Sci. Technol.*, **63**, 1349.
93 Pipes, R.B. and Hubert, P. (2003) *Compos. Sci. Technol.*, **63**, 1571.
94 Liao, K. and Li, S. (2001) *Appl. Phys. Lett.*, **79**, 4225.
95 Wong, M., et al. (2003) *Polymer*, **44**, 7757.
96 Qian, D., Liu, W.K., and Ruoff, R.S. (2003) *Compos. Sci. Technol.*, **63**, 1561.
97 Gou, J.H., et al. (2004) *Comput. Mater. Sci.*, **31**, 225.
98 Gou, J.H., et al. (2005) *Compos. Part B: Eng.*, **36**, 524.
99 Li, C.Y. and Chou, T.W. (2003) *J. Nanosci. Nanotechnol.*, **3**, 423.
100 Gao, X.L. and Li, K. (2005) *Int. J. Solids Struct.*, **42**, 1649.
101 Tsai, J. and Sun, C.T. (2004) *J. Compos. Mater.*, **38**, 567.
102 Pryamitsyn, V. and Ganesan, V. (2006) *Macromolecules*, **39**, 844.
103 Salaniwal, S., Kumar, S.K., and Douglas, J.F. (2002) *Phys. Rev. Lett.*, **89**, 258301.
104 Sen, S., et al. (2007) *Macromolecules*, **40**, 4059.
105 Chauve, G., et al. (2005) *Biomacromolecules*, **6**, 2025.
106 Yoon, P.J., Fornes, T.D., and Paul, D.R. (2002) *Polymer*, **43**, 6727.
107 Fornes, T.D. and Paul, D.R. (2003) *Polymer*, **44**, 4993.
108 Li, X.D., et al. (2007) *J. Nanosci. Nanotechnol.*, **7**, 2309.
109 Buxton, G.A. and Balazs, A.C. (2002) *J. Chem. Phys.*, **117**, 7649.
110 Kojima, Y., et al. (1993) *J. Polym. Sci., Part A: Polym. Chem.*, **31**, 983.
111 Kojima, Y., et al. (1993) *J. Polym. Sci., Part A: Polym. Chem.*, **31**, 1755.
112 Lordi, V. and Yao, N. (2000) *J. Mater. Res.*, **15**, 2770.
113 Zheng, Q.B., et al. (2007) *J. Phys. Chem. C*, **111**, 4628.
114 Minisini, B. and Tsobnang, F. (2005) *Compos. Part A: Appl. Sci. Manuf.*, **36**, 539.
115 Aleperstein, D., et al. (2005) *J. Appl. Polym. Sci.*, **97**, 2060.
116 Chabert, E., et al. (2004) *Compos. Sci. Technol.*, **64**, 309.
117 Raos, G., Moreno, M., and Elli, S. (2006) *Macromolecules*, **39**, 6744.
118 Liu, H. and Brinson, L.C. (2006) *J. Appl. Mech.*, **73**, 758.
119 Guz, A.N., Rodger, A.A., and Guz, I.A. (2005) *Int. Appl. Mech.*, **41**, 233.
120 Guz, A.N. (2006) *Int. Appl. Mech.*, **42**, 19.
121 Xia, Z.H., Guduru, P.R., and Curtin, W.A. (2007) *Phys. Rev. Lett.*, **98**, 245501
122 Zhang, W., Picu, R.C., and Koratkar, N. (2007) *Appl. Phys. Lett.*, **91**, 193109.
123 Zhang, W., Picu, R.C., and Koratkar, N. (2008) *Nanotechnology*, **19**, 285709.
124 Mallick, P.K. and Zhou, Y.X. (2003) *J. Mater. Sci.*, **38**, 3183.
125 Boo, W.J., Liu, J., and Sue, H.J. (2006) *Mater. Sci. Technol.*, **22**, 829.
126 Song, M. and Chen, L. (2006) *Macromol. Theory Simul.*, **15**, 238.
127 Jeong, B.W., Lim, J.K., and Sinnott, S.B. (2007) *Nanotechnology*, **18**, 485715.
128 Seshadri, M. and Saigal, S. (2007) *J. Eng. Mech.*, **133**, 911.
129 Fyta, M.G., et al. (2006) *Phys. Rev. Lett.*, **96**, 185503.
130 Zimmermann, A., Hoffman, M., and Rodel, J. (2000) *J. Mater. Res.*, **15**, 107.
131 Ward, D.K., Curtin, W.A., and Qi, Y. (2006) *Acta Mater.*, **54**, 4441.
132 Ward, D.K., Curtin, W.A., and Qi, Y. (2006) *Compos. Sci. Technol.*, **66**, 1151.
133 Friedrich, K. and Schlarb, A.K. (2008) *Tribology of Polymeric Nanocomposites: Friction and Wear of Bulk Materials and Coatings*, Elsevier, Oxford.
134 Dasari, A., Yu, Z.-Z., and Mai, Y.-W. (2009) *Mater. Sci. Eng. R*, **63**, 31.
135 Yang, J.L., et al. (2006) *Polymer*, **47**, 6745.
136 Ohji, T. (2003) *J. Ceram. Soc. Jpn*, **111**, 793.
137 Perez, C.J., Alvarez, V.A., and Vazquez, A. (2008) *Mater. Sci. Eng. A*, **480**, 259.

15
Morphology Generation in Polymer Nanocomposites Using Various Layered Silicates

Kenji Tamura and Hirohisa Yamada

15.1
Introduction

Following the pioneering studies of the Toyota research groups, who demonstrated the first practical application of a clay/polyamide 6 nanocomposite in the automobile industry [1–3], numerous other groups have reported the preparation of exfoliated clay/polymer nanocomposites in a variety of polymer systems, including polyimides [4], epoxies [5–8], elastomers [9, 10], polypropylene [11–13], polyester [14], polylactide [15, 16], and hydrogels [17]. More often than not, the clay (layer silicate) component in conventional polymer nanocomposites is smectite clay (e.g., montmorillonite; MMT). In an *exfoliated* clay/polymer nanocomposite (one particular type of nanocomposite), the layers of the clay component are individually dispersed in the polymer matrix. The resulting nanocomposite exhibits a high tensile strength, modulus and heat distortion temperature, as well as excellent barrier properties and flame resistance, even with a low clay content. It is therefore of great interest to understand the factors underlying this high performance. Numerous reports have identified the aspect ratio (the ratio of lateral length to thickness) of the dispersed clay layers as a key influence in this improvement [18]. An individual clay nanolayer has the highest possible aspect ratio when it is in its fully exfoliated state. For further improvements in the properties of the nanocomposite (through an increase in the aspect ratios), new layered materials with higher aspect ratios are needed, besides the commonly used clay.

In the following section, the various structures and parameters of layered silicates that are important when controlling the aspect ratio of the dispersed silicate platelets are highlighted. Later sections in the chapter will details the specific types of nanocomposite that have been prepared using various layered silicates. Recent studies extending this research into controlling techniques for nanocomposite preparation are also described.

Optimization of Polymer Nanocomposite Properties. Edited by Vikas Mittal
Copyright © 2010 WILEY-VCH Verlag GmbH & Co. KGaA, Weinheim
ISBN: 978-3-527-32521-4

15.2
Aspects of Layered Silicates

15.2.1
General Structure [19]

The structure of layered silicates consists of sheets of tetrahedra (T) and a sheet of edge-sharing octahedra (O) (Figure 15.1). The junction plane between the tetrahedral and octahedral sheets consists of shared apical oxygen atoms of the tetrahedrons and unshared hydroxyls (or fluorides). The 1:1 TO layer type (kaolin) shown in Figure 15.1a has one tetrahedral sheet which shares corners with an octahedral sheet. The thickness of this two-sheet unit is about 0.7 nm. The 2:1 TOT layer type (Figure 15.1b), with no interlayer cation (composed of pyrophyllite and talc), consists of two sheets of tetrahedra, with an octahedral sheet sandwiched in between. These 2:1 layers are electrostatically neutral, in the ideal case, with no interlayer ion present. The thickness of this 2:1 unit is about 0.91~0.94 nm. Figure 15.1c shows the 2:1 layer type with interlayer cations (mainly composed of smectite, vermiculite, illite, and mica groups). It also contains 2:1 layers (similar to Figure 15.1b), but it differs from the latter which uses pyrophyllite-talc, in that a more significant amount of isomorphous substitutions occurs, sufficient to raise the charge per unit cell. The net negative charge that arises from this is compen-

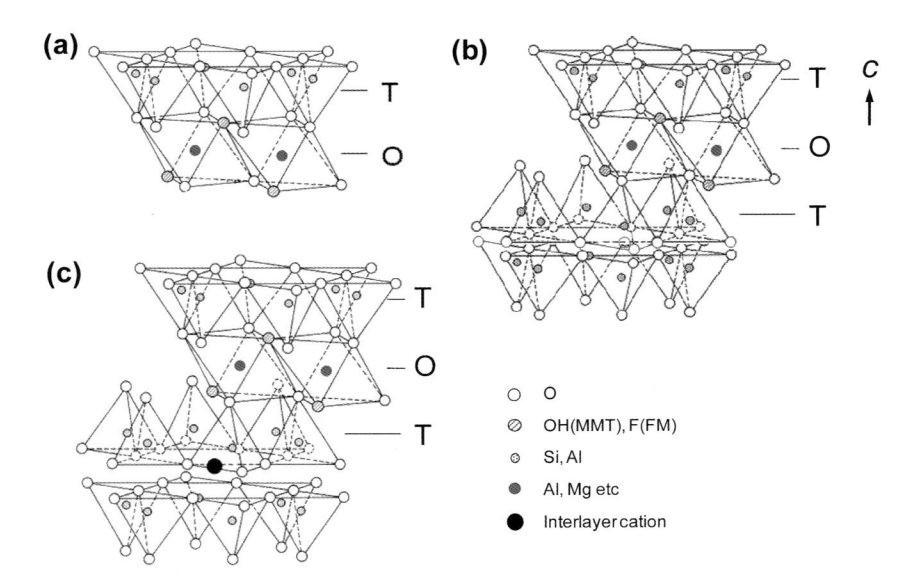

Figure 15.1 Typical structures of layered silicates. (a) 1:1 layer type, kaolin group; (b) 2:1 layered type (e.g., pyrophyllite, talc); (c) 2:1 layer type with interlayer cation (primarily K, Na, Ca, and Mg).

sated for by interlayer cations (primarily K, Na, Ca, and Mg). The thickness of this platelet is about 1 nm.

The concepts of "layer," "particle," "aggregate," and "association" in layered silicates are dependent upon the size of the unit cell, the three-dimensional (3-D) arrangement, and the chemical composition. The various properties of layered silicates due to the above criteria are reflected in their morphology and their physico-chemical properties. In the figures that follow, lateral extensions in the (a, b) plane are measured using images of objects within the sheet that are normal to the optic axis. The thickness, perpendicular to the (a, b) plane, is measured using images obtained from objects within the sheet that are oriented parallel to the optic axis of the microscope. After measuring the parameters of individual objects, they can be categorized by grain size or selection criteria (e.g., loose or aggregate). The features of typical layered silicates are introduced below. It is important that attention is always paid to the size and shape of the layered silicates that are used, so that the different species of layered silicates can be related to the varying properties of the prepared nanocomposites.

15.2.2
Various Types of Layered Silicates

Most previous reviews have dealt with layered silicate/polymer nanocomposites based on smectite clays. However, smectite clay is just one of many possible bases that can be used. Typical species of layered silicates are shown below.

- **1:1 TO Layer Silicates: Kaolin:** The kaolin group consists of 1:1 layer structures with the general composition of $Al_2Si_2O_5(OH)_5$. Kaolinite (Figure 15.2a), dickite, and nacrite are polytypes [19]. Halloysite is a hydrated polymorph of kaolinite with curved layers and a basal spacing of about 1 nm due to the hydration.

- **2:1 TOT Layer Silicates (with no interlayer cation): Pyrophyllite and Talc:** The ideal 2:1 layer structures of pyrophyllite, $Al_2Si_4O_{10}(OH)_2$ and talc, $Mg_3Si_4O_{10}(OH)_2$, are electrically neutral, and hence no charge balancing cation is present in the interlayer space. Continuous layers are held together by van der Waals interactions.

- **2:1 TOT Layer Silicates (with Interlayer Cation): Smectite:** Smectite is a 2:1 type layered silicate with a total negative charge of between 0.2–0.6 per unit cell. Typical smectites are MMT [$E_{0.33}(Al_{1.67}Mg_{0.33})Si_4O_{10}(OH)_2$], beidellite [$E_{0.33}Al_2(Si_{3.67}Al_{0.33})O_{10}(OH)_2$], nontronite [$E_{0.33}Fe_2^{3+}(Si_{3.67}Al_{0.33})O_{10}(OH)_2$], saponite [$E_{0.33}Mg_3(Si_{3.67}Al_{0.33})O_{10}(OH)_2$], hectorite [$E_{0.33}(Mg_{2.67}Li_{0.33})Si_4O_{10}(OH)_2$] and stevensite [$E_{0.16}Mg_{2.92}Si_4O_{10}(OH)_2$] (here, E denotes exchangeable interlayer hydrated cations such as Na, Li, Ca). Although smectite is a typical clay mineral, it has a very fine grain size (Figure 15.2b). In general, although particle size is a key parameter when defining clay, there is no generally accepted upper limit. In contrast, in pedology, for example, the clay fraction refers to a class of materials, the particles of which are smaller than 2 μm (equivalent spherical

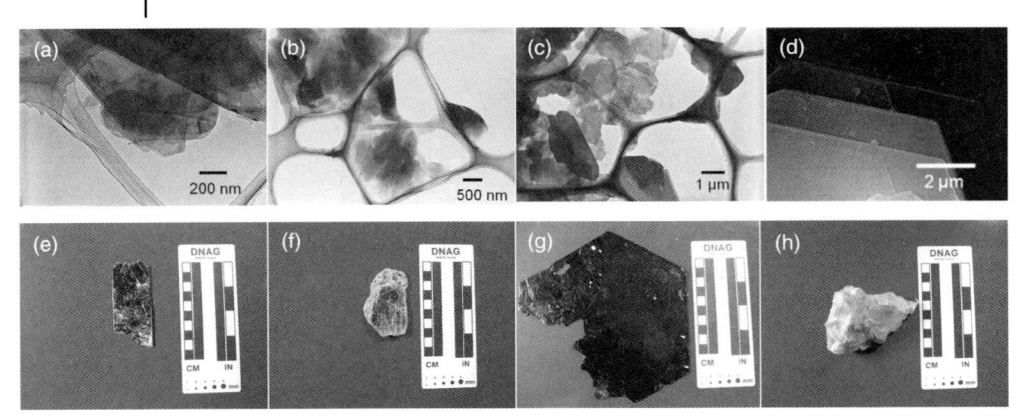

Figure 15.2 (a–c) Transmission electron microscopy, (d) scanning electron microscopy, and (e–h) digital camera images of various layered silicates. (a) Kaolinite; (b) Montmorillonite; (c) Synthetic expandable fluoromica (in this picture, sodium tetrasilicic mica; Coop Chemical); (d) Sericite; (e) Vermiculite; (f) Muscobite; (g) Biotite; (h) Synthetic nonexpandable fluoromica (in this picture, fluorophlogopite; Topy Ind.).

diameter). The ideal formula of synthetic expandable fluoro-mica (sodium tetrasilicic mica) is $Na_{1.0}Mg_{2.5}Si_4O_{10}F_2$. The estimated chemical composition of the sample shown in Figure 15.2c was $Na_{0.66}Mg_{2.68}(Si_{3.98}Al_{0.02})O_{10}F_2$. The layer charge of the synthetic expandable fluoro-mica is 0.66; therefore, it is categorized here with the smectite group.

- **Illite (Sericite):** Illite (sometimes referred to as sericite) also is a 2:1 TOT micaceous clay mineral, but is nonexpandable. This is because the charge per half unit cell (approximately 0.9) arises primarily from tetrahedral substitutions, and the interlayer cations are mainly K ions. The planes are packed parallel to each other, with K ions embedded in their surfaces, with strong electrostatic forces [20]. Illite contains less K and more water than true mica. The illite particle size is of the order of approximately several dozen micrometers (Figure 15.2d).

- **Vermiculite:** These minerals have a 2:1 TOT structure with layer charge densities of 0.6–0.9 per half unit cell (due to greater levels of isomorphous substitution). Vermiculites occur frequently as large crystals with a platey morphology, similar to micas, although they are softer and contain interlayer water (Figure 15.2e shows a larger specimen).

- **Mica:** Macrocrystalline micas are not considered to be smectite clay minerals (due to their larger size and greater layer charge). They are also 2:1 TOT minerals. True micas normally have K or Na interlayer cations, whereas in brittle mica the main interlayer cation is Ca. The high layer charge (~1.0 per half unit cell) of micas leads to electrostatic forces that are so strong that

15.2 Aspects of Layered Silicates

interlayer polar molecules are not present. Consequently, they do not swell under ordinary conditions. Furthermore, the crystal size can become large (Figure 15.2f–h; approximately several dozen centimeters).

- **Mixed-Layer Clay Minerals:** Mixed-layer clay minerals or interstratified clay minerals can be built up using two or more different components [21]. Interstratified clay minerals can have: (i) ordered (or regular) mixed-layer structures if different layers alternate along the c* direction in a periodic pattern (e.g., the stacking of generic type of A and type B layers can be ... ABABAB ... or ... AABAABAA ... or ... AAABAAABAAA ..., etc.); and (ii) disordered (or irregular) mixed-layer structures, wherein the stacking along the c* direction of type A and type B layers is random (e.g., ... ABBAAABABB ...).

Figure 15.3 shows an example of interstratification between illite (I): anhydrous layers, with periodicity of about 1.0 nm, and smectite (S): hydrated layers with periodicity of about 1.2~1.4 nm. Regular sequences are identified by special names. For example, the name "rectorite" is attributed to a regular interstratification of a mica and smectite; "tosudite" is a regular interstratification of chlorite and smectite [21]. Irregularly stacked structures are identified using the names of the two components, such as illite–smectite, smectite–chlorite, and kaolinite–smectite.

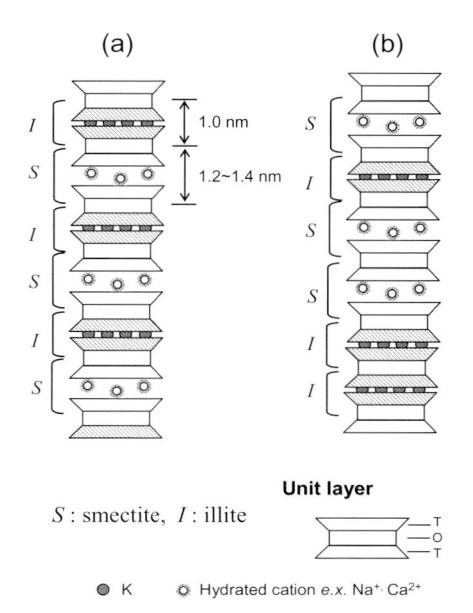

Figure 15.3 Structures of interstratified layer silicate (illite/smectite system). (a) Regularly interstratified structure (1:1 type); (b) Randomly interstratified structure.

15.3
Conventional Layered Silicate Polymer Nanocomposites using Smectite and Expandable Synthetic Fluoro-Mica

15.3.1
Relationship Between Morphology and Properties

Generally, clay/polymer nanocomposites are classified according to the level of intercalation and exfoliation (Figure 15.4). At one end of this classification scheme are well-ordered and stacked multilayers that consist of intercalated polymer chains resident between the host layers. At the other end are exfoliated silicate layers in a continuous polymer matrix. Some exfoliated clay/polymer nanocomposites have already been used in practical applications because of their acceptable rigidity, strength, and barrier properties, despite having a far lower silicate content than conventional composites that are filled with other minerals. Three main methods have been used to prepare these nanocomposites, namely, *solvent intercalation, in-situ polymerization,* and *melt blending.* The melt blending method has been identified as the most desirable for the preparation of clay/polymer nanocomposites that are to be used in the creation of thermoplastics, in part because it conforms to standard industrial methods for polymer processing.

The best results are obtained if the layered silicate (TOT type) is fully exfoliated into single layers with a thickness of approximately 1 nm. During exfoliation, the layered silicate not only becomes much smaller, but the shape changes simultaneously from rigid platelets to thin, flexible nanolayers. The shape of a layered silicate platelet is usually expressed by its aspect ratio. An important factor in achieving excellent performance from layered silicate/polymer nanocomposites is a large aspect ratio of the dispersed silicate platelets.

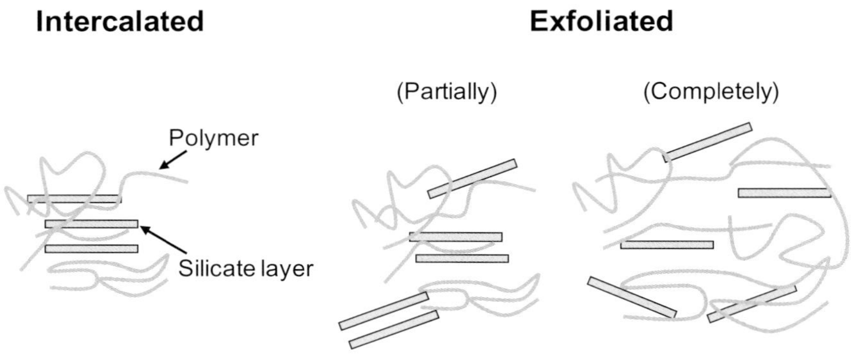

Figure 15.4 Dispersion conditions of layered silicate in a nanocomposite. (a) Well-stacked intercalated layered silicates with polymer chains; (b) Partially exfoliated layered silicate/polymer nanocomposite; (c) Completely exfoliated layered silicate/polymer nanocomposite.

15.3.2
Properties of Conventional Layered Silicate/Polymer Nanocomposites

Today, there is increasing demand for a reduction in the weight and cost of industrial materials used in the automotive market. Plastic products used in automobiles are required to perform well in terms of rigidity and strength (compared to conventional products such as mineral-filled composites and fiber-reinforced plastics), but at the same time are also expected to show the benefits of a lower weight and reduced cost. Figure 15.5 shows the specific rigidities of two different polyamide-66 (PA66) composites, one using a distealyldimethylammonium-modified synthetic expandable fluoromica (DSDMA-SEFM), and the other a conventional glass fiber-reinforced PA66 (so-called FRP) [22]. The nanocomposite using synthetic expandable flouromica clearly shows high rigidity, and also has a significantly lighter weight compared to the conventional FRP.

Figure 15.6 shows the relationship between the heat distortion temperature (HDT) and the flexural modulus for three materials: polyamide 6 (PA6) containing DSDMA-SEFM; PA6 containing DSDMA-modified synthetic smectite (DSDMA-SSm); and PA6 containing pristine SEFM [22]. The nanocomposite based on the DSDMA-SEFM had a significantly higher modulus and HDT than both the DSDMA-SSm-based nanocomposite and the pristine SEFM-based composite.

Figure 15.7 shows the transmission electron microscopy (TEM) images of PA6 composites containing DSDMA-SEFM and DSDMA-SSm (silicate content 5 wt%), respectively; the images show the cross-sections of the dispersed silicates. The image in Figure 15.7a (showing layers with larger aspect ratios compared to those in Figure 15.7b) shows a correlation between the improved properties of DSDMA-SEFM and the larger aspect ratios.

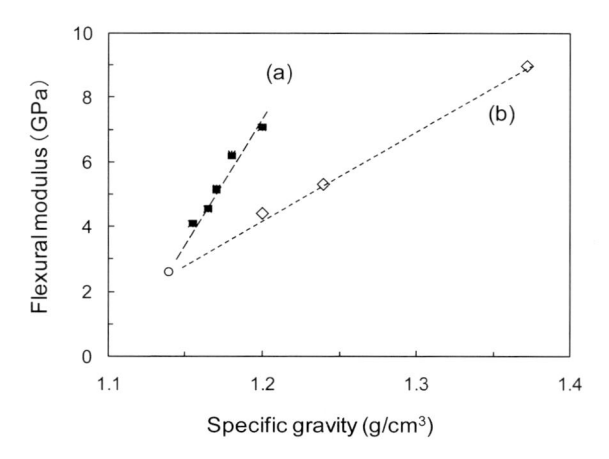

Figure 15.5 Specific rigidity of (a) DSDMA-SEFM/PA66 nanocomposite and (b) glass fiber-reinforced PA66 (FRP). The nanocomposite shows a high specific stiffness. Reprinted with permission from Ref. [22]; copyright Plastics Age Co. Ltd.

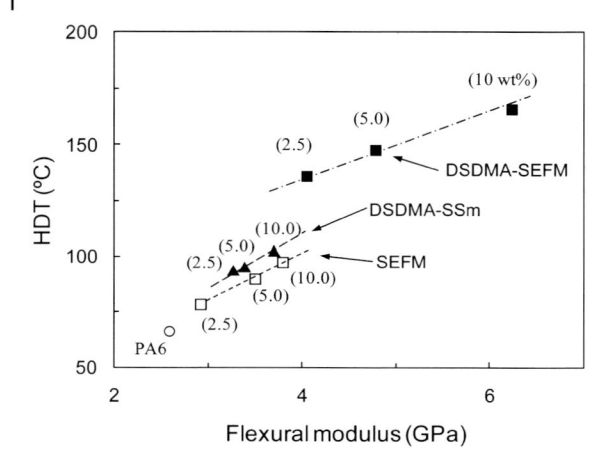

Figure 15.6 The relationship between heat distortion temperature (HDT) and flexural modulus for neat PA6 (○), PA6 nanocomposites containing DSDMA-SEFM (■), DSDMA-SSm (▲), and SEFM/PA6 composites (□). The figures in parentheses indicate silicate content (wt%). Reprinted with permission from Ref. [22]; copyright Plastics Age Co. Ltd.

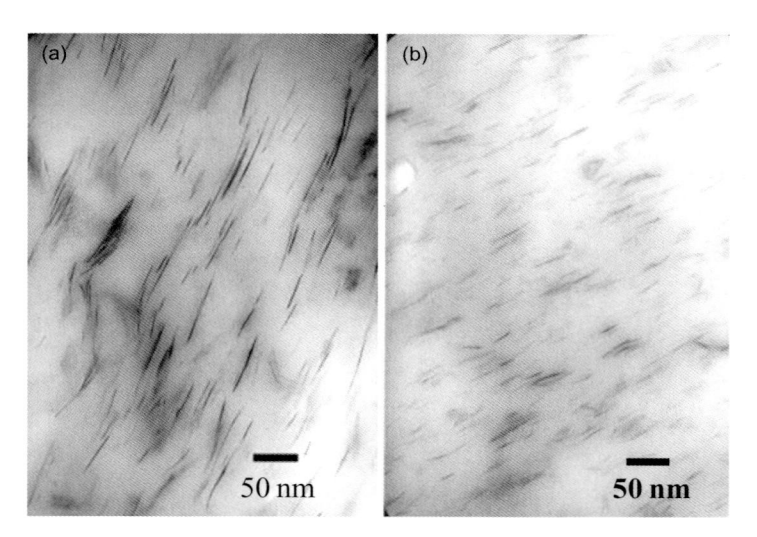

Figure 15.7 Transmission electron microscopy images of (a) DSDMA-SEFM/PA6 and (b) DSDMA-SSm/PA6 nanocomposites (silicate content = 5 wt%).

The aspect ratio of the silicate layers varies greatly, depending on the species and physico-chemical properties of the layered silicate, as well as the manufacturing process used. From the perspective of development of nanocomposite technology, therefore, it is very important to know the aspect ratios of the silicate platelets that are dispersed in a polymer, and to clarify the relationship between the mor-

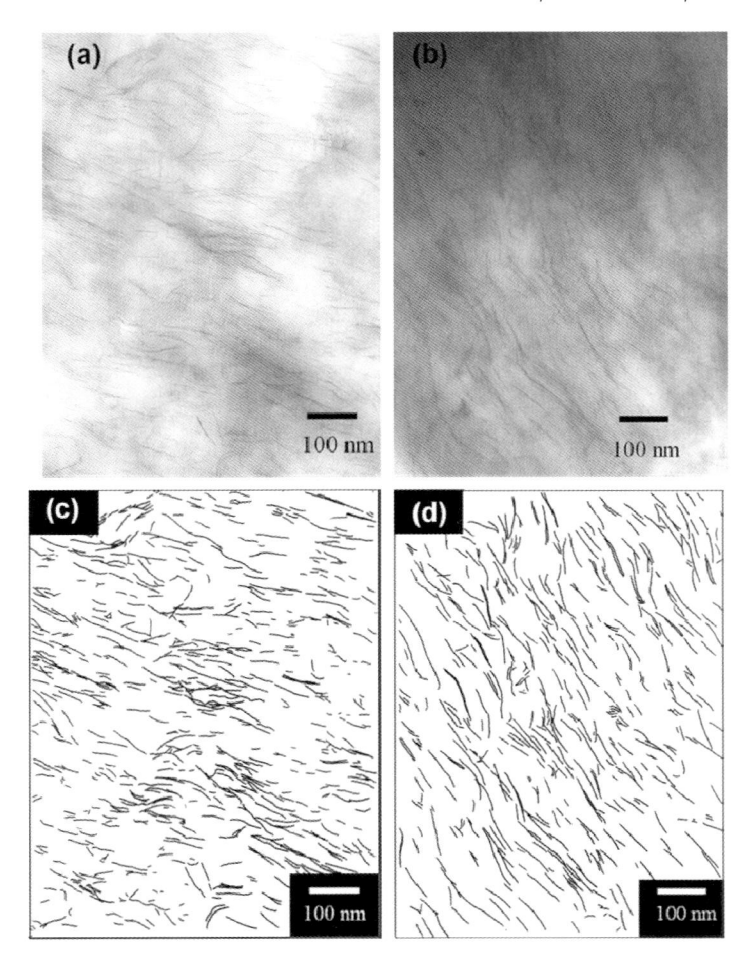

Figure 15.8 Transmission electron microscopy (TEM) images and binary images from TEM images of DDA-MMT/PA6 and DDA-SEFM/PA6 nanocomposites containing 5 wt% silicate. (a, c) DDA-MMT/PA6 nanocomposite; (b, d) DDA-SEFM/PA6 nanocomposite. Reprinted with permission from Ref. [24]; copyright John Wiley & Sons, Ltd.

phology (e.g., particle shape and size) and properties of the nanocomposite (e.g., rigidity, strength, and heat resistance). Fornes *et al.* have developed a technique of binarizing the TEM images, and then measuring the length and width of the dispersed particles from them [23]. In this way it is to estimate statistically the average aspect ratio of particles dispersed in nanocomposites.

Figure 15.8a and b show TEM images of PA6 nanocomposites containing dodecylamine hydrochloride modified-MMT (DDA-MMT) and -synthetic expandable fluoromica (DDA-SEFM), both of which were prepared using a melt blending method [24]. The observed TEM images were binarized using the thresholding procedure described by Fornes *et al.*, and the length (l), thickness (t) and aspect ratio (α) of the dispersed particles subsequently calculated (Figure 15.8c and d) [24]. The silicate platelets in PA6 were classified as either single-layer or laminated, and the sheet length ($l_{platelet}$) and thickness ($t_{platelet}$) were measured individually. The aspect ratio (α) was then calculated using the relationship:

$$\alpha = l_{platelet}/t_{platelet} \tag{15.1}$$

where $t_{platelet}$ is estimated to be 0.94 nm (the thickness may be calculated by adding the center-to-center distance between the outer oxygen atoms in the outer tetrahedral layers of the expandable silicate, to twice the sum of the atomic radius oxygen) [23, 25]. The number of layers (*n*) was counted from the TEM images, and the particle thickness ($t_{particle}$) calculated using

$$t_{particle} = d_{001}(n-1) + t_{platelet} \tag{15.2}$$

The morphological parameters (length [*l*] and thickness [*t*]) of the dispersed particles were measured from cross-sectional TEM images, and plotted in the histograms shown in Figure 15.9 [24]. The results of the morphological parameters

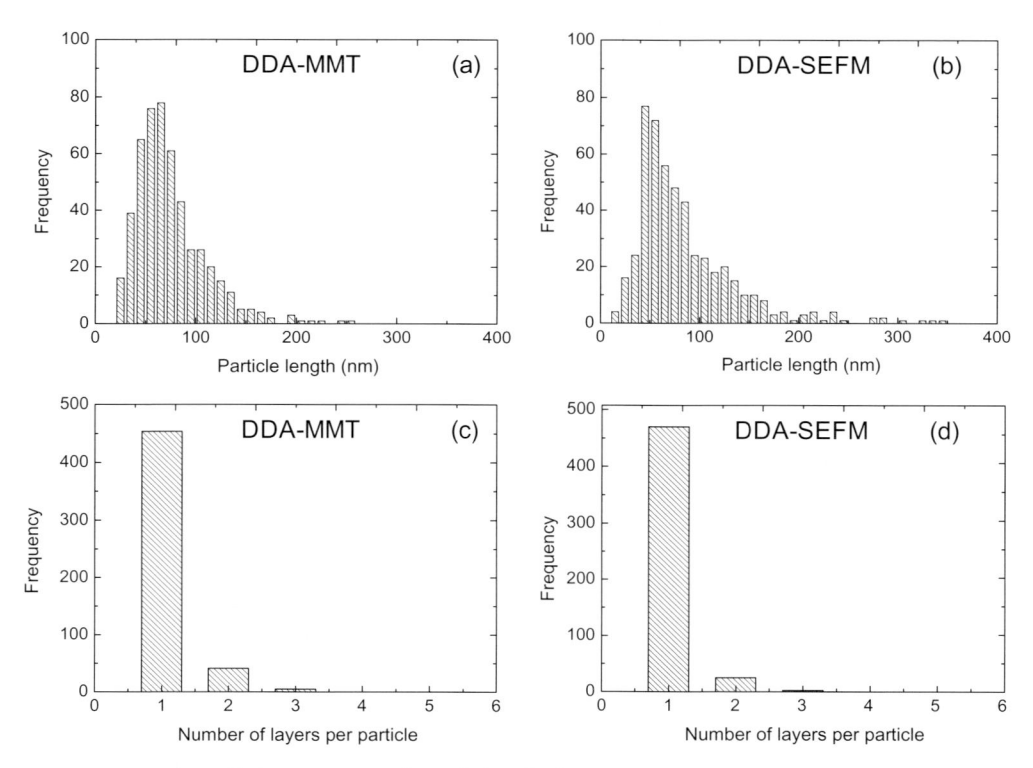

Figure 15.9 Histogram of particle length of platelets (a and b) and the number of layers (c and d) per particle data obtained by transmission electron microscopy images of PA6 nanocomposites containing DDA-MMT and DDA-SEFM. Reprinted with permission from Ref. [24]; copyright John Wiley & Sons, Ltd.

Table 15.1 Image analysis results obtained from transmission electron microscopy images of injection-molded PA6 nanocomposites containing DDA-MMT and DDA-SEFM. The values are mean ± standard deviation.

Parameter	DDA-MMT/PA6	DDA-SEFM/PA6
Number average particle length, l_n (nm)	76 ± 38	85 ± 50
Number average particle thickness, t_n (nm)	1.2 ± 0.3	1.2 ± 0.5
Number average particle aspect ratio, α	76 ± 39	85 ± 47

Reprinted with permission from Ref. [24]; copyright John Wiley & Sons, Ltd.

Table 15.2 Mechanical properties of PA6 nanocomposites containing DDA-MMT and DDA-SEFM. The figures in parentheses indicate the standard deviation.

Nanocomposite	Silicate content (wt%)	Flexural modulus (MPa)	Tensile modulus (MPa)	Elongation at break (%)	HDT (°C)
PA6	0	2300 (57)	2580 (120)	34 (7.3)	54
DDA-MMT/PA6 (2.5%)	1.6	3030 (50)	3250 (74)	6 (4.0)	65 (4.2)
DDA-MMT/PA6 (5.0%)	4.0	3880 (48)	3960 (48)	2 (0.2)	106 (2.3)
DDA-SEFM/PA6 (2.5%)	2.2	3350 (38)	3310 (287)	11 (3.2)	79 (0.3)
DDA-SEFM/PA6 (5.0%)	4.5	4210 (24)	4260 (89)	4 (0.2)	108 (4.7)

Reprinted with permission from Ref. [24]; copyright John Wiley & Sons, Ltd.

obtained by TEM analysis are summarized in Table 15.1 [24]. The average thickness of particles in both samples was approximately 1.2 nm, indicating that most were exfoliated to single layers. The number-average aspect ratio α for the DDAMMT-PA6 and DDASEFM-PA6 nanocomposites was estimated to be 76 and 85, respectively.

On comparing the two nanocomposites, it was found that DDA-SEFM/PA6 had a greater rigidity and heat resistance than DDA-MMT/PA6 (Table 15.2) [24]. Since the rigidity of nanocomposites is well known to be highly sensitive to the degree of exfoliation, it was of interest to know if the modulus was correlated with the aspect ratio of the dispersed particles. As can be seen from the image analysis data in Table 15.1, the pronounced reinforcement effects seen in DDA-SEFM/PA6 may have resulted from their high aspect ratios.

Figure 15.10 displays the permeability coefficients of the composites when tested using methanol vapor together with films of the composites, in a 40 °C atmosphere [24]. It is clear from these results that the nanocomposites showed larger improvements in barrier properties as the silicate content increased; this suggested that a higher silicate content posed a more tortuous path for the gas during its movement through the polymer [26, 27]. These studies indicated that three factors – the degree of exfoliation, the aspect ratio of the platelets, and the

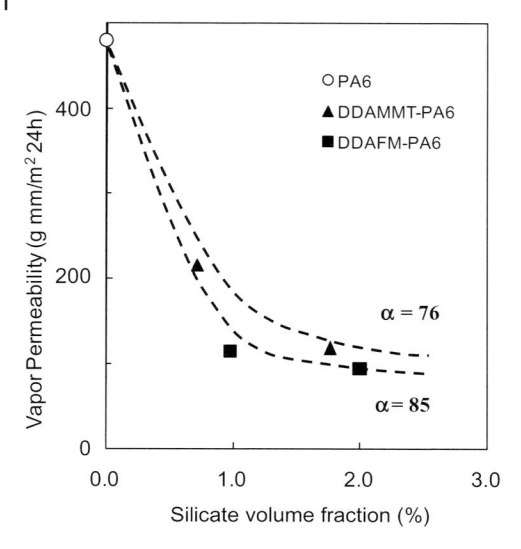

Figure 15.10 Experimental methanol vapor permeability versus silicate volume fraction at 40 °C. Reprinted with permission from Ref. [24]; copyright John Wiley & Sons, Ltd.

dispersion homogeneity of the platelets – played an important role in determining the properties of these high-performance nanocomposites. Thus, control over these three factors would represent keys to the future development and improvement of polymer nanocomposite technology. In the next section, attention is focused on recent investigations into the control of one these factors, namely, the aspect ratio of the platelets.

15.4
Aspect Ratio Variation Using Various Layered Silicates

15.4.1
Exfoliation of High Crystallinity Nonexpandable Mica

Generally, micaceous clay minerals and mica minerals have a relatively higher negative layer charge than smectite. Their interlayer K^+ ions are believed to be nonexchangeable because they are strongly bound to the interlayer surfaces [20], and for these reasons micaceous clay minerals and mica minerals have rarely been used in nanocomposites. If their layered silicates were to be fully exfoliated, however, the aspect ratio of the individual layers would greatly exceed that of the smectite clays (Figure 15.11). Recently, new polymer nanocomposites of this type were successfully prepared for the first time using a conventional process [28, 29]. The nonexpandable micas used were a natural potassium sericite (K-SE) and a synthetic potassium fluorophlogopite (K-FPH),

15.4 Aspect Ratio Variation Using Various Layered Silicates

(a) Clay (smectite)

Small Particle   Exfoliation ------▶   Polymer / Silicate nanolayer

~2 μm

(b) High Crystallinity Non-Expandable Mica

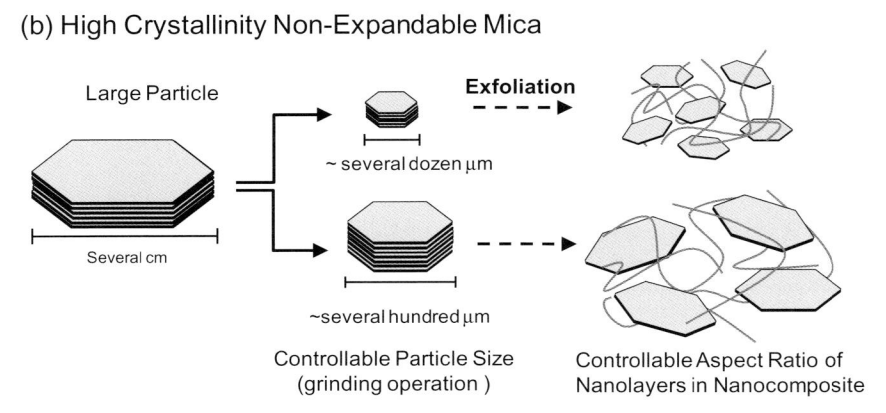

Large Particle (Several cm) → ~ several dozen μm   Exfoliation ----▶

→ ~several hundred μm ----▶

Controllable Particle Size (grinding operation)   Controllable Aspect Ratio of Nanolayers in Nanocomposite

Figure 15.11 Comparing the dimensions of silicate nanolayers derived from (a) smectite clay and (b) high-crystallinity nonexpandable mica.

both of which are commonly used, well-known micas. The respective compositions of these micas are $(K_{0.77}Na_{0.03})(Al_{1.67}Fe^{2+}_{0.18}Mg_{0.10})(Si_{3.24}Al_{0.76})O_{10}(OH)_2$ and $(K_{0.89}Mg_{0.02}Al_{0.03})Mg_{3.00}(Si_{2.97}Al_{1.03})(O_{10}F_2)$. In order to prepare exfoliated mica/polymer nanocomposites, it is very important to form a completely organically modified sample (for a much easier preparation of the intercalated stage) [28]. Thus, organically modified micas were prepared using an ion-exchange reaction between mica and an aqueous solution of dodecylamine (DDA) or octadecylamine (ODA) hydrochloride.

Figure 15.12 shows the TEM images of: (a) DDA-SE/epoxy; (b) ODA-SE/epoxy; and (c) DDA-FPh/PA6 nanocomposites. The samples in Figure 15.12 a and b were cured at 180 °C for 6 h, whilst the sample in Figure 15.12c was prepared using a melt-mixing process at 260 °C. The morphologies of the DDA-SE/epoxy and ODA-SE/epoxy nanocomposites showed silicate nanolayers with extremely high aspect ratios; in fact, they were on the order of several dozen to hundreds of times greater than that of conventional exfoliated clay/polymer nanocomposites. The DDA-FPh/PA6 image shows FPh platelets that are thinner than pristine FPh platelets. Typically, these platelets had a cross-sectional length of several tens to several thousands of nanometers, and a thickness of one to a few nanometers. The degree of exfoliation, however, was significantly lower than that of the more

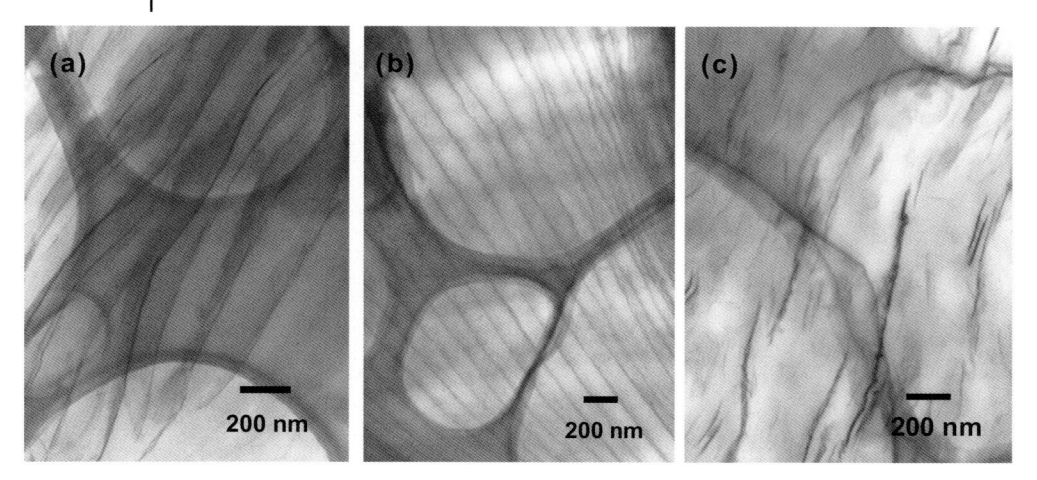

Figure 15.12 Transmission electron microscopy (TEM) images of thin sections of nanocomposites. (a) DDA-SE/epoxy system; (b) ODA-SE/epoxy system cured at 180 °C for 6 h; (c) DDA-FPh/PA6 system prepared by extruding process. Well-dispersed individual silicate layers (dark lines) are found in all images. For the DDA-SE/epoxy system (panel a), the nanolayer thickness of approximately 2–3 nm deduced from the TEM images is thicker than a single silicate layer; this may have been caused by tilted layers being inclined obliquely to the sectioned plane.

conventional DDA-MMT/PA6 nanocomposite (a typical smectite nanocomposite). For example, it has been recently reported [24] (and also discussed above) that approximately 90% of silicate platelets in the case of DDA-MMT/PA6 were dispersed into single layers of MMT. Similarly, DDA-SEFM/PA6 (an expandable silicate nanocomposite, which is categorized with the smectite group (as noted in Section 15.2.2) showed a level of exfoliation comparable to that of DDA-MMT/PA6. These differences between these two smectite-based nanocomposites and the FPh-based PA6 nanocomposite (as introduced in this section) are due to differences in the particle sizes and/or electronic interactions resulting from differences in the layer charges.

When comparing the differences in polymer processing between the DDA-SE/epoxy and DDA-FPh/PA6 systems, the shear stress used by the extruding machine had a significant influence on the dispersed particle size during exfoliation of the silicate layer. Due to this influence, as the dispersed silicate layers became thinner, the shear stress caused the particle length to be reduced. It was shown clearly in these studies that, for polymer nanocomposite systems, both the degree of exfoliation and the aspect ratio of the silicate platelets can play important roles in determining mechanical performance. Another important finding here was that the dispersion homogeneity (the distribution of particle parameters including particle length, thickness, and aspect ratio) of the silicate platelets can greatly influence the mechanical properties of the polymer nanocomposites.

15.4.2
Controlling the Number of Nanolayers (in the Dispersed Platelets): Interstratified Layered Silicate/Polymer Nanocomposites

In this section, a method used to control the number of layers of the dispersed silicate platelets is described. In order to achieve control of the number of layers, the first step was to prepare a layered silicate with alternating layers possessing antagonistic properties (i.e., swelling and nonswelling properties) [21]. Some natural clays consist of layers in which different types of silicate layer are stacked together (interstratification), with each type having either swelling or nonswelling properties; these silicates are termed "mixed-layer" or "interstratified layer" silicates. By using an interstratified layer silicate, nanolayers of silicates consisting of two to three layers per group could be dispersed in a polymer matrix. The use of new techniques such as these might represent an important step towards the creation of a new generation of polymer nanocomposite materials.

Recently a new, regularly interstratified layer silicate composed of talc/talc/smectite (TTS) in an ordered sequence was synthesized using a hydrothermal technique. An organically modified TTS was then prepared via an ion-exchange reaction from TTS and an aqueous solution of ODA hydrochloride; the resultant organically modified TTS was designated as ODA-TTS. The ordered interstratified TTS and organically modified TTS (ODA-TTS) structures are shown schematically in Figure 15.13 [30]. The final nanocomposite was prepared using the organically modified TTS (ODA-TTS) structures with intercalated compounds, obtained via a common described procedure.

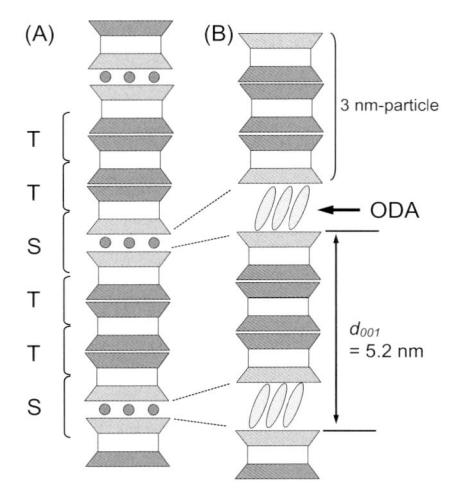

Figure 15.13 Schematic of orderly interstratified talc/talc/smectites (TTS) (A) and organically modified TTS (ODA-TTS) structures (B). Reprinted with permission from Ref. [30]; copyright Wiley-Blackwell.

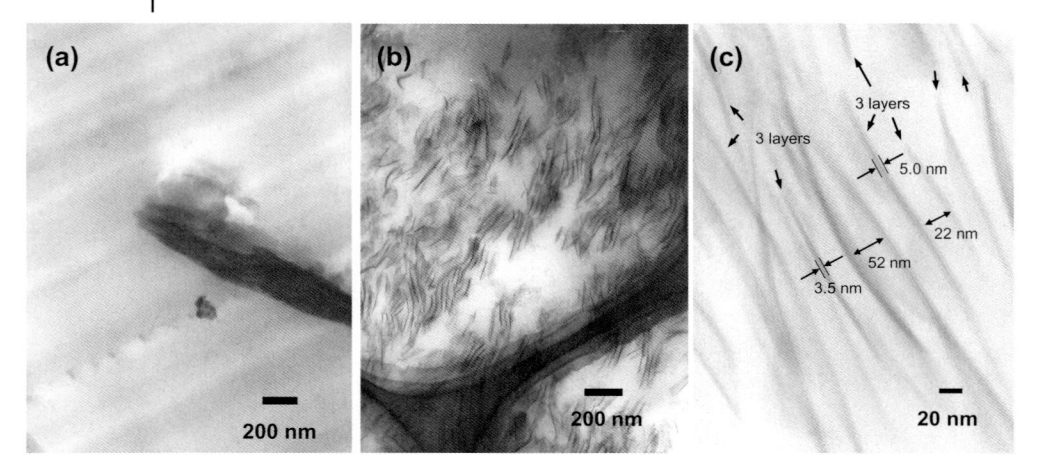

Figure 15.14 Transmission electron microscopy images of thin sections of glassy epoxy composites cured at 150 °C for 6 h. (a) TTS/epoxy system; (b, c) ODA-TTS/epoxy nanocomposite at low and high magnification, respectively (dispersion of three layers per group). Each TTS loading was 2.5 wt% on a silicate basis. Reprinted with permission from Ref. [30]; copyright Wiley-Blackwell.

The TEM images of conventional TTS/epoxy composite, and ODA-TTS/epoxy nanocomposite (at low and high magnification) are shown in Figure 15.14a–c [30]. In Figure 15.14a, the TTS is aggregated in the epoxy matrix with a cross-sectional length of approximately 1–2 μm and a thickness of 200–300 nm. In Figure 15.14b and c, the ODA-TTS/epoxy nanocomposite has an homogeneous dispersion of ODA-TTS platelets in the epoxy matrix. A closer examination of the nanocomposite at high magnification (Figure 15.14c) confirms a rather rigid platelet (cross-section), in which the length of the cross-section is a few hundred nanometers, and the thickness is approximately 3–5 nm. Most of the dispersed nanolayers shown in Figure 15.14c are actually comprised of three nanolayers in one group. The techniques used to achieve this provide a means of controlling the number of dispersed layers in polymer nanocomposite technology.

15.5
Summary

Several examples of polymer nanocomposite using different types of layered silicate have been introduced in this chapter, and the relationship between the aspect ratio of the dispersed silicates in the nanocomposite and the properties of the nanocomposite has been shown to have great significance. Generally, smectites and other expandable layered silicates are versatile materials that, in a practical sense, are well adapted to yield a wide variety of nanocomposites due to their ion-exchange, expandability, and colloidal properties. In contrast, the so-called "nonexpandable layered silicates" (e.g., micaceous clay minerals and mica) have not been used to create new nanocomposites, despite their high crystallinity and large particle size (which is over several dozen times that of smectite particles). Clearly, this lack of effort is based on firmly held notions regarding the nonexpandability or "difficult ion-exchangeability" of these materials. Likewise, regularly interstratified clay minerals have never been used in nanocomposites because it is very difficult to obtain an homogeneous structure (regular sequence of platelets).

The morphology of the mica/polymer nanocomposites demonstrated silicate nanosheets with an extremely high aspect ratio, at levels which generally were several dozens to hundreds of times greater than that of conventional exfoliated clay/polymer nanocomposites, and consequently their applications would be expected to extend over a wide range. Yet, other new composite systems may extend the field of high-performance materials beyond traditional applications, so as to encompass unexpected new functionalities. It would be significant – from the point of view not only of materials development but also from the environment – to elucidate the exfoliation behavior of nonexpandable layered silicates and to apply them in the field of nanotechnology.

References

1 Fukushima, Y., Okada, A., Kawasumi, M., Kurauchi, T., and Kamigaito, O. (1988) *Clay Miner.*, **23**, 27–34.
2 Kurauchi, T., Okada, A., Nomura, T., Nishio, T., Saegusa, S., and Deguchi, R. (1991) *SAE Tech. Pap.*, **910584**, 1–7.
3 Usuki, A., Kawasumi, M., Kojima, Y., Fukushima, Y., Okada, A., Kurauchi, T., and Kumigaito, O. (1993) *J. Mater. Res.*, **8**, 1179–1184.
4 Yano, K., Usuki, A., and Okada, A. (1997) *J. Polym. Sci., Part A: Polym. Chem.*, **35**, 2289–2294.
5 Messersmith, P.B. and Giannelis, E.P. (1994) *Chem. Mater.*, **6**, 1719–1725.
6 Wang, M.S. and Pinnavia, T.J. (1994) *Chem. Mater.*, **6**, 468–474.
7 Koerner, H., Jacobs, D., Tomlin, D.W., Busbee, J.D., and Vaia, R.D. (2004) *Adv. Mater.*, **16**, 297–302.
8 Ma, J., Yu, Z.Z., Zhang, Q.X., Xie, X.L., Mai, Y.W., and Luck, I. (2004) *Chem. Mater.*, **16**, 757–759.
9 Wang, Z. and Pinnavaia, T.J. (1998) *Chem. Mater.*, **10**, 3769–3771.
10 Lee, H.S., Fasulo, P.D., Rodgers, W.R., and Paul, D.R. (2005) *Polymer*, **46**, 11673–11689.
11 Hasegawa, N., Kato, M., Usuki, A., and Okada, A. (1997) *Macromolecules*, **30**, 6333–6338.
12 Garcés, J.M., Moll, D.J., Bicerano, J., Fibiger, R., and MacLeod, D.G. (2000) *Adv. Mater.*, **12**, 1835–1839.
13 Sun, T. and Garcés, J.M. (2002) *Adv. Mater.*, **14**, 128–130.
14 Tsai, T.Y. (2000) *Polymer-Clay Nanocomposites*, (eds T.J. Pinnavaia and G.W. Beall), John Wiley & Sons, Inc., New York.
15 Ray, R.R., Maiti, P., Okamoto, M., Yamada, K., and Ueda, K. (2002) *Macromolecules*, **35**, 3104–3110.
16 Maiti, P., Yamada, K., Okamoto, M., Ueda, K., and Okamoto, K. (2002) *Chem. Mater.*, **14**, 4654–4661.
17 Haraguchi, K. and Takehisa, T. (2002) *Adv. Mater.*, **14**, 1120–1124.
18 Fornes, T.D., Hunter, D.L., and Paul, D.R. (2004) *Polymer*, **45**, 2321–2331.
19 Bailey, S.W. (1980) *Crystal Structures of Clay Minerals and Their X-Ray Identification* (eds G.W. Brindley and G. Brown), Mineralogical Society, London.
20 van Olphen, H. (1977) *An Introduction to Clay Colloid Chemistry: For Clay*

21 Reynolds, R.C. (1980) *Crystal Structures of Clay Minerals and Their X-Ray Identification* (eds G.W. Brindley and G. Brown), Mineralogical Society, London.
22 Tamura, K. and Nakamura, J. (1999) *Plastics Age*, **45**, 106–113.
23 Fornes, T.D. and Paul, D.R. (2003) *Polymer*, **44**, 4993–5013.
24 Tamura, K., Uno, H., Yamada, H., and Umeyama, K. (2009) *J. Polym. Sci., Part B: Polym. Phys.*, **47**, 583–595.
25 Tsipursky, S.I. and Drits, V.A. (1984) *Clay Miner.*, **19**, 177–193.
26 Bharadwaj, R.K. (2001) *Macromolecules*, **34**, 9189–9192.
27 Nielsen, L.E. (1967) *J. Macromol. Sci. Chem.*, **A1**, 929–942.
28 Tamura, K., Yokoyama, S., Pascua, C.S., and Yamada, H. (2008) *Chem. Mater.*, **20**, 2242–2246.
29 Yokoyama, S., Tamura, K., and Yamada, H. (2009) *Chem. Lett.*, **38**, 48–49.
30 Tamura, K., Yamada, H., Yokoyama, S., and Kurashima, K. (2008) *J. Am. Ceram. Soc.*, **91**, 3668–3672.

16
Thermomechanical Properties of Nanocomposites
Lucia Helena Innocentini-Mei

16.1
Introduction

Today, nanocomposites continue to find novel applications as investigations prove their efficiency in different areas. Indeed, nanocomposites are equally popular among researchers from academia and industry, due to their excellent properties that have been described as being superior to those of both virgin polymers and conventional composites [1].

The application areas of nanocomposites include sectors such as automotive, medical, packaging and aerospace among others, as they have far superior thermal, mechanical and barrier properties than conventional composites.

During the past few years, nanocomposites based on silicates filled polymers have been the target of numerous research studies to identify "new materials," but using "old" raw materials. In order to produce nanocomposites, research groups are faced with the challenge of creating a more compatible material using an organic hydrophobic polymeric matrix and a hydrophilic nanostructured filler to obtain superior properties. Yet, by using knowledge acquired with "old" composite materials, a great deal of success has been achieved in preparing nanocomposites with good miscibility and compatibility so as to guarantee the desired properties, based largely on the chemical modification of the silicates that are used to introduce an organic character to a hydrophilic ceramic. To date, many reports have been produced on organic–inorganic composites that combine the desirable properties of a ceramic phase (heat resistance, retention of mechanical properties at elevated temperatures, low thermal expansion) with those of organic polymers (toughness, ductility, processability); the result has been the development of novel properties that are not demonstrated by regular composites [2].

For regular use in engineering applications ceramics must present, among others, with a low density, a high Young's modulus, a high resistance to temperature or a high heat distortion temperature (HDT), a low thermal conductivity, a high resistance to shearing and abrasion, a low expansion coefficient or good dimensional stability, high hardness, and magnetic and electrical properties. During the past 20 years, many investigations have been conducted to develop

Optimization of Polymer Nanocomposite Properties. Edited by Vikas Mittal
Copyright © 2010 WILEY-VCH Verlag GmbH & Co. KGaA, Weinheim
ISBN: 978-3-527-32521-4

advanced ceramics for use in areas such as industrial sensors, in recovery systems, electronics materials, and filters. As a consequence of these studies, *nanoceramics* have acquired huge industrial importance, as shown in a recent report from BCC Research. Amazingly, for the year 2006 this market totaled US$ 2.2 billion in the United States alone, with projections indicating a value of US$ 3.4 billion in 2011 (equivalent to a total annual growth rate of 8.9%). Moreover, it has been predicted that the nanoceramics market will grow by 23.4%, and the advanced ceramics market by 6.9% during this time (www.fapesp.br).

Among the ceramics used extensively to develop nanocomposites, montmorillonite (MMT) is probably the most widely known, it being regarded as a special member of the group of four dozen or so "clay minerals." These materials are silicates that contain Al, Fe, and Mg hydrates, and which have a layered crystalline structure that consists of continuous sheets of SiO_4 tetrahedron ordered in an hexagonal geometry, condensed with octahedral sheets that are made from metal hydroxide and are bi- and tri-positively charged. Consequently, each structural unit has two tetrahedral sheets, with one edge-shared octahedral sheet of aluminum or magnesium hydroxide between the two. The clays most often used in the preparation of polymeric layered silicate nanocomposites belong to the structural family of 2:1 phyllosilicates (smectites), of which MMT is the main representative. This class of clays presents layers formed by sheets with a thickness of about 1 nm and lateral dimensions from 100 nm to 1 μm. Depending on the nature of the intercalated cation among layers, the polymer used and the method of preparation, different composites can be obtained. These natural clays have hydrophilic characteristics due to the presence of Na^+ and Ca^{2+} cations housed in galleries, which link the layers by weak van der Waals forces. When Al^{3+} is replaced by Mg^{2+} or Fe^{2+}, or Mg^{2+} by Li^+ and Si^{4+} by Al^{3+}, the negative charges resultants are counterbalanced by the gallery cations, Na^+ and Ca^{2+} located in the interlayer space. Due to the hydrophilic character of the natural clays, their dispersion in organic polymers is not straightforward. In fact, this behavior led to the development of chemical modifications of clays in order to make them chemically compatible with the organic polymer matrices normally used in engineering [3].

16.2
Thermomechanical Analysis [4]

Polymers are viscoelastic materials that have properties intermediate between those of an *elastic material*, which instantly recover its original size and shape after deformation, and a *viscous material* that flow under the effect of an external force.

The viscoelastic behavior of polymer can be described as complex, as it depends on the chemical composition and molecular structure, temperature, frequency of solicitation, and also the stress to which the polymer is subjected.

The viscoelastic (or rheological) properties of a polymer can be characterized with a rheometer, which enables the analysis of effects of temperature, frequency, strain or tension on the response viscoelasticity of polymer. In dynamic mode, the sample is subjected to an alternating shear stress $\sigma(\omega)$, and the magnitude of the amplitude γ and phase angle δ from the resultant deformation $\gamma(\omega)$ is measured. With all these parameters, it is possible to calculate the complex shear modulus $E^*(\omega) = E' + jE''$ of the polymer. The real part E', which is called the *storage modulus*, is characteristic of the stiffness of the material, and means that the elastic energy stored and then returned by the polymer. In turn, the imaginary part E'' is the loss modulus; this is specific to the "sticky part" of the material; that is, the energy dissipated by macromolecular friction or the viscous dissipation within the polymer.

Mathematically, it is possible to understand how the elastic and viscous contributions of a material can be evaluated. If a material with an ideal elastic behavior is submitted to a sinusoidal deformation, then:

$$\gamma = \gamma_0 \sin(\omega t) \tag{16.1}$$

where γ_0 is the amplitude at maximum deformation; the response of the material will also be sinusoidal and in phase with the deformation:

$$\sigma = \sigma_0 \sin(\omega t). \tag{16.2}$$

Since $E = \sigma_0/\gamma_0$, the material answer may be written as:

$$\sigma = E\gamma_0 \sin(\omega t) \tag{16.3}$$

For a material with an ideal viscous behavior, the stress will present a sinusoidal response, but will be 90° out of phase with the deformation. For fluids, the stress and the rate of deformation are related by viscosity as:

$$\eta = \sigma/\dot{\gamma} \tag{16.4}$$

Substituting in Equation 16.1 for the applied deformation, Equations 16.5 and 16.6 become:

$$\dot{\gamma} = d\gamma/dt = \omega\gamma_0 \cos(\omega t) \tag{16.5}$$

$$\sigma = \eta\omega\gamma_0 \cos(\omega t) \tag{16.6}$$

For a material which exhibits a linear viscoelastic behavior, the stress response will be sinusoidal and will be out of phase with the deformation by an angle δ ($0 \leq \delta < \leq \pi/2$). As in Equation 16.1, if $\gamma = \gamma_0 \sin(\omega t)$, then σ may be expressed as:

$$\sigma = \sigma_0 \sin(\omega t + \delta) \tag{16.7}$$

In order to obtain the stress response, it should be remembered that sin (A + B) = sin A cos B + sin B + cos A. So, it is possible to deduce Equation 16.8:

$$\sigma = \sigma_0 \sin(\omega t)\cos\delta + \sigma_0 \cos(\omega t)\sin\delta \tag{16.8}$$

This last equation shows that the stress response may be resolved into two components:

- $\sigma_0 \sin(\omega t) \cos \delta$, when the component is in phase with the deformation and is related to the elastic energy stored;
- $\sigma_0 \cos(\omega t) \sin \delta$, when the component is 90° out of phase with the deformation and is related to the viscous energy dissipated.

16.3
Dynamic Mechanical Analysis and the Principle of Time-Temperature Superposition

Dynamic mechanical tests can be carried out as function of either temperature or time. Whilst some instruments are unable to register the parameters as a function of time, all can operate at several frequencies, which indirectly provides the dynamic mechanical behavior of a material as a function of time, as the frequency is the inverse function of time ($\omega = 1/t$). A typical curve of modulus as a function of time or frequency, also called a relaxation spectra, is shown in Figure 16.1. The point to note in this figure should be the similarity of these curves with that of modulus versus temperature curve. In fact, if the chains achieve mobility as a function of temperature (molecular relaxation under the action of temperature), then the same may be assumed for the effect of time (molecular relaxation under mechanical action).

The principle of time–temperature superposition is, quite simply, based on two points:

- That the curves of modulus as a function of temperature are similar to the curves of modulus as a function of time.
- That the strong variation of the modulus value as a function of time is closely related to the value of the T_g for a particular material.

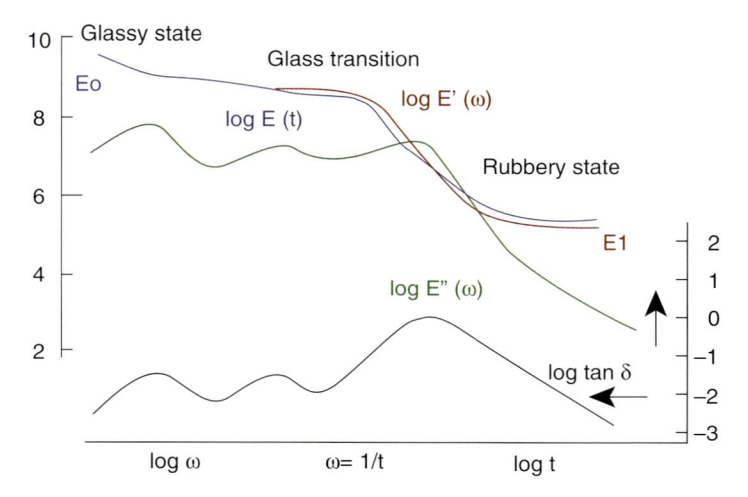

Figure 16.1 Typical curve of modulus (E) variation as function of time (t) or frequency (ω).

This method allows, over a short period of analysis at high temperature, information to be obtained concerning the properties of a material that would be obtained over a very long period if the test were to be conducted at low temperature (it is valid when the time of relaxation has the same temperature dependence).

As noted by Ray *et al.* [1, 5], dynamic mechanical analysis (DMA) is a tool that permits evaluation of the amount of elastic and dissipated energy stored in nanocomposites under mechanical strain. In turn, the geometrical characteristics, the level of dispersion of filler in the matrix and the degree of interaction between the matrix and filler's surface will strongly influence the results [1].

16.4
Nanoclays and Their Influence on the Thermomechanical Properties of Polymer Composites: Some Case Studies

Several reports have shown that, among other parameters, the materials processing and the type of nanoclay used are very important when creating a final product with good properties [43]. Several groups [6–8] have suggested that the best thermomechanical properties are achieved with exfoliated nanocomposites, as the clay is dispersed homogeneously in the polymer matrix. An understanding of how the dispersion of clay in the final composites can alter the thermomechanical properties is provided in the following chronological review of studies relating to this topic, the results of which are highlighted according to the opinion of the authors.

The thermomechanical properties of poly(ethylene oxide) (PEO)/clay nanocomposites were investigated by Ratna *et al.* [9]. For this, intercalated nanocomposites were prepared using a solution of PEO in chloroform, after which the effects of clay incorporation on the viscoelastic response of PEO/clay nanocomposites and pure PEO were observed by preparing loss tangent versus temperature and dynamic modulus versus temperature plots for the PEO- and clay-containing nanocomposites. It was noted that, when the temperature increased, the damping passed through a maximum in T_g. Although the damping was low below T_g (because the chain segments in that region were frozen), when the system reached the T_g region the damping rose due to an initiation of micro-Brownian motion of the molecular chain segments. It was suggested, however, that not all segments would take part in such relaxation together, and that the frozen segments could store much more energy for a given deformation than could a rubbery segment that was free to move. Thus, whenever a stressed frozen-in segment becomes free to move, it dissipates its excess energy. In a polymer chain, the micro-Brownian movement is dependent on the cooperative diffusional motion of the other main chain segments. A maximum damping is observed at T_g when most of the chain segments take part in this cooperative motion under harmonic stress. The loss factor (E'') decreases above T_g because, beyond this temperature, the molecular segments will gain movement and offer little resistance to flow.

A slight shift in loss tangent towards a higher temperature was detected for the nanocomposite compared to pure PEO; this was considered evidence of the chains' confinement as a result of polymer intercalation between the interlayer gallery of the clay. The increase in stiffness, which was reflected in a reduction in the loss peak height, was also reported by Ratna et al. [9] for an epoxy/clay nanocomposite using the same clay. Here, the loss peak was wider due to incorporation of the clay, which suggested that the clay introduced some form of relaxation modes to the PEO chain, as proposed by Ogata, Kawakage and Ogihara [10]. The storage modulus for the whole temperature range was higher for the nanocomposite compared to the pure PEO. Likewise, the change in storage modulus at T_g became smaller as clay was added to the matrix, which reconfirmed the intercalation of the PEO chains.

Other studies with ternary nanocomposites were conducted by Asif et al. [11], who used an organo-montmorillonite (OMMT), named Cloisite 25A. These studies were conducted in stepwise fashion, the first step being to prepare a hydroxyl-terminated poly(etheretherketone) oligomer with a pendant methyl group (PEEKMOH). This oligomer was blended with an epoxy resin based on diglycidyl ether of bisphenol-A (DGEBA) in the presence of Cloisite 25A, and then cured using 4,4'-diamino-diphenyl sulfone as the crosslinking agent. The thermomechanical properties were studied in a thermomechanical analyzer, where the coefficient of thermal expansion (CTE) and T_g were evaluated. The values of CTE were seen to decrease when the clay content was increased to 3 phr (parts per hundred), but recovered beyond this point. The explanation for this phenomenon was based on the nanoparticle distribution in the polymer matrix; this was homogeneous up to 3%, but higher amounts of filler most likely caused agglomerates to form in the matrix. As a consequence, the nanofiller was not distributed homogeneously in the matrix, and this in turn compromised the efficiency of the matrix to transfer stress to the clay layers and to control its thermal expansion. When measured by thermomechanical analysis (TMA), the T_g of this system did not change significantly with the clay concentration, although when measured by DMA the T_g showed a considerable decrease. This behavior may have resulted from the plasticizing effect of the organic moiety of the clay particles, which affected the crosslink density of the epoxy network.

Asif et al. [11] also studied the viscoelastic behavior of these materials using dynamic mechanical thermal analysis (DMTA). The results of the dynamic mechanical spectra showed only one T_g, instead of two corresponding to the epoxy and thermoplastic phases, as observed earlier [12]. According to Asif et al. [11], this result could be attributed to both a low PEEKMOH concentration and a restriction of the thermoplastic chain mobility in the presence of rigid nanoclay layers. It was also noted that the T_g of the epoxy phase decreased with the addition of PEEKMOH, most likely due to a small amount of this component being present in the bulk of the epoxy matrix, hindering its complete crosslinking. In addition, a decrease in the T_g of the nanocomposites was observed with an increase in clay content, most likely due to the presence of an organic modifier in the clay layers that functioned as a plasticizer, pushing down the value of T_g. Nonetheless, the measured values of E' and E'' modulus which were higher for values below and above T_g when compared to the pristine resin. The increases in E' and E'' were about 92% and

33%, respectively, for the 5:5 phr proportion of PEEKMOH and nanoclay, below T_g. The authors suggested that this was supported by the increase in interfacial interaction between the epoxy matrix, PEEKMOH, and the organolayers of the exfoliated clay. In contrast, when the nanoclay content was increased to 8 phr (and also PEEKMOH), the values of E' and E'' were each decreased, to 45% and 20%, respectively, due to the presence of clay aggregates that contributed to a smaller interfacial interaction between the components. The fact that E' was lower for nanocomposites at T_g, in comparison to the net epoxy resin, provided evidence of a low crosslink density when PEEKMOH and nanoclay were added. When analyzing the plot of E'' versus temperature, two peaks corresponding to the T_g of PEEKMOH-rich phase and epoxy-rich phase were observed. Finally, a lowering of T_g was observed with an increase in nanoclay content.

Thermomechanical properties were also investigated by Ray, Bandyopadhyay and Bousmina [5], who worked with nanocomposites of poly[(butylene succinate)-co-adipate)] (PBSA), an environmentally friendly synthetic aliphatic polyester. The nanocomposites were prepared by melt-mixing, using a twin-rotor Thermo Haake mixer (Polylab), at 135 °C, with a rotor speed of 60 rpm, for 8 min. In these studies another type of nanofiller was utilized, namely an organomodified synthetic layered fluorine mica (OSFM) that belonged to the phyllosilicates family and differed from MMT due to the F groups on its surface (these are absent from MMT). The amount of OSFM was fixed at 6 wt% in the matrix, and samples of nanocomposites were dried to remove any residual water, under vacuum at 65 °C for 7 h. The molded samples were cooled at room temperature and then annealed at 60 °C for 5 h to crystallize isothermally before being subjected to investigations.

When plots of storage (E') and loss modulus (E'') and their variation with temperature, as well as tan δ, were obtained [5], it was observed that a significant increase in E' for the nanocomposites occurred when compared to PBSA over the temperature range investigated. Since, at low temperature, both the PBSA matrix and the nanocomposite were in a glassy state, the T_g of the matrix had clearly not been affected significantly by OSFM incorporation. However, when the behavior of E' of the nanocomposites was observed over the entire temperature range, it was always higher than that of the neat PBSA. The authors suggested that this behavior might be due to the high degree of intercalation of polymer chains into the silicate layers of the OSFM, as confirmed by X-ray diffraction (XRD) patterns and transmission electron microscopy (TEM) images [5], which exhibited a large surface area for the favorable interactions between silicate layers and the polymer matrix. Based on these results, it was concluded that immobilization of the polymer chains inside the silicate galleries may be the main factor responsible for this substantial increase in E'. As was noted, at room temperature (25 °C) the increase in E' for the nanocomposite was 140% compared to that of neat PBSA [5]. The same behavior was noted for E'' with the incorporation of OSFM into the PBSA matrix [5], and this was attributed to the presence of a strong internal friction between the homogeneously dispersed intercalated silicate particles.

In a later study, Kontou and Anthoulis [13] investigated the effect of silica (SiO_2) on the thermomechanical properties of polystyrene (PS) nanocomposites, their aim being to identify the optimum silica content for the PS matrix, using a conventional system of melt mixing to prepare the PS nanocomposites. The degree of reinforcement and the viscoelastic response was explored via DMA testing, among other measurements. The curves for E' and E'' were plotted following the principle of time–temperature superposition, with a very low data fluctuation. Subsequently, it was observed that E' exhibited a very slight shift to higher frequencies for higher silica contents, and a similar behavior was observed for E''. Kontou and Anthoulis [13] observed a plateau region at high frequencies, followed by a dissipative transition that indicated a very strong viscoelastic behavior. In contrast, at low frequencies a slight leveling-off was observed, with $E'' > E'$ and with a clear crossover point.

As noted by Zhao, Morgan and Harris [14], the particle distribution in the bulk of the polymer matrix represents an important factor that seems to be responsible for the rheological behavior of the nanocomposites. Kontou and Anthoulis [13] observed that, for a higher nanofiller concentration, the plateau at higher frequencies was slightly increased, as evidence of the reinforcing clay. However, at low frequencies this effect was more substantial, with two crossover frequencies for PS nanocomposites with 4 wt% SiO_2 and three frequency regions being noted, where the higher frequency region was attributed to the glassy plateau. The intermediate region showed that E' and E'' were similar, while the low-frequency region revealed a secondary plateau.

In another series of studies [14], a similar relationship between E' (or G') and E'' (or G'') it was obtained, though this was explained in terms of a good nanofiller dispersion that led to a solid-like behavior.

When Kontou and Anthoulis [13] examined PS nanocomposites with a silica content of 8 wt%, E' was noted always to be higher than E'', this behavior being attributed to the presence of more filler in the matrix, resulting in a dominant solid-like behavior. However, when PS nanocomposites with 8 wt% and 4 wt% filler were compared, a slight plateau was noted for the former nanocomposite at low frequency [13]. This was attributed to the existence of more and larger filler agglomerates, which do not contribute to the mechanical properties. An examination of a 10 wt% filled nanocomposite showed the relationship E'/E'' to be very close to that for pure PS. In the same study, it was noted that the T_g was not affected by the silica nanoparticles for all PS silica loaded formulations investigated, although others reported that changes in T_g with the filler content were common [15–18]. Elsewhere, investigations of the mechanical properties (e.g., tensile versus strain as a function of different temperatures, e.g., 20, 85, and 100 °C) led to some very interesting results. Notably, if the tests were developed in the glassy region, the PS and PS/SiO_2 nanocomposites demonstrated a brittle behavior. However, all types of nanocomposite studied showed a higher tensile strength and a lower elongation at break when compared to the pure PS. According to Shah et al. [19], the mobility of nanoparticles in the bulk of the polymer is very important with regards to the mechanism of energy dissipation, which results in an enhanced toughness of the nanomaterial and is also related to matrix mobility. It appears that, in the glassy state, the mobility of the matrix is restricted and does not contribute to the mechanism of energy dissipation. An examination of results, obtained at 85 °C, for the same systems [13], showed the material to behave very differently, with both PS and its silica nanocomposites presenting a sharp yield stress, and with a resultant strain softening and subsequent strain hardening in PS and in a PS nanocomposite with a 4 wt% filler content. Both, the yield stress and Young's modulus showed a similar trend at 20 °C, which denoted an enhanced mechanical response when nanoparticles were present in the PS matrix. At temperatures close to T_g, the polymer chains and particles gain mobility and, as a consequence, their interfacial interaction increases so as to influence the macroscopic response of the material. With an 8 wt% silica loading, the modulus increment was more pronounced and the elongation at break higher; yet, with a 10 wt% silica loading the behavior reversed and the properties of the nanocomposites began to decay, most likely due to the presence of agglomerates [13]. An examination of the materials at 100 °C showed a viscoplastic response, which resulted in a maximum yield stress followed by a smooth decrease at higher strain values. Although it was not possible to evaluate the elongation at break due to limitations of the equipment, it could be seen that the tensile properties presented the same trend as that for 85 °C. Consequently, it was concluded [13] that the optimum silica content was 4 wt%, as this showed the best particle dispersion, whereas above 8 wt% the composite's macroscopic properties were in fact worsened. By conducting tensile stress–strain tests, it was also possible to follow increases in the mechanical response of the nanocomposites, compared to the pristine PS. Likewise, raising the temperature caused the materials' behavior to change, from brittle at 20 °C to viscoplastic at 85 °C.

Another interesting concept was developed by Harada and coworkers [20, 21], who studied the effect of a layered titanate filler on the flame retardancy and thermomechanical properties of a poly(glycidyloxypropyl) phenyl silsesquioxane (PGPSQ) matrix, and of poly(glycidyloxypropyl) silsesquioxane (PGSQ). Although PGSQ showed a high thermal stability and flame retardancy due to the presence of an inorganic component, when the formulation was loaded with 5% of an organolayered titanate it exhibited a self-extinguishing property. Based on these findings, the same authors investigated the effect of a phenyl moiety in PGSQ, named PGPSQ, cured with 4,4′-diaminodiphenylmethane (DDM), the aim being to improve the flame retardancy of the resultant material, but with a lower content of organolayered titanate. The epoxy resin used was the diglycidyl ether of bisphenol A (DGEBA), and the target was to enhance the high thermal stability and gas-barrier properties of cured PGSQ loaded with layered titanate. In order to characterize the thermomechanical properties, the authors first observed the temperature dependence of such properties for the DGEBA, PGSQ, and PGPSQ systems. Notably, the E'-value of the DGEBA system suffered a significant decrease due to a glass-rubbery transition near 180 °C, while the PGSQ and PGPSQ systems each showed E'-values 10-fold higher than DGEBA over the same temperature range (30–100 °C). This was attributed to the relaxation peak of the aliphatic chains between the inorganic moiety and the epoxy group. As expected, the PGPSQ system showed a slight shift in this relaxation peak due to the presence of the phenyl group, which was less mobile than the glycidyloxypropyl group [20]. The ductility of PGPSQ loaded with 27.4 wt% silica demonstrated a serious limitation of poor fracture toughness for practical applications of this material. However, this observation was opposed to the findings of other reported studies, which indicated an improvement of this characteristic not only following the incorporation of clay, but also when a high-aspect ratio filler was used [22–24]. According to Harada *et al.* [20], it may be possible to improve the ductility of the silsesquioxane by altering the organolayered titanate content; in other words, the toughness of the PGPSQ could be improved by adding a filler that would also act as a flame retardant. Based on these results, it was concluded that PGPSQ possessed much better flame-retardancy properties than PGSQ, with only 1 wt% of organolayered titanate being required to provide the self-extinguishing characteristic of PGPSQ (as measured with the UL-94 method). Such properties were considered to be the result of a multiplication effect of the high thermal stability of PGPSQ and the gas-barrier property of the titanate-layered filler.

Another group compared the microstructure and nanostructure of poly(L-lactic) acid (PLLA)/layered silicate hybrids [25], by preparing micro- and nanocomposites of PLLA; for this, a casting method was used, and the loadings of natural and hexadecylamine-modified MMT varied. The inorganic nanosized particles, typically with less than 5 wt% content, provided the polymer matrix with characteristics that included enhanced mechanical properties, an enhanced barrier resistance, decreased flammability, and an increased biodegradability when added to biodegradable polymers [26–30]. The main aim of these studies [25] was to produce hybrid nanocomposites with properties superior to those of pure PLLA, as many questions regarding the composite's structure–properties relationship have not yet been resolved. It was suggested that an understanding of this relationship, with attention focused on the composite's morphology and its effect on the physical properties, would provide opportunities for designing materials with tailor-made properties. Based on TGA results for the pure and hybrid PLLA [25], carried out at $10\,°C\,min^{-1}$ and in the absence of humidity, it was noted that the PLLA volatilized completely in a single step, starting at 270 °C. The introduction of an organophilic clay caused an improvement in the thermal stability of the matrix, raising its temperature of decomposition, especially when filler content was low. This behavior was attributed to the tortuous barrier effect generated by the high aspect ratio of the clay platelets, when dispersed in the polymer matrix; this caused a delay in the volatilization of the products generated during the degradation process [31]. As the clay loading was increased, the polymer matrix became destabilized due to the presence of more ammonium cations in the bulk of the matrix; these would decompose to produce ammonia and the corresponding olefin, according to the Hoffmann mechanism [32], as shown in Scheme 16.1.

As noted above, the TGA thermograms of pure PLLA and PLLA matrix filled with natural (NaMMT) and organomodified filler (C_{16} MMT) [25] suggested that the presence of C_{16} MMT and NaMMT had caused a delay in the onset of degradation temperature. However, when the concentration of organophilic clay in the matrix was raised to >9 wt%, the corresponding hybrid nanocomposites showed a

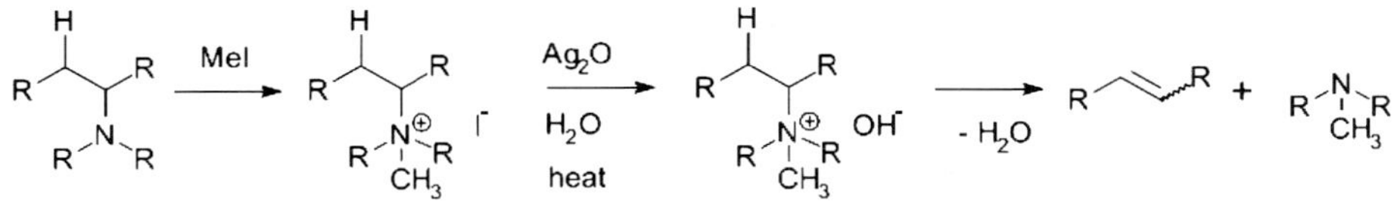

Scheme 16.1 Hoffmann elimination mechanism.

decrease in the onset of decomposition temperature compared to the pure PLLA that was attributed to organic nanofiller degradation. In order to validate their results, Marras, Zuburtikudis and Panayiotou [25] employed the same analysis conditions for microcomposites as were used for the hybrid nanocomposites. Here, the incorporation of a natural clay (NaMMT) had a much less significant effect on the thermal degradation of the polymer matrix than did the organomodified filler. Subsequently, to determine the effect of the analysis atmosphere on polymer/clay nanocomposites, a series of experiments was performed under oxidative conditions. Not unexpectedly, oxygen encouraged PLLA degradation at a lower temperature than did nitrogen. In turn, the addition of organoclay was shown to cause a significant delay in the mass loss, and also raised the temperature of degradation of the hybrid material. The presence of an organophilic inorganic material seemed to prevent the diffusion of volatile products, presumably by increasing the carbonaceous material during the initial stage of combustion, which would in turn protect the nanocomposites [28]. It was also proposed that the silicate platelets would act as an additional barrier, most likely by preventing the diffusion of oxygen into the bulk of the polymer matrix [31]. Based on thermogravimetric analyses of the composites with NaMMT, the introduction of NaMMT to the polymer matrix was seen to result in a small increase in the thermal stability of the microcomposite, which was attributed to the poorer dispersion of the inorganic unmodified filler. When these experiments were continued in an inert atmosphere, it was observed that only nanocomposites filled with nanoclay would showed an enhanced thermal stability, when compared to pure PLLA [25]. Similar investigations of PLLA/MMT nanocomposites in the presence of air were also performed by Pluta et al. [33].

When experiments were performed under isothermal conditions at 280 °C in the presence of air, the PLLA nanocomposites with a 3–5 wt% filler loading delayed their rate of weight loss (to reach 50% mass loss) almost threefold compared to pure PLLA. This considerable delay in material volatilization may involve the formation of a char structure on the nanocomposite's surface during burning, as noted previously [34]. Contact between the clay and matrix in an oxidative atmosphere favored the formation of char, which would then act as an efficient insulator and mass transport barrier so as to delay the diffusion of oxygen from the surroundings to the polymer mass. This in turn would reduce the rate at which volatile products could escape as the material decomposed thermally. An effective contact between the matrix and the organoclay silicate (which was well distributed in the matrix) resulted in the thermal stability of these materials. This effect was confirmed by comparing the thermal behavior of microcomposites studied under the same conditions as were used for nanocomposites, namely in the presence of oxygen at 280 °C. For microcomposites, the lack of any delay in volatilization owing to the presence of NaMMT in the matrix was indicative of the inhomogeneous clay dispersion and a weak interfacial interaction between the matrix and the inorganic filler. Further evidence of this was provided when nanocomposites and microcomposites with the same MMT content (3 wt%) demonstrated similar results [25].

One very interesting characteristic to be explored was the effect of nanofillers on the crystallization (T_c) and melting (T_m) temperatures of nanocomposites, these being important parameters that affected the materials' properties, notably their thermomechanical responses [25]. The differential scanning calorimetry (DSC) heating curves demonstrated an exothermic peak for all tested samples, over the range of 95–120 °C (which included the temperature range where pure PLLA would crystallize, ~107 °C). In the presence of nanoparticles the T_c of PLLA was lowered; this was considered to be due to the large surface area of the nanosized filler and to its high aspect ratio, and caused the nanofiller to become a good nucleating agent by enhancing the rate of PLLA crystallization [35–37]. At a higher organoclay content, the number of nucleation sites was increased, and at 5 wt% the lowest T_c for PLLA was observed. At higher contents there was a delay in crystal growth which led to an increase in T_c [38]. The T_m of PLLA was also affected by the incorporation of MMT nanoparticles, and showed a slight decrease when the clay content was enhanced. In turn, the T_g was not influenced, as was also observed by Pluta et al. [33], who showed T_g to be independent of the filler concentration. Others [25] reported that, for unmodified MMT filled composites, the thermal parameters of the system would not be affected significantly, as the interfacial interaction between the matrix and filler would be insufficient to cause major changes. The same authors observed the coexistence of both exfoliated and intercalated organo-MMT nanoparticles at filler concentrations exceeding 5 wt%. When compared to the microcomposites and to pure PLLA, the nanocomposites showed a higher thermal stability, even within an oxidative environment. However, if submitted to much higher temperatures, the inorganic platelets created a barrier to mass transport and functioned as an insulator, thereby retarding the rate of matrix decomposition.

Recently, it has been shown that physical properties of polyamides can be improved by the addition of both inorganic micro- and nanofillers to a poly(trimethylhexamethylene terephthalamide) matrix [2]. For this, the composite samples were prepared using a sol–gel technique, with the incorporation in situ of a silica precursor (tetraethyl orthosilicate; TEOS), using diethylamine as the catalyst. After evaporation of the solvent (dimethylformamide; DMF), the films obtained were submitted to mechanical, dynamic mechanical and morphological analyses. The DMA of these films was carried out using the bending mode, and the curves were reported as tan δ variation versus temperature. It was observed that, on increasing the temperature, some segments of the polyamide chains would achieve mobility, and the corresponding tan δ curve was reported at high

16.4 Nanoclays and Their Influence on the Thermomechanical Properties

values around 102 °C. This phenomenon was attributed to the temperature of α-relaxation for the pure polymer. The sharp peak obtained then shifted to high temperatures, due to the nanofiller being introduced into the polyamide matrix, up to 20 wt% silica content. Not unexpectedly, as more inorganic filler was added a greater reduction of chain motion could be observed. It was also reported that greater interactions between the organic and inorganic phases led to an increase in T_g, a point also detected by others. All of these reported studies were in agreement, that limitations of motion of the polymer chains due to the presence of nanofiller was mainly responsible for the higher T_g values that were observed.

It has also been reported that, although the storage modulus, E', is initially increased with increasing silica content, further additions of silica hinder the movement of the chains segments, such that the system loses flexibility [2]. Moreover, there is a limit to the effect of silica concentration on the properties, since at higher concentrations silica will form agglomerates and contribute to the deterioration of E'. Layered clay/epoxy nanocomposites were also investigated, and their thermomechanical properties analyzed, by Kaya, Tanoglu and Okur [3]. Here, the nanocomposites were prepared by stirring an epoxy resin with different amounts (1, 3, and 10 wt%) of OMMT and MMT, at room temperature for 1 h. The OMMT was obtained using hexadecyltrimethylammonium chloride (HTCA) as the modifier organic agent. The blends were then poured into silicon molds and allowed to cure at room temperature.

Maiti et al. [39] studied the influence of crystallization on the intercalation, morphology, and mechanical properties of composites based on polypropylene/nanoclay (PPCN), using several characterization techniques that included DMA. For this, the variation of G' with temperature was followed for a PP-maleic anhydride (PP-MA) matrix and nanocomposites with 4 wt% nanofiller, crystallized using two different temperatures (i.e., 70 °C and 130 °C). By selecting a determined condition of strain frequency and amplitude, the absolute value of G' was obtained at 50 °C for different systems, but with a fixed strain amplitude (0.05%) and a frequency of 6.28 rad s^{-1}, in a strain-controlled rheometer. The results showed that G' for crystallized PP-MA was increased only marginally at 130 °C when compared to the same material at 70 °C. However, in nanocomposites with 4 wt% nanofiller the increase in G' was significant. It was also observed that, for a particular value of T_c, the value of G' would rise as the amount of clay in the system was increased. The results showed that, for PP-MA crystallized at 130 °C, G' showed an increase of about 10% when compared to the same procedure at 70 °C. Subsequently, when PP nanocomposites containing 7.7 wt% and 4 wt% of nanoclay were investigated, G' was seen to increase by 13.3% and 30.6%, respectively. Based on the results of these studies, the effect of T_c on G' for these systems could be presented in the following order: PP-MA < PP-MA nanocomposite 7.5 wt% < PP-MA nanocomposite 4 wt%. It was concluded that a higher efficiency of intercalation, providing good reinforcement, should be achieved with nanocomposites with a 4 wt% filler content. These proposals were also based on the results of other studies, including clay segregation and crystallization, both of which are related directly to the nanocomposite's mechanical and morphological properties.

In order to confirm segregation of the dispersed clay particles between the spherulite and interspherulitic regions, a series of TEM investigations was carried out on the spherulite of PPCN with 4 wt% filler, crystallized at 140 °C for five days (annealing). When using this technique, a higher density of dispersed clay particles was observed in the interspherulitic region. However, when PPCN with 4 wt% filler was crystallized at $T_c < 100$ °C, the excess of clay particles in the interspherulitic region was suppressed, since at low T_c the rate of crystallization would be higher and the system would solidify very quickly. For high T_c values, the crystallization rate would be very low, and this would permit the chains to intercalate the nanofiller galleries. Such intercalation would be promoted by the strong hydrophilic interaction between the maleic anhydride group and the polar groups on the clay surfaces.

It was also reported that the interlayer spacing $d_{(001)}$ of the clay would increase with T_c for the nanocomposites, in the following order: PP nanocomposite 7.5 wt% filler < PP nanocomposite 4 wt% filler. In other words, intercalation would proceed at T_c and increase when the clay content decreased. However, as the clay content was lower in nanocomposites with a 2 wt% filler loading, the system would crystallize at high T_c and exhibit a higher degree of intercalation compared to systems with a high clay content, that would crystallize at any T_c. It should also be noted here that crystallinity was observed not to increase significantly when T_c was increased from 70 °C to 130 °C, for either PP-MA or for nanocomposites with 4 wt% filler. Based on these observations, it was concluded that the degree of intercalation would affect the storage modulus more than the crystallinity. Furthermore, the segregation of nanofillers that occurred at high T_c would increase the nanofiller concentration in the interspherulitic region and, as a consequence, a higher modulus would be obtained for PPCNs crystallized at high T_c.

In a later study, a clear explanation was provided for the action of the nanoparticles on the elasticity modulus (E') [40]. Here, the authors verified that the increase in temperature would promote a decrease in the storage modulus (E'') for epoxy nanocomposites loaded with different weight ratios of SiO_2 nanoparticles (1, 3, 5, and 7 wt%). The curve of E' as a function of temperature showed three stages: stage 1 was related to the glassy behavior of the material in the range of −100 to 10 °C; stage 2, at 10–60 °C, related to the elastic fraction of the material; and stage 3, at 60–100 °C, corresponding to the viscous fraction. In stage 1, E' was in the range of 4000–8000 MPa, and was attributed to the extreme stiffness of the material due to gelation of the chains movement. Following this, in stage 2, E' was significantly decreased (to 500 MPa) due to deformation of the material caused by free movement of the polymer chains. Finally, in stage 3, as the chains slid over each other, the viscous flow would lead to an irreversible deformation, forcing the E' value to fall to zero. The results obtained by Yao et al. [40] also showed that the presence of SiO_2 nanoparticles improved the reinforcement of the material, which presented a high value of E' in the glassy stage, this being attributed to interaction of the nanoparticles with the epoxy polymer chains. This may be confirmed by comparing the values of E' for pure epoxy (6000 MPa) and nanocomposites (8000 MPa). When the polymer chains achieve mobility, their Brownian movement

16.4 Nanoclays and Their Influence on the Thermomechanical Properties

increases during the viscous stage. As a consequence, the action of the nanofillers is weakened and this results in a value of E' which is close to that for pure epoxy.

When monitoring the behavior of tan δ, which was the same for all samples (T_g ~60 °C), a higher T_g was obtained for 3 wt% SiO_2; this was attributed to the good compatibility of the epoxy and nanoparticles, which reduced the free volume of the system. This was confirmed using positron annihilation lifetime spectroscopy (PALS), where the 3 wt% SiO_2-loaded nanocomposites presented a lower free volume compared to other nanocomposites. An interesting behavior was reported in a low-temperature environment, when a weak loss peak was seen to be present at about −60 °C for the nanocomposites; this correlated with a slight relaxation of the chains in this region, and showed that the systems had a good impact resistance. When analyzing E'' for all samples, these materials were seen to have a maximum E'' in the glassy state, which meant that a great degree of damping was present in this type of nanocomposite. When the temperature was raised, however, such damping was decreased due to limitations in the movement of the chains. The reason for the increased damping as a function of the amount of nanoparticles within the T_g region could be attributed to the interaction between the matrix and the nanofiller, which would hinder the motion of the polymer chains and impose a greater rigidity to the material. On the other hand, at T_g, whereas the epoxy chains would achieve Brownian movement, the SiO_2 nanoparticles would restrict molecular motion, such that the resultant loss in modulus would be very similar to that for pure epoxy.

Samal, Nayak and Mohanty [41] reported PP nanocomposites using PP-g-MAH as compatibilizer and MMTs organomodified with alkyl ammonium [Cloisite 15A; OMMT-I (modified by hexadecyloctadecyltrimethyl ammonium chloride) and OMMT-II (modified by octadecylammonium)]. The nanocomposites were prepared using a melt-blending technique, with the aim of investigating the effect of Na^+-MMT and OMMT on the mechanical, thermal, and morphological properties of the materials developed. With regards to dynamic mechanical properties, it was noted that the value of E' was increased with the incorporation of nanoclays, compared to that of pure PP. PP–Na^+-MMT showed a slightly higher E' than virgin PP, most likely due to the absence of any interfacial compatibility between both components. However, when OMMT was added the E'-values were higher for the entire temperature range considered (−100 to 150 °C). This provided evidence that the nanoclay had acted as a reinforcing agent since, according to Chiu et al. ([42], cited in [41]), its high aspect ratio permits stress transfer with a greater efficiency at the interface. In addition, restricting the movement of chain segments at the apolar (organic)–polar (inorganic) matrix–filler interface, and exfoliation of the clay that takes apart the nanoplatelets, may lead to an increase in E'. In these studies, it was noted that the OMMT modified with the long alkyl tallow (Cloisite 15A) exhibited the best exfoliation of the galleries compared to others, and was responsible for the highest E'. Still in relation to E', two regions of decrease were observed for the nanocomposites, the more pronounced region being in the range of −10 to ~20 °C, and the other region over 70 °C. In the first case, the decrease was attributed to a relaxation of the amorphous phase (α-relaxation); following this, E'

continued to fall gradually such that, beyond the range of 70 to 80 °C the decrease in E' was very slow until fusion occurred.

An analysis of E'' for pure PP and its nanocomposites led to the proposal that nanoclay incorporation would cause increases in the modulus, according to the following order: Cloisite 15A > OMMT-I > OMMT-II > Na$^+$-MMT > PP. All of these nanocomposites presented similar relaxation peaks to the polymeric matrix, independent of the presence of nanoclay. It was also noted that the T_g values for the nanocomposites varied with the type of clay, and were lower than the value for PP. Amongst all nanocomposites, the OMMT-II-filled material showed the lowest T_g, at about −2 °C. An explanation for this was that the T_g decreased because the amorphous molecules which intercalated the nanoclays had a low molecular weight that allowed them to move. Instead, the softening point (T_s) for OMMT-II nanocomposites was higher than for pure PP, which may be due to limitations in the mobility of the rigid amorphous molecules around the crystalline regions. In addition, the difference between T_g and T_s for the different clays was attributed to the type of alkyl ammonium used to modify the nanoclay surfaces.

16.5
Conclusions

It is impossible to draw a unique conclusion for the thermomechanical behavior properties of the nanocomposites, as each system has its own particularity and is composed of different resins, additives, and nanoclay modifications, and is also processed under different conditions and with different equipments. Despite few types of nanoclay being available, each must be investigated separately in order to ascertain the influence and interaction of each component, and its concentration, on the thermomechanical properties of the nanocomposites. It is possible, however, to draw some conclusions based on the behavior of these properties, as reported in the literature:

- In general, the more inorganic filler that is added to a system, the higher the reduction of chains motion that may occur. However, if the nanoclay is organomodified, this may be reverted up to a certain concentration of clay, due to the presence of the organic modifier in the clay layers that may function as a plasticizer.

- The particle distribution in the bulk of a polymer matrix is a very important factor next to the rheological behavior of the nanocomposites, as it may influence their thermomechanical properties. Hence, the way in which every system is processed is very important in terms of their final properties.

- There is, for each system, a certain amount of nanoparticles that will be well dispersed in the polymer matrix, and result in good thermomechanical properties. Above this amount, however, the nanocomposite's macroscopic properties may be worsened.

- If exfoliation of the nanoclay is achieved owing to an interfacial interaction between the polymer matrix and clay layers, then E' and E'' will probably increase as a consequence, though only up to a critical nanoclay concentration. Beyond this critical concentration, however, the values of E' and E'' will be reduced due to the formation of clay aggregates.

- An improvement in the ductility of the materials may be observed when organoclays are added, and also when a high aspect ratio filler is used.

- Up to a certain concentration, the nanofillers may reduce the T_c of the material; however, above this concentration such behavior may be reversed.

- The rate of weight loss of a nanocomposite material in an air atmosphere may be delayed compared to the pure base resin. This may be caused by the formation of a char structure on the material's surface, that acts as an efficient insulator and mass transport barrier to prevent oxygen diffusing from the surroundings into the polymer mass.

- In crosslinked resins, the presence of a modified clay (OMMT) may reduce the degree of crosslinking, which will result in a lower T_g value.

References

1 Ray, S.S. and Okamoto, M. (2003) *Prog. Polym. Sci.*, **28**, 1539–1641.
2 Sarwar, M.I., Zulfiqar, S., and Ahmad, Z. (2008) *Polym. Int.*, **57**, 292–296.
3 Kaya, E., Tanoglu, M., and Okur, S. (2008) *J. Appl. Polym. Sci.*, **109**, 834–840.
4 Lucas, E.F. and Soares, B.G. (2001) *Caracterização de Polímeros*, Editora e-papers, Rio de Janeiro, Brasil.
5 Ray, S.S., Bandyopadhyay, J., and Bousmina, M. (2007) *Polym. Degrad. Stab.*, **92**, 802–812.
6 Giannelis, E.P. (1996) *Adv. Mater.*, **8**, 29.
7 Carrado, K.A. and Xu, L. (1998) *Chem. Mater.*, **10**, 1440.
8 Kawasumi, M., Usuki, A., Okada, A., and Karauchi, T. (1996) *Mol. Cryst. Liq. Cryst.*, **281**, 91.
9 Ratna, D., Divekar, S., Samui, A.B., Chakraborty, B.C., and Banthia, A.K. (2006) *Polymer*, **47**, 4068.
10 Ogata, N., Kawakage, S., and Ogihara, T. (1997) *Polymer*, **38**, 5115.
11 Asif, A., Leena, K., Lakshmana Rao, V., and Ninan, K.N. (2007) *J. Appl. Polym. Sci.*, **106**, 2936.
12 Francis, B., Thomas, S., Jose, J., Ramaswamy, R., and Rao, V.L. (2005) *Polymer*, **46**, 12372.
13 Kontou, E. and Anthoulis, G. (2007) *J. Appl. Polym. Sci.*, **105**, 1723.
14 Zhao, J., Morgan, A.B., and Harris, J.D. (2005) *Polymer*, **46**, 8641.
15 Hergeth, W., Steinau, U., Bittrich, H., Simon, G., and Schmutzler, K. (1989) *Polymer*, **30**, 254.
16 Iisaka, K. and Shibayama, K. (1978) *J. Appl. Polym. Sci.*, **22**, 3135.
17 Kotsilkova, R., Fragiadakis, D., and Pissis, P. (2005) *J. Polym. Sci., Part B: Polym. Phys.*, **43**, 522.
18 Arrighi, V., McEwen, I.J., Qian, H., and Prieto, M.B. (2003) *Polymer*, **44**, 6259.
19 Shah, D., Maiti, P., Jiang, D.D., and Batt, C.A. (2005) *Adv. Mater.*, **17**, 525.
20 Harada, M., Minamigawa, S., Tachibana, K., and Ochi, M. (2007) *J. Appl. Polym. Sci.*, **106**, 338–344.
21 Harada, M., Minamigawa, S., and Ochi, M. (2008) *J. Appl. Polym. Sci.*, **110**, 2649–2655.
22 Liu, W., Hoa, S.V., and Pugh, M. (2005) *Compos. Sci. Technol.*, **65**, 307.

23 Wang, K., Chen, L., Wu, J., Toh, M.L., He, C., and Yee, A.F. (2005) *Macromolecules*, **38**, 788.
24 Liu, W., Hoa, S.V., and Pugh, M. (2004) *Polym. Eng. Sci.*, **44**, 1178–1186.
25 Marras, S.I., Zuburtikudis, I., and Panayiotou, C. (2007) *Eur. Polym. J.*, **43**, 2191–2206.
26 Cho, J.W. and Paul, D.R. (2001) *Polymer*, **42**, 1083.
27 Lu, C. and Mai, Y.-W. (2005) *Phys Rev. Lett.*, **95**, 88303.
28 Gilman, J.W., Jackson, C.L., Morgan, A.B., Harris, R., Manias, E., Giannelis, E.P., et al. (2000) *Chem. Mater.*, **12**, 1866.
29 Jang, B.N., Costache, M., and Wilkie, C.A. (2005) *Polymer*, **46**, 10678.
30 Ray, S.S., Bousmina, M., and Okamoto, K. (2005) *Macromol. Mater. Eng.*, **290**, 759.
31 Zanetti, M., Camino, G., Reichert, P., and Mulhaupt, R. (2001) *Macromol. Rapid Commun.*, **22**, 176.
32 March, J. (1985) *Advanced Organic Chemistry, Reactions, Mechanisms and Structure*, 3rd edn, John Wiley & Sons, Ltd.
33 Pluta, M., Galeski, A., Alexandre, A., Paul, M.A., and Dubois, P. (2002) *J. Appl. Polym. Sci.*, **86**, 1497.
34 Gilman, J.W., Jackson, C.L., Morgan, A.B., Harris, R.H., Jr, Manias, E., Giannelis, E.P., Wuthenow, M., Hilton, D., and Phillips, S.H. (2000) *Chem. Mater.*, **12** (7), 1866.
35 Di, Y.W., Iannace, S., Di Maio, E., and Nicolais, L. (2005) *J. Polym. Sci., Part B: Polym. Phys.*, **43**, 689.
36 Nam, J.Y., Ray, S.S., and Okamoto, M. (2003) *Macromolecules*, **36**, 7126.
37 Lee, J.H., Park, T.G., Park, H.S., Lee, D.S., Lee, Y.K., Yoon, S.C., and Nam, J.D. (2003) *Biomaterials*, **24**, 2773.
38 Fornes, T.D. and Paul, D.R. (2003) *Polymer*, **44**, 3945.
39 Maiti, P., Nam, P.H., Okamoto, M., Hasegawa, N., and Usuki, A. (2002) *Macromolecules*, **35**, 2042.
40 Yao, X.F., Yeh, H.Y., Zhou, D., and Zhang, Y.H. (2006) *J. Compos. Mater.*, **40**, 4.
41 Samal, S.K., Nayak, S.K., and Mohanty, S. (2008) *J. Thermoplastic Compos. Mater.*, **21**, 243.
42 Chiu, F.C., Lai, S.-M., Chen, J.-W., and Chu, P.-H. (2004) *J. Polym. Sci., Part B: Polym. Phys.*, **42** (22), 4139–4150.
43 Pollet, E., Paul, M.-A., and Dubois, P. (2003) Layered silicate as nanofiller, in *Biodegradable Polymers and Plastics* (eds E. Chiellini and R. Solaro), Kluwer Academic/Plenum Publishers, New York, p. 329.

17
Effect of Processing Conditions on the Morphology and Properties of Polymer Nanocomposites

Michele Modesti, Stefano Besco, and Alessandra Lorenzetti

17.1
Introduction

The processing of hybrid polymer nanocomposites has been accomplished by several routes, including the *in situ* polymerization of monomer/filler mixtures, solution intercalation, and melt compounding. In this chapter, attention will be paid to aspects involved with the melt intercalation of thermoplastic polymer-based nanocomposites, in particular dealing with the analysis and comprehension of the factors that affect filler dispersion during the process, and with the relationships that exist between these parameters and the final structure and properties of the composites. Layered silicates and carbon nanotube (CNT)-based nanocomposites will be examined separately because of the partially different mechanisms that affect the dispersion of these two types of filler within polymer matrices. Moreover, solution blending of polymer nanocomposites will be considered, despite it having been applied to a lesser degree than melt blending; interesting results related to the high flexibility of this technique, involving intensive mixing devices that allow a very effective dispersion of filler particles, will be described. Progresses and perspectives concerning innovative melt blending processing will also be discussed, notably with regards to fluid (water, supercritical CO_2) and ultrasound-assisted compounding techniques that have recently been developed, and the effectiveness of which has been proven for several systems. The processing of thermosetting polymer-based nanocomposites will also be considered, analyzing in particular different examples of solvent-aided processes, high-shear stirring, cryo-compounding, sonication, and microwave-assisted mixing.

17.2
Melt-Intercalation of Polymer Nanocomposite Systems

17.2.1
Melt Intercalation of Polymer/Clay Systems

Melt-intercalation is by far the most versatile and widely employed procedure for synthesizing silicate hybrids from thermoplastic resins, as it does not require organic solvents and can be realized with processing operations commonly adopted for the parent polymers. The thermal mixing of clay/polymer mixtures at temperatures above the polymer softening or melting point is involved, as according to first studies conducted by Vaia *et al.* [1–4], which showed that nanocomposites could be obtained also by direct polymer melt-intercalation, where the polymer chains diffused into the space between the clay layers or galleries. In particular, Vaia *et al.* [1] prepared intercalated polystyrene (PS)/organoclay nanocomposites by statically heating powder mixtures of the components, with a mechanism that was predominantly enthalpic. Although the entropy loss associated with confinement of the macromolecules inside the galleries can be compensated by the increase of conformational freedom of surfactant organic chains, these authors stated that the total entropy change would not grant intercalation without the enthalpic contribution of polar interactions between the polymer and filler. During the preparation of nanocomposites by melt-compounding techniques, such as batch mixing or extrusion, the molten polymer can be mechanically mixed with fillers within a heated mixing chamber. This is possible because of the polymer molecules' increased mobility through the input of thermal energy, while filler particles are mechanically dispersed and mixed under the influence of shear forces due to polymer viscosity. The dispersion of filler agglomerates can be achieved when the cohesive forces of the agglomerates are exceeded by the hydrodynamic separating forces applied by the fluid polymer [5]. In the case of the intercalated layers, the extent of exfoliation appears to be strongly affected by the conditions of mixing, while the degree of dispersion is generally governed by the matrix viscosity, the average shear rate, and the time of mixing.

17.2.1.1 Effects of Temperature, Shear, and Residence Time
A high temperature may help to improve the mobility of the polymer but, at the same time, it will reduce the viscosity and thus the mechanical force being applied to the nanofillers; this, in turn, will make it more difficult to break up the original agglomerates. In contrast, long residence times may improve mixing but may also enhance degradation and increase the cost of the process. Despite the importance of processing parameters on the final nanocomposite properties, few reports have considered their effects, and it is difficult to compare and generalize observations deriving from different sources. Therefore, no summaries of optimal processing conditions actually exist in the literature, and relatively little can be found about the relationship between the degree of mixing, processing parameters and com-

posite properties, with the objective of optimizing the degree of mixing in terms of such parameters. If different processing techniques may be used for mixing operation, then co-rotating twin-screw extrusion remains the most popular device, due mainly to its great flexibility in terms of screw profile design and mixing configurations. When dealing with the twin-screw extruder processing of polymer-based nanocomposites, it is generally admitted that an increase in screw speed leads to a better dispersion [6–13], because a higher shear rate will increase the agglomerates breakup while intercalation and exfoliation are largely accelerated in the presence of a flowing fluid [14]. However, it should be noted that a negative effect of the screw speed on exfoliation was also observed in some specific cases [15]. The influence of other parameters, including the feed rate and temperature, has also been reported; in particular, increasing the feed rate can induce a major reduction in residence time and specific energy. Several authors [12, 16] have found that feed rate had no effect on the state of intercalation, whereas the state of exfoliation was significantly improved by decreasing the feed rate. This improvement in exfoliation was assumed to be related to the corresponding increase in residence time as the feed rate decreased [12, 17–19]. With regards to the effect of barrel temperature, the reported results have been relatively contradictory, because this parameter cannot be considered independently of the screw speed and feed rate [6, 10, 19]. As the temperature increases, the viscosity decreases with the shear required to break the clay aggregates. In addition, diffusion is improved, which helps polymer intercalation and the migration of exfoliated platelets. On the other hand, a high temperature may cause degradation of the organoclay surfactants, leading to a collapse of the interlayer galleries and to a loss of organically modified layered silicate (OMLS) efficiency, thus leading to a dispersion of unintercalated tactoid particles [20].

An investigation into the optimization of processing parameters (temperature, residence time, shear) for the melt blending of PA6/clay nanocomposites was reported by Paul *et al.* [6]. For these studies, two different extruders were used. First, a Haake intermeshing co-rotating twin-screw extruder (with screw diameter $D = 30$ mm) having a screw configuration containing two kneading blocks consisting of one right-handed medium-pitched, one left-handed medium-pitched kneading disc element and one mixing ring was used, setting up a barrel temperature of 240 °C and screw speed of 180 rpm; this led to a mean residence time for neat PA6 of 5.3 min. For comparative purposes, a Killion single-screw extruder was used ($D = 25.4$ mm) with two intensive mixing heads and compounding was carried out at 240 °C, using a screw speed of 40 rpm; this led to a mean residence time for neat PA6 of 2.35 min.

A morphological analysis showed clearly that, for the composite prepared by single-screw extrusion, full exfoliation was not achieved, mostly because the amount of shear was insufficient and the residence time too short. Thus, the final product was a mixture of exfoliated and unintercalated clay particles. In contrast, the tensile strength and modulus of the organoclay composite produced after a second pass through the extruder were slightly improved, whereas the elongation at break and impact strength were not.

When dealing with the composites obtained with a twin-screw extruder, the organoclay appeared uniformly dispersed and exfoliated within the polymer matrix. An absence of the characteristic clay d_{001} diffraction peak in X-ray diffraction (XRD) patterns provided further evidence of the formation of exfoliated nanocomposites. A processing parameters analysis showed that PA6 nanocomposites with good properties could be obtained over a broad range of conditions by using a twin-screw extruder. Typically, the nanocomposite's properties were almost independent of the barrel temperature over the range of typical PA6 processing (230–280 °C), even if the mechanical properties could be slightly improved, either by increasing the screw speed or with a second pass through the extruder (hence increasing the total residence time).

Vergnes and Lertwimolnun [11] investigated the influence of processing conditions on clay dispersion for polypropylene (PP) -based formulations prepared by direct melt-intercalation. For this, an internal mixer (Haake Rheomix 600, equipped with roller rotors) was used, with the rotor speed being varied from 10 to 150 rpm, the mixing time from 5 to 30 min, and the mixer temperature from 180 to 200 °C. Melt mixing of the PP/OMLS containing maleic anhydride grafted PP (PP-g-MA) composites was also studied using a self-wiping, co-rotating twin-screw extruder (Clextral BC45; $D = 50$ mm; length $(L)/D$ ratio = 24). The screw profile was composed of a left-handed element for the melting, followed by three mixing zones, with different configurations of the kneading blocks. The polymer pellets and nanoclay were dry-mixed and introduced simultaneously through the main hopper, setting the barrel temperature at 180 °C and a screw speed of 200 rpm. Only the influence of feed rate, which was varied between 4 and 30 kg h^{-1}, was investigated. When the effect of mixing time was evaluated for composites produced using the internal mixer, it was observed (by means of morphological and rheological analyses) that the degree of dispersion was improved by increasing the mixing time. With regards to XRD characterization, intercalation appeared not to depend significantly on the shear rate, while the degree of resultant intercalation was substantially independent of temperature within the range investigated. Notwithstanding, for the specific system, the temperature appeared to have a significant effect on the degree of exfoliation; in particular, a low mixing temperature led to a higher degree of exfoliation because of the level of stress applied during mixing. Twin-screw extruder tests showed that the value of interlayer spacing was not affected by varying the feed rate, which suggested that complete intercalation could occur, despite the mixing time in the twin-screw extruder being very short at high feed rates (ca. 63 s at 30 kg h^{-1}). In addition, the value of interlayer spacing obtained by extrusion was comparable to that obtained via internal mixer blending producing the same formulation. Whilst it was reasonable to state that the intercalation was independent of the processing conditions, the authors also showed that the feed rate heavily influenced clay exfoliation. Notably, the best dispersion was obtained at a low feeding rate, and hence the longer residence times, which ranged from 290 s at 4 kg h^{-1} to 63 s at 30 kg h^{-1}. Further evidence confirming these observations was provided by Modesti et al. [9, 10], who investigated the effects of processing conditions on the properties of PP/OMLS nanocomposites prepared by melt-intercalation. For this, an intermeshing co-rotating twin-screw extruder ($D = 42$ mm, $L/D = 40$) was used, with a medium-shear configuration modular screw. The melt temperature (and hence the barrel temperature profile) and screw speed were varied respectively within 170 to 200 °C and 250 to 350 rpm on the basis of an experimental design, while maintaining a constant feed rate. A consideration of the different processing conditions showed that the shear stress exerted on and by the polymer, depending on the screw speed and melt viscosity, had a much greater influence on the extent of intercalation and delamination than did the residence time. In fact, the best results in terms of clay dispersion and properties enhancements were obtained using processing conditions that maximized the shear stress exerted on the polymer, by using higher screw speeds and lower melt temperatures (Figure 17.1). On the other hand, even when using optimized processing conditions, a high degree of intercalation or exfoliation could be achieved only when a high compatibility existed between the polymer and clay. In their studies, Kim *et al.* [21] assessed the effects of melt temperature, screw speed, and feed rate on the degree of mixing of a melt-compounded alumina poly(ethylene terephthalate) (PET) nanocomposite

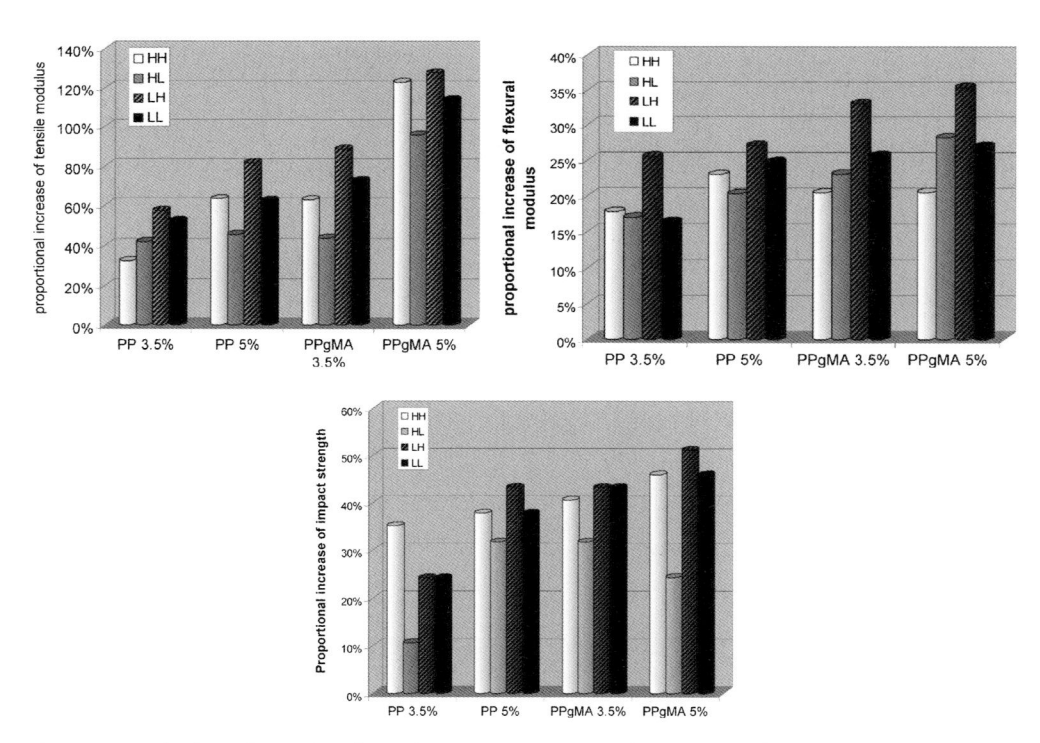

Figure 17.1 The dependence of mechanical properties on processing conditions. HH = high screw speed and high melt temperature; HL = high screw speed and low melt temperature; LH = low screw speed and high melt temperature; LL = low screw speed and low melt temperature [9]. Reprinted with permission from Elsevier.

system. Here, the overall effects of each parameter were evaluated through a full-factorial, two-level experimental design. A Leistritz co-rotating, intermeshing twin-screw extruder ($L/D = 40$; $D = 27$ mm) and operating under nitrogen was used for the compounding experiments. The screw speed was found to have the greatest effect on the degree of mixing, with better mixing observed at higher screw speeds. The feed rate also had a significant effect, with a lower feed rate providing better mixing, whereas a lower melt temperature led to an improved degree of dispersion.

In addition, the specific energy input (SEI) to the mixer was shown to be a good indicator of the degree of mixing; indeed, the latter improved with increasing SEI up to a limiting critical value which corresponded to the optimum dispersion level for a given system. When the critical SEI value was maintained, this approach allowed for variations in the process parameters whilst maintaining a good dispersion. Taking together these and other evidences [2, 14, 18, 22, 23] it is clear that, whilst the shear forces may considerably expedite the break-up of clay agglomerates, penetration of the polymer chains within the clay galleries, and exfoliation of the silicate layers stacks, are mainly driven by residence or mixing times and, of course, by the chemical compatibility between the clay and polymer. Thus, processing conditions may be tailored to achieve the greatest intercalation and exfoliation of clay platelets even if, unfortunately, the same effects of processing parameters on nanocomposite structure are not universally observed. It follows that each system must be specifically studied in order to optimize filler dispersion.

17.2.1.2 Effects of Extruder Configuration and Screw Profiles

Besides processing conditions, the screw profile has an important effect on the dispersion of organoclays. This situation occurs because the final morphology of the nanocomposite is not only a question of shear stress or residence time, but also a complex result of the thermal and mechanical history supported by the material when flowing along the screw profile [24]. Very few studies have focused on the influence of screw profile on clay dispersion and on the evolution of nanocomposite microstructure along the extruder [12, 25–27].

In their investigations, Dennis et al. produced PA6/montmorillonite (MMT) nanocomposites while comparing the effectiveness of several single- and twin-screws extruders set up with several types of modular screw design [18]. For this, two screw configurations were evaluated with a Japan Steel Works (JSW) co-rotating twin-screw extruder: (i) a low-shear intensity configuration with one kneading block positioned early in the screw; and (ii) a medium-shear intensity configuration that incorporated three kneading blocks. The kneading discs block sections could include different paddles (the wider the paddle, the more elongational field and dispersive mixing), while the discs in the kneading blocks could be characterized by different staggering angles promoting forward movement of the polymer. A forward-pumping kneading section typically preceded a reverse or neutral element in order to increase the mixing intensity and mean residence time, and to broaden the residence time–distribution curve (Figure 17.2). Three screw

Killion Single Screw

Japan Steel Works Co-Rotating Twin Screw

Low Shear

Medium Shear

Figure 17.2 Screw configurations used for single screw and co-rotating twin-screw extruders [18]. Reprinted with permission from Elsevier.

Low Shear

Medium Shear

High Shear

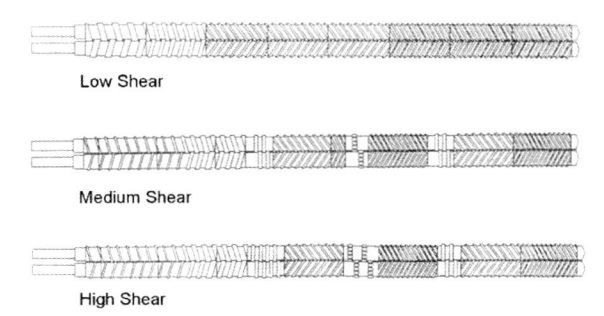

Figure 17.3 Screw configurations used in counter-rotating, intermeshing twin-screw extruders [18]. Reprinted with permission from Elsevier.

configurations were used in a Leistritz counter-rotating extruder in both intermeshing and nonintermeshing mode. For the intermeshing mode, the low-shear intensity configuration was roughly a closed-flighted positive displacement pump, while the medium-shear intensity configuration was a closed-flighted pump to which two series of shearing elements and a turbine distributive mixer were added. The high-shear intensity configuration was a closed-flighted pump to which two sets of four shearing elements (two shearing elements and a set of two turbine distributive mixers) were added (Figure 17.3). When dealing with the nonintermeshing extruder, the low-shear intensity configuration was a simple, open-flighted distributive mixer, while the medium-shear intensity system had a shearing element and a reverse element added, and the high-shear intensity system was a distributive mixer with a set of four shearing elements and a set of two turbines (Figure 17.4). The processing conditions were held constant, with a feed rate of $5\,\text{kg}\,\text{h}^{-1}$, a mean temperature of 240 °C and a screw speed of 200 rpm being used for the twin-screw extruders. For the single-screw extruder a lower feed rate and a lower screw speed were chosen. These conditions produced residence time–distribution curves that were measured with spiking pulses of 1–2 mm aluminum pellets feeding the hopper [28]. The single-screw extruder with an intensive mixing head

Low Shear

Medium Shear

High Shear

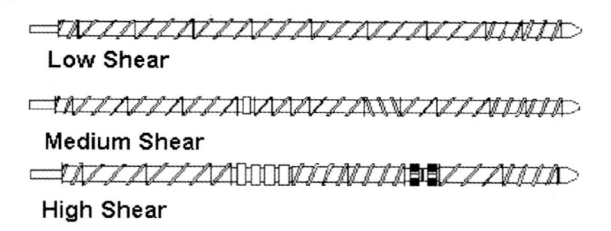

Figure 17.4 Screw configurations used in counter-rotating, nonintermeshing twin-screw extruders [18]. Reprinted with permission from Elsevier.

showed a relatively narrow residence time–distribution, whilst, as expected, each of the twin-screw extruders provided much more back-mixing, measured as the value of normalized variance of the curves. In summarizing these findings, the co-rotating intermeshing extruder provided the greatest degree of back-mixing, and the counter-rotating intermeshing extruder the least. The counter-rotating nonintermeshing extruder gave intermediate degrees of back-mixing that were significantly affected by the screw configuration. A subsequent examination of the materials, using transmission electron microscopy (TEM), with data expressed as the number of platelets or intercalates per unit area, showed that the process conditions had a strong effect on the dispersion of the clay. All of the twin-screw extruders gave a good delamination of OMLS, with the best results being obtained when using the medium-shear intensity screw configurations. In particular, the shear intensity was seen to be increased for a given extruder type (from the medium- to the high-shear intensity configuration). Typically, a degree of shear intensity was reached where delamination and dispersion showed no further increases; hence, it appeared that very high shear was not the key to delamination, at least for this system. With regards to residence times, although the data did not follow a clear trend, delamination was clearly improved as the residence time was increased for all samples prepared in twin-screw extruders. Notably, delamination was poor for the single-screw extruder, despite the relatively long residence time, mainly because of the poor backmixing. Vergnes et al. [24] also studied the effects of screw profile on the microstructure of PP/OMLS nanocomposites prepared with an industrial self-wiping co-rotating twin-screw extruder (Clextral BC45; $D = 50\,\text{mm}$; $L/D = 24$). For this, three different screw profiles were used in order to obtain a progressive increase in the mixing and residence times, by including increasing numbers of kneading blocks and left-handed elements. An evolution of morphology was observed along the screw in order to explain and understand the mechanisms of filler break-up, intercalation, and exfoliation. Intercalation appeared as a rapid mechanism, and was generally observed as soon as the matrix was melted; the level of intercalation seemed to be quite independent of the processing conditions. In contrast, exfoliation may be partially controlled by the processing conditions and screw profile; indeed, for a fixed screw geometry the level of exfoliation was proportionally related to the N/Q ratio (the screw speed/

feed rate). It also seemed to be controlled by the specific energy experienced by the material.

With regards to screw profile effects, based on previously reported results [18] the most severe profile was not necessarily the most efficient for enhancing clay dispersion, and this indicated that other parameters must be taken into account in the exfoliation mechanism. Rather, the evolution of exfoliation along the screws was largely influenced by the feed rate and, in some cases, an unexplained decrease in exfoliation after the melting zone was observed.

17.2.1.3 Effect of Processing Route

Another factor affecting the structure and properties of polymer/OMLS composites involves the processing route adopted. In particular, for melt blending there are two generally feasible approaches:

- A one-step process (direct melt blending) in which the polymer and clay are dry-premixed or fed separately to the extruder in the correct proportions, and which represent the final concentrations required for the composite.

- A two-step process in which the final nanocomposite is obtained from the dilution of a concentrate masterbatch containing more filler than the correct amount required for the final polymer/clay composite.

Paul and Shah [29] investigated the melt-mixing masterbatch process by preparing PA6 nanocomposites, the aim being to obtain well-exfoliated nanocomposites with low melt viscosities that were particularly suited to applications such as injection molding. It has long been known that high-molecular-weight (HMW) grades of PA6 lead to higher levels of exfoliation of organoclays than do low-molecular-weight (LMW) grades; this effect is due to the higher melt viscosity providing major particles break-up and a more efficient shear transfer during mixing. Thus, a novel two-step process to prepare PA6/clay-based nanocomposites was explored. In the first step, a HMW PA6/OMLS masterbatch was prepared by melt processing to give intercalation and exfoliation, while in the second step the concentrate was diluted with LMW PA6 to the desired MMT content. Concentrate masterbatches containing 20, 14, and 8.25 wt% MMT were produced using a Haake co-rotating, intermeshing twin-screw extruder ($D = 30$ mm, $L/D = 10$) operating with a barrel temperature of 240 °C, a screw speed of 280 rpm, and a feed rate of 1200 g h^{-1}. Each of the masterbatches was then diluted with LMW PA6, using the same apparatus and processing conditions, in order to produce nanocomposites containing 2.0, 4.0, and 6.5 wt% MMT. For comparison, the same nanocomposites were prepared directly (one-step process) from HMW and LMW PA6, using the same organoclay and processing conditions as noted above. Mechanical tests showed that, although there was a minimal difference between the stiffness of virgin HMW PA6 and of LMW PA6, the moduli of the nanocomposites based on HMW PA6 were 10–15% higher than those of their LMW PA6-based counterparts. Moreover, wide-angle X-ray diffraction (WAXD) scans of the HMW PA6 composites indicated a well-exfoliated character of these systems, whereas those nanocomposites prepared from LMW PA6 showed a distinct broad peak that indicated the presence of intercalated clay tactoids. The mechanical stiffness values of all samples prepared using the masterbatch approach fell between those of the equivalent HMW and LMW nanocomposite samples produced in one step. With regards to the structural characterization of polymer/clay composites, WAXD patterns obtained for 20 wt% and 14 wt% masterbatches revealed a low, broad peak that suggested these systems had a mixed morphology which consisted of intercalated clay tactoids and exfoliated clay platelets. However, the WAXD pattern of the masterbatch containing 8.25 wt% MMT showed an extremely broad peak which indicated that the system was almost exfoliated; this was in good agreement with the mechanical characterization results. Subsequent TEM images of the HMW PA6 nanocomposite revealed a well-exfoliated structure, whereas the LMW PA6 nanocomposite revealed a partial exfoliation along with areas containing exfoliated platelets and unintercalated tactoids. Whilst the masterbatch-based nanocomposites showed a mixed morphology, the unexfoliated clay stacks were fewer in number and smaller in size than those found in images of the LMW nanocomposite; this was in good agreement with data obtained in WAXD and mechanical analyses.

17.2.2
Melt-Intercalation of Polymer/CNT Systems

The outstanding performances of CNTs, which combine unique electrical and thermal properties with high mechanical strength [30–47], support their promising use as fillers for polymer matrices over a wide range of industrial applications. Although the large-scale melt-mixing of nanotubes into thermoplastic polymers ha been reported [48–50], no systematic investigations have yet been identified relating to the influence of melt-mixing conditions on nanotube dispersion. As with OMLS, the key challenge regarding these systems is the suitable distribution and dispersion of the filler inside the polymer matrix, with the formation of an electrically conductive path (termed the percolation threshold), combined with an enhancement of physical and mechanical performances at low filler contents. During the past decade, several methods have been proposed and even applied for the synthesis of CNT/polymer nanocomposites, including physical melt-mixing with polymers [51], electrospinning [52], *in situ* polymerization in the presence of CNTs [53], surfactant-assisted processing of CNT/polymer nanocomposites [54], mechanochemical pulverization processes [55], latex fabrication methods [56], coagulation spinning [57], and solid-state shear pulverization [58]. Both, high shear forces during the melt-mixing process and relatively long processing times, have been found to be suitable for the successful dispersion of multiwall nanotubes (MWNTs), as reported for several systems [48, 50, 59–70]. For small-scale mixing [66], the influence of rotation speed on the state of MWNT dispersion and on their electric/dielectric properties, at concentrations below and above the percolation threshold, has been widely investigated. Generally, below the percolation threshold an increased screw speed has resulted in a better dispersion, whilst above the threshold a decrease in electric conductivity can be observed; the suggested reason

17.2 Melt-Intercalation of Polymer Nanocomposite Systems

for this was breakage of filler particles. Furthermore, as in the case of polymer/clay nanocomposites, it has been reported that an increase in the mixing time generally improves the MWNT dispersion, independently of their concentration. Similar results have been reported by Takase [67], who showed that increasing the rotation speed of the twin-screw extruder caused the agglomerate size to be decreased. Moreover, as indicated by Andrews *et al.* [65], a certain decrease in the aspect ratio during melt-mixing may improve the dispersibility of these fillers, due to a decreasing number of physical entanglements. In particular it was reported that, during the melt-mixing of MWNTs with PS in a laboratory mixer, although a shortening of MWNTs occurred with increasing mixing time, a better MWNT dispersion was generated when the mixing energy input was increased.

Masterbatch dilution processes have also been described aimed at dispersing CNTs in polymer melts, in order to reduce human contact with nano- and microstructured powders that might, potentially, be hazardous to health [48, 50, 61–63, 71–73]. Recently, Villmow *et al.* [74] used a twin-screw extruder with a modular assembling to study systematically the effects of the screw profile, temperature profile and rotation speed on the dispersion and distribution of MWNTs within a PLA polymer matrix, via a masterbatch processing route. The greatest influence was identified for the rotation speed, whereby an increase from 100 rpm to 500 rpm led to a significant enhancement of MWNT dispersion with a lower number of residue agglomerates. This finding was in agreement with former reports of polycarbonate/CNT composites [66], and also with the data reported by Takase [67] which described a decreasing agglomerate size with increasing rotational speed during twin-screw extrusion. The level of shear stress acting on the primary agglomerates proved to be an important factor for the individualization of MWNTs during melt-mixing [74]. In the studies conducted by Villmow *et al.*, the masterbatch characterized by the best dispersion was obtained using a screw profile mainly with mixing elements, a temperature profile with a rising temperature towards the extruder die, and a high rotational speed. These experiments also showed that the potential for further MWNT dispersion during the masterbatch dilution process was limited. Consequently, primary MWNT agglomerates from the masterbatches would remain in the final composites if their number and dimension in the masterbatches were to exceed certain critical values. Although not all MWNT agglomerates of the masterbatch could be dispersed in the dilution step, the rotation speed was seen to have a major impact in the dilution step.

Chen *et al.* [75] reported details of the fine dispersion of CNTs in a polyvinylidene fluoride (PVDF) matrix, using a novel ultrahigh-shear extruder in which a feedback-type screw system was applied operating at high rotation speed (1000 rpm). In this peculiar device, the raw materials fed to the top of the screw were able to return to the root of the screw via a feedback path. This allowed the PVDF/MWCNT composites to be prepared by melt-compounding of PVDF with MWCNT at 220 °C for 4 min, and to demonstrate an apparent effect on the electrical conductivity of the screw rotation speed. A significant increase in conductivity was achieved with a percolation threshold of about 2.5 wt% CNTs under low-shear processing, whereas a percolation threshold of only 1.5 wt% CNTs was required for high-shear processing. Using the same device, Li and Shimitzu [76] assessed the effects of screw rotation speed on a different system, namely poly(styrene-*b*-butadiene-*co*-butylene-*b*-styrene) (SBBS)/MWCNT, and found the CNT dispersion to be significantly improved when the mixing speed was increased. A homogeneous dispersion was achieved with a very high screw rotation speed (2000 rpm); this was confirmed by the fact that the tensile modulus and tensile strength of the composites with a fine CNT dispersion (processed at a high shear rate) were much higher than those of composites obtained using a low shear rate. This is a further indication that high-shear processing might represent a general – but effective – method for improving the dispersion of unmodified CNTs in polymer matrices.

17.3
Solution-Intercalation of Polymer Nanocomposites

In solution-intercalation, a solvent capable of dissolving the polymer and swelling the clay is first selected, after which a homogeneous, three-component mixture of an appropriate composition is prepared by using heating and mechanical and/or ultrasonic stirring. The solvent is then removed by evaporation, or the polymer is precipitated in a nonsolvent. The entropy gain associated with desorption of the solvent molecules is thought to compensate for the entropic loss of the intercalated polymer chains [5]. Thus, with this procedure, an entropy-driven intercalation might be expected to occur, even in the absence of an enthalpy gain, due to favorable interactions between the macromolecules and the surface of the clay layers. This technique was used with water-soluble polymers, such as poly(vinyl alcohol) [77–79], poly-(acrylic acid) [79], and poly(ethylene oxide) [78, 80–83], the polarity of which most likely contributes to an enthalpy gain that assists the intercalation. Several examples involving organic solvents and hydrophobic polymers have also been reported [83–91]. For example, Jeon *et al.* [86] dissolved high-density polyethylene (HDPE) and a MMT modified with dodecylamine in a mixture of xylene and benzonitrile, and recovered the composite after precipitation with an excess of tetrahydrofuran (THF). The filler was found to be well dispersed in the form of relatively thick stacks of silicate layers, the *d*-spacing of which, as measured using XRD, was only few Angstroms larger than that of pristine organoclay. This expansion was most likely too small to be ascribed to intercalation of the HDPE chains [92, 93]. A definite intercalation was achieved by the same group, by using a nitrile copolymer in place of HDPE [86], while Xu *et al.* [94] prepared master blends from polyethylene-grafted maleic anhydride (PE-*g*-MA) and a MMT modified with trimethylhexadecylammonium ions either in the melt, or from xylene solution, and diluted these compounds with HDPE by melt-compounding in a roller mixer. In terms of morphological results (via WAXD, TEM), solution-intercalation appeared to be more effective than melt-intercalation, most likely because of kinetic (enhanced mobility of the macromolecules in solution) and/or thermodynamic (entropic gain associated with desorption of solvent molecules) factors.

17.3 Solution-Intercalation of Polymer Nanocomposites | 381

Qiu et al. [91] suspended a classic OMLS in boiling xylene for 12 h, and added the appropriate amount of linear low-density polyethylene (LLDPE) required to obtain composites with 5.0 or 10.0 wt% organoclay. The mixture was then stirred for 6 h and the composite precipitated with an excess of ethanol. An XRD analysis of the products showed that a high level of intercalation with partial exfoliation was apparently obtained for the LLDPE/organoclay composites prepared from solution. The results of Qiu et al. [91], when compared with those of Hotta and Paul [92] and others [94–98], showed that an organic treatment of the clay would be insufficient to enable PE intercalation by melt-compounding, and that enhanced levels of intercalation or exfoliation of PE nanocomposites would be granted by the use of a solution-intercalation procedure.

Following these guidelines, Filippi et al. [99] prepared ethylene acrylic acid (EAA)/clay composites under different processing conditions in order to explain the thermodynamic and kinetic aspects of the intercalation process. Both, solution-intercalation and thermal annealing of clay/polymer mixtures in the absence of shear have been used, in addition to melt-compounding. For solution-intercalation, EAA was first dissolved under stirring in a suitable solvent (mainly toluene or xylene) at 110–140 °C, after which an appropriate volume of a dispersion of the organoclay in the same solvent was added to the polymer solution under stirring for 2 h. The polymer nanocomposites were then recovered by solvent evaporation or precipitation with the addition of an excess of nonsolvent (acetone, ethyl alcohol, pentane, n-hexane, or mixtures of THF). For comparison, EAA/clay nanocomposites were prepared using a Brabender mixer that had been preheated at 120 °C, and of which the rotor speed was maintained at 30 rpm for approximately 3 min, but then increased gradually to 60 rpm. Static melt mixing was investigated starting from powder mixtures of EAA and clay that, after being compressed at room temperature to prepare 2 mm-thick tablets, were placed in a mold and heated at the selected temperature for different times before being cooled. The dispersion of the clay particles in the composites prepared from solution was generally lower than that achieved by the melt-compounding procedure. However, although solution blending does not grant the same level of dispersion which can be achieved by the application of shear forces during melt-compounding, it may certainly help to disperse the clay particles within the polymer matrices as individual platelets and/or small tactoids. Moreover, microscopic analyses, together with X-ray characterization, indicated solution blending would allow a high level of intercalation and partial exfoliation to be achieved in all cases, independent of the preparation method (i.e., the solvent/nonsolvent couple employed). Static melt-mixing molding of the powders obtained via a low-temperature solution method led easily to the intercalation of EAA, and yielded nanocomposites that displayed XRD patterns very similar to those obtained with analogous materials produced by melt-compounding. Subsequently, Modesti et al. [100] evaluated a dimethyl-hexadecyl-imidazolium (DMHDIM) -modified MMT (which also contained Nile Blue A as a fluorescent optical probe) for the preparation of polymer nanocomposites, the aim being to compare the respective effectiveness of melt-blending and solution-intercalation techniques. Samples obtained using solution-intercalation, assisted by ultrasonication, were first prepared by dissolving and stirring acrylonitrile butadiene styrene (ABS) into refluxing acetone, and then mixing the resultant solution with a dispersion of OMLS in the same solvent. Ultrasonic mixing (Bransons 1510; 70 W, 72 kHz) was applied for 6 h to disperse the OMLS in acetone, and also to the polymer solution after addition of the OMLS. The solvent was removed by evaporation at 50 °C under vacuum. Melt-blended composites were instead produced using a Brabender mixer preheated at 190 °C and with screws rotating at 4.8 rad s^{-1}, for a mixing time of about 5 min.

Fluorescence has long been used for monitoring the intercalation/exfoliation of the clay [101]. Indeed, it has been shown that for PA6/DMHDIM-MMT/Nile Blue nanocomposites, the spectroscopic emission at about 560 nm can be related to intercalation, while fluorescence effects at 610 nm are indicative of mixed intercalated/exfoliated structures. Emission wavelengths depend on the nanoconfinement of the optical probe and on the polar character of the local environment; thus, fluorescence spectra can also be used to investigate the preferential localization of the clay in one of the phases composing the polymer structure (i.e., styrene-acrylonitrile and butadiene). Hence, a modified dyed clay was dispersed in three different solvents at a concentration of 5 wt%; in this case, each solvent was chosen to mimic the polarity of a different ABS phase (heptane for butadiene, acetonitrile for acrylonitrile, and toluene for styrene).

Subsequent morphological and structural analyses (comparative TEM images are shown in Figure 17.5) showed that, with both melt-blending and solution processes, mixed intercalated/exfoliated structures were obtained. The sonication process further reduced the size of these aggregates (compared to melt-processing results) and improved the degree of dispersion. Hence, for this system, solution-intercalation appeared to be more effective at dispersing and improving the mechanical properties than did melt-blending. The fluorescence spectra for the ABS/OMLS composites (see Figure 17.6) revealed two peaks at about 600 nm and 480 nm, which appeared to be a combination of the peaks obtained with toluene and acetonitrile. This observation suggested that clay would reside in the styrene-acrylonitrile (SAN) rigid phase, as observed previously by Stretz et al. [102]. In the composite images (Figure 17.7) (generated by superimposing 20 individual confocal images), nonfluorescent aggregates (the clay will not fluoresce until the effective distance between fluorophores is at least 3–5 nm) of intercalated tactoids with a maximum dimension of about 50 μm are observed for the melt-blending sample. Moreover, aggregates are much smaller in the ultrasound composites, further supporting a more homogeneous dispersion for ultrasound samples. An investigation into parent systems was recently reported by Pourabas and Raeesi [103] which involved the development of a solvent/nonsolvent method for the preparation of ABS/clay nanocomposites. For this, ABS was first dissolved at 60 °C in THF under stirring. In this method, a special mechanical homogenizer was used which consisted of two co-axial cylinders with a small gap between them, rotating with a relative speed of 10 500 rpm. A dispersion of OMLS in ethanol was added dropwise to the ABS/THF solution under intensive stirring, and the polymer/clay composite was precipitated. For comparative purposes, and in order to investigate the homog-

17.3 Solution-Intercalation of Polymer Nanocomposites | 383

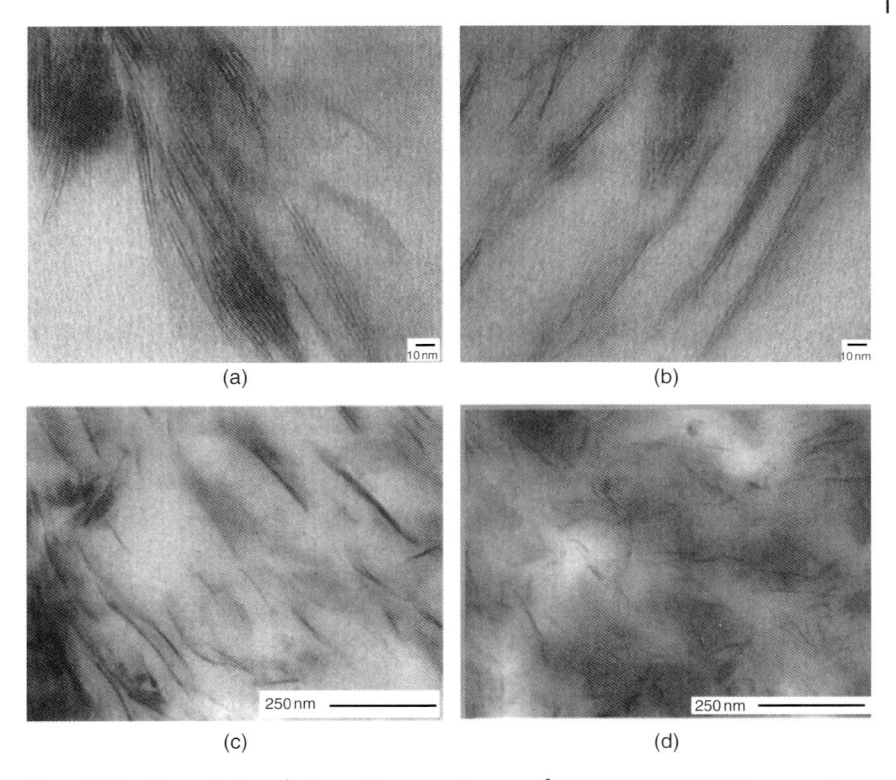

Figure 17.5 Transmission electron microscopy images of ABS/DMHDIM–MMT composites obtained with melt-blending (a, c) and ultrasonic solution blending (b, d) at different magnifications [100]. Reprinted with permission from Wiley.

enizer effect, the same procedure was repeated using a traditional mechanical stirrer rotating at 200 rpm. This led to intercalated structures with a uniform interlayer spacing of the clay silicate layers, as detected from the XRD peak shape and position and TEM images. With respect to thermal stability, the silicate layers appeared to act as thermal shields for the intercalated polymer chains, which in turn led to an enhanced thermal resistance. Moreover, these investigations confirmed that the solution-intercalation of a polymer between the silicate layers was a thermodynamically controlled process that involved two steps: (i) a preliminary intercalation step; and (ii) a subsequent delaminating process. The application of severe mixing or shear stresses during the first step caused an acceleration in attaining the pseudo-equilibrium. Following the initial intercalation, further shear stresses, whether applied with a twin-roller mill or via internal mixer processing, may simply lead to exfoliation of the particles [18].

Morgan and Harris [104] also focused their interests on the synthesis of an exfoliated PS nanocomposite by solvent blending, in order to consider the effects of solvent type and blending conditions. The specific processing conditions for

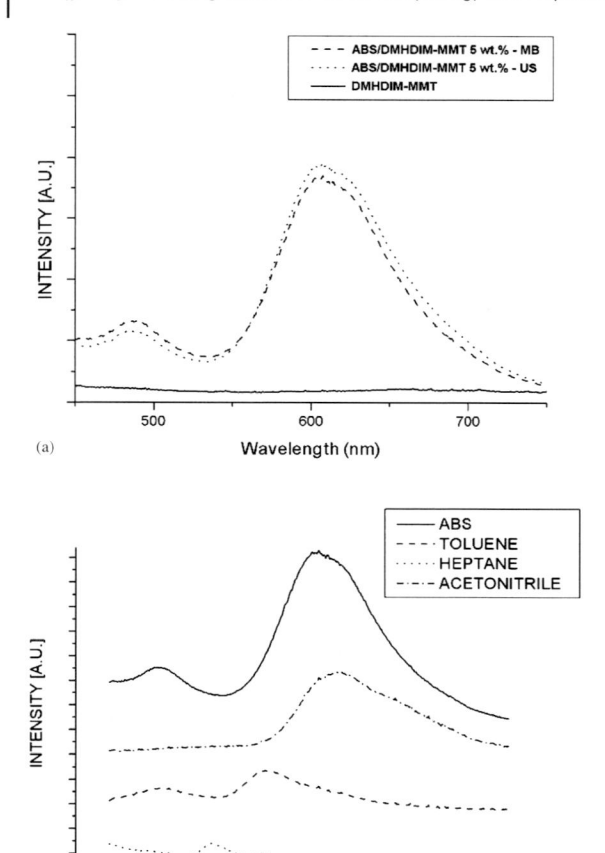

Figure 17.6 (a) Fluorescence spectra for ABS/DMHDIM–MMT composites obtained with melt-blending (MB) and ultrasonic solution (US) blending. The solid line indicates pristine OMLS; (b) Fluorescence spectra for ABS/DMHDIM–MMT composite obtained with solution-intercalation and for OMLS solutions using representative solvents [100]. Reprinted with permission from John Wiley & Sons. Ltd.

this experiment were selected based on previous discussions [104], the aim being to determine how solvent blending, with and without high-energy mixing, could disperse a natural MMT and a synthetic clay (fluorinated synthetic mica; FSM) in a PS matrix. The organic treatment for both clays was a 1,2-dimethyl-3-hexadecyl imidazolium (DMHDI) cation, the use of which had been reported previously [105]. The solvent selected was chlorobenzene, whilst high-energy mixing was achieved with a sonic horn; this led to a PS nanocomposite with high levels of clay exfoliation. The one clear key factor in these experiments appeared to be the high-energy mixing accomplished by sonication since, in its absence, the XRD and TEM

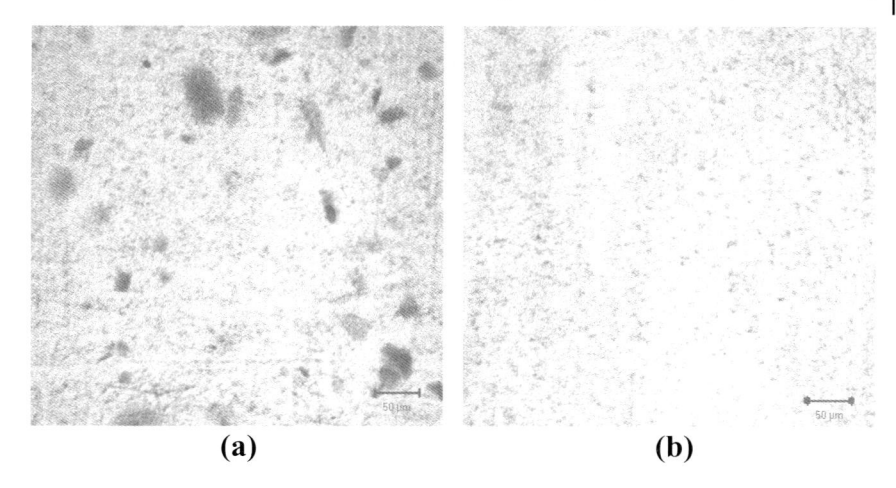

Figure 17.7 Composite images using confocal microscopy of: (a) ABS/DMHDIM-NB–MMT composites produced by melt-blending, and (b) ultrasonic solution blending [100]. Reprinted with permission from John Wiley & Sons. Ltd.

data indicated that these levels of clay dispersion had not led to the creation of well-dispersed exfoliated clay nanocomposites. Clearly, when solvent blending was used, sonication had a major effect on the formation of an exfoliated PS nanocomposite.

17.4
Progress in Polymer Nanocomposites Processing

17.4.1
Water Injection-Assisted Melt-Compounding

The intercalation of polymers into unmodified sodium-MMT represents a major challenge for avoiding the use of OMLS. It has been shown that the presence of large amounts of LMW surfactants (ca. 30 wt%) may adversely affect the reinforcing efficiency of the clay layers, and the thermal stability and durability of the final products, even if all of the clay layers are fully exfoliated [106, 107]. Recently, much effort has been directed towards developing alternative routes to produce exfoliated nanocomposites, without using organically modified clay [108–113]. Hasegawa *et al.* [108] reported a novel process for preparing PA6 nanocomposites from an aqueous slurry of unmodified MMT, that has been melt-blended with PA6 by using a laboratory extruder; the water released was removed by evacuation. Both, WAXS and TEM investigations clearly indicated exfoliation of the filler, while the mechanical properties appeared to be very similar to those obtained with conventional organomodified nanocomposites. The major drawback of this method was the need to pump very large amounts of water into the extruder so as to reduce the viscosity of the slurry. Subsequently, Korbee and Van Geneen reported and patented a similar process where water was injected directly into the extruder melt while MMT was introduced with the polymer (PA6) pellets [114]. This concept was further developed recently by Yu *et al.* [115], whereby the clay did not require any pretreatment in water before being blended in the extruder, which in turn meant that much smaller amounts of water could be used. Devaux *et al.* further developed this technique [116] by using a laboratory-scale extruder set in scale-up conditions. In this case, a combination of high-shear compounding and very high-throughput and reactive processing technology was used to improve the dispersion of natural MMT in PA6. Unfortunately, to date most of the above-described methods have been described only for polyamides and hydrophilic polymers [117].

A recent report [111] described the preparation of PP/clay nanocomposites using a modified clay slurry method, where the slurry was obtained by pumping water into the extruder containing the PP melt and clay mixture. Although, in this case, exfoliated structures were obtained, organic compatibilizers had to be used in order to stabilize the nanomorphology.

Other groups obtained a MMT/polyamide-based masterbatch by first producing a water injection-assisted melt blending, and then diluting this into a polyolefin [118]. In this case, polyamide was used as a compatibilizer and placed in direct contact with a natural MMT; compatibilization with the polyolefin was then performed in a second step, using a maleic anhydride-functionalized polymer. For this, a co-rotating extruder (ZSK25 WLE; Coperion) was used ($D = 1000$ mm; L/D ratio = 40). The PA6 pellets and clay were introduced through the feeder at throughputs which varied between 10 and 50 kg h^{-1}, while the screw speed varied between 200 and 1200 rpm. Water was pumped into the extruder, from 0 to 20 wt% of the total throughput, and compounded with the PA6 and clay at a barrel temperature of 240 °C. The water modified the viscosity and polarity of molten polymer and, under these processing conditions of temperature and pressure, was completely miscible with the polyamide. Water diffusing between the silicate layers was adsorbed onto the surface, causing the interlayer distance to be increased as the result of a well-known swelling phenomenon (see Figure 17.8). Grim [27] showed that, by increasing the relative water pressure, the existence of one to four successive water layers in the interlayer would be possible. Consequently, when more than one molecular layer of water has been adsorbed, the d-spacing of the hydrated MMT would be in the same range as that of organomodified clay. Moreover, the combination of these two effects would create ideal conditions for the diffusion and adsorption of PA6 chains onto the top-MMT surface and between the platelets. While diffusing between the silicate layers, the polyamide chains were adsorbed on the surface, causing previously adsorbed water molecules to be desorbed.

Polyamide 6/pristine MMT nanocomposites were obtained [118] which exhibited an exfoliated structure similar to that of nanocomposites prepared with organomodified clay and produced by simple melt intercalation. In this way, it would be possible to compare the morphology and properties of these materials to those

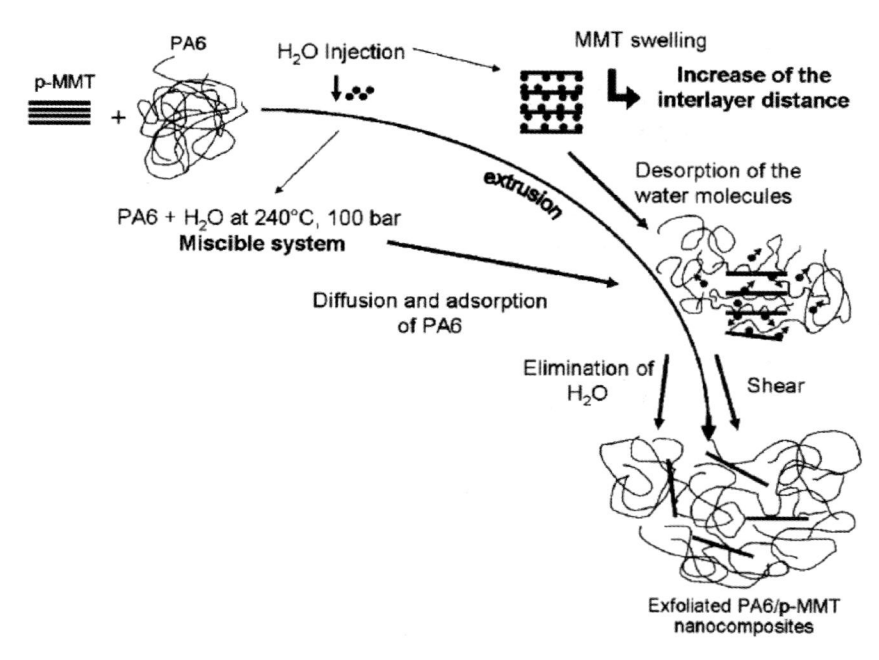

Figure 17.8 Schematic description of the dispersion of natural MMT in a PA6 matrix during water injection-assisted extrusion [118]. Reprinted with permission from Elsevier.

of the PA6/MMT composites. The morphologies of PP-based dilutions indicate that, whichever processing and sequence has been used, the clay will always reside within the polyamide phase domains, and water assistance will be valuable only when the clay is dispersed into the polyamide. Dasari *et al.* [119] prepared PA6 composites at a barrel temperature of 200 °C using a Clextral twin-screw extruder ($D = 28$ mm; L/D ratio = 36). Here, the PA6 pellets and pristine clay powder were fed to the hopper of the extruder at a throughput of 40 g min^{-1}, while water was injected into the extruder barrel downstream with a flow rate of 9 g min^{-1}. The water was then removed from the composite via a venting gate located before the extruder die. From the TEM images it could be seen clearly that the use of water to disperse the clay layers was very effective, such that an exfoliated morphology was obtained (Figure 17.9).

Following the removal of water from the extruder further downstream, the exfoliated morphology obtained would not re-agglomerate, as might be expected based on the favorable interactions between the polar PA6 and clay layers. This was despite the suggestion being made that, for nonpolar matrices, re-agglomeration of the finely dispersed clay layers was likely. Most importantly, the addition of water during the extrusion process did not cause the resultant nanocomposites to deteriorate. Such stability was confirmed by gel-permeation chromatography (GPC), which showed clearly that the molecular weight of PA6 had not been reduced by the addition of water. However, this may have been due to the very

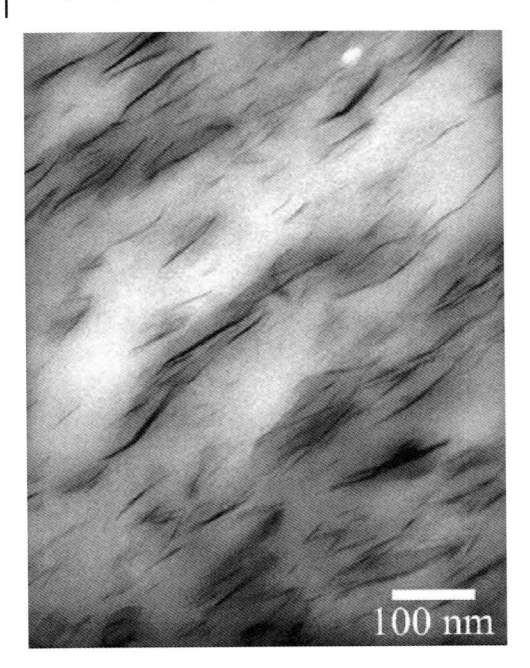

Figure 17.9 Transmission electron microscopy image of PA6-based nanocomposites obtained with a water-assisted melt-intercalation process, showing the well-intercalated structures [119]. Reprinted with permission from Elsevier.

short contact time between the water and polymer melt, to the relatively low processing temperature (compared to the melt temperature of PA6), and/or because of a water-plasticizing effect.

17.4.2
Supercritical CO_2-Assisted Melt-Compounding

During recent years supercritical carbon dioxide (scCO_2) technology has been widely investigated for its applications in polymer processing. The addition of scCO_2 into a polymer phase generally reduces the chain–chain interactions and increases the interchain distance and free volume. As a result, the polymer would be plasticized, its physical properties would be modified (in line with a depression of its glass transition temperature; T_g), its interfacial tension would be lowered, and its melt viscosity reduced. In the case of mechanical compounding, although the role of scCO_2 has not been yet investigated systematically, initial studies have indicated clear improvements in the dispersion of clay within a polymer matrix [120]. Bilotti et al. [121] previously reported the preparation of PP/sepiolite nanocomposites using a traditional melt-compounding method, and investigated the advantages of their preparation with scCO_2.

The $scCO_2$ processing method involves the contact of nanoclays or polymer/nanoclay mixtures with $scCO_2$ at high pressure (above the critical point for CO_2); after soaking for an appropriate time, the system is then rapidly depressurized to atmospheric pressure. Under these selected processing conditions, the mixture of CO_2 and polymer would diffuse between the clay layers; then during depressurization, an expansion of the $scCO_2$ between the layers would push the latter apart, and this would result in delaminated nanocomposites or nanoclays. A method of preparing polymer nanocomposites using $scCO_2$ in a melt-intercalation process was recently reported by Lesser et al. [122], who used a modified hopper that allowed the polymer and clay to interact with $scCO_2$ before blending. The $scCO_2$ promoted a significant increase in the basal spacing of the clay, thus enhancing polymer intercalation. Elsewhere, Mielewski et al. [123] directly injected $scCO_2$ into a melt mixture of silicate particles and polymer in an extruder, although no evidence of any exfoliated morphology was presented. Alternatively, Manke et al. [124] developed a process that allowed clay particles to exfoliate after first being pretreated with $scCO_2$ in a pressurized vessel, and then rapidly depressurized at atmospheric pressure. However, no mechanism was suggested that would ensure that the exfoliated particles would remain exfoliated when they were combined with the polymer via a conventional melt-blending.

Another innovative process, developed by Nguyen and Baird [125], involved the use of a pressurized CO_2 chamber to assist in the exfoliation and delivery of the clay into a stream of polymer melt; this allowed contact to be made only between the clay and $scCO_2$. In this case, the mixture of exfoliated clay and $scCO_2$ was fed into the extruder in a one-step process, rather than a two-step process (e.g., Manke's method). As usual, for comparison the same materials were produced using a direct melt-blending technique with a single-screw extruder, and with conventional twin-screw extrusion using PP-g-MA. Samples were extruded at a melt temperature of about 190 °C and a screw speed of 15 rpm using a single, two-stage screw extruder (Killion KL-100; $D = 25.4$ mm; L/D ratio = 30). A capillary die was attached to one end of the extruder, and a chamber inserted between the CO_2 pump and the injection port at the start of the second stage of the screw (Figure 17.10). When using the pressurized chamber, the clays were allowed to come into direct contact with $scCO_2$ at high pressure and 80 °C for a certain period of time. The pressure was then released before the nanoparticles and $scCO_2$ were injected into the molten polymer stream.

By using $scCO_2$, more exfoliated clay particles may be produced compared to traditional melt-blending. The presence of exfoliated clay greatly enhanced the mechanical and rheological properties, and outstanding improvements were observed with the technique when using the pressurized CO_2 chamber. Both, WAXD and TEM data showed a good degree of exfoliation for clay concentrations up to 6.6 wt%, while mechanical properties such as modulus were increased by up to 54%. When Ma et al. [126] used $scCO_2$ to prepare PP/sepiolite nanocomposites, PP and sepiolite – with and without PP-g-MA as compatibilizer – were placed in an autoclave into which $scCO_2$ that had been chilled to −6 °C was pumped (Isco model 260D syringe pump). The autoclave was maintained at 15 MPa and 200 °C

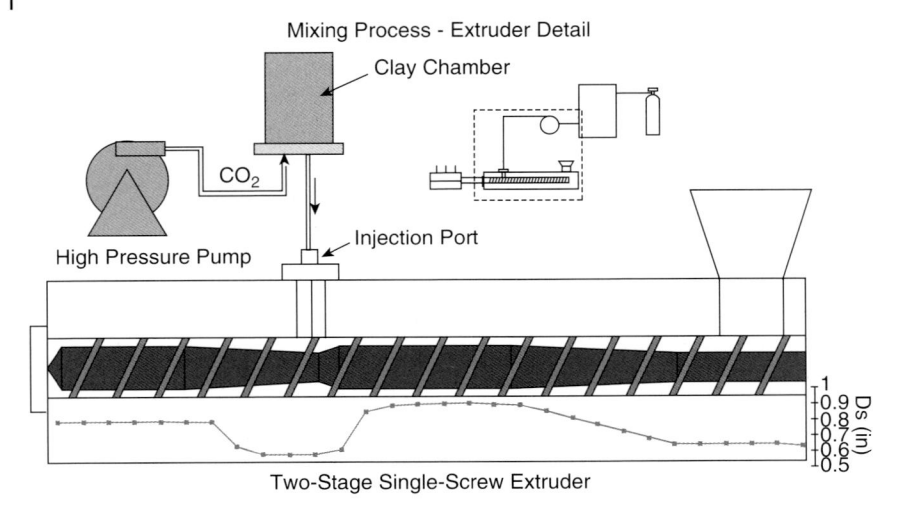

Figure 17.10 Schematic description of the overall process showing the CO_2 chamber and the two-stage single-screw extruder [125]. Reprinted with permission from Elsevier.

under stirring (using a pitched-blade turbine) for 30 min, and then cooled in water to room temperature after having vented off the CO_2. In this way, nanocomposites with clay contents of 1.0, 2.5, 5.0, and 10 wt% were obtained by selecting different weight ratios between PP and sepiolite. For comparison, traditionally melt-compounded nanocomposites were prepared using a two-step blending process in a mini twin-screw extruder at 200 °C and 200 rpm. For this, a masterbatch of 10 wt% of filler was prepared with a 5 min processing time, and subsequently diluted with pure PP with a 10 min processing time. The large aggregates of sepiolite fibers observed in the melt-compounded PP were not seen in the $scCO_2$/PP nanocomposites, thus confirming the better dispersion of clay with $scCO_2$-assisted mixing, as suggested by others [120]. In particular, it was noted that PP-g-MA was not required to achieve a good dispersion under $scCO_2$ conditions. This result is especially relevant as compatibilizers are generally regarded as essential for the creation of well-dispersed nanoclay composites using traditional melt-compounding methods [121].

Although previous studies have reported significant reductions in sepiolite length due to the compounding process [121], TEM images recorded after $scCO_2$-PP nanocomposite production showed the sepiolite needles to be separated one from another. The mechanical properties of PP nanocomposites prepared in $scCO_2$ were also significantly improved, without a need for PP-g-MA, again confirming that $scCO_2$ can provide significant benefits in the preparation of PP/sepiolite nanocomposites. When Horsch et al. [127] investigated $scCO_2$-based dispersion in polydimethylsiloxane (PDMS) nanocomposites, they concluded that nanoclays could be delaminated by the process and that the extent of dispersion would depend on the CO_2-philicity of the nanoclay. A significant dispersion was achieved with OMLS, whereas relatively little dispersion was achieved for the unmodified natural clay. In particular, it seemed that the presence of an acidic hydrogen on the ammonium might improve the CO_2-philicity of the clay, thus increasing its dispersion. In addition, the results showed that natural clay can be dispersed by combining CO_2-philic PDMS with the clay, prior to processing.

17.4.3
Ultrasound-Assisted Melt-Compounding

Over the past decade, extensive studies have been conducted to develop a novel ultrasound-assisted extrusion process. Consequently, the results of several studies of the effect of ultrasound on polymers have been reported [128–132]. For example, it has been shown that ultrasonic oscillations can break down the three-dimensional (3-D) network of vulcanized rubber within seconds, as well as improve the compatibilization of immiscible polymer blends [133]. The use of ultrasound to disperse nanofillers within a polymer matrix has attracted attention because it seems to help the rapid intercalation and partial exfoliation of the nanoclay in the polymer matrices [134]. Although, for the preparation of nanocomposites ultrasound has been used widely to promote the dispersion of nanoparticles in polymer solution, very few studies have been conducted to investigate the ultrasound-assisted melt-intercalation of clay in polymer matrices. Previous studies [135, 136] have shown that ultrasound could uniformly disperse nanoclays in low-viscosity monomers for *in situ* polymerization, as well as for the direct dispersion of fillers prior to melt mixing [137–139]. Lee *et al.* [140] extended the application of ultrasound in the preparation of polymer-based nanocomposites by introducing the power ultrasound directly into the polymer melt. Such "*in situ* ultrasound" of the polymer melt phase proved to be an effective method for enhancing the dispersion, intercalation and exfoliation of nanoclays in thermoplastic-based nanocomposites. In a later study, Zhao *et al.* [141] successfully prepared PP/OMLS nanocomposites with an intercalated structure by using an ultrasonic extrusion technology in conjunction with a special compounding system aided by ultrasonic oscillations in a direction parallel to the flow of the polymer melt. This not only improved the dispersion effect of OMMT particles in the PP matrix, but also diminished the spherulite size of PP in the nanocomposites. As a consequence, the mechanical properties of the nanocomposites were greatly improved, especially the elongation at break and the impact strength.

Several reports have involved the application of ultrasound to the preparation and characterization of nickel–PS nanocomposites [142], conductive polyaniline–nanosilica particle composites [143] under static ultrasonic treatment conditions, and silica agglomerate breakdown in a continuous ultrasonic extruder [144], the general finding being that ultrasound will increase the dispersion of a filler. A continuous method to achieve rapid intercalation within short residence times and a partial exfoliation of PP/clay nanocomposite without chemical modification of the matrix, was also developed [145]. Polyolefin matrices (both PP and PE) have been further investigated for the fabrication of clay-based nanocomposites using a continuous ultrasound-assisted process. In particular, the influence of the extruder feed rate (and hence the residence time of the polymer in the ultrasonic treatment zone) on the properties of composites has been studied. The results have indicated that intercalation was higher for a low feed rate, thus enhancing the residence time under sonication. Notably, PE appeared to form well-intercalated and partially exfoliated nanocomposites, whereas two competing processes, such as intercalation/exfoliation and degradation of the polymer and clay under the influence of ultrasound, have been encountered. In this case, the elongation at break and toughness were substantially improved, in particular for the PE/clay nanocomposites. Ryu et al. [146] reported the details of a sonication process that had been used to enhance nanoscale dispersion during the melt-mixing of polymer blends and organically modified clay, and showed that ultrasound-assisted processes could successfully generate exfoliated nanocomposites. Lee [140] also developed a process which used ultrasound to enhance the exfoliation and dispersion of clay platelets in PP-based nanocomposites, and concluded that the ultrasonic processing of polymer nanocomposites in the melt state provided an effective means of improving the exfoliation and dispersion of nanoclays.

Notably, as none of these processes was continuous, this influenced the potential application of ultrasound-assisted melt-blending on an industrial scale. In order to avoid this problem, a technique of electromagnetic dynamic extrusion was proposed by Luo et al. [147], which introduced a vibration force field into the extrusion process by axial vibration of the screw [148]. Subsequently, several groups [149–151] proved that this method could reduce the die pressure and the apparent viscosity of the melt, and also improve filler dispersion. In particular, a report was made [147] detailing the production HDPE/$CaCO_3$ (calcium carbonate) nanocomposites by using an electromagnetic dynamic extruder, the aim being to investigate the effects of a vibration force field on their structure and properties. For this, a pre-dried mixed polymer/filler blend was first fed into a twin-screw extruder and melt-mixed at 200 °C, with a rotational speed of 90 rpm; the melted material was then pelletized. After drying, the filled pellets were fed into an electromagnetic dynamic extruder with a split die and made into sheets for characterization.

The results showed that the introduction of a vibration force field could effectively improve the dispersion of nanoparticles within the HDPE matrix, leading to a smaller particle size and a narrower particle size distribution when compared to the commonly extruded sample. In addition, the dynamically extruded nanocomposites had a higher melting temperature and a smaller average crystal size, and consequently exhibited both greater strength and toughness. In extending these investigations, Kim et al. [152] applied a high-intensity ultrasonic wave to enhance the nanoscale dispersion of polymer blends and organically modified clay in the PP/PS/clay nanocomposites, and compared this process to the conventional compounding method. In order to impose the ultrasonic wave during melt extrusion, an intermeshing co-rotating twin-screw extruder (model TEK 25; SM Platek Co., Korea) was specially designed (Figure 17.11) with a horn which vibrated longitudinally at a frequency of 20 kHz and an ultrasonic power of 100 W. The extruder had 10 barrel sections, and the ultrasonic waves were transmitted through the

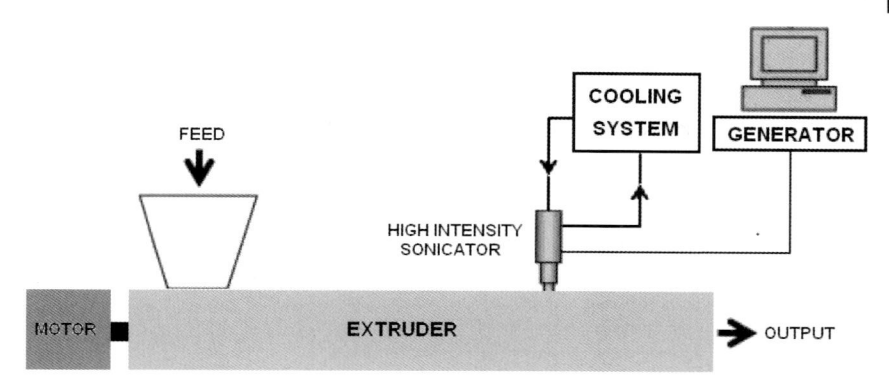

Figure 17.11 Schematic description of the twin-screw extruder with high-intensity ultrasonic system [152]. Reprinted with permission from John Wiley & Sons. Ltd.

middle part of the barrel. The screw diameter (D) was 25 mm, and the L/D ratio 41. Before irradiation with ultrasound, a preliminary extrusion was carried out for 3 min so as to reach a stable state whilst, for durable operation, an additional cooling accessory was fitted to the instrument. The sonic wave, when applied during the extrusion process, provided an effective means of accomplishing the nanoscale dispersion of layered silicates in the PP/PS/clay nanocomposites. The technique was especially effective in enhancing the interfacial interaction between the immiscible polymer components, the break-up of the clay agglomerates and, as a result, the exfoliation of the clay layers.

In a later report, Isayev et al. described a novel method and apparatus for the continuous dispersion of CNTs by means of an ultrasound-assisted twin-screw extruder, the effectiveness of which was monitored by carrying out the compounding of a polyetherimide (PEI)/MWNT system [153]. For this, a commercially available, continuously co-rotating intermeshing twin-screw microextruder (PRISM USA LAB 16; Thermo Electron Corp., UK) with a diameter (D) of 16 mm and an L/D ratio of 25, was modified by the addition of ultrasonic treatment sections at the die, as shown in Figure 17.12. A pair of 800 W ultrasonic power supplies (Branson Ultrasonics Corp., CT, USA) generated ultrasound at 40 kHz frequency. Each power supply was connected in series to a converter (Branson Ultrasonic Corp.) and a 1:1 booster (Branson Ultrasonics Corp.), while the water-cooled titanium horns, which were in direct contact with the polymer melt, provided the longitudinal vibrations in a direction perpendicular to the flow. Both horns were mounted symmetrically on the slit die, and both were operated simultaneously for all ultrasonically treated samples. The horn tips were calibrated in air for an ultrasonic amplitude in the range of 0 to 8 mm. A range of ultrasonic amplitudes, from 0 to 6 mm, was applied to the molten compound in order to study the effect of increasing ultrasonic power on the dispersion of CNTs in a polymer melt. The temperature in the barrel section was set in the die zone as 360 °C, while the screw speed was set at 50 rpm at a feed rate of 500 g h^{-1}; this led to a mean residence time in the ultrasonic treatment zone of 29 s. The ultrasonically treated nanocom-

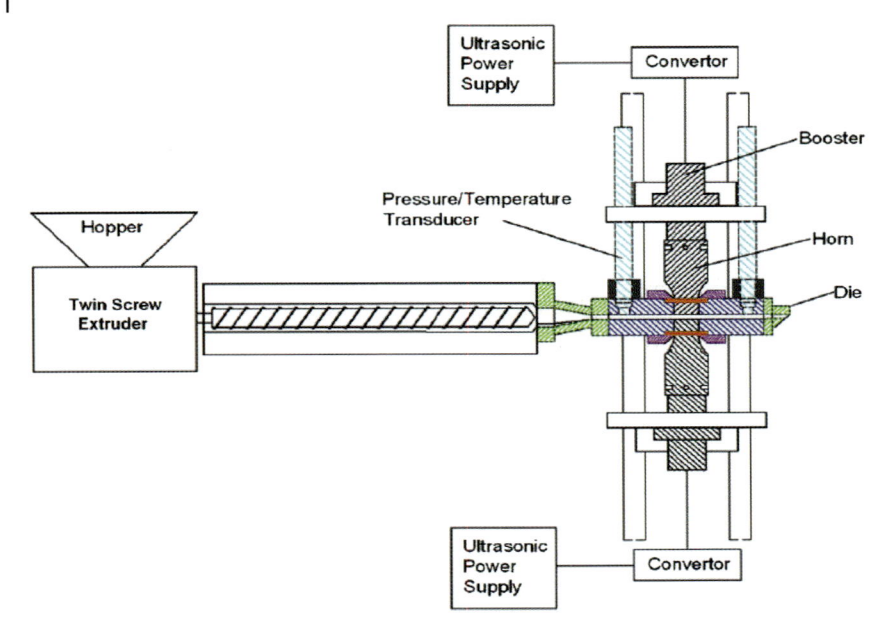

Figure 17.12 The ultrasonic twin-screw microextruder [153]. Reprinted with permission from Elsevier.

posites showed an increase in viscosity and storage modulus, but reduced damping characteristics, as compared to the untreated nanocomposites, indicating the presence of a better dispersion of nanotubes under ultrasonic treatment. Data relating to the mechanical properties showed the Young's modulus and tensile strength of the nanocomposites to have increased from the untreated to the treated sample. Rheology was found to be more sensitive than electrical conductivity to MWNT dispersion, while high-resolution scanning electron microscopy (HRSEM) images confirmed that the ultrasound-assisted dispersion of nanotubes had resulted from an effective breaking-up of the MWNT bundles.

17.5
Processing of Thermoset Nanocomposites

Nanocomposites based on thermoset polymers are mainly obtained via *in situ* intercalative polymerization methods, where the layered silicate swells within the liquid monomer or prepolymer, such that polymer formation can occur between the intercalated sheets. During swelling, the monomer diffuses from the bulk into the galleries between the silicate layers. Depending on the degree of penetration of the monomer into the layered silicate structure, different types of nanocomposite can be obtained, ranging from intercalated to exfoliated or delaminated. Polymer penetration resulting in a finite expansion of the silicate layers produces intercalated hybrids that consist of well-ordered multilayers composed of alternating polymer/silicate layers. Extensive polymer penetration, resulting in disorder and eventual delamination of the silicate layer, produces exfoliated hybrids that consist of individual nanometer-thick silicate layers suspended in the polymer matrix [4].

Thus, multicomponent thermoset systems require a preloading of the inorganic into one component, or an incorporation of the inorganic into the component mixture. In order to obtain the benefits of a large surface area of the dispersed phase in nanocomposites, nanoparticles must be fully dispersed at the nanolevel. Homogenization of the dispersion necessitates Brownian motion of the individual layers, which is hindered by: (i) an interlayer orientational coupling, arising from the original low-dimensional crystallite structure; and (ii) an extra-gallery polymerization which increases the medium's viscosity and ultimately leads to gelation. The disturbance of the local orientational correlations between nanoparticles must occur in a pre-processing step, well before the onset of gelation [154]. Preloading requires a high concentration of inorganic in one component, and this may lead to compounding processing difficulties, arising from a high viscosity medium. However, these problems may be overcome in several ways. For example, the use of low-boiling-point solvents (e.g., acetone, ethanol, isopropyl alcohol) which enhance mixability by lowering the viscosity, may be useful in achieving an homogeneous mixing of the components at temperatures much lower than those at which the reactions take place. However, the solvents must be removed before polymerization takes place, by either evaporation or vacuum stripping [155]. The pre-intercalation of monomers into the clay galleries prior to addition of the curing agent may be further favored by increasing the temperature so as to further reduce the viscosity and thus facilitate diffusion of the monomers into the galleries. Under these latter conditions, however, care must be taken to maintain the solvation temperature below the point where gelation of the resin will occur.

Besides solvent-aided processing, the dispersion of silicate in the liquid monomer may be helped by using other more environment-friendly techniques, either alone or in combination. Examples include high-speed stirrers, high-shear mixers and, more commonly, sonication [154, 156–159].

The differences between the various dispersion techniques, such as high-shear mixing, high-speed stirring and sonication, were reported for elastomeric polyurethane (PU) by Rhoney *et al.* [156]. These authors showed that both high-shear and sonication could provide a fairly good silicate dispersion, though sonication gave the better results. Gintert *et al.* [157] compared three techniques for the dispersion of clay in polyimide resin solution, the aim being to obtain the smallest possible tactoids before the resins were subjected to curing. (It is important to reduce the size of tactoids as much as possible before the polymer reaches its gel point and ceases the molecular motion, which in turn prevents exfoliation.) In a "baseline" method, a magnetic stirring bar was used to disperse the clay; a high-intensity mixing blade and an ultrasonic bath were also used, with the expectation that a better clay dispersion would be produced. Subsequent TEM images showed that, although the magnetic stirring bar provided sufficient stirring for clay gallery intercalation by the resin, the clay tactoids were typically so large that many clay layers were seen to be stacked within them. Although the use of a high-shear mixing blade for clay dispersion proved superior to a stirring bar, it was still unsatisfactory as large numbers of clay layers were still present in the tactoids. Otherwise, blades used were designed for the mixing, blending, and dispersion of paints, adhesives, and cements – that is for liquids which typically are more viscous than the monomer solution used in such a role. Neither was the viscosity increased by very much, due to a low loading of clay in the solution. On the other hand, sonication produced the best results as the number of clay layers in the stacks was significantly reduced. When Lekakou *et al.* [158] investigated a combination of solvent-aided mechanical mixing and solvent-free sonication for processing silica–epoxy and silica–polyester nanocomposites, the solvent-aided approach proved to be superior since, in the absence of a solvent, an inhomogeneous dispersion of the silica nanoparticles occurred at the microscale.

Occasionally, in order to increase monomer viscosity for transferring shear stress more effectively to a secondary phase, the advanced technology of *cryo-compounding* has been used [154]. In these studies, three different processing techniques were compared to assess the effect of cryo-compounding on silicate dispersion in epoxy resins:

- Method I, which comprised high-shear mixing alone.
- Method II, which comprised shear mixing and sonication.
- Method III, in which some material was separated from method II following the addition of a curing agent to undergo a compounding step at sub-ambient temperatures (cryo-compounding).

The key concept of cryo-compounding is to maximize the thermoset viscosity by halting the cure before gelation, and by compounding at sub-ambient temperatures close to the resin's T_g. High shear forces, due to the very high viscosity of the system, may facilitate homogenization of the layered silicate nanocomposite in the thermoset. The results obtained via TEM imaging for nanocomposites by method III and suitable OMLS, show an extremely uniform dispersion of the particles (even if apparently contradicted by the XRD data), in spite of a more agglomerated morphology of nanocomposite being achieved by using method II.

Recently, Modesti *et al.* (unpublished results) identified a new processing technique for improving clay dispersion in monomers. It had been shown that microwave irradiation may be usefully employed not only to initiate polymerization but also to promote the intercalation and exfoliation of clays in PU raw materials (polyols and isocyanates). Data acquired with both XRD and TEM showed clearly that the same degree of dispersion of the final polymer nanocomposite could be obtained either through ultrasonication (requiring about 2 h of processing time) or by using microwave processing, which is applied for a much shorter time (the full cycle takes about 5 min). Because of the greater viscosity and polarity, better results are obtained by using polyol as the swelling monomer, although good results have also been obtained with isocyanate as the swelling agent. Based on this success, it is likely that microwave irradiation will be extended to the processing of other thermosetting materials.

From a more chemical viewpoint, pioneering studies conducted by Pinnavaia *et al.* [160, 161] with MMT–thermoset systems (e.g., epoxy resins, elastomeric polyurethanes) established the initial conceptual methodology. Interfacial modifiers, such as primary ammonium alkyls, are intercalated between the MMT layers to not only compatibilize the inorganic aluminosilicate and organic resin, but also to accelerate the crosslinking reaction between the layers through acid-catalysis. In fact, a suitable balance between the relative rate of reagent intercalation, chain formation, and network crosslinking must be maintained in order to achieve exfoliation before network formation and matrix gelation. Another critical point is to enhance the intragallery polymerization rate so that it is comparable or greater than the extragallery polymerization. The reason for this is to increase monomer consumption within the swollen low-dimensional crystallite, and to enhance layer separation due to mass flow of the monomer into the interlayers. In this instance, the preferred sequence of component mixing (organoclay/monomers/curing agent), as well as the temperature and time of curing, might have a significant effect on the quality of the final nanocomposite material. The curing temperature may be a key feature in obtaining a well-dispersed morphology, as it influences both the reactivity and diffusion rates (mobility and flexibility) of the curing agent, thus affecting the extragallery and intragallery reaction rates. However, whilst in some cases improved exfoliation has been reported for increased cure temperatures, other systems have been found not to be affected within a certain temperature range [162]. The curing time also plays a significant role in that, by using a low-reactivity curing agent (e.g., 4,4′-diaminodiphenyl sulfone; DDS), a greater degree of exfoliation may be achieved, since the slow curing of the epoxy resin will delay extragallery gelation and provide sufficient time for intergallery polymerization [163].

For the mixing sequence, Vaia *et al.* [155] reported that the addition of a curing agent (diamine) to a mixture of OMLS and a low-molecular-weight diglycidyl ether of bisphenol A (Epon828) led to a nanocomposite with a smaller gallery spacing than was found in a nanocomposite obtained by adding the diamine directly to the OMLS. This suggests that, once intercalated, Epon828 resides preferentially in the interlayer and is not displaced by the diamine, and also that the order of reagent addition is critical. Clearly, the reagents must be added in such an order that a substantially greater intragallery reactivity is achieved that, in turn, will create a gradient to drive reagents from the exterior regions into the interlayer.

A similar conclusion relating to the sequence of component mixing was also drawn for other thermoset polymeric nanocomposites such as PUs. For example, Ma *et al.* [164] investigated the effect of the route of clay compounding with polyols on the mechanical properties of the elastomeric PU nanocomposites. For this, three types of polyol with different molecular weights and functionalities – that is, two polypropylene glycols (PPGs) – PPG1 and PPG2, with different molecular weights – and glycerol propoxylate (GPO), were used for preparing PU and PU/

clay nanocomposites. The ratio of the polyols was held constant to ensure the same chemical structure of the PU and PU/MMT nanocomposites. The resultant nanocomposites showed better tensile properties when MMT was first dispersed in GPO and then mixed with PPG, rather than dispersing clay within the PPG and mixing the composite with GPO.

In a later study, Woo et al. [165] investigated the effect of different preparation routes on the structure and properties of rigid PU-layered silicate nanocomposites, by using an organoclay containing a surfactant with reactive hydroxyl groups that could react with the isocyanate; here, the hydroxyl groups of the Si OH bonds of the clay may also have participated in the reaction. The synthesis route analyzed comprised clay dispersions in both isocyanates (route A) and polyols (route B). It appeared that, even when using identical compositions, the route of synthesis greatly affected the structure of the nanocomposites, with TEM images showing many more tactoids in samples produced via route B than route A. This difference was due to the level of the polymerization reaction taking place in the interlayer under various synthesis routes. In synthesis route B, the degree of intragallery polymerization was less than for route A, due to encapsulation of the stacks of swollen clay platelets that resulted primarily from a faster polymerization reaction in the bulk phase (extragallery reaction). In contrast, when the organoclay was first dispersed in the isocyanate precursors (route A), the excess isocyanate would begin to react with the surfactants of the clay (intragallery reaction). This, in effect, reduced the difference between reaction rates of the interlayer and the bulk phase, such that exfoliation was encouraged while tactoid formation was reduced.

The importance of the component mixing sequence was also reported for unsaturated polyester (UP) resins by Suh et al. [166], who used two different methods of mixing. In *simultaneous mixing*, the unsaturated polyester chains, styrene monomers and OMLS were simply mixed together for 3 h at 60 °C, whereas in *sequential mixing* the OMLS was first dispersed in the unsaturated polyester chain, after which the styrene monomer was added, with the mixing time being varied between 15 and 180 min, at 60 °C. Finally, both mixtures were cured at 80 °C for 3 h. In the simultaneous mixing system, as the styrene monomer moved more easily than uncured UP chains, this generated a higher styrene monomer concentration in the OMLS gallery than in any other part. If polymerization were to occur under these conditions, then the total crosslinking density of the sample would be decreased due to the low concentration of styrene in the uncured UP linear chains. In contrast, in sequential mixing, as the clay interlayers are intercalated with UP, the styrene monomers would be more dispersed inside and outside of the silicate layers, and the dispersion would improve as the mixing time increased. Consequently, the crosslinking reaction would take place homogeneously both inside and outside of the silicate layers, and the crosslinking density would reach that of the cured pure UP.

Another operating parameter that may affect thermoset nanocomposite structure is the *preconditioning time* – that is, the time between clay swelling in the monomer and the addition of a curing agent. When Tolle et al. [167] studied such an effect in epoxy nanocomposites, preconditioning (or aging) was carried out at room temperature for periods which ranged from a few hours to several weeks. Subsequently, it was shown (using TEM) that mixture-aging resulted in a more-expanded or more-ordered exfoliated morphological state, with smaller colonies of tactoids, whereas the un-aged material showed an ordered, intercalated morphology with a longer range registry, both within and between the tactoids. During preconditioning, a minor network formation may be produced in the interphase region of the mixture-aged materials; as a result, participation in the subsequent cure-agent-induced network formation would be minimized, and the balance between network formation within and around the tactoids would favor OMLS mobility prior to gelation. These subtle variations in the interface and interphase regions may alter the balance of intragallery and extragallery network formation, leading to differences in morphology development.

17.6
Conclusions

In this chapter, attention was focused on the processing of polymer-based nanocomposites, notably the analysis and understanding of those factors that affect filler dispersion during the process, and the relationships that exist between such parameters and the final structure and properties of the composites.

In the melt-intercalation of thermoplastic-based hybrids, several factors have been identified that influence the composites' properties, including temperature, shear, mixing time and extruder configuration (notably screw profile design). It is clear that shear forces may expedite the break-up of clay agglomerates, whilst penetration of the polymer chains within the clay galleries and exfoliation of the silicate layers stacks is mainly driven by temperature, residence or mixing times and, of course, by any chemical compatibility between the clay and polymer. In this respect, the processing conditions may be tailored to achieve the highest degrees of intercalation and exfoliation of the clay platelets even if, unfortunately, the same effects on nanocomposite structure are not universally observed and each system must be studied specifically so as to optimize filler dispersion. As severe screw profiles are generally unnecessary to enhance clay dispersion, this indicates that other parameters must be taken into account with regards to exfoliation mechanisms.

Although strong shear stresses are required to break up CNT aggregates within polymer melts, nanotube shortening can occur under high-shear conditions or with increasing mixing times. In contrast, a decrease in the aspect ratio during melt mixing may improve the dispersibility of these fillers as the result of a decreased number of physical entanglements.

In solution-intercalation processes, one key factor was the need for high-energy mixing that often was accomplished with sonication; without sonication, a clay dispersion generally does not produce well-dispersed, exfoliated structures.

The intercalation of polymers into unmodified sodium-MMT also represents a major challenge to avoid the use of OMLS, as the presence of large amounts of

LMW surfactants may adversely affect the reinforcing efficiency of the clay layers and the thermal stability and durability of the final product, even if all the clay layers are fully exfoliated. Much effort has been directed towards developing alternative routes to produce exfoliated nanocomposites without using organically modified clay, notably by means of water injection, $scCO_2$ and ultrasound-assisted melt-intercalation. Although the use of water has been shown very effective, and exfoliated morphologies can be obtained, it is important to note that adding water during the extrusion process does not cause deterioration of the resultant nanocomposites, due not only to the short contact time between the water and polymer melt but also to the lower processing temperatures as the result of a plasticizing effect. The use of $scCO_2$ leads to the production of more exfoliated clay particles than with traditional melt-blending. Indeed, although the use of compatibilizers can be avoided, their role is seen as essential for creating well-dispersed nanoclay composites via traditional melt-compounding. More recently, the use of ultrasound to disperse nanofillers in a polymer matrix has attracted attention, mainly because, when applied to melt-intercalation, it aids in rapid intercalation and partial exfoliation of the fillers, and the successful generation of exfoliated nanocomposites.

In thermosetting nanocomposite processing, the dispersion of silicate in the liquid monomer can be achieved by several methods that may be applied either alone or in combination, and include high-speed stirring, high-shear mixing, sonication, and cryo-compounding. In cryo-compounding, the thermoset viscosity is maximized by halting the cure before gelation, and by compounding at temperatures close to the T_g of the resin. High shear forces, due to the very high viscosity of the system, may in fact facilitate homogenization of the layered silicate nanocomposite in thermoset polymers. Microwave irradiation has also been shown to be successful, not only to initiate polymerization but also to promote intercalation and exfoliation of the clays in raw materials. The curing temperature is a key feature for achieving a well-dispersed morphology, as it influences the reactivity and diffusion rates of the curing agent, which in turn affects extragallery and intragallery reaction rates. The importance of the sequence in which components are mixed has also been recognized as important, as has the preconditioning time.

References

1 Vaia, R.A., Ishii, H., and Giannelis, E.P. (1993) *Chem. Mater.*, **5**, 1694–1696.
2 Vaia, R.A., Jandt, K.D., Kramer, E.J., and Giannelis, E.P. (1995) *Macromolecules.*, **28**, 8080–8085.
3 Vaia, R.A. and Giannelis, E.P. (1997) *Macromolecules*, **30**, 7990–7999.
4 Vaia, R.A. and Giannelis, E.P. (1997) *Macromolecules*, **30**, 8000–8009.
5 Bolen, W.R. and Colwell, R.E. (1958) *SPE J.*, **14**, 24–28.
6 Cho, J.W., and Paul, D.R. (2001) *Polymer*, **42**, 1083–1094.
7 Incarnato, L., Scarfato, P., Russo, G.M., Di Maio, L., Iannelli, P., and Acierno, D. (2003) *Polymer*, **44**, 4625–4634.
8 Kwak, M., Lee, M., and Lee, B.K. (2002) Proceedings SPE ANTEC, San Francisco, p. 224.

9 Modesti, M., Lorenzetti, A., Bon, D., and Besco, S. (2005) *Polymer*, **46**, 10237–10245.
10 Modesti, M., Lorenzetti, A., Bon, D., and Besco, S. (2006) *Polym. Deg. Stab.*, **91**, 672–680.
11 Lertwimolnun, W. and Vergnes, B. (2005) *Polymer*, **46**, 3462–3471.
12 Lertwimolnun, W. and Vergnes, B. (2006) *Polym. Eng. Sci.*, **46**, 314–323.
13 Peltola, P., Välipakka, E., Vuorinen, J., Syrjälä, S., and Hanhi, K. (2006) *Polym. Eng. Sci.*, **46**, 995–1000.
14 Homminga, D., Goderis, B., Hoffman, S., Reynaers, H., and Groeninckx, G. (2005) *Polymer*, **46**, 9941–9954.
15 Scatteia, L., Scarfato, P., and Acierno, D. (2006) *e-Polymers*, **23**, 1.
16 Nassar, N., Utracki, L.A., and Kamal, M.R. (2005) *Int. Polym. Proc.*, **20**, 423–431.
17 Fasulo, P.D., Rodgers, W.R., Ottaviani, R.A., and Hunter, D.L. (2004) *Polym. Eng. Sci.*, **44**, 1036–1045.
18 Dennis, H.R., Hunter, D.L., Chang, D., Kim, S., White, J.L., Cho, J.W., and Paul, D.R. (2001) *Polymer*, **42**, 9513–9522.
19 Tillekeartne, M., Jollands, M., Cser, F., and Bhattacharya, S.N. (2003) Proceedings of Polymer Processing Society 19th Annual Meeting, Melbourne, CD ROM.
20 Médéric, P., Razafinimaro, T., and Aubry, T. (2006) *Polym. Eng. Sci.*, **46**, 986–994.
21 Kim, D., Lee, J.S., Barry, C.F., and Mead, J.L. (2008) *J. Appl. Polym. Sci.*, **109**, 2924–2934.
22 Vaia, R.A., Jandt, K.D., Kramer, E.J., and Giannelis, E.P. (1996) *Chem. Mater.*, **8**, 2628–2635.
23 Ratinac, K.R., Gilbert, R.G., Ye, L., Jones, A.S., and Ringer, S.P. (2006) *Polymer*, **47**, 6337–6361.
24 Lertwimolnun, W. and Vergnes, B. (2007) *Polym. Eng. Sci.*, **47**, 2100–2109.
25 Wang, H., Zeng, C.C., Elkovitch, M., Lee, L.J., and Koelling, K.W. (2001) *Polym. Eng. Sci.*, **41**, 2036–2046.
26 Read, M.D., Liu, L., Harris, J.D., and Samson, R.R. (2004) Proceedings SPE ANTEC, Chicago.
27 Lew, C.Y., Moore, I., McNally, G.M., Murphy, W.R., Abe, K., Yanai, S., and Tshai, K.Y. (2005) Proceedings of Polymer Processing Society 21st Annual Meeting, Leipzig.
28 Usuki, A., Koiwai, A., Kojima, Y., Kawasumi, M., Okada, A., Kurauch, T., and Kamigaito, O. (1995) *J. Appl. Polym. Sci.*, **55**, 119–123.
29 Rhutesh, K., Shah, K., and Paul, D.R. (2004) *Polymer*, **45**, 2991–3000.
30 Iijima, S. (1991) *Nature*, **354**, 56–58.
31 Iijima, S. (1993) *Mater. Sci. Eng. B*, **19**, 172–180.
32 Iijima, S., Ichihashi, T., and Ando, Y. (1992) *Nature*, **352**, 776–778.
33 Charlier, J.C. and Michenaud, J.P. (1993) *Phys. Rev. Lett.*, **72**, 1858–1861.
34 Saito, R., Dresselhaus, G., and Dresselhaus, M.S. (1993) *J. Appl. Phys.*, **73**, 494–500.
35 White, C.T., Robertson, D.H., and Mintmire, J.W. (1993) *Phys. Rev. B*, **47**, 5485–5488.
36 Dresselhaus, M.S., Dresselhaus, G., and Pimenta, M. (1999) *Eur. Phys. J. D*, **9**, 69–75.
37 Ebbesen, T.W., Lezec, H.J., Hiura, H., Bennett, J.W., Ghaemi, H.F., and Thio, T. (1996) *Nature*, **382**, 54–56.
38 Dai, H.J., Wong, E.W., and Lieber, C.M. (1996) *Science*, **272**, 523–526.
39 Hone, J., Llaguno, M.C., Biercuk, M.J., Johnson, A.T., Batlogg, B., Benes, Z., et al. (2002) *Appl. Phys.*, **74**, 339–343.
40 Berber, S., Kwon, Y.K., and Tomanek, D. (2000) *Phys. Rev. Lett.*, **84**, 4613–4316.
41 Che, J.W., Cagin, T., and Goddard, W.A. (2000) *Nanotechnology*, **11**, 65–69.
42 Ruoff, R.S. and Lorents, D.C. (1995) *Carbon*, **33**, 925–930.
43 Li, F., Cheng, H.M., Bai, S., Su, G., and Dresselhaus, M.S. (2000) *Appl. Phys. Lett.*, **77**, 3161–3163.
44 Pan, Z.W., Xie, S.S., Chang, B.H., Wang, C.Y., Lu, L., Liu, W., et al. (1998) *Nature*, **394**, 631–632.
45 Walters, D.A., Ericson, L.M., Casavant, M.J., Liu, J., Colbert, D.T., Smith, K.A., et al. (1999) *Appl. Phys. Lett.*, **74**, 3803–3805.

46 Salvetat, J.P., Bonard, J.M., Thomson, N.H., Kulik, A.J., Forro, L., Benoit, W., et al. (1999) *Appl. Phys. A*, **69**, 255–260.
47 Demczyk, B.G., Wang, Y.M., Cumings, J., Hetman, M., Han, W., Zettl, A., et al. (2002) *Mater. Sci. Eng. A*, **334**, 173–178.
48 Meincke, O., Kaempfer, D., Weickmann, H., Friedrich, C., Vathauer, M., and Warth, H. (2004) *Polymer*, **45**, 739–748.
49 Schartel, B., Pötschke, P., Knoll, U., and Abdel-Goad, M. (2005) *Eur. Polym. J.*, **41**, 1061–1070.
50 Pötschke, P., Fornes, T.D., and Paul, D.R. (2002) *Polymer*, **43**, 3247–3255.
51 Chen, G.X., Kim, H.S., Park, B.H., and Yoon, J.S. (2005) *J. Phys. Chem. B*, **109**, 22237–22243.
52 Sen, R., Zhao, B., Perea, D.E., Itkis, M.E., Hu, H., Love, J., et al. (2004) *Nanoletters*, **4**, 459–464.
53 Chen, G.X., Kim, H.S., Park, B.H., and Yoon, J.S. (2006) *Carbon*, **44**, 3373–3375.
54 Vigolo, B., Poulin, P., Lucas, M., Launois, P., and Bernier, P. (2002) *Appl. Phys. Lett.*, **81**, 1210–1212.
55 Xia, H., Wang, Q., Li, K., and Hu, G.H. (2004) *J. Appl. Polym. Sci.*, **93**, 378–386.
56 Regev, O., El Kati, P.N.B., Loos, J., and Koning, C.E. (2004) *Adv. Mater.*, **16**, 248–251.
57 Vigolo, B., Penicaud, A., Coulon, C., Sauder, C., Pailler, R., Journet, C., et al. (2000) *Science*, **290**, 1331–1334.
58 Furgiuele, N., Lebovitz, A.H., Khait, K., and Torkelson, J.M. (2000) *Macromolecules*, **33**, 225–228.
59 Pötschke, P., Häußler, L., Pegel, S., Steinberger, R., and Scholz, G. (2007) *Kaut. Gummi Kunst.*, **7**, 28–33.
60 Villmow, T., Pegel, S., Pötschke, P., and Wagenknecht, U. (2008) *Comp. Sci. Technol.*, **68**, 777–789.
61 Li, C.Y., Thostenson, E.T., and Chou, T.W. (2007) *Appl. Phys. Lett.*, **91**, article no. 223114.
62 Lin, B., Sundararaj, U., and Pötschke, P. (2006) *Macrom. Mater. Eng.*, **291**, 227–238.
63 Pötschke, P., Bhattacharyya, A.R., and Janke, A. (2004) *Eur. Polym. J.*, **40**, 137–148.
64 Donder, W.E. and Gorga, R.E. (2006) *J. Polym. Sci. B*, **44**, 864–878.
65 Andrews, R., Jacques, D., Minot, M., and Rantell, T. (2002) *Macrom. Mater. Eng.*, **287**, 395–403.
66 Pötschke, P., Dudkin, S.M., and Alig, I. (2003) *Polymer*, **44**, 5023–5030.
67 Takase, H. (2005) High-conductive PC/CNT composites with ideal dispersibility, in: K. Schulte (ed.), Polymer Composites International Conference, TUHH, Hamburg, Germany.
68 Shiga, S. and Furuta, M. (1985) *Rubber Chem. Tech.*, **58**, 1–22.
69 Rwei, S.P., Manas-Zloczower, I., and Feke, D.L. (1990) *Polym. Eng. Sci.*, **30**, 701–706.
70 Li, Q., Feke, D.L., and Manas-Zloczower, I. (1995) *Rubber Chem. Tech.*, **68**, 836–841.
71 Pötschke, P., Bhattacharyya, A.R., Janke, A., Pegel, S., Leonhardt, A., Täschner, C., et al. (2005) *Fullerenes Nanotubes Carbon Nanostruct.*, **13**, 211–224.
72 Pötschke, P., Brünig, H., Janke, A., Fischer, D., and Jehnichen, D. (2005) *Polymer*, **46**, 10355–10363.
73 Pötschke, P., Bhattacharyya, A.R., Janke, A., and Goering, H. (2003) *Compos. Interfaces*, **10**, 389–404.
74 Villmow, T., Pötschke, P., Pegel, S., Häussler, L., and Kretzschmar, B. (2008) *Polymer*, **49**, 3500–3509.
75 Chen, G., Li, Y., and Shimizu, H. (2007) *Carbon*, **45**, 2334–2340.
76 Li, Y. and Shimizu, H. (2007) *Polymer*, **48**, 2203–2207.
77 Greenland, D.J. (1963) *J. Colloid Sci.*, **18**, 647–664.
78 Ogata, N., Kawakage, S., and Ogihara, T. (1997) *J. Appl. Polym. Sci.*, **66**, 573–581.
79 Billingham, J., Breen, C., and Yarwood, J. (1997) *Vib. Spectrosc.*, **14**, 19–34.
80 Aranda, P. and Ruiz-Hitzky, E. (1992) *Chem. Mater.*, **4**, 1395–1403.
81 Lemmon, J.P. and Lerner, M.M. (1994) *Chem. Mater.*, **6**, 207–210.
82 Malwitz, M.M., Lin-Gibson, S., Hobbie, E.K., Butler, P.D., and Schmidt, G. (2003) *J. Polym. Sci. B*, **41**, 3237–3248.

83 Shen, Z., Simon, G.P., and Cheng, Y.-B. (2002) *Polymer*, **43**, 4251–4260.
84 Jimenez, G., Ogata, N., Kawai, H., and Ogihara, T. (1997) *J. Appl. Polym. Sci.*, **64**, 2211–2220.
85 Ogata, N., Jimenez, G., Kawai, H., and Ogihara, T. (1997) *J. Polym. Sci., Part. B: Polym. Phys.*, **35**, 389–396.
86 Jeon, H.G., Jung, H.-T., Lee, S.W., and Hudson, S.D. (1998) *Polym. Bull.*, **41**, 107–113.
87 Ren, J., Silva, A.S., and Krishnamoorti, R. (2000) *Macromolecules*, **33**, 3739–3746.
88 Li, Y. and Ishida, H. (2003) *Polymer*, **44**, 6571–6577.
89 Wu, T.-M. and Wu, C.-W. (2005) *J. Polym. Sci., Part. B: Polym. Phys.*, **43**, 2745–2753.
90 Chiu, F.-C. and Chu, P.-H. (2006) *J. Polym. Res.*, **13**, 73–78.
91 Qiu, L., Chen, W., and Qu, B. (2006) *Polymer*, **47**, 922–930.
92 Hotta, S. and Paul, D.R. (2004) *Polymer*, **45**, 7639–7654.
93 Mainil, M., Alexandre, M., Monteverde, F., and Dubois, P. (2006) *J. Nanosci. Nanotechnol.*, **6**, 337–344.
94 Liang, G., Xu, J., Bao, S., and Xu, W. (2004) *J. Appl. Polym. Sci.*, **91**, 3974–3980.
95 Wang, K.H., Choi, M.H., Koo, C.M., Choi, Y.S., and Chung, I.J. (2001) *Polymer*, **42**, 9819–9826.
96 Morawiec, J., Pawlak, A., Slouf, M., Galeski, A., Piorkowska, E., and Krasnikova, N. (2005) *Eur. Polym. J.*, **41**, 1115–1122.
97 Xu, Y., Fang, Z., and Tong, L. (2005) *J. Appl. Polym. Sci.*, **96**, 2429–2434.
98 Bergaya, F., Mandalia, T., and Amigouët, P. (2005) *Colloid Polym. Sci.*, **283**, 773–782.
99 Filippi, S., Mameli, E., Marazzato, C., and Magagnino, P. (2007) *Eur. Polym. J.*, **43**, 1645–1659.
100 Modesti, M., Besco, S., Lorenzetti, A., Zammarano, M., Causin, V., Marega, C., Gilman, J.W., Fox, D.M., Trulove, P.C., De Long, H.C., and Maupin, P.H. (2008) *Polym. Adv. Tech.*, **19**, 1576–1583.
101 Maupin, P.H., Gilman, J.W., Harris, R.H. Jr, Bellayer, S., Bur, A.J., Roth, S.C., Murariu, M., Morgan, A.B., and Harris, J.D. (2004) *Macromol. Rapid Commun.*, **25**, 788–792.
102 Stretz, H.A., Paul, D.R., and Cassidy, P.E. (2005) *Polymer*, **46**, 3818–3830.
103 Pourabas, B. and Raeesi, V. (2005) *Polymer*, **46**, 5533–5540.
104 Morgan, A.B. and Harris, J.D. (2004) *Polymer*, **45**, 8695–8703.
105 Gilman, J.W., Awad, W.H., Davis, R.D., Shields, J., Harris, R.H. Jr, Davis, C., Morgan, A.B., Sutto, T.E., Callahan, J., Trulove, P.C., and DeLong, H.C. (2002) *Chem. Mater.*, **14**, 3776.
106 Xie, W., Gao, Z., Pan, W.P., Hunter, D., Singh, A., and Vaia, R. (2001) *Chem. Mater.*, **13**, 2979–2990.
107 Morgan, A.B. and Harris, J.D. (2003) *Polymer*, **44**, 2313–2320.
108 Hasegawa, N., Okamoto, H., Kato, M., Usuki, A., and Sato, N. (2003) *Polymer*, **44**, 2933–2937.
109 Chen, B., Liu, J., Chen, H., and Wu, J. (2004) *Chem. Mater.*, **16**, 4864–4866.
110 Wang, K., Chen, L., Wu, J., Toh, M.L., He, C., and Yee, A.F. (2005) *Macromolecules*, **38**, 788–800.
111 Kato, M., Matsushita, M., and Fukumori, K. (2004) *Polym. Eng. Sci.*, **44**, 1205–1211.
112 Strawhecker, K.E. and Manias, E. (2000) *Chem. Mater.*, **12**, 2943–2949.
113 Lee, D.C. and Jang, L.W. (1996) *J. Appl. Polym. Sci.*, **61**, 1117–1122.
114 Korbee, R. and Van Geneen, A. (1999) WO99/29 767, DSM.
115 Yu, Z.Z., Hu, G.H., Varlet, J., Dasari, A., and Mai, Y.W. (2005) *J. Polym. Sci., Part B: Polym. Phys.*, **43**, 1100–1112.
116 European Research Project FP5 (2000–2005) Melt Made Polymer Nanocomposites, G5RD-2000-00348.
117 Fedullo, N., Sclavons, M., Bailly, C., Lefebvre, J.M., and Devaux, J. (2006) *Macromol. Symp.*, **233**, 235–245.
118 Fedullo, N., Sorlier, E., Sclavons, M., Bailly, C., Lefebvre, J.-M., and Devaux, J. (2007) *Prog. Org. Coat.*, **58**, 87–95.
119 Dasari, A., Yu, Z.-Z., Mai, Y.-W., Hu, G.-H., and Varlet, J. (2005) *Comp. Sci. Technol.*, **65**, 2314–2328.
120 Garcia-Leiner, M. and Lesser, A.J. (2003) *Polym. Mater. Sci. Eng.*, **89**, 649–650.

121 Bilotti, E., Fischer, H.R., and Peijs, T. (2008) *J. Appl. Polym. Sci.*, **107**, 2572–2577.
122 Zerda, A.S., Caskey, T.C., and Lesser, A.J. (2003) *Macromolecules*, **36**, 1603–1608.
123 Mielewski, D.F., Lee, E.C., Manke, C.W., and Guari, E. (2004) US Patent 6, 753,360.
124 Manke, C.W., Gulari, E., Mielewski, D.F., and Lee, E.C. (2002) US Patent 6, 469,073.
125 Nguyen, Q.T. and Baird, D.G. (2007) *Polymer*, **48**, 6923–6933.
126 Ma, J., Bilotti, E., Peijs, T., and Darr, J.A. (2007) *Eur. Polym. J.*, **43**, 4931–4939.
127 Horsch, S., Serhatkulu, G., Gulari, E., and Kannan, R.M. (2006) *Polymer*, **47**, 7485–7496.
128 Isayev, A.I., Wong, C.M., and Zeng, X. (1990) *Adv. Polym. Technol.*, **10**, 31–45.
129 Isayev, A.I. and Chen, J. (1994) US Patent 5, 284,625.
130 Isayev, A.I., Chen, J., and Tukachinsky, A. (1995) *Rubber Chem. Technol.*, **68**, 267–280.
131 Isayev, A.I., Yushanov, S.P., and Chen, J. (1996) *J. Appl. Polym. Sci.*, **59**, 815–824.
132 Tukachinsky, A., Schworm, D., and Isayev, A.I. (1996) *Rubber Chem. Technol.*, **69**, 92–103.
133 Feng, W. and Isayev, A.I. (2004) *Polymer*, **45**, 1207–1216.
134 Lapshin, S. and Isayev, A.I. (2006) *J. Vinyl Additive Technol.*, **12**, 78–82.
135 Chen, G., Guo, S., and Li, H. (2002) *J. Appl. Polym. Sci.*, **84**, 2451–2460.
136 Kim, H. and Lee, J.W. (2002) *Polymer*, **43**, 2585–2589.
137 Jianling, Z., Zhimin, L., Buxing, H., Tao, J., Weize, W., Jing, C., et al. (2004) *J. Phys. Chem. B*, **108**, 2200–2204.
138 Gul, R.J., Wan, P.S., Hyungsu, K., and Wook, L.J. (2004) *Mater. Sci. Eng. C*, **24**, 85–88.
139 Artzi, N., Nir, Y., Narkis, M., and Siegmann, A. (2002) *J. Polym. Sci., Part B: Polym. Phys.*, **40**, 1741–1753.
140 Lee, E.C., Mielewski, D.F., and Baird, R.J. (2004) *Polym. Eng. Sci.*, **44**, 1773–1782.
141 Zhao, L., Li, J., Guo, S., and Du, Q. (2006) *Polymer*, **47**, 2460–2469.
142 Kumar, R., Koltypin, Y., Palchik, O., and Gedanken, A. (2002) *J. Appl. Polym. Sci.*, **86**, 160–165.
143 Xia, H. and Wang, Q. (2003) *J. Appl. Polym. Sci.*, **87**, 1811–1817.
144 Isayev, A.I., Hong, C.K., and Kim, K.J. (2003) *Rubber Chem. Technol.*, **76**, 923–947.
145 Lapshin, S. and Isayev, A. (2005) *SPE ANTEC*, **2**, 1911–1915.
146 Ryu, J.G., Kim, H., and Lee, J.W. (2004) *Polym. Eng. Sci.*, **44**, 1198–1204.
147 Luo, W., Zhou, N., Zhang, Z., and Wu, H. (2006) *Polym. Test.*, **25**, 124–129.
148 Qu, J.P. (1992) *J. South China Univ. Technol.*, **20**, 1–8.
149 Qu, J.P. (2002) *Polym. -Plastics Technol. Eng.*, **41**, 115–132.
150 Lui, Y.J., Qu, J.P., Xu, B.P., and Peng, X.F. (2004) *Polym. Mater. Sci. Eng.*, **20**, 18–21.
151 Qu, J.P., Qing, Y.M., Tian, Y.C., and Zhou, N.Q. (2000) *China Plast. Ind.*, **28**, 23–24.
152 Kim, K.Y., Ju, D.U., Nam, G.J., and Lee, J.W. (2007) *Macrom. Symp.*, **283**, 249–250.
153 Isayev, A.I., Kumar, R., and Lewis, T.M. (2009) *Polymer*, **50**, 250–260.
154 Koerner, H., Misra, D., Tan, A., Drummy, L., Mirau, P., and Vaia, R. (2006) *Polymer*, **47**, 3426–3435.
155 Brown, J.M., Curliss, D., and Vaia, R.A. (2000) *Chem. Mater.*, **12**, 3376–3384.
156 Rhoney, J. (2004) *Appl. Polym. Sci.*, **91**, 1335–1343.
157 Gintert, M.J., Jana, S.C., and Miller, S.G. (2007) *Polymer*, **48**, 4166–4173.
158 Lekakou, C., Kontodimopoulos, I., Murugesh, A.K., Chen, Y.L., Jesson, D.A., Watts, J.F., and Smith, P.A. (2008) *Polym. Eng. Sci.*, **48**, 216–222.
159 Yasmin, A., Abot, J.L., and Daniel, I.M. (2003) *Scripta Mater.*, **49**, 81–86.
160 Lan, T. and Pinnavaia, T.J. (1994) *Chem. Mater.*, **6**, 2216–2219.
161 Wang, Z. and Pinnavaia, T.J. (1998) *Chem. Mater.*, **10**, 3769–3771.
162 Becker, O., Cheng, Y.B., Varley, R.J., and Simon, G.P. (2003) *Macromolecules*, **36**, 1616–1625.

163 Kong, D. and Park, C.E. (2003) *Chem. Mater.*, **15**, 419–424.
164 Ma, J., Zhang, S., and Qi, Z. (2001) *J. Appl. Polym. Sci.*, **82**, 1444–1448.
165 Woo, T., Halley, P., Martin, D., and Kim, D.S. (2006) *J. Appl. Polym. Sci.*, **102**, 2894–2903.
166 Suh, D.J., Lim, Y.T., and Park, O.O. (2000) *Polymer*, **41**, 8557–8563.
167 Tolle, T.B. and Anderson, D.P. (2004) *J. Appl. Polym. Sci.*, **91**, 89–100.

Index

a

abrasion resistance 67
abrasiveness 211f
activation energy of polymer/MWNT
    nanocomposites 163f
AFM, see atomic force microscopy
agglomerate of nanoparticles 231
alkyl ammonium salt 2, 6ff, 79
allylamine 61
3-aminopropyl triethoxy silane
    (KH570) 62ff
amorphous thermoplastic polymer 145,
    156f
analysis of polymer nanocomposites 6ff
analytical modeling 131
– of nanocomposite properties 305
– three-phase approach 127f, 318
– two-phase (particle/matrix) approach
    125ff, 149
application of nanocomposites 302
aspect ratio (AR) 126f, 189, 263, 267ff, 333,
    340, 367, 379
atomic force microscopy (AFM) 143, 148,
    152, 167
attapulgite (ATP) 281, 283, 287
average strain of nanocomposites 306
average stress of nanocomposites 306

b

barium titanate 280, 282, 286
barrier effect 196
barrier generation
– in nonpolar nanocomposites 183ff
– in polar nanocomposites 176ff
barrier properties, modeling of 189ff
barrier resistance 173, 360
basal plane spacing 6f, 11, 179ff, 185
– influence of supercritical carbon
    dioxide 389

batch mixing 370
BEM, see boundary element method
Bingham plastic behavior 98
biodegradability 360
– influencing properties of PNC 257f
biodegradable polyester/layered silicate
    nanocomposite 236
bisphenol A diglycidyl ether (DGEBA) 22f,
    356f, 359, 397
bisphenol expoxy/nanoclay composite 24,
    34
bisphenol F diglycidyl ether (DGEBF) 24
boehmite 282, 292
boundary element method (BEM) 312,
    319
Brownian dynamics (BD) 309f
buckling of PNC 324f
bulk polymerization 3

c

Cahn-Hillard-Cook (CHC) nonlinear
    diffusion equation 311
calcium carbonate 282, 285, 289, 292,
    392
carbon black as nanofiller 205
carbon nanofiber 202ff
carbon nanotube (CNT) 24, 93, 139, 281
– chitosan-induced dispersion 105f
– effect on tribological behavior of
    PNC 224f
– influence of processing parameters on
    nanotube dispersion 378
– load transfer 140
– modification 102
– percolation 149
– properties 102, 104, 140, 223f, 312
– stabilizing effect 202f
– stress transfer 139, 320
– stress-induced fracture 141

carbonaceous nanofiller 202ff
cation exchange 2, 62
cation-exchange capacity (CEC) 6, 16, 185
ceramic-matrix nanocomposite 302
char formation 203
chemorheology 28
chitosan 105
choc resistance 134
clay 281, 287, 295, 352
– *in situ* modification 62
Cloisite 24, 30, 35, 114, 356
CNT, *see* carbon nanotube
CNT/thermoset polyester nanocomposite 31
coagulation spinning 378
co-coagulation 58
compatibilizer 8, 44, 71, 75f, 77, 79, 184
corrosion protection 263
coupling agent 44, 64, 216
crack blunting 158f, 167
crack initiation 158f
crack propagation behavior 145, 156ff, 165f
crack resistance behavior 156, 160
crack tip opening displacement (CTOD) 158f, 165f
creep resistance 212
cristallinity of polymers 13
crossover frequency 115f
crossover point 358
cryo-compounding 400
crystal growth 23
crystalline lamellae 85
crystalline orientation 86f
crystalline structure 84
crystallization 279f, 303
– effect of nanofiller incorporation 296
– effect of organic modifier 287
– effect on the intercalation morphology 363
crystallization of matrix polymers 83f
crystallization temperature 77, 83, 364, 367
CTOD, *see* crack tip opening displacement
curing reaction 22f, 24f, 397
– effect of organically modified clay 27
curing temperature 54f
Cussler-Aris model 175

d

d-spacing, *see* interlayer distance
damping degree 365
DDS, *see* diaminodiphenyl sulfone
deintercalation 44
delamination 10f
– acidic onium ions 22
– of OMLS 376
dendritic polymer 33f
density of nanocomposites 306
DENT, *see* double-edge notch in tension
DGEBA, *see* bisphenol A diglycidyl ether
DGEBF, *see* bisphenol F diglycidyl ether
diaminodiphenyl sulfone (DDS) 34
differential scanning calorimetry (DSC) 13f, 23, 125, 141
1,2-dimethyl-3-hexadecyl imidazolium (DMHDI) 381, 384
dispersion technique 395
dissipative particle dynamics (DPD) 310, 324
distortion temperature 340
DMA, *see* dynamic mechanical analysis
Doi-Hess hydrodynamic theory 317
double-edge notch in tension (DENT) specimen 134, 144, 156
DPIM, *see* dynamic packing injection molding
DSC, *see* differential scanning calorimetry
ductile-to-brittle transition 156
ductility 367
dynamic density functional theory (dynamic DFT) 311f
dynamic flow property of polycarbonate/MWNT nanocomposites 117ff
dynamic mechanical analysis (DMA) 23, 141, 148, 163, 354f
dynamic mechanical thermal analysis (DMTA) 356f
dynamic modulus, *see* storage modulus
dynamic shear modulus 28, 108
dynamic viscoelasticity 73, 76, 242, 249, 251
dynamic viscosity 75

e

electron tomography 36f, 266
electrospinning 378
encapsulation of healing agents 267ff
enhancement of load transfer 146
epoxy nanocomposite 9f, 22f, 397ff
– anhydride cured 25f
– effects of processing and aging 27f, 398
– nanoreinforcement 25, 226
– SCF-filled, tribological behavior 226f
– thermomechanical properties 363f
– tribological improvement by grafting of nanoparticles 217

– viscosity   26, 29
epoxy/amine nanocomposite   22f
epoxy/multiwall nanotube nanocomposite   25
epoxy/quaternary organoammonium-modified MMT nanocomposite   178ff
epoxy/silica nanocomposite   24
epoxy/titanium dioxide nanocomposite   23
epoxy/vermiculite nanocomposite   27
epoxy-HBP nanostructured system   33f
equivalent-continuum approach   307, 317
essential work of fracture (EWF) concept   134f, 144f, 156, 162
– preconditions   162
ethylene propene diene monomer rubber/OMC nanocomposite   43
EWF, see essential work of fracture
exfoliated graphite (xGnP)   281f
exfoliation   10f, 67, 69f
– effect on polymer crystallization behaviour   291
– effect on tensile strength   125
– influence of screw speed   371
– influencing factors   74, 372
– layer-by-layer mechanism   100
– of nanoclays   72f, 99f, 125, 291, 338, 367, 372
exfoliation/dispersion process
– of layered silicate   76, 79, 346
– rate   73f
extensional flow mixer (EFM)   80
extragallery (bulk phase) polymerization   398
extruder   372, 387
– co-rotating twin-screw extruder   374ff, 392f
– intermeshing co-rotating twin-screw extruder   373
– single, two-stage screw extruder   389f
– single-screw extruder   375
– ultrahigh-shear extruder   379
– ultrasonic twin-screw microextruder   393f
extrusion   370
– co-rotating twin-screw extrusion   371
– screw speed   371, 373
– single-screw extrusion   371
– water injection-assisted   387

f
fatigue of PNC   325f
FEM, see finite element method
filler delamination   10
filler dimensions   123, 342
filler dispersion   58, 123ff, 173, 176f
– by application of a vibration force field   392
– dispersion of agglomerates   370
– dispersion of unintercalated tactoid particles   371
– effect of MWD   77f
– effect of OMMT content   78
– effect on tensile strength   131
– influence of the screw profile   374
filler phase   1
filler platelet alignment   175
filler surface modification   9ff
filler volume fraction   123f, 129f
– correlation with permeation of polymer matrices   174
– effect on yield stress   130f
– models for effect on mechanical properties of PNC   153ff
filler-bound polymer chain   187
filler-filler interaction   112
filler-matrix interaction   125, 155
film extrusion casting   90
finite element method (FEM)   312, 318f
3-D finite element modeling   271
flame resistance   201
flame retardancy   149, 203, 359f
flexural modulus   340
flocculant   57
Flory-Huggins interaction parameter   183
flow-induced crystallization   84, 86
fluorescent optical probe   381f
fracture mechanics of nanotubes   141, 144f
Frederickson-Bicerano model   175
friction coefficient   67, 211, 222
friction principles   211
fullerenes   205
fumed silica   282, 285

g
gas diffusivity   306
gas-barrier properties   54f, 67, 173
Gauchy stress tensor   133
gel-permeation chromatography (GPC)   387
grafted nanoparticle   217
graphite as nanofiller   205, 212, 226, 285, 291
Guth-Smallwood equation   153, 318

h
halloysite nanotube (HNT)   207
Halpin-Tsai model   148f, 153, 270f, 274f, 306, 318
hardener   34
HBP, see hyperbranched polymer

HDPE, *see* high-density polyethylene
heat distortion temperature (HDT)   351
heat insulation effect   197
helical conformation of polymer nanocomposites   139, 168
Hershel-Bulkley model   98
heterogeneous nucleation   283, 287, 290, 295
high-density polyethylene (HDPE)   83, 90
high-shear mixing   395f
high-speed stirring   395
Hoffmann elimination mechanism   361
hydroxyapatite   290
hydroxyl-terminated PEEK   356f
hyperbranched polymer (HBP)   33f

i

illite   336, 344
in situ intercalative polymerization   25, 394
in situ polymerization   24, 41, 57, 94f, 104, 114, 291, 369, 378
in situ ultrasound   391
infrared spectroscopy (IR)   11f, 30
inhomogeneous clay dispersion   362
injection molding
– conventional injection molding   87
– dynamic packing injection molding (DPIM)   88
– shear-controlled orientation injection molding (SCORIM)   88
inorganic filler effect on polymer chain mobility   366
integral procedure decomposition temperature (IPDT)   24
interaction
– between surface-modified filler and polar polymer matrix   176
– chain-chain interaction   388
– in CNT-based nanocomposites   140f, 146
interaction energy   323
intercalation degree   4
intercalation process   74, 376
– entropy-driven intercalation   380
intercalation/exfoliation process of OMLS   382
interface enhancement   60f
– with silane coupling agent   62ff
interfacial adhesion   215, 217, 303
interfacial area   303
interfacial bonding, *see* interfacial interaction
interfacial interaction   60, 62, 110, 202, 359
– between organoclay surface and polymer matrix   114, 359

– prediction of, by molecular mechanics   323
interfacial tension   75
interlayer cation   335
interlayer distance   7, 11f, 31, 71ff, 76ff, 238, 247, 364
interlayer orientational coupling   395
interlayer spacing, *see* interlayer distance
internal lubricant   212
interspherulite region   364
interstratified layered silicate   347
intragallery polymerization   397
IPDT, *see* integral procedure decomposition temperature
IR, *see* infrared spectroscopy
isothermal crystallization   282, 285, 291

j

J-integral concept   133, 156

k

kaolin   335
kinetics of curing reaction   23

l

latex compounding
– mechanism   57
latex fabrication method   378
latex-silicate interaction   59
lattice Boltzmann (LB)   310f
layer aggregation   51
layered alumosilicate   1
– structure   1, 94, 280f
– surface modification   2
layered silicate
– aspect ratio variation using   344f
– degree of intercalation/exfoliation   338, 344
– dimensions of silicate nanolayers   345
– dispersion behavior   28
– expandability   345, 348f
– for reinforcement of biodegradable polyesters   235
– imidazolium-stabilized   26, 381, 384
– interstratified-layer silicates   347
– organic (OLS)   28, 35, 385f
– structural classification   335f
– structural principles   334f
– surface tension   76
limited intercalation   70
liquid healing agent   262

Index | 411

– encapsulation 267ff
living polymerization 187
load transfer between nanotubes and polymer matrix 140, 145f
load-displacement curve 133, 156
loss modulus 107ff, 111, 148
– temperature effect on 357
lubricating filler 212
Lusti-Gusev model 175

m

maleated polypropylene (MAPP) 70f, 76
MAPP, see maleated polypropylene
mass transport barrier 361, 367
mass transport barrier mechanism 197
masterbatch process 78, 377, 379
material stiffness 124f, 212
matrix material 302
matrix mobility 262
matrix viscosity 370
matrix/nanoparticles stress transfer 213, 217
MD, see molecular dynamics
mechanical failure of PNC 324f
mechanical properties of PNC 316
– effect of nanoparticle dispersion 301f
– effect of processing conditions 373f
– model for prediction 304
mechanical reinforcement 150, 321f
mechanochemical pulverization process 378
melt blending, see melt compounding
melt compounding 3, 41, 67, 73, 94, 125, 131, 142, 369
– comparision with latex compounding 57
– conditions 44
– direct melt blending (one-step process) 377
– supercritical carbon dioxide-assisted 388ff
– synthesis of polymer/CNT nanocomposites 104
– synthesis of RCN 42f
– thermodynamic principle 42
– two-step process 377
– ultrasound-assisted 391ff
– water injection-assisted 385ff
melt elasticity 77, 144
melt intercalation, see also melt compounding
– effect of average shear rate 370f
– effect of residence time 370f
– temperature effect 370f
melt mixing, see melt compounding

melt rheometry 71
melt-compounding technique 370
metal-matrix nanocomposite 302
mica 3, 336f, 344ff
micro-Brownian motion 355, 364
micromechanical model 16, 125ff, 140
microscale modeling 308ff
microstructure
– correlation with rheological behavior 109
– exfoliated 4f, 68
– intercalated 4f, 68
– microcomposite 4f
– of clay particles 44, 46
– relation to ultimate mechanical properties of PCN 129
– unintercalated 4f
microwave irradiation 400
mixed-layer clay minerals 337
mixing condition
– feed rate 82
– mixing time 82, 370
– rotation speed 82
– specific energy input (SEI) 374
– temperature effect 82
mixing in elongational flow 80f
mixing method 78f
mixing sequence 398
mixing under an electric field 81
MMT, see montmorillonite
MMT/Nylon 6 nanocomposite 9f
mobility of polymer chains 168, 359, 364
– reduction 363, 365f
modulus of elasticity 306
molecular dynamics (MD) 308f, 316
molecular modeling 307, 316
molecular weight (of polymers) 77f
molecular weight distribution (MWD) 77f
molybdenum disulfide 212
Monte Carlo (MC) technique 309
montmorillonite (MMT) 1, 67
– basal plane spacing 7
– in expoxy/diamine nanocomposites 23
– intercalation 68
– modification with amines 237ff, 341f, 360
– particle size 72, 76
– properties 2, 352
– SEM 280
– swelling 386
– unmodified 359ff, 385f
Mori-Tanaka model 271, 306f, 317

morphology
- development, effect of injection-molding technique 87f
- effect of nanofiller incorporation on polymer morphology 296
- of base polymer 82f
- of fracture surfaces 221, 223, 230f
multiscale composite 34, 36
multiscale fiber-reinforced nanocomposite 25
multiscale modeling 315
- application in prediction of PNC properties 315
- concurrent approach 314
- sequential approach 313
multi-specimen fracture evaluation (MSFE) 161
multi-walled nanotube (MWNT) 25, 31f, 34, 139, 146f, 225
- as nucleating agent 149, 284, 287f
- SEM 280
- stabilizing effect 202f
MWD, see molecular weight distribution
MWNT, see multi-walled nanotube
m-xylylenediamine 61

n

NaMMT, see MMT
nano-alumina 215, 280, 282, 287
nanoceramic 352
nanoclay 69
- dispersion by ultrasound 391
- dispersion state 69f, 338
- effect on rheology of nanocomposites 28
- exfoliation process 72f
- organophilized 69
nanocomposite type 302
nanocontainers 263
nanofiller material 302
nanofiller properties 314
nanofiller-induced polymer degradation 200
nanoparticle
- as active corrosion protection 263
- coexistance of exfoliated and intercalated OMMT nanoparticles 362
- effect of nanoparticle content on storage modulus 367
- effect of nanoparticle content on thermomechanical properties 366
- effect on nucleation and crystallization of polymers 296
- effect on the elasticity modulus 364, 367

- for improvement of wear performance 214
- in transfer film formation 214
- segregation of nanofillers 364
nanoparticle dispersion 213, 362
nanoparticle-filled silane film 263
nanoparticle-reinforced polymer 213f
- tribological performance 214f, 220
nanoreinforcement 25, 93, 140, 146, 213f
nano-silica 218f, 227
nanotube dispersion
- influencing processing parameters 378
- ultrasound-assisted 393
nanotube shortening 399
natural clay, see MMT
Nicolais-Narkis model 154
Nielsen model 154
Nile Blue A 381f
nitrile-butadiene rubber/OMC nanocomposite 43
nonisothermal crystallization 83, 282, 286, 290
nonlinear mechanical properties 123
nonpolar polymer system 9f, 12, 76
non-self-healing polymer system 273
nucleation 283
numerical modeling method 307f
nylon 10,10/MMT nanocomposite, crystallization 294
nylon 11/MMT nanocomposite, crystallization 293
nylon 6 nanocomposite 322
- crystallization behavior 291
- polymorphism 292
nylon 66 nanocomposite
- crystallization behavior 291
- polymorphism 291

o

octadecyl triethylammonium chloride (OMT) 79
OLS, see layered silicate, organic
OMC, see organically modified clay
OMLS, see OLS
OMM, see organo-montmorillonite, Cloisite
OMMT, see organophilized montmorillonite
OMT, see octadecyl triethylammonium chloride
organolayered titanate 359
organomodified synthetic layered fluorine mica (OSFM) 357
organo-montmorillonite (OMM) 30

organophilized montmorillonite (OMMT)   71f, 237, 356
– as nucleating agent for PP   85, 87
– content, effect on crystallization enthalpy   85
– dispersion of, in injection-molded PP nanocomposite   88
orientation order parameter   89
oxygen permeation   16f, 177f, 185, 189f, 361, 367

p

PALS, *see* positron annihilation lifetime spectroscopy
PB nanocomposite   285f
PB, *see* poly-1-buten
PBAT, *see* poly(butylene adipate-ω-terephthalate)
PBAT/organomodified MMT nanocomposite
– biodegradability   244ff
– mechanical properties   241f
– morphology   237ff
– thermal properties   243f
PBS, *see* poly(butylene succinate)
PBS/MMT nanocomposite
– biodegradablility   253f
– interlayer spacing   247
– mechanical properties   248f
– preparation   245f
– tensile properties   250
– thermal properties   252f
PBT, *see* polybutylen terephthalate
PBT/clay nanocomposite, nonisothermal crystallization   286
PCH, *see* polymer-clay hybris
PCL, *see* poly(ε-caprolactone)
PCN, *see* polymer/clay nanocomposite
PDT, *see* polymerization-clay delamination temperature
PE nanocomposite   197
PE, *see* polyethylene
PEEK, *see* poly(etheretherketone)
PEEK/nano-alumina nanocomposite   290f
PEEK/nano-silica nanocomposite   219f, 291
PEN nanocomposite   288f
PEN, *see* polyethylen naphthalate
percolation network   73, 116, 167
percolation treshold   159, 161, 378f
permeability of polmer membranes   174f
PET nanocomposite   287
PET, *see* polyethylene terephthalate   287
PHB nanocomposite   290
PHB, *see* poly (3-hydroxybutyrat)
PHBV, *see* poly(3-hydroxybutyrate-ω-3-hydroxyvalerate)
phyllosilicate, *see also* smectite   352, 357
Piggot-Leidner model   154
PLA, *see* polylactide
plastic flow   221f
plastic resistance   163
PLLA nanocomposite
– crystallization behavior   289f
– thermomechanical properties   360f
PLLA, *see* polylactic acid
PLSN, *see* polymer/layered silicate nanocomposite
PMR, *see* polyimide thermoset resin
PNC, *see* polymer nanocomposite
Poisson's ratio   306
polar PNC   176
polar polymer system   9f
polarity of filler interlayer   176
polarized micrograph   84
poly (3-hydroxybutyrat)   290
poly trimethylene terephthalate (PTT)
poly(3-hydroxybutyrate-ω-3-hydroxyvalerate) (PHBV)   235, 290
poly(butylene adipate-ω-terephthalate) (PBAT)   235f, 237
poly(butylene succinate) (PBS)   235f
poly(butylene succinate)-*co*-adipate (PBSA)   357
poly(etheretherketone) (PEEK)   214ff, 227f
poly(ethylenoxide)/clay nanocomposite   355f
poly(glycidyloxypropyl) phenyl silsesquioxane (PGPSQ) matrix   359
poly(glycidyloxypropyl) silsesquioxane (PGSQ) matrix   359
poly(methyl methacrylate) (PMMA)   214ff, 227f
poly(tetrafluoroethylene) (PTFE)   212
poly(ε-caprolactone) (PCL)   235
poly-1-buten (PB)   285
polyamide nanocomposite   362, 386
polybutylen terephthalate (PBT)   286
polycarbonate/MWNT nanocomposite   148
– fracture behavior   157ff
polydimethylsiloxane polymer   9
polyester/organo-clay nanocomposite   30f
polyester/silica nanocomposite   396

polyethylene (PE)   3
polyethylene naphthalate (PEN)   288
polyethylene terephthalate (PET)
polyhedral oligomeric silsesquioxane   206
polyimide polymer   9, 395f
polyimide thermoset resin (PMR)   32
polylactic acid (PLLA)   289
polylactide (PLA)   235
polymer chain density   185
polymer crosslinking   367
polymer nanocomposite (PNC)
– characterization   36
– crystallization   279f
– definition   1
– manufacturing process   68f
– material stiffness
– mechanical properties   123, 145, 343
– mechanical property enhancement   123f
– microstructure classification   4
– modeling of barrier properties   189f
– processing   369ff
– properties   15f, 67
– properties at break   133f
– requirements for application as packaging material   173
– rheological behaviour in melt state 107ff
– rheological behaviour in solution state   95ff
– stress transfer   145f, 320f
– synthesis   3, 67, 78f, 187, 378
– synthesis of exfoliated PNC without using OLS   385f
– thermo-oxidative degradation   197, 200
– ultimate mechanical properties   129
polymer sheathing   151, 157
polymer toughness   67, 150, 212
polymer wrapping   139, 157, 168
polymer/clay nanocomposite, see polymer/layered silicate nanocomposite
polymer/clay/CNT nanocomposite   93
polymer/CNT nanocomposite   93
– application   93
– deformation-morphology correlation   147
– fracture mechanics   156f
– interactions   140f
– mechanics   141
– melt-intercalation   378ff
– molecular modeling   316f
– solution-intercalation   380ff
– stress transfer   147, 320
– surfactant-assisted processing   378
– synthesis   94

– tribological performance   224f
polymer/layered silicate nanocomposite (PLSN)   41f, 338
– conventional, properties   339f
– degree of intercalation/exfoliation   338
– melt intercalation   370ff
– preparation   338
– with controlled number of nanolayers   347
polymer-clay hybris (PCH), see PCN   41f
polymerization-clay delamination temperature   41f
polymer-matrix nanocomposite, see polymer nanocomposite
polymer-nanoparticle interaction   270
polymer-nanotube interaction   140, 148, 168
polymorphism in nanoreinforced polymers   285, 289, 292f
– effect of nanofiller incorporation on polymer morphology   296
polyol/clay dispersion   96f, 396f
polyol/MWNT dispersion   100ff
polyol/SWNT dispersion   100ff
polyolefine nanocomposite   183
– correlation between gas permeability and CEC   186
– decompositions mechanisms   201
polyolefine/MMT nanocomposite   68f
– melt characteristics   91
polyoxymethylene/MMT nanocomposite   196f
polypropylene (PP)   13, 16
– containing ammonium-groups   77
polystyrene (PS)   3, 7f
– effect of clay on crystallization of syndiotactic polystyrene (sPS)   295
– polymorphism in syndiotactic polystyrene (sPS)   296
positron annihilation lifetime spectroscopy (PALS)   365
POSS, see polyhedral oligomeric silsesquioxane   206
post-yield fracture mechanics (PYFM) concept   163
poylvinylidene fluoride (PVDF)
PP nanocomposite   3, 12 f, 16, 184, 388
– crystalline morphology   84f
– crystalline orientation   86f
– crystallization   283f
– influence of crystallization on the intercalation morphology   363
– oxygen permeation   13

– polymorphism of PP 285
– preparation by modified clay slurry method 386
PP, see polypropylene
PP/MAPP/OMMT nanocomposite 70, 77f, 83, 88
PP/MMT nanocomposite 83f
– injection-molded 87
– thermomechanical properties 365
– viscoelastic behavior 109
PP/MWNT nanocomposite
– characterization 142ff
– crystallization 284
– fracture behavior 163ff
– mechanical reinforcement 150
– melt rheological behavior 150
– morphology 152
– nanotube deformation 159, 161
– synthesis by melt extrusion 142
PP/OMMT nanocomposite 70, 365
preconditioning (aging) time 398
pre-intercalation of monomers 395
processing effects on nanocomposite formation 3
processing of PNC 369ff
– influence of processing parameters on clay dispersion 372
– optimization of processing parameters 371
processing technique 28f
properties of nanocomposites 302
– effect of component mixing sequence 398
– effect of processing parameters 373f
– effect of processing route on 377f
– influence of filler properties 314
– properties predictable by analytical modeling 306
– properties predictable by multiscale modeling
– properties predictable by numerical modeling 307
PS, see polystyrene 28f
PS/clay nanocomposite 109ff, 179, 392
– effect of silica on thermomechanical properties 357f
– melt-intercalation 370ff
– solvent-blending of exfoliated PS nanocomposite 383f
– stress relaxation behaviour 116
PS-nickel nanocomposite 391
PTT nanocomposite 288
PTT, see poly trimethylen terephthalate

PU/clay nanocomposite 95f, 395ff
PU/CNT nanocomposite 100f, 147
Pukanszky model 154
PVDF nanocomposite 294f
PVDF, see polyvinylidene fluoride 294
pyrophyllite 335

q
quiescent crystallization 83

r
radical trapping 202
RCN, see rubber-clay nanocomposite 3
reaggregation 57f
rectorite 281
reinforcement of biodegradable polyesters 235
reinforcing filler 211
α-relaxation 365
release improve factor (RIF) 267
representative volume element (RVE) 305
residence time (in a mixing apparatus) 370f
residence time-distribution curve 375
restriction of thermal motion 198
rheological behavior, see also viscoelastic behavior 352ff
– effect of nanoparticle dispersion 366
– of MWNT/water-based PU nanocomposites 104
– of poly(ethylene oxide)/surfactant-modified MMT nanocomposites 114
– of polyol/clay dispersion 97
– of PP/MMT nanocomposites 107ff
– of PS/clay nanocomposites 111ff
– of PU/CNT nanocomposites 100
rotation speed 378
rubber latex 42f, 57f
rubber/pristine nanocomposite 57
rubber-clay nanocomposite (RCN) 41f, 58ff
– effects on morphology 44f
– interface-enhanced 64
– properties 53
– synthesis 42f, 57
rule of mixture 305f
RVE, see representative volume element

s
saponite 281, 292
SAXS, see small-angle X-ray scattering
scanning electron microscopy (SEM) 151, 158, 161, 164f, 203, 219, 226f, 230f, 256, 280, 284

SCF *see* short carbon fiber
screw profile 374f
screw speed 371
segregation of nanofillers 364
self-extinguishing property 359f
self-healing nanoparticle-reinforced polymer system 263
self-healing polymer system
– boric acid-modified polydimethylsiloxane (Silly-Putty) nanocomposite 275f
– encapsulation in liquid-based systems 265
– glass fiber-reinforced nanocomposite 275f
– liquid-based thermoplastic polymers 265
– liquid-based thermosetting polymers 264
– microstructured 264ff
– nanoparticle-reinforced 270, 274f
– principles 261f
– reversible supramolecular polymer nanocomposite 275
self-similar approach 307
SEM, *see* scanning electron microscopy
semi-crystalline thermoplastic polymer 149f, 162f, 262
sepiolite 281, 292, 388ff
sericite, *see* illite 336f
shear force 79, 320
shear rate 96f, 370
shear stress 98, 373
shear thickening 98
shear thinning 28, 98f
sheet extrusion 89f
short carbon fiber (SCF) 211
– effect on tribological behavior of PNC
silica
– as nanofiller 205f, 287, 289
– *in situ* incorporation of silica precursor 362
silica-loaded nanocomposite 365
silicon carbide 215, 217
silicon nitride 214
silicon nitride ceramic 282
silver nanoparticle 291, 295
single-specimen fracture evaluation (SSFE) 145, 161, 166
single-walled nanotube (SWNT) 102, 126, 139, 146, 225, 318
– polyurethane-grafted (SWNT-g-PU) 102ff
sinusoidal deformation 353
skin/core structure 87
small-angle X-ray scattering (SAXS) 23, 30, 36, 86, 90, 182
smectite 335ff, 344, 347, 352

sol-gel technique 362
solide-state shear pulverization 378
solution casting 94, 104
solution-intercalation 369, 380ff
solvent blending 383
solvent intercalation, *see also* solution-intercalation 338
solvent-aided mechanical mixing 396
sonication 395
spherulite growth 279
spherulite region 364
spherulite structure 85f, 279
stiffness of PNC 316, 353, 378
storage modulus 73, 75f, 107, 111, 148, 353f, 356
– correlation with the elasticity modulus 358, 364
– effect of nanoclay incorporation 366
– temperature effect 357
strain localization 157, 166
strength of PNC 316
stress relaxation behavior 116
stress response 353f
stress transfer 139, 320
structure-property experiment 28, 31
structure-tribological property relationship 215f
styrene butadiene rubber/OMC nanocomposite 43
sub-micro filler 226f
supercritical carbon dioxide 80, 388ff
supercritical fluid-assisted mixing 80
surface modification
– by alkyl ammonium salts 11ff, 177ff, 184
– by esterification 14
– by physical adsorption of organic molecules 188
– by polymerization 187f
– carbon nanotube surface modification 24
– of alumosilicate 2
– of filler, effect on oxygen permeation 178
surface modification 43
surface reaction
– of clay by polymerization 14 f
surfactant 43f
surfactant-modified clay 61
SWNT, *see* single-walled carbon nanotube
Szazdi model 155

t
talc 335
tearing modulus 161
TEM, *see* transmission electron microscopy

temperature resistance 351
tensile modulus 16, 126, 241
tensile strength 146, 220, 241
tensile test 144
ternary nanostructured system 34
TGA, *see* thermogravimetric analysis
thermal annealing of clay/polymer mixtures 381
thermal decompositon process of polymers 195
thermal expansion coefficient (CTE) 351, 356
thermal expansion coefficient of nanocomposites 306
thermal stabilization of polymers 208, 359
– by carbonaceous nanofillers 202ff
– by clay minerals 196ff
– by metal oxides 206f
– by metals 206f
– by physical adsorption of organic molecules 202
– by radical trapping with metal cations 201
– by restriction of thermal motion 198
– by silica-based nanofillers 205f
– catalytic effects 198f
– char formation 198f
thermogravimetric analysis (TGA) 6ff, 23, 50, 188, 197, 360
thermomechanical analysis (TMA) 352ff
thermomechanical properties 351ff
– effect of nanoclays 355ff
– effect of nanoparticle content 366
– effect of silica 357f
thermoplastic polymer-based nanocomposite 369ff
thermoplastic polyolefin (TPO) 83, 145
thermoset nanocomposite, *see also* epoxy nanocomposite 21f, 30f, 145
– modeling 36
– processing 394ff
thermosetting polyurethane/ MWCNT nanocomposite 31f
time-dependent Ginzburg-Landau method (TDGL) 311
time-temperature superposition 354f, 358
titanium dioxide
– as filler for epoxy resin 23
– effect on wear resistance 215
– stabilizing effect in PNC 207
– sub-micro particles in SCF/graphite-filled PNC 227f
total fracture work 134

tough-to-brittle transition 156, 159
TPO, *see* thermoplastic polyolefin
transfer film 214f
transmission electron microscopy (TEM) 4f, 10, 22ff, 27, 32, 42, 45, 48, 52f, 56f, 60f, 70, 80, 85, 88, 102, 142f, 175, 180, 186, 240, 336, 340f, 364
transparency of PNC 173
tribological behavior of polymers 211, 224
tribological system 211
triisopropanolamine 61

u

ultra-small-angle X-ray scattering (USAXS) 24
ultrasonic mixing 382ff
ultrasound for analysis of PNC 391
USAXS, *see* ultra-small-angle X-ray scattering

v

vacuum-assisted resin infusion molding (VARIM) 25
van der Waals interaction 141, 146, 167, 204
VARIM, *see* vacuum-assisted resin infusion molding
vermiculite 3, 27, 336
vibration force field 392
viscoelastic behavior, *see also* rheological behavior 352ff, 366
viscoelastic response 359
viscoelasticity 353f
viscosity
– influence of shear rate 101, 104, 106
– of chitosan/modified CNT dispersions 106
– of polymer/clay dispersion 95f
– of polyol/clay dispersion 96f, 101ff
– thermoset viscosity 396
viscous energy 354
vulcanization of RCN 44f
– chemical effects of curing agents 45ff
– morphological effects 45
– pressure effects 50ff
– temperature effects 49f, 54
vulcanization temperature 51, 54f

w

water injection-assisted melt-compounding 385ff
water-assisted mixing 78f
WAXD, *see* wide-angle X-ray diffractometry
wear of PNC 326f

wear performance 214
wear rate 211, 216, 222
weight loss rate 367
wide-angle X-ray diffractometry
 (WAXD) 42f, 46ff, 71, 76ff, 86ff, 143

x
X-ray diffractometry (XRD) 4, 6ff, 10,
 71, 99, 110, 179f, 187, 238f, 247,
 295

y
yield strength 147
yield stress 131, 154f, 359
Young's modulus 124ff
– influence of aspect ratio 272
– normalized 140
– theoretical prediction 129

z
zinc oxide 282